SAUNDERS GOLDEN SUNBURST SERIES IN ENVIRONMENTAL STUDIES

Introduction to Environmental Studies
by Jonathan Turk

Ecology, Pollution, Environment
by Jonathan Turk, Janet T. Wittes, Robert Wittes, and Amos Turk

Physical Science with Environmental and Other Practical Applications
(Second Edition) by Jonathan Turk and Amos Turk

Jonathan Turk, Ph.D.
Dillingham, Alaska

Amos Turk, Ph.D.
City College of the City University of New York

with contributions by

Karen Arms, Ph.D.
Cornell University

ENVIRONMENTAL SCIENCE

THIRD EDITION

SAUNDERS COLLEGE PUBLISHING

Philadelphia New York Chicago
San Francisco Montreal Toronto
London Sydney Tokyo Mexico City
Rio de Janeiro Madrid

Address orders to:
383 Madison Avenue
New York, NY 10017

Address editorial correspondence to:
West Washington Square
Philadelphia, PA 19105

Text Typeface: Palatino
Compositor: Clarinda Company
Acquisitions Editor: John Vondeling
Developmental Editor: Lloyd Black
Project Editor: Patrice L. Smith
Copyeditor: Ruth Low
Managing Editor & Art Director: Richard L. Moore
Art/Design Assistant: Virginia A. Bollard
Text Design: Lawrence R. Didona
Cover Design: Lawrence R. Didona
Text Artwork: Tom Mallon
Production Manager: Tim Frelick
Assistant Production Manager: Maureen Iannuzzi

Cover credit: El Capitan, Yosemite National Park, California. Copyright 1983 by David Stoecklein.

ENVIRONMENTAL SCIENCE/3e   ISBN 0–03–058467-1

CBS COLLEGE PUBLISHING
Saunders College Publishing
Holt, Rinehart and Winston
The Dryden Press

PREFACE

Environmental science is an integration of material from the disciplines of biology, the physical sciences, and the social sciences. A prime objective of this book is to emphasize this interdependence and to maintain a balanced presentation of environmental issues important to the student. To this end, we have developed a principles-oriented book that emphasizes concepts rather than the memorization of large amounts of raw data. This text is designed to explain how the underlying science, economics, and politics affect important environmental issues. Adequate facts are given to illustrate and support the principles developed, and, more importantly, facts and principles are integrated throughout the text.

The publication of the third edition of ENVIRONMENTAL SCIENCE marks the 10th year in print of this title. In the past 10 years many environmental questions, once of interest mainly to scientists, have become serious issues of public policy and have sustained a steadily growing public awareness. Concern about the environment is now worldwide, and policies set in the United States can have far-reaching global effects.

It is with these and other changes in mind that we have revised ENVIRONMENTAL SCIENCE. To provide a better basis for the discussion of environmental problems, the biological and geological material in the third edition have been considerably improved. Karen Arms, noted author of biology textbooks (Arms and Camp, BIOLOGY, 2nd edition, Saunders College Publishing, 1982; Camp and Arms, EXPLORING BIOLOGY, 2nd edition, Saunders College Publishing, 1984), has strengthened greatly the chapters on the biology of natural ecosystems. There

is increased emphasis on the technical and social issues regarding global resources—fuels, fiber, food, water, soil, and minerals. This greater international overview is evident in the number of examples of environmental problems and solutions cited from other countries. Contemporary areas of concern—energy use, land management, and pollution and waste control—have been expanded and updated.

In order to incorporate all the changes of the last decade, it was quickly discovered that a major restructuring was necessary. Thus, the organization has been revised so that environmental science is presented as a unified, integrated study. The stage for this study is set in the very first unit, in which both legal and technical limitations on environmental improvement are introduced.

Following is an annotated list of noteworthy features contained in the third edition.

(1) *Organization.* It has been our aim to offer logical, orderly sequences of presentation and to avoid awkward breaks in the continuity. At the same time, we recognize that the instructor requires flexibility and may choose to emphasize certain topics more than others. Therefore, the material has been arranged so that the presentation can be individualized to suit the particular course objectives. It is for this reason that the text is organized into units that are individually self-contained and need not be presented in the sequence in which they appear in the text. The units are as follows:

Unit 1. Environmental Science: An Overview
This unit introduces environmental science and its relation to the human con-

dition, discusses the classification of environmental problems, shows how environmental improvement is tied to economic and legal concerns, and discusses how the laws of nature limit our response to environmental challenges.

Unit 2.   The Ecological Background
This unit provides a comprehensive look at the biology of natural ecosystems, populations, speciation, extinction and genetic resources, and an overview of the biosphere.

Unit 3.   Human Impact on the Earth
This unit discusses the relationship of human beings and the environment, beginning with the origins of the human species and the growth of human populations and ending with discussions of the effects of chemical and hazardous wastes on human health.

Unit 4.   Energy
This unit, extensively revised and expanded in this edition, discusses sources of energy; nuclear energy and its environmental effect; the search for and use of energy and its environmental, economic, and political consequences; and the need for both short-term and long-term planning.

Unit 5.   Soil, Land, and Minerals
This unit deals with soil and agriculture; the control of pests and weeds and the debate over insecticides; food production, the Green Revolution, and the consequences of world hunger; land use and encroaching urbanization; and nonrenewable mineral resources.

Unit 6.   Air, Water, and Wastes
This unit discusses water resources; water and air pollution and their environmental effects; the social, legal, and economic aspects of water and air pollution; and the disposal of solid wastes.

Epilogue.   Planning for a Sustainable Society
This coda discusses some of the realistic options we do have and stresses the need for planning.

(2)   *Enrichment Material.* Various case histories and special topics (which appear in boxes) are

interspersed throughout the text. These are presented, for the most part, because they are interesting. Many offer a human outlook that is not necessarily conveyed in a purely didactic approach. However, they are not required for a reading or understanding of the chapter, nor are questions at the end of the chapter based on them.

(3)   *Pedagogy.* Several pedagogical elements are incorporated into the text.

a.   Chapter summaries. These summaries highlight the principles of each chapter and the key words that the student should know.

b.   Questions. End-of-chapter questions function in two ways. First, they help students review important points covered in the chapter. Second, they present real-world situations that challenge students to apply what they have learned as they develop solutions for environmental problems. The questions are usually grouped by major subtopic.

c.   Suggested readings. The references at the end of the each chapter are selected to include classic sources, readings at both introductory and more advanced levels, and the most recent contributions to the field. They are not intended to be exhaustive in scope.

d.   Appendix. The appendix is a reference section on units, various physical concepts, chemical formulas, and calculations of growth rates and of decibel levels.

e.   Glossary. An extensive glossary of terms used in the text is provided at the end of the book.

f.   Use of the metric system. Metric units are given first, with the United States Customary or British units sometimes following. Measures of heat are shown in calories. The confusion between the different kinds of tons is avoided by using only the designation **tonne,** which means metric ton and is intermediate between the customary long and short tons. All needed conversions are provided in the Appendix.

(4)   *"Take-Home" Experiments.* The various "take-home" experiments that appeared in previous editions, as well as new ones, have been gathered into one supplementary laboratory section at the end of the book. These experiments require no more, or very little more, equipment and supplies than are commonly available in a grocery, hobby, or hardware store. Some of the experiments can be used as classroom projects or can serve as springboards for class discussion. They should all be interesting and informative and serve to reinforce the principles of environmental science.

(5) *Supplements.* Three supplementary items are available to adopters of ENVIRONMENTAL SCIENCE. An Instructor's Manual provides teaching suggestions and other useful information. A set of 100 Overhead Transparencies containing about 100 illustrations from the text is also available. An environmental simulation entitled "WORLD: The Fate of Civilization" is available on disk for microcomputers and is accompanied by a manual explaining the use of the software.

Since ENVIRONMENTAL SCIENCE is an intro-

ductory text, only a minimal science background is required by the student. Also, no mathematics is needed to comprehend the concepts presented. Even though the text offers a broad-spectrum view of environmental science, we have written a book that tries to explain the true complexity of environmental issues without over-simplifying or providing simplistic solutions that really don't work. We have always attempted to be realistic in discussing the true complexity of the world in which we live; yet we have, at the same time, strived to convey a picture of humans in their environment that is understandable and relevant to scientists and nonscientists alike.

ACKNOWLEDGMENTS

The following reviewers read the entire manuscript and offered many valuable and incisive suggestions. We benefited from their expertise and are deeply grateful for their help.

John D. Cunningham
  Keene State College
David E. Kidd
  University of New Mexico
Peter W. Frank
  University of Oregon
Martha C. Sager
  The American University
Susan Uhl Wilson
  Miami-Dade Community College
Robert R. Churchill
  Middlebury College
H. Gray Multer
  Fairleigh Dickinson University
David A. Stewart
  Ferris State College
C. B. Coburn, Jr.
  Tennessee Tech University

In addition, Professor Stanley S. Wecker of The City College of The City University of New York offered helpful comments on particular sections related to the ecological background.

Mrs. Pearl Turk helped gather and classify source material and photographs.

The editorial and production staff at Saunders College Publishing did their usual superlative work: John Vondeling, associate publisher; Lloyd Black, developmental editor; Patrice Smith, project editor; Richard Moore, managing editor and art director; Virginia Bollard, design assistant; Tim Frelick, production manager; and Maureen Iannuzzi, assistant production manager.

JONATHAN TURK
AMOS TURK

CONTENTS

x   Contents

UNIT 3
Human Impact on the Earth   133

UNIT 4
Energy   201

UNIT 5
Soil, Land, and Minerals   295

Environmental Science: An Overview

This view of the head of Yellowstone Falls shows
the fresh water, clear skies, and lush vegetation
that a protected natural environment can provide.
(Photo courtesy of Richard L. Moore.)

Studying the Environment

During the 1960s and 1970s people began to listen seriously to predictions that human existence would be threatened unless humans adjusted their relationship with their environment. Scientists said that the earth was becoming polluted and that soil, mineral, and fuel resources were being used faster than they were being replaced. Demographers said that the world's human population was growing faster than its food supply. Economists pointed out that the gap between the rich and the poor was growing wider. These concerns were the incentive for the development of the discipline of environmental science—the study of the relationship between human beings and their environment.

The human environment is the Earth we live on. It includes all the physical parts of the Earth such as air, soil, minerals, rocks, and water, and all its biological inhabitants such as animals, including human beings, bacteria, fungi, and plants. Environmental science provides an approach toward understanding the human environment and the impact of human life upon that environment. It is also a search for solutions to the environmental problems that confront us.

This chapter provides an overall look at the topics that will be covered in more depth in succeeding chapters. This overview includes consideration of the worldwide problems that initiated the study of environmental science and of how those problems are related to each other.

1.1 The Human Condition

Consider the state of humankind on Earth, the planet we call home. In the past, life was hard and short for most people. Then, at the beginning of the twentieth century, rapid progress in medicine, agriculture, and industrial techniques seemed to promise that everyone might soon be able to enjoy long life, decent food, satisfying employment, and adequate housing. This promise has not been realized. In 1982 there were about 5 billion people on Earth. One quarter of them had a greater life expectancy and lived in greater luxury than anyone a hundred years ago would have believed possible. But at the same time three quarters of all people had inadequate or unsatisfactory water and shelter, and more than one third suffered from malnutrition and hunger (Fig. 1–1). More

2

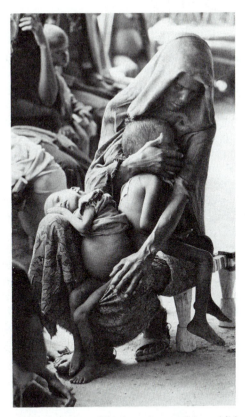

Figure 1–1 A malnourished mother and her children in India. Millions of people starve to death every year and many more are permanently crippled by malnourishment. (Photo courtesy of Henri Bureau-Sygma.)

Our vast and swiftly growing population consumes the Earth's resources of agricultural land, minerals, water, and fuel faster than natural processes can replace them. This is a serious problem in itself. But global conditions are further endangered because these resources are not distributed evenly. One quarter of the population, those that live in the developed nations, use nearly 80 percent of the resources consumed by humankind in any one year. People in the United States alone, although they comprise only 5 percent of the Earth's population, use about 35 percent of all raw materials. The other three quarters of the population consume only about 20 percent of the resources used in a year. The gap between the haves and the have-nots is growing wider. Today, when transistor radios are to be found even in the most remote African villages, the world's poor know how underprivileged they are. This knowledge produces political instability. Years ago, political upheaval in one nation meant little to the rest of the world. But times have changed, because modern technology ensures that nearly all nations possess weapons that can wreak havoc far beyond their own borders. As a result, no nation can afford to ignore the problems of another.

For many years most people forgot that the Earth is largely a closed system, with only finite supplies of resources and a limited ability to absorb wastes. Because only a thoughtful few re-

people starved to death in 1982 than in any year since the beginning of time.

Most of the people who starve to death live in poverty-stricken developing nations. But now, at the end of the twentieth century, the wealthy nations are in trouble too. According to most estimates, the average standard of living in North America and Western Europe peaked in about 1967. Even the wealthiest nations are running out of fuel, hardwoods, and some minerals. As a result, necessities such as housing, food, and fuel are demanding more and more of the family budget, leaving less available for luxuries. Pollutants (Fig. 1–2) contaminate cities, towns, and even rural environments. Sewage or poisonous pesticides in waterways, smog-laden air, and garbage in streets or parks lower the standard of living of everyone, no matter how wealthy.

Figure 1–2 Air pollutants from electrical generating stations. (Courtesy of EPA.)

alized what was happening, little was done about the problems, and the situation continued to deteriorate. Some experts now think that everyone on Earth can have enough to eat and a reasonable standard of living only if the human population is reduced considerably. Some even think that our time on Earth is nearly done. These pessimists believe that the disparity between the rich and the poor and the destruction of the environment will lead to social unrest that can end only in a nuclear war that will destroy us all. Others think that the human plight is serious but not desperate. They believe that technology can be used to solve environmental problems as it has solved so many others, and that the developed nations will see the wisdom of spending more of their wealth on improving the human condition. But surely everyone must agree that a better understanding of the relationship between humankind and the environment is one positive step forward. To this end, the rest of this book is a beginning.

1.2 Classifying Environmental Problems

Environmental problems are always interrelated. Sometimes a solution to one problem actually creates another problem. For example, when people are sick and dying from disease, it is natural to want to improve human health. When health is improved and infant mortality is reduced, a population explosion may result. To feed this growing population, natural habitats are often destroyed by turning them into farmland. As natural habitats are destroyed, the wild plants, predatory animals, and parasites that once lived there are killed as well. Because of the lack of predators and parasites, outbreaks of insect pests become more common. Farmers use pesticides to control the pests and protect the crops, but in the process the environment becomes polluted. The development of this entire cycle in itself consumes fossil fuel supplies that are becoming scarce. In addition, when fuels are burned, air pollutants are generated.

How does a person begin to study such a network of interlocking problems? To make the task a bit more manageable, we will divide environmental disruptions into five main types.

(a) Overpopulation. Overpopulation may be defined as the presence in a given area of more people than can be supported adequately by the resources available in that area. Many people argue that the population explosion that has taken place in the twentieth century is now the most important problem we face (Fig. 1–3). It is important first because overpopulation is a major cause of all other environmental problems: Fewer people would use less oil, chop down fewer trees, and pour less sewage into rivers. Second, overpopulation and the starvation that accompanies it are generally higher on our list of priorities than other environmental concerns. It is hard to argue that an area should be set aside as parkland to preserve a vanishing forest or savanna when that land might be used to raise crops that would prevent fellow human beings from starving to death.

Figure 1–3 Some densely populated nations view overpopulation as one of their most devastating problems and encourage small families. Here a vendor displays the *Family Planning* publicity on his cart.

BOX 1.1 POLLUTION

For many years, the word pollute has meant "to impair the purity of," either morally* or physically.† The terms "air pollution" and "water pollution" refer to the impairment of the normal compositions of air and water by the addition of foreign matter, such as sulfuric acid. Within the past few years two new expressions, "thermal pollution" and "noise pollution," have become common. Neither of these refers to the impairment of purity by the addition of foreign matter. Thermal pollution is the impairment of the quality of environmental air or water by raising its temperature. The relative intensity of thermal pollution cannot be assessed with a thermometer, because what is pleasantly warm water for a man can be death to a trout. Thermal pollution must therefore be appraised by observing the effect of a rise in temperature. Similarly, noise pollution has nothing to do with purity: foul air can be quiet, and pure air can be noisy. Noise pollution (to be discussed in Chapter 10) is the impairment of the environmental quality of air by noise.

---

*(1857.) Buckle, *Civilization.* I., viii, p. 526: "The clergy . . . urging him to exterminate the heretics, whose presence they thought polluted France."

†(1585.) T. Washington. trans. *Nicholay's Voyage,* IV, ii; p. 115: "No drop of the bloud should fall into the water, least the same should thereby be polluted."

(b) Pollution. Pollution is a reduction in the quality of the environment by the introduction of impurities (Box 1.1). Smoke pollutes the air; sewage pollutes the waters; junk cars pollute the land. We know that such contamination exists; it can be seen, smelled, or even tasted. The effects of pollution on human welfare or on the economy, however, may be matters of considerable disagreement (Fig. 1–4).

There are two distinctly different types of pollution.

1. *Concentration of Organic Wastes.* All living organisms produce waste products; wastes are associated with the act of living. Upon death, the entire organism becomes a waste product. Before modern civilization, most organic wastes did not accumulate in the environment because they were consumed by other organisms and thereby recycled. In modern times, the natural decomposition of organic wastes does not always operate efficiently. For example, a city may house several million people in a small area of land. Organic wastes from such a city are not spread evenly about the countryside but instead

are concentrated in a few locations. Sewage that is dumped into a river decays naturally, but the process takes time. If the volume or the concentration of sewage is high, the water may not be purified by the time it reaches the next site downstream where pure water is needed. If several cities line the river and each city discharges its wastes, then the river cannot cleanse itself. Many plants and animals that depend on the river and are beneficial to people die. Organisms that take their place are often less useful to humans or may even carry disease.

2. *Introduction of Synthetic Chemicals into the Environment.* Everything is made of chemicals—people, eagles, trees, lakes, plastic—everything. Although many natural chemical compounds have existed for billions of years, people have recently learned to make new chemical compounds, called **synthetic chemicals.** The quantity and variety of new synthetic chemicals are staggering. They are present in paints, dyes, food additives, drugs, pesticides, fertilizers, fire retardants, building materials, clothes, cleaning supplies, cosmetics, plastics, and so on. In 1980, some 70,000 different synthetic chemicals were produced in quantity for common use. About 2000 new compounds enter the environment every year. Just before the beginning of World

Figure 1–4 The effect and ultimate cost of pollution are often hard to assess. This is Copper Basin at Copperhill, Tennessee, where a luxuriant forest flourished until all vegetation was killed by the fumes from copper smelters. In such cases the rate and degree of recovery cannot be accurately predicted. (U.S. Forest Service. From Odum: *Fundamentals of Ecology.* 3rd ed. Philadelphia, W. B. Saunders Co., 1971.)

War II, production of synthetic chemicals in the United States totalled less than ½ billion kg per year. By 1980, that total had jumped to some 80 billion kg per year.

Synthetic chemicals are noted for the variety of their properties. Some of them are drugs that save millions of lives every year, and others are poisons. But because most of them are new to the environment, the traditional patterns of decay and recycling do not necessarily apply. Some synthetic chemicals break down rapidly in the environment by the action of sunlight, air, water, or soil, and some are eaten by living organisms. Such processes may take place over a span of minutes, hours, or days. A material that decomposes in the environment as a result of biological action is said to be **biodegradable.**

Many compounds, however, do not disappear so readily. Synthetic plastics, for example, remain in the environment for a long time because organisms that feed on them and break them down are rare. (Scientists have bred bacteria that do decompose a few plastics.) Plastic shampoo bottles may produce unsightly litter, but they are not biologically active. Many other synthetic chemicals, however, are harmful. For example, DDT was developed as an insecticide, but experience with it has shown that it has undesirable environmental effects as well. It kills nontarget insects and interferes with the life cycle of birds, by making their egg shells so thin that they break easily, causing many of the young to die prematurely. DDT decomposes so slowly that appreciable quantities remain in the soil 15 years after a single spray application (see Chapter 15).

Natural chemicals can also cause environmental problems. For example, arsenic is found in minerals that occur naturally in small concentrations in many rocks and soils. Arsenic compounds are poisonous, but when small amounts are chemically bound into rock formations they do not present a serious hazard to human health. Now, enter a mining operation. Rocks are dug up, crushed, and treated chemically to separate valuable minerals such as gold from the less valuable ones. In the process, arsenic is inadvertently concentrated. In many cases, the semiconcentrated arsenic is thrown away. In its new form it enters the environment as a real hazard to health.

(c) Depletion of Resources. A **resource** is any source of raw materials. Fuels, minerals, water, soil, and timber are all resources. A material is depleted, or used up, as it becomes less available for its intended function. Material resources can become depleted in three different ways. First, a substance can be *destroyed*, that is, converted into something else. Fuels are destroyed when they are used: Coal is converted to ashes and gas; uranium is converted to radioactive waste products. The ashes or waste products are no longer fuels.

Second, a substance can be lost by being *diluted*, or by being *displaced* to some location from which it cannot easily be recovered. If you open a helium-filled balloon, the gas escapes to the atmosphere. Not one atom of helium is destroyed, but nevertheless the gas is lost because it would be impossible, as a practical matter, to recover it. The same concept of loss by dilution applies to minerals. An **ore** is considered to be a rock mixture that contains enough valuable minerals to be mined profitably with currently available technology. Today, iron is profitable to mine when the ore contains about 40 percent iron. However, iron is also widely dispersed in the soil, in many rocks, and even in the ocean, where its concentration is about one millionth of 1 percent. These sources of iron are useless to humans as minerals because it would take far too much energy and equipment to recover them. When products containing iron, such as automobiles or paper clips, are thrown away, the metal rusts. The rust is iron oxide, which could be reprocessed to produce iron again. However, many of the old auto hulks (and all of the paper clips) are scattered so widely over the countryside that it is uneconomical to find and collect them. It is in this way that our reserves of iron are being depleted.

Third, a substance can be rendered unfit for use by being *polluted*. In this way pollution and depletion are related to each other. If industrial or agricultural wastes are discharged into a stream, or if they percolate down through soil and porous rock to reach a supply of groundwater, then these water resources become less fit for drinking or, in the case of the stream, for recreation or for the support of aquatic life.

Depletion of resources can be slowed by conservation, recycling, and substitution. Fuels,

Figure 1–5 Electric cars use energy that can be generated at a plant that burns coal, which will still be available after petroleum reserves are depleted. (United Press International photo.)

however, can be saved only by conservation (using less) and substitution (using other energy sources, see Fig. 1–5). Fuels cannot be recycled. Political decisions about depletion of resources, and especially about policies of conservation, are often difficult because there is generally little agreement on the amount of resources available, on future rates of consumption, or on alternative sources of energy and materials. It would be unrealistic, for example, to expect agreement on energy policy between one group who expects oil to run out by the end of the century, coal to last for only a few hundred years, and nuclear energy to be unavailable, and another group who believes that most fossil fuel reserves have not yet been discovered and that nuclear energy will be plentiful for thousands of years.

Finally, conservation is often seen as a measure whose benefits will be realized later, perhaps only by our children or grandchildren, and not all makers of policy are equally concerned about future generations.

(d) Changes in the Global Condition. Scientists have begun only recently to wonder whether human activities might affect the global environment. Aerosol sprays and aircraft exhaust may be destroying the ozone layer in the atmosphere that filters out ultraviolet radiation. Burning fossil fuels releases carbon dioxide that could affect planetary weather patterns. Pollution of the oceans destroys plant life that produces oxygen, and such pollution might eventually reduce the oxygen content of the air we breathe. Throughout much of the world, forests, jungles, shrublands, and other natural systems are being converted to farmland. In many areas, this process is depleting the fertility of the soil, altering the climate, and causing the extinction of literally thousands of species of plants and animals. Except in emotional terms, people often do not know precisely what has been lost when a species becomes extinct. Scientists are convinced that many endangered species of plants or animals should be saved because they may be essential in breeding valuable crops or domestic animals.

(e) War. In many ways, war is a combination of all environmental problems rather than a separate category. From time immemorial, overpopulation and want have led human groups into wars over food, land, or some other coveted resource. In modern times war and the preparation for war have led to pollution and depletion of resources that are far more extreme than any single peacetime activity. War reduces population, although the effect is trivial: more people were born in 6 months in 1982 than were killed throughout the first and second world wars. Finally, a nuclear war places the global systems of the Earth, human civilization, and even the human species itself at risk.

1.3 The Tragedy of the Commons

Some people were aware of environmental problems a long time ago. In 1789 Thomas Robert Malthus, an English clergyman and economist, published an *Essay on the Principle of Population*. In this essay he argued that human populations must invariably outgrow their food supplies and will then be reduced by starvation, disease, and war. The English poet William Wordsworth was well aware of air pollution. In his poem "On Westminster Bridge" he celebrated one of the rare days in the nineteenth century when the air was clear and the London skyline was visible. If we have known about the problems for centuries, then, why have we done so little about them? Different people would give different answers to this question.

An American living in poverty would probably argue that it is outrageous to spend public money cleaning up the environment when the money would be better spent on housing, medical care, and education for the many who still lack such amenities, even in the United States.

A family living on the verge of starvation in Africa would find it hard to believe that anyone could spend money on anything but food. As for cutting down trees faster than they can grow, what other fuel is there to cook with? It is heart breaking not to have enough food for starving children, but many children must be conceived if any are to survive because half the babies die before they are one year old. In that part of the world children are vital for such tasks as tending the goats and for supporting their aging parents.

These arguments against controlling population and pollution illustrate the conflict between the short-term welfare of individuals and the long-term welfare of society. Garrett Hardin has called this phenomenon "the tragedy of the commons." He illustrates it with the case of commons in medieval Europe. A common was grazing land that belonged to a whole village. Any member of the village could graze cows and sheep there. It was in the interest of each individual to put as many animals on the common as possible to take advantage of the free animal feed. However, if too many animals grazed the common, they eventually destroyed the grass. Then everyone suffered because no one could raise any cattle on it at all. For this reason, common land was eventually replaced by individually owned, enclosed fields. The owner of a field is careful not to put too many cows on one patch of grass, because overgrazing one year means that fewer cows can be raised the following year.

In the same way, our air, water, and land are commons. Today, many people are "overgrazing" or, more precisely, depleting many natural resources because it is profitable to do so. But what is profitable in the short term may be disastrous in the long run. Similarly, pollution is profitable if anyone can dump wastes into the

Figure 1–6 *Pure mountain stream.*

air and water, thereby saving the cost of disposing of the pollutants in other ways. Then every smoky factory can avoid the cost of air pollution controls and every household can discharge its wastes directly into the local stream. The difficulty is that polluting, like overgrazing, often benefits individuals or corporations in the short run. It is cheaper for a firm to operate its factory without pollution controls than with them, even though pollution is expensive to society in the long run. If the "tragedy of the commons" is to be avoided, this situation must be reversed by increasing short-term incentives for protecting the environment (Fig. 1–7).

1.4 The Human Condition— Looking into the Future

It is relatively easy to determine the condition of the environment today. Pollution levels

Figure 1–7 It takes time and effort to bring cans to a recycling center, and so most people don't do it. Paying consumers even a small amount to bring cans in enormously increases the number of cans that are recycled.

can be monitored and quantified precisely. Go down to the service station and try to buy some gasoline. If you can buy as much as you want at an affordable price, then gasoline is readily available. If there is some form of rationing, or if you must wait in long lines for service, or if the price is exorbitant, then it is not. But what about the future? What will our world be like 10, 25, 50, or 100 years from now? This is a much more difficult question to answer.

During the latter part of his term in office, President Carter directed a group of scientists to write a report predicting what general living conditions for the human race would prevail by the year 2000. This report, entitled the *Global 2000 Report,* gives cause for alarm. The following excerpts are from the introduction:

> If present trends continue, the world in 2000 will be more crowded, more polluted, less stable ecologically, and more vulnerable to disruption than the world we live in now. Serious stresses involving population, resources, and environment are clearly visible ahead. Despite greater material output, the world's people will be poorer in many ways than they are today.

> . . . real prices for food are expected to double.

> . . . The Study projects that the richer industrialized nations will be able to command

enough oil and other commercial energy supplies to meet rising demands through 1990. With the expected price increases, many less developed countries will have increasing difficulties meeting energy needs. For the one-quarter of humankind that depends primarily on wood for fuel, the outlook is bleak. Needs for fuelwood will exceed available supplies by about 25 percent before the turn of the century.

> . . . Regional water shortages will become more severe. In the 1970–2000 period population growth alone will cause requirements for water to double in nearly half the world.

> . . . Significant losses of world forests will continue over the next 20 years as demand for forest products and fuelwood increases. Growing stocks of commercial-size timber are projected to decline 50 percent per capita. The world's forests are now disappearing at the rate of 18–20 million hectares a year (an area half the size of California), with most of the loss occurring in the humid tropical forests of Africa, Asia, and South America.

> Serious deterioration of agricultural soils will occur worldwide, due to erosion, loss of organic matter, desertification, salinization, alkalinization, and waterlogging. Already, an area of cropland and grassland approximately the size of Maine is becoming barren wasteland each year, and the spread of desert-like conditions is likely to accelerate.

> Atmospheric concentrations of carbon dioxide and ozone-depleting chemicals are expected to increase at rates that could alter the world's climate and upper atmosphere significantly by 2050. Acid rain from increased combustion of fossil fuels (especially coal) threatens damage to lakes, soils, and crops. Radioactive and other hazardous materials present health and safety problems in increasing numbers of countries.

> Extinctions of plant and animal species will increase dramatically. Hundreds of thousands of species—perhaps as many as 20 percent of all species on earth—will be irretrievably lost as their habitats vanish, especially in tropical forests.

These statements sound grim. Yet all hope is not lost. Re-read the first phrase of the first sentence of the report carefully. It says, "If present trends continue. . . ." Optimists claim that present trends are not at all likely to continue.

They say that human ingenuity and resourcefulness will rise to the challenge and engineer a future environment that is even better than the one we have today. This solution is sometimes called the **technological fix,** because it claims that technology will solve our problems. The *Global 2000 Report* recognizes the power of technology when it says:

> . . . It must be emphasized that the Global 2000 Study's projections are based on the assumption that national policies regarding population stabilization, resource conservation, and environmental protection will remain essentially unchanged through the end of the century. But in fact, policies are beginning to change.

But in the next paragraph it warns:

> Encouraging as these developments are, they are far from adequate to meet the global challenges projected in this Study. Vigorous, determined new initiatives are needed if worsening poverty and human suffering, environmental degradation, and international tension and conflicts are to be prevented. There are no quick fixes.

Some people claim that there always have been and always will be technologically feasible solutions to the problems we face. They point to historical successes. For example, in the 1950s, people feared that as the richest iron mines were

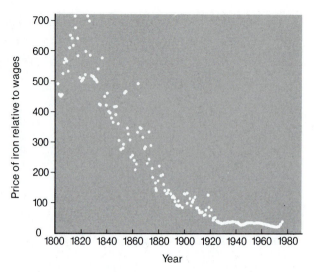

Figure 1–8 The price of pig iron relative to wages in the United States. (From J. Simon: *The Ultimate Resource.* Princeton, N.J., Princeton University Press, 1982.)

being rapidly depleted, iron shortages would develop, thereby threatening our entire civilization. In fact, there has been no shortage of iron. Technological improvements in mining and refining have made it possible to extract metals from ores that were uneconomical to mine 35 years ago. As a result, the real price of iron in the United States has actually decreased in recent years (see Fig. 1–8). Similarly, the real price of many other metals has declined as well.

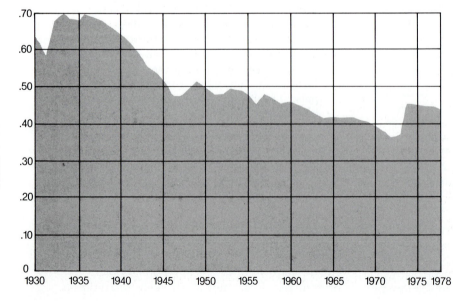

Figure 1–9 Cost of gasoline at the service station in constant 1972 dollars. Notice that the price of gasoline has gone down since 1930 relative to the buying power of average wages. (From J. Simon: *The Ultimate Resource.* Princeton, N.J., Princeton University Press, 1982.)

SPECIAL TOPIC A
Environmental Impact Statement

It is better to prevent environmental damage than to control it after the harm is done. It is therefore helpful to try to predict the environmental effect of a new project *before* the project is undertaken. Such an evaluation is called an Environmental Impact Statement. A typical study attempts to provide answers to the questions outlined here. Chapter numbers in parentheses deal with the particular question.

1. Description of proposed action.

2. What are the anticipated adverse effects of the proposed project on
   (1) The natural ecosystem?   (Chapter 3)
   (2) Air quality?   (Chapter 21)
   (3) Water quality?   (Chapter 20)
   (4) Solid waste management?   (Chapter 22)
   (5) Human health and safety?   (Chapter 10)
   (6) Parks, and historical or aesthetic features?
                                     (Chapter 17)
   (7) Energy supplies?   (Chapters 11, 12, 13)
   (8) Existing land use and resource conservation strategies?   (Chapter 17)

3. What is the anticipated impact of the proposed project on
   (1) Population growth?   (Chapters 8, 9)

   (2) Economic growth?   (Chapter 2)
   (3) Social conditions?   (Chapters 2, 8, 9)

4. Does the proposed project have any beneficial environmental effects? (Describe.)

5. To what extent does the proposed action constitute an *irreversible* commitment?

6. Will this action create a need for additional related actions, and what is the anticipated impact of the latter?

7. What are the alternatives to this proposal, and what are their respective impacts? (Use additional space as necessary.)

8. Which alternative(s) would significantly conserve energy?

9. Are there related proposals that could offset the adverse environmental effects of the present proposal?

10. What, if any, overriding concerns justify the adverse environmental effects of the proposed action?

11. Additional comments.

What about fuels? Many of us remember the precipitous increase in fuel prices in 1973. However, the real price of petroleum in the United States has declined during the past century, despite short-term fluctuations (Fig. 1–9). This decline has resulted from technological improvements in the extraction, processing, and distribution of fuel. Some claim that when fossil fuel reserves become scarce, there will be an increased economic incentive to develop alternative sources of energy such as solar, wind, tidal, and nuclear energy. These sources will replace conventional fuels as soon as the need for them becomes urgent enough.

Thus, this argument continues, economic need is a driving force for technological change. In a free economy, scarcity is automatically countered by change and substitution. For example, when the price of gasoline rose in the United States in 1973, there was a sudden shift toward the production and sale of smaller, more

fuel-efficient cars. As a result, gasoline consumption has declined in the United States during the last decade, and the fuel bills for many families have not increased substantially.

Does all this mean that environmental concern is unnecessary and superfluous? No, not at all. Two points are important to remember. First of all, although it is true that technology has led to increases in global availability of a number of resources, many people in local areas continue to suffer. For example, even though world food productivity has increased steadily over the years, millions have starved to death in recent times, and millions more are hungry today. Global statistics of increased food productivity are meaningless to an individual who has nothing to eat.

The second point to remember is that the world is different today than at any other time in history. It is much more densely populated. Technology, especially weaponry, is much more sophisticated. The unequal distribution of wealth threatens unprecedented social upheavals. As a result, human activities pose a greater threat to the environment than they have ever before (see Special Topic A). Thus, history may not be a model for the future.

Some people predict that technology will save us from environmental disruption only if we are willing to implement drastic social change as well. For example, it is argued that we must completely restructure the distribution of resources. At the present time, people in developed countries are adapting to the use of small cars, but in the meantime hundreds of millions of people in the less developed countries cannot even afford bicycles!

Summary

In the second half of the twentieth century, people became aware that their interactions with the environment threaten the future of human life on Earth. From this realization, the discipline of environmental science was born.

Although the major problems are interrelated, they can be divided into five main categories: **overpopulation** causes human suffering directly and exacerbates all other environmental problems; **pollution** makes vital resources less useful and reduces the quality of life; **depletion** of resources makes things that are vital to human existence more expensive or impossible to obtain; **global changes** that result from human activities may permanently alter the Earth in unpredictable ways; **war** may be caused by or may cause all other environmental problems, and modern warfare threatens the survival of the human species.

Environmental problems are among the hardest for human societies to solve because individuals seldom have to pay directly for their contribution to these problems. Individuals act in ways that promote their own short-term welfare, which often conflicts directly with the long-term environmental interests of present and future generations. This "tragedy of the commons" is the main factor that limits the effectiveness of social solutions to environmental problems.

According to the *Global 2000 Report*, the human condition will be bleak by the year 2000 if present trends continue. Some observers believe that technological solutions will arise to solve our problems as they have in the past.

Questions

Environmental Issues

1. What types of disruptions have the potential for long-range degradation of the environment?

2. Suggest one action you personally might take to help preserve the environment. How do you assess the importance of your own action? How would you estimate the number of people who would have to join you before the action would have a significant environmental impact?

3. Today people talk about the depletion of soil and timber resources. What is meant by the word "depletion" in these two instances?

4. The five major types of environmental disruptions were listed as a series of isolated topics. Explain how each one of them may be related to one or more of the other four in certain circumstances.

5. Apply the "tragedy of the commons" principle to the problem of limiting human family size.

6. Prepare a class debate with one side arguing the "doomsday" forecast that the human condition will be much worse in the year 2000 than it is today and the other side taking the opposite view that conditions on Earth will improve from now on.

Suggested Readings

A comprehensive book filled with data on environmental problems and very useful for reference is:

P. R. Ehrlich, A. H. Ehrlich, and J. P. Holdren: *Ecoscience: Population, Resources and Environment.* San Francisco, W. H. Freeman, 1977.

The thought-provoking argument that the environmentally ethical individual is doomed:
G. Hardin: The tragedy of the commons. *Science, 162:* 1243, 1968.

One forecast for the future is given in:
Gerald O. Barney (director): *The Global 2000 Report.* New York, Pergamon Press, 1980. (Volume I, *Summary Report;* Volume II, *Technical Report;* Volume III, *Government Global Model.*)

A more recent viewpoint is:
State of the Environment 1982 by Conservation Foundation. Dr. Edwin H. Clark II, editor, 1982, Washington, D.C.

A defense of the technological fix is given by:
Julian L. Simon: *The Ultimate Resource.* Princeton, N.J., Princeton University Press, 1981. 415 pp.

The fascinating and influential book that has led many people to rethink traditional ideas about what kinds of technology are appropriate for our overpopulated planet is:
E. F. Schumacher: *Small Is Beautiful: Economics as if People Mattered.* New York, Harper & Row, 1975.

An operating manual for the citizen lobbyist is:
George Alderson and Everett Sentman: *How You Can Influence Congress.* New York, Dutton, 1981.

Environmental Improvement—
An Interplay of Social and Technological
Problems

Perhaps the single most important generalization in environmental science is that there are no simple solutions to the important problems of our times. All the issues are complex, and nearly every perceived "answer" invariably leads to yet other complications.

One reason for this complexity is that neither technical nor social solutions can be entirely effective by themselves. Instead, scientific, legal, economic, and political realities must all be taken into account. This chapter attempts to relate these different aspects of environmental science. Even though subsequent chapters deal with separate topics, the reader should bear in mind that environmental issues always involve such complexities.

2.1 Environmental Improvement and the Economy

In 1980 a large ore smelter located in Anaconda, Montana, closed down (Fig. 2–1). Hundreds of workers lost their jobs, and the entire economic base of the town was disrupted. The owners of the smelter, Atlantic Richfield Company, claimed that the cost of meeting new air pollution regulations had become prohibitive. Since they couldn't simultaneously process the copper ore, comply with the regulations, and

Figure 2–1 Anaconda, Montana, with the smelter, now shut down, in the background.

make a profit, they had to shut the doors.* Anaconda had always been a mining and refining town. Throughout its history mining and related activities had provided the main support for the region. There had been little other industry and only a moderate amount of farming and ranching. People were stunned to learn that the Anaconda Company had pulled out. Many local residents felt that jobs were more important than clean air and that the regulations should be eased. They pointed out that people must work to survive. Compared to this economic necessity, clean air is a luxury that cannot always be afforded.

This argument warrants close analysis. First of all, it is important to understand that the situation cannot be analyzed fairly by isolating one single smelter from the rest of the environment. Think back to Garrett Hardin's tragedy of the commons (Chapter 1). If many cows are grazing a commons, a person learns little by studying the impact of one cow; the impact of the entire herd must be considered. Similarly, if many factories are polluting a common resource such as air, it is unreasonable to isolate one factory from the others. Therefore, it is necessary to widen our focus and consider not just the Anaconda smelter but all sources of air pollution. Within this framework, what are the costs and benefits of air pollution control?

Even within such a large framework, it is relatively easy to estimate the cost of installing and maintaining pollution control devices. Engineering and construction firms can make reasonable estimates of the cost of pollution controls on factories, automobiles, and home furnaces. On the other hand, what would be the cost to the community as a whole if there were no air pollution controls anywhere? The additional expenses incurred by individuals as a consequence of living in a polluted environment are not paid for by manufacturers. They are therefore separate or

outside the direct cost of manufacturing and are called external costs, or **externalities.** Externalities are generally not so obvious or easy to quantify as direct, internal manufacturing costs, but they are nevertheless quite real (Fig. 2–2). For example, pollution can affect human health. Consider medical bills, loss of work because of illness, decreased productivity, and death of a wage earner at an early age. These are just some of the external costs of pollution. If the smoke from manufacturing (or from waste disposal) darkens nearby houses and soils the clothes of local residents, the costs of more frequent repainting and laundering are also externalities. Many pollutants are reactive chemicals such as corrosive acids. Some of these acids collect in the atmosphere, dissolve in water droplets, and fall to the Earth in the form of acid rain which destroys buildings, kills fish, and slows the growth of agricultural crops. These costs, too, must be added to the cost of manufactured goods. In many areas, severe pollution has resulted in loss of tourist trade. In these instances, factory and smelter owners may gain by ignoring the expense of pollution control devices, but motel and restaurant owners and many other local businesses must pay the cost instead. Note that externalities include only factors that can be measured in monetary terms. Human pain or suffering cannot be valued in this manner and is therefore not accounted for in this type of analysis.

External costs are sometimes omitted from economic analysis, but they must be paid for nevertheless. They are often difficult to estimate because they are so spread out, but some attempts have been made. According to the Environmental Protection Agency, the cost of meeting the regulations of the Clean Air Act in the United States in 1977 was about $6.7 billion. But the savings in external costs was $8 billion. Thus, the American people saved $1.3 billion by regulating air pollution in 1977. In a more recent report, the Council on Environmental Quality (CEQ) analyzed the total cost of all pollution controls in the United States. According to this report, the people in the United States spent a total of $37 billion for all forms of pollution control in 1979. This amounted to an average of $164 per person per year. Does this figure sound like a lot or a little? Think for a moment of what

---

* It is not certain that the requirements of the Clean Air Act forced Atlantic Richfield to shut down the Anaconda smelter. There were some implications that the shutdown was a move against organized labor, because a strike was in progress at the time of the plant closure. Still other reports state that the plant was old and outdated anyway. These factors remind us that economic analysis is not always as clearcut as it might appear to be at first glance.

Figure 2–2 An example of an economic externality. A home in Appalachia collapses as the overburden on which it stands subsides into an abandoned, unsupported deep mine shaft. (Courtesy of the Bureau of Mines, U.S. Department of the Interior.)

it would be like to live in a world with no pollution controls at all. In such an environment, untreated sewage would be dumped directly into the rivers, black oily smoke would pour out of factory chimneys, and garbage would be allowed to rot and fester in unsanitary open dumps. Seen from this perspective, pollution control is a necessity, not a luxury, although its cost of implementation falls more heavily on some than on others.

Cost/Benefit Analysis

Pollution control is not a yes–no, on–off affair. Imagine that a copper smelter with no pollution control devices releases a certain quantity of sulfur dioxide into the air every month. Equipment can be designed to remove any portion of the pollutant, from a minor amount to practically all of it. In general, the more pollution that is removed, the more expensive the process becomes.

It is roughly accurate to say that the cost of pollution control is constant for each reduction of pollution by 50 percent. That means that if it costs, say, $1 million per year to remove 50 per-

cent of a given pollutant, it will cost another $1 million to remove 50 percent of the remaining pollution. Therefore, $2 million will accomplish a 75 percent (50 percent + 25 percent) removal, $3 million will achieve 87.5 percent, and so on. Thus, limited pollution control can be relatively inexpensive, but an essentially pollution-free environment is very costly.

How much pollution control can be justified? Some people suggest that pollution control measures should be applied only when it can be shown that there is a positive economic return on the investment. This approach, known as **cost/benefit analysis,** (Fig. 2–3), can be illustrated by the example of air pollution from a copper smelter as previously described. If no control devices are used at all, air pollution is so severe that the external costs are very high. Thus, the total cost of the smelter to society is quite large.

Suppose that some modest pollution control devices are installed. At first a moderate outlay of money will reduce the pollution by a significant amount. The small increased cost of manufacturing will eventually be passed on to the consumer who will have to pay a little more for

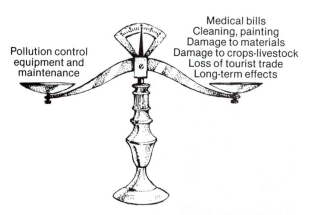

Figure 2–3 *The principle of cost/benefit analysis. This accounting system does not recognize psychological trauma, pain of illness, or emotional needs for a clean environment.*

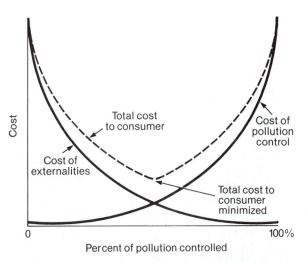

Figure 2–4 *Schematic graph showing the relationship between the cost of pollution control and the cost of externalities. The actual numbers will vary with the situation.*

various goods. But the same consumer will save a lot more money because the cost of externalities will decline dramatically. This level of pollution control is economically beneficial. Examine Figure 2–4 and note the two solid curves that show the relationship between pollution control costs and externalities. Imagine that more and more money is spent on pollution control. At first, a small amount of money causes a large drop in external costs (left side of the graph). As complete pollution control is approached, a large increase of manufacturing expenditure yields only a small decrease in the cost of externalities (right side of the graph). The dotted line in Figure 2–4 is the sum total of all costs to an individual consumer who must pay both manufacturing costs and external costs. This line is roughly U-shaped, and the low point in the curve represents the level of pollution control at which the total cost to an individual is at a minimum. Proponents of cost/benefit analysis argue that pollution control should be regulated only to this point.

Opponents of this proposal pose questions about the nonquantifiable cost of living in a polluted world. How can a dollar value be placed on the annoyance of a vile odor, or of unclear air, or of industrial noise? Such annoyance involves not the destruction of property but the deterioration of the quality of human life. What about recreational opportunities? Fifty years ago, for instance, on the south shore of Lake Erie,

people fished, boated, swam, and explored the beaches and marshes. Today, swimming is unsafe in many parts of the Lake, the number of tasty game fish has declined dramatically, the wild secluded beaches have disappeared, and boats ply polluted waters. Alternative recreational opportunities are now available. Swimmers can find artificial swimming pools, children can play in approved parks, people who want to see turtles can drive to the Cleveland Zoo, and lovers of nature can visit a museum. For many, these recreational changes involve poignant losses, but their dollar value is difficult to assess (Fig. 2–5).

Going beyond annoyance, how can you measure the dollar value of human suffering and misery caused by illness, or the value of a human life that ends too soon? Many people feel that such costs are beyond our right to judge.

Let us return again to the plight of the unemployed worker at the smelter in Anaconda. Many people argue that they have paid too high a personal price for a clean environment for others. On a national scale, however, the development, design, installation, and maintenance of pollution control devices provide more jobs than they eliminate (Fig. 2–6). Between 1971 and 1978 an estimated 22,000 people in the United States lost their jobs when plants were forced to close because they could not meet pollution control

Figure 2–6 Pollution control creates jobs. (Photo courtesy of E. D. Switala, Owens-Corning Corp.)

Figure 2–5 Swimming in a pool is different from swimming in a natural environment. Who can place a dollar value on such differences?

requirements. Many of these plants, like the one in Anaconda, were old and inefficient and might have closed anyway. On the other hand, some 680,000 jobs were created by pollution control work. Yet these national figures do not help un-

employed smelter workers who cannot sell their homes in a depressed local real estate market.

A study reported by the Conservation Foundation at the end of 1982 found that the environmental laws passed in the 1970s have not blocked construction of new oil refineries, steel mills, and other industrial plants. Moreover, there is no evidence that environmental and land-use laws have caused a major movement of industry from states perceived to be environmentally strict to those that are thought to be more permissive. The most important factors that determine locations of plants are still the traditional ones: access to markets, to labor, and to raw materials. If these factors are favorable, the strictness of environmental regulations is not a handicap.

2.2 Environmental Degradation and the Law

In many situations, environmental degradation results from one form of the tragedy of the commons. It is cheaper for an individual to pollute the environment than not to pollute it, but if we all polluted our surroundings the external costs would be prohibitive. Therefore, a legitimate role of governments is to regulate pollution. Legal systems of government have always recognized the need to prevent one individual from harming another. Criminal statutes punish direct assaults on the well-being of others such as robbery or murder. A second type of wrongful act, called a **tort,** is a noncriminal action that results in personal injury or damage to property.

Legal Actions in Tort Law

1. Nuisance. In legal usage, a nuisance is the "substantial, unreasonable interference with the reasonable use and enjoyment of property." In an early form of environmental legislation, the Romans interpreted pollution as a form of nuisance, and anyone caught polluting the city aqueduct system was subject to severe penalties. Today, air pollution, water pollution, and excess noise are also considered nuisances and therefore belong in the jurisdiction of legal concern.

However, the legal issues are not always simple or straightforward. Jet planes (Fig. 2–7) are noisy, and as a consequence, some of the people who live near Kennedy International Airport in New York City filed suit claiming that the airport was a nuisance and should be closed permanently. The courts recognized that such a closure would disrupt the economy of the city of New York and provide a dangerous precedent for other airport closures. In addition, defense lawyers pointed out that many of the people who had filed the suit had moved to the area after the airport had been in operation for many years. They argued that it is unreasonable for a

person to move to an industrial area and then claim at a later date that their personal rights have been infringed. In cases like this one, the court is forced to balance competing interests. The airport agreed to adopt certain noise control procedures, and the local residents agreed to drop the suit.

2. Trespass. Medieval English law was strict in prohibiting people from intruding upon or invading other people's property. It was illegal for peasants to hunt in a forest belonging to a noble, and one of the first illegal acts committed by the legendary figure Robin Hood and his followers was simple trespass. Trespass has become important in environmental law because it has been recognized that an object as well as a person can trespass. For example, if the pollutants from a factory fall on a forest and kill the trees, the intrusion of pollutants on another person's property is a form of trespass.

3. Negligence. An act carelessly performed or carelessly omitted when it should have been performed is called negligence. An example is the faulty construction of an automobile so that carbon monoxide can leak into the passenger compartment, or neglect by a factory manager on one day to turn on the air pollution control system. Many lawsuits, some amounting to millions of dollars, have been successfully prosecuted on the basis of negligence.

General Types of Legal Actions for Pollution Control

Within the broad philosophy outlined by Western law, several general types of legal actions are commonly used.

1. Restriction. Some pollutants are so poisonous that even small concentrations can kill wildlife, domestic stock, agricultural crops, or people. Here the concepts of cost/benefit analysis or economic evaluation lose their validity, and the government has the power to restrict or abolish the harmful practices. For example, in the United States, the use of several pesticides, including DDT, has been banned (see Chapter 15). Violators in some cases are subject to criminal prosecution. It should be noted, however, that criminal prosecution is used only rarely.

Rationing is a form of restriction. For example, if fuel is rationed, every person is allotted a certain amount and no more. Rationing was

Figure 2–7 *Jet planes use the most power, and make the most noise, during take-off. (Photo by David F. Hall; courtesy of the Connecticut Lung Association.)*

used during World War II but would be very un-popular at the present time.

2. Penalties. Many pollutants are not acute poisons. For example, sulfur dioxide is an air pollutant released in various concentrations whenever fossil fuels are burned. It is harmful in many ways, but a small dose is certainly not lethal. In addition, it would be economically impractical to prohibit the release of all sulfur dioxide. One approach to such a problem is to permit each factory to release a specific quantity of the gas. If more than the legal limit is released, fines can be levied, or criminal prosecution can be instituted.

3. Subsidies. Regulations that simply prohibit pollution ignore the resulting hardships that are imposed on individual companies or communities. An alternative approach is to offer economic incentives that will persuade polluters not to despoil the environment. Subsidies operate as follows. An industrial firm declares that its normal operating procedures would yield a certain amount of pollutants. However, by installing antipollution devices, the amount of pollution would be reduced. Then the government rewards the firm with a subsidy that depends directly on this reduction. Subsidies can be very effective because businesses respond favorably to economic incentives. However, subsidies are expensive for the government. Moreover, although it is relatively easy to measure the total amount of pollution, it is difficult to estimate the amount that would have occurred without controls. Therefore, the calculation of subsidies is open to abuse and corruption.

A second type of subsidy is given in the form of **tax deductions** for environmentally favorable practices. Under the Carter administration, special deductions were allotted for adding insulation, solar heating, or other conservation devices to older homes.

4. Pollution tax or residual charge. Another approach is the so-called **pollution tax,** or **residual charge.** Each polluter is charged an amount proportional to the quantity of pollutant emitted. Such a scheme would encourage environmentally sound manufacturing because it penalizes processes that pollute. Surveys of industries have generally found that this approach is favored by most firms because it internalizes pollution costs for all companies simultaneously. Of course, standards for particularly hazardous or toxic materials would still need to be established and enforced. The main disadvantage of this approach is that, like sales taxes and value-added taxes, a pollution tax is passed on to the consumer in a regressive fashion. Thus, under this system, the poor ultimately pay a higher proportion of their income for pollution control than the rich.

Ideally, residual charges present the polluter with a choice—either dump waste and pay the fine, or deal with the waste in some other way—treat it, recycle it, store it, or minimize it in manufacturing. If the residual charge is too low, there will be little economic incentive to prevent waste, and pollutants will continue to be dumped untreated into the environment.

Pollution taxes can be charged either to the manufacturer of an item or directly to the consumer. For example, an electric generating utility could be charged for the sulfur dioxide emitted, or, alternatively, individuals could be encouraged to use less by taxes on their consumption. One suggestion is a steeply increasing tax. The tax should allow a reasonable amount of usage at very low rates, so that the poor are not charged exorbitantly for necessities. Then, according to this system, heavy users would be taxed severely.

Pollution Problems and the Courts. To many people, especially those who live in a relatively unpolluted environment, the pollution problem may seem to be remote. But to others, pollution may be a very real and obvious problem. Imagine, for example, that a factory in your neighborhood is polluting the air and water around you and making your life unpleasant. What can you do as an individual? One recourse is to take the managers of the factory to court. The Anglo-American legal system has developed judicial rules and procedures governing controversies in which one party, the **plaintiff,** alleges that he has been harmed by another party, the **defendant.** If the court judges that harm has indeed occurred, the defendant is often ordered to pay the plaintiff money to offset the damage. Perhaps, if you were the plaintiff you would want not money but an **injunction,** a legal order to the defendant requiring him to stop doing the

Sierra Club v. Morton. Supreme Court of the United States 405 U.S. 727 (1972). Dissent by Justice Douglas

Inanimate objects are sometimes parties in litigation. A ship has a legal personality, a fiction found useful for maritime purpose. The corporation sole—a creature of ecclesiastical law—is an acceptable adversary and large fortunes ride on its cases. The ordinary corporation is a "person" for purposes of the adjudicatory processes, whether it represents proprietary, spiritual, aesthetic, or charitable causes.

So it should be as respects valleys, alpine meadows, rivers, lakes, estuaries, beaches, ridges, groves of trees, swampland, or even air that feels the destructive pressures of modern technology and modern life. The river, for example, is the living symbol of all the life it sustains or nourishes—fish, aquatic insects, water ouzels, otter, fisher, deer, elk, bear, and all other animals, including man, who are dependent on it or who enjoy it for its sight, its sound, or its life. The river as plaintiff speaks for the ecological unit of life that is part of it. Those people who have a meaningful relation to that body of water—whether it be a fisherman, a canoeist, a zoologist, or a logger—must be able to speak for the values which the river represents and which are threatened with destruction.

wrongful act. Classic English examples are cases in which a plaintiff sought an injunction against a defendant to prevent the latter from hunting on the plaintiff's land.

A plaintiff who fulfills all necessary requirements to pursue a case in court is said to have **legal standing** (see Special Topic A). Standing for damage suits is granted only if an individual's harm is distinguishable from that of the public in general, or, as the United States Supreme Court* has stated, one is required to have a "personal stake in the outcome of the controversy." The question of standing has rendered the courts inaccessible to many environmental suits. For example, an individual man in Santa Barbara could not have sued Union Oil Company for the nearby oil spill simply because of his own personal anguish over the killing of birds. How-

ever, if he had owned waterfront land near the spill, he would have been able to sue for damage to his own property. Nor can an individual woman sue the Department of the Interior for destroying a wilderness area by commercializing it, even though part of her taxes are being used and she is thus being economically "hurt" by the development. She lacks standing, for her harm cannot be distinguished from that of the public in general.

Sometimes a group is granted standing when an individual is not. For example, a city or a town can file a case representing all its citizens, or an organization (such as the Sierra Club) can file a case on behalf of all its members. A factory spewing noxious gases into the atmosphere might be sued by a town and legally declared a "public nuisance." Certain legal scholars declare that a pollution-free and healthy environment is one of the unenumerated rights of the Ninth Amendment, which says, "The enumeration in the Constitution, of certain rights, shall not be construed to deny or disparage others retained by the people."

An adversary approach to environmental protection is fraught with problems. Litigation is very expensive and slow. Another disadvantage is that lawsuits may be addressed to a very small problem. In addition, a given decision is not generally binding on future offenders. Each violation requires a new case. Environmentalist groups may be pitted against an array of giant corporations and government agencies, and an individual case may cost as much as half a million dollars. Often, however, the possibility of a suit and adverse publicity will discourage the defendant. Furthermore, the action of the court may be tempered by provisions in federal regulations permitting a judge to take the cost of compliance into account. Thus, various laws specify that a company must exercise "economically feasible" control of emissions and need not install the most effective "technologically feasible" control measures.

Environmental law is complex and constantly changing. Each country has its own particular environmental legislation. Even within a country, national policy may fluctuate, and is affected in part by local ordinances and by other factors such as the vigilance of the government in enforcing existing legislation.

---

* Flast *v.* Cohen. 392 U.S. 83 (1968).

LAW	PURPOSE OF LAW	FOR FURTHER READING, SEE CHAPTER
National Environmental Policy Act (NEPA)	(a) To "use all practicable means, consistent with other essential considerations of national policy, to . . . fulfill the responsibilities of each generation as a trustee of the environment for succeeding generations" (b) Any person or corporation planning to engage in any act that affects the environment must file an "Environmental Impact Statement"	1
Federal Water Pollution Control Act	To regulate the discharge of toxic pollutants into the natural waterways	19
Clean Air Act (CAA)	To protect and enhance air quality in order to protect the public health and welfare	20
Resource Conservation and Recovery Act (RCRA)	To initiate a federal role in overseeing solid wastes, hazardous wastes, and resource recovery	10, 22
Noise Control Act	To protect all Americans from "noise that jeopardizes their health and welfare"	10
Occupational Safety and Health Act (OSHA)	To protect workers from on-the-job injury or exposure to harmful pollutants	10
Food and Drug laws	To insure that the foods and drugs available on the market are not harmful to health	10
Federal Insecticide, Fungicide, and Rodenticide Act (FIFRA)	To regulate poisonous agricultural chemicals	15
Toxic Substances Control Act (TSCA)	To regulate toxic substances that are not under the jurisdiction of the Food and Drug Administration (FDA), FIFRA, or (OSHA)	10

A brief summary of some important United States environmental legislation is listed above. Specific topics will be discussed in more detail in appropriate places throughout this book.

2.3 Case History: Reserve Mining—A Lengthy Legal Battle

Reserve Mining Company is one of the largest iron mining and processing corporations in the world. Every *day* the company digs nearly 100,000 tonnes of ore out of the ground (Fig. 2–8). This ore contains iron mixed with less valuable minerals and rocks. Of the 100,000 tonnes mined, about 30,000 tonnes of concentrated iron ore pellets are produced, and 70,000 tonnes of solid waste material are thrown away. In the

1950s the company was dumping these wastes into the waters of Lake Superior (Fig. 2–9).

Reserve Mining operations are economically important to regions of northern Minnesota as well as to the country in general. The company employs 3000 people and supplies over 15 percent of the iron ore used in the United States. If the plant were shut down, the local economy would be destroyed. Not only would the 3000 employees be out of jobs, but many of the business people in the neighborhood would also be out of work. There might be a national shortage of iron ore, and the price of steel could rise. That, in turn, would lead to increased inflation across the country.

The waste that is dumped into Lake Superior contains many different types of minerals. One of these is a fibrous material called **asbestos.** Some of these fibers float along the lake to Duluth, Minnesota, and to Superior, Wisconsin,

Figure 2–8 The iron ore processed at Silver Bay, Minnesota, is mined with the aid of huge power shovels. Each shovel removes over 10 tonnes of ore in a single scoop. (Photo courtesy of Reserve Mining Company.)

residents, various government agencies have tried to force the company to stop the pollution. The legal battle has been long and complex.

In 1969 the federal government recommended that Reserve stop polluting Lake Superior. Nothing was done. Two years later the United States Environmental Protection Agency (EPA) ordered the company to develop a water pollution control program within six months. Reserve submitted a plan that was unacceptable, so the United States government took the company to court. The first court battle lasted two years. On April 20, 1974, a district judge ruled that the air and water pollution endangered the health of the people living in the area. He ordered the plant to shut down. Rather than clean up their operations, Reserve immediately appealed the legal case to a higher court. Two days later a court of appeals judge overruled the district judge and said that more time was needed to review the case. The plant was allowed to resume operations immediately. The company continued to dump 70,000 tonnes of waste into Lake Superior every day. One year later the higher court ruled that Reserve must stop water pollution "within a reasonable time." Nothing was done. In 1976, the court ordered that all water pollution must cease by July 7, 1977, unless "circumstances in the case had changed before that time."

and enter the drinking water in these cities. It is known that asbestos fibers in the air cause cancer in humans. Experts feel that it is likely that asbestos fibers in water also cause cancer, but no one is sure. Since there is a strong possibility that Reserve is endangering the health of local

Figure 2–9 Aerial view of the Reserve Mining Company plant at Silver Bay, Minnesota. The white area in the photograph consists of taconite tailings poured into Lake Superior. (Wide World Photos.)

By now seven years had gone by since the United States government had demanded that the pollution stop. During that seven years the mining had continued without any change in operation. After 1976 the company finally agreed to start construction of a water pollution control facility. This agreement was considered to be a "change in circumstances." In May, 1977, the court stated that as long as construction of the water pollution control facilities was proceeding normally the company would be allowed to continue to pollute until April 15, 1980.

The company did build a waste disposal facility, completed in March, 1980, and the dumping site at the lake was closed down. The legal battle ended on April 23, 1982, with a final agreement by Reserve to pay $1.84 million to Duluth, Minnesota and three smaller communities for the costs of filtering their drinking water.

The 12½ year dispute is a classic example of a struggle between economics and the desire for a clean, healthy environment. The government finally forced Reserve to build a waste disposal site that does not pollute Lake Superior. But from 1969, when the case first went to court, to 1980, when the construction was completed, people lived in a polluted and perhaps poisonous environment for 11 years. The men and women who work at Reserve have held their jobs. Pay at the mine is high, and most people have been able to live a "good life." But the external price may be enormous. In the years ahead many may die of cancer contracted from mine pollutants.

An important issue has been raised by the legal battle. Article V of the Bill of Rights guarantees that "no person . . . shall be . . . deprived of . . . property without due process of law." A corporation is a "person" in the eyes of the law, and a working factory is property. Therefore, a factory must be judged to be in violation of the law before action can be taken against it. But the law may move slowly, and a corporation that can afford the necessary millions of dollars can make the court battle long indeed. Meanwhile, the affected population must await the legal outcome. If the government wins its case, the pollution can be stopped. Affected individuals may sue for damages. But if their health has been impaired, no court can turn the clock back.

2.4 Environmental Problems and Laws of Nature

A few incentives for reducing or preventing environmental degradation have been presented. But the will to confront environmental problems is only a first step. How are specific problems actually solved? As mentioned in Chapter 1, environmental solutions can be divided into two general categories—social and technical. To illustrate the difference between the two, consider the problem of oil reserves being depleted to provide large amounts of fuel for automobiles. One social solution would be to encourage people to drive less. With fewer cars on the road, there would be less demand for gasoline. On the other hand, a technical solution would be to design automobiles that burn fuel more efficiently so that a car could go further on a gallon of gasoline.

Although both types of solutions are helpful, each has its limitations. In a country like the United States, it is reasonable to expect that people can drive less and still maintain a high standard of living. Fuel consumption did, in fact, decline during the early 1980s. No one expects, however, that the entire population will abandon the automobile in favor of buses, bicycles, walking, or riding horses to work. Thus this social answer, like most others, can solve the problem only partially. Technical solutions, powerful as they are, are also limited. There are certain laws of nature that govern the way matter behaves. These laws can never be broken, and they set boundaries on how far technical solutions can help us. Two of the most important of these laws are the First and Second Laws of Thermodynamics.

The Laws of Thermodynamics are laws about energy. *"Energy"* is the capacity to do work or to transfer heat. In ordinary speech we refer to "physical work" or "mental work" to describe activities that we think of as energetic. To the physicist, however, the word "work" has a very specific meaning: *work is done on a body when the body is forced to move*. Merely holding a heavy weight requires the application of a force, but it is not work because the weight is not being moved. Lifting a weight, however, is work.

A moving freight train has kinetic energy.

A stone perched high above the floor of the Grand Canyon has potential energy.

This campfire is producing heat energy.

This pile of coal can burn to produce heat. It has chemical potential energy.

Figure 2–10 *There are many different forms of energy.*

Climbing a mountain or a flight of stairs is work. So is stretching a spring or compressing the gas in a cylinder or in a balloon because all of these activities force something to move.

Heat is another form of energy. When a lump of coal is burned, the energy can be used to warm a room, and this is a form of heat transfer. Heat can also change the chemical state or composition of a substance. For example, when ice is heated, it melts; when sugar is heated enough, it decomposes. All of these changes require energy (Fig. 2–10).

Now imagine that you see an object lying on the ground. Does it have energy? If it can do work or transfer heat, the answer is yes. If it is a rock lying on a hillside, then you could nudge it with your toe and it would tumble down, doing work by hitting other objects during its descent and forcing them into motion. Therefore, the rock must have had energy by virture of the *potential to do work* that was inherent in its hillside position. If it is a lump of coal, it could burn and be used to heat water or cook food, and therefore it too has energy. If it is an apple, you could eat it, and it would enable you to do work and to keep your body warm, and so an apple also has energy (Table 2–1).

Energy can be changed from one form to another. For example, you eat an apple and then walk to school. The chemical **potential energy** of

TABLE 2–1   Energy: What It Can Do

	Heat a Body Above the Surrounding Temperature	Work
Apple	Eating it helps you to maintain your body temperature at 37°C (98.6°F), even when you are surrounded by air at 20°C (68°F)	Eating it helps you to be able to pull a loaded wagon up a hill
Coal	Burning it keeps the inside of the house warm in winter	Burning it produces heat that boils water that makes steam that drives a piston that turns a wheel that pulls a freight train up a hill

the apple is converted to **kinetic energy,** or energy of motion. Many examples come to mind. An automobile engine, like your body, also converts the potential energy of a fuel to kinetic energy, energy of motion.

The First Law of Thermodynamics

The ability to convert energy from one form to another was an important discovery, but what happens to the energy during its conversion? Repeated experimentation has shown that it is always conserved. This **Law of Conservation of Energy** is also known as the **First Law of Thermodynamics.** The following example illustrates the importance of this concept:

Since the beginning of civilization, people have been faced with the problem of how to lift heavy loads. Buildings have been traditionally built of wood or stone, and engineers have had to move these heavy items from the ground onto the walls and roofs. In the early days, we can imagine that people worked together to lift stones and timbers. Later, clever inventors developed simple machines such as the lever, the wheel, and the inclined plane. A person who can lift 50 kg unaided can lift hundreds of kilograms with a lever and even more with a well-designed system of gears or pulleys. In the early days, engineers believed that machines actually reduced the amount of work required to lift an object.

Scientists now understand that such devices do not actually reduce the amount of work. Instead, they spread it out over a longer period of time and smooth out the effort. However, this difference can easily be overlooked, and a device

such as a lever can be mistakenly thought of as a "work-saver." Since many "work-savers" had been invented, people reasoned that if you were clever enough you could build a machine that would do all the work for free. They called this imaginary device a **perpetual motion machine.**

Figure 2–11 Drawing by Pieter Pourbus the Elder of a man-powered crane in use in the sixteenth century. The development of clever machines led some early inventors to believe that it would be possible to build a perpetual motion machine. However, we are now certain that this is impossible.

If you owned a perpetual motion machine you could turn it on, and it would lift all the boulders you wanted while you sat and watched. It may be difficult for the modern reader to appreciate that the search for the perpetual motion machine seemed entirely reasonable. Many very clever people looked for a solution (Fig. 2–11). However, all attempts failed. The failures have been so consistent that we are now convinced that the effort is hopeless. We believe that it is a fundamental law of nature that it is impossible to build a perpetual motion machine.

The First Law can be stated in another way: **Energy cannot be created or destroyed.** The perpetual motion machine fails because it cannot create the energy needed to keep itself running forever. A real machine works because energy is constantly being supplied to it. When that energy is used up (converted to another form) the machine stops. When your gasoline tank is empty, your engine can't run and your car won't move. All of the energy in the gasoline has been converted by the engine into other forms of energy, such as heat and the motion of the car.

Don't ask for a formal proof of the First Law; there is none. Its truth comes from a broad range of experience. Careful experiments repeated again and again support the truth of the First Law.

It is immediately obvious that the First Law limits technical solutions to environmental problems. If a person must move from a home in the suburbs to a job in the city, some form of energy is required. The energy may be gasoline to drive a car. It may be coal to produce electricity to operate an electric trolley, or potatoes and apples to provide the muscle energy to ride a bicycle, but some form of energy is absolutely essential. There is no magic way to perform work without a supply of energy.

The Second Law of Thermodynamics

If energy cannot be found for free, it would seem desirable at least not to waste any. But unfortunately, this too is impossible. Whenever energy is transformed from heat to work, some must always be wasted. Concentrated, or high-temperature, heat energy has the ability to perform useful work. However, as a machine per-

forms work some of the energy escapes in the form of low-temperature heat. It is this low-temperature heat, which has little ability to do useful work, that is wasted. The inability to convert all heat energy into useful work is another fundamental law of nature and is called the **Second Law of Thermodynamics.**

The Second Law, like the First, arose out of a long series of observations. If a hot iron bar is placed on a cold one, the hot bar always cools while the cold one becomes warmer, until both pieces of metal are at the same temperature. No one has ever observed any other behavior. Similarly, if a small quantity of blue ink is dropped into a glass of water, the ink will disperse until the solution becomes uniformly light blue. The ink does not stay concentrated in one section of the clear water (Fig. 2–12).

Thus there appears to be a natural drive toward sameness or disorder. If we have two blocks of iron at different temperatures in a system, or a spot of ink in a glass of water, there is a differentiation of physical properties. Such dif-

Figure 2–12 The Second Law of Thermodynamics states that any undisturbed system will naturally tend toward maximum disorder. If a drop of ink is placed in a glass of water, the ink will always disperse until it is evenly distributed.

ferentiation results from some kind of *orderly* arrangement among the individual parts of the system. This is a subtle but important point. In your experience, how is *order* different from *disorder?* The answer is that *order* is characterized by repeated *separations.* Your room is orderly if all the books are separated from your socks; books on the shelves, socks in the drawer. It is disorderly if books and socks are all mixed up in both places. Similarly, if a small spot of blue ink is *separated* from clear water, the system is orderly. In all cases disordered systems are more natural (that is, more probable) than ordered ones. In your own common experience, you know that if you neglect your room, it naturally becomes disorderly. The reason is that there are always more ways to be disorderly than to be orderly (or, in other words, there are more ways

to break rules than to follow them). Therefore, any isolated system, if left alone, will tend toward disorder. **Entropy** is a thermodynamic measure of disorder. It has been observed that *the entropy of an undisturbed isolated system always increases during any spontaneous process;* that is, the degree of disorder always increases. Thus, if you drop a spot of ink in water, the ink will spread out evenly throughout the liquid. It will become disorderly. If you don't clean your room regularly, it will become messy (Figure 2–13).

Superficially, living organisms may appear to run counter to the Second Law by creating order out of disorder. Consider a frog developing from a tiny egg. The embryo builds large, highly ordered, high-energy molecules, such as carbohydrates from small building blocks such as sugar molecules. However, living things are not really exempt from the Second Law of Thermodynamics. They are not isolated systems and must use an outside supply of energy—food—to synthesize the molecules they need, to move substances around, and to combat the universal tendency toward increasing disorganization.

Now consider a heat engine that could burn any organic fuel such as gasoline, wood, starch, or sugar. As the fuel burns, the engine can do work. It can supply energy to a generator that can make electricity. Imagine that the electricity is used to light up sunlamps that can promote photosynthesis to make plants grow and produce new plant tissue (Fig. 2–14). Could the plant tissue (sugars, starches, and so on) then be burned in the heat engine, using the oxygen supplied by photosynthesis, to recycle all the energy? In other words, could the system operate indefinitely, continuously recycling its energy and producing new fuel for the engine? The answer is no, such a system could *not* work, even it it were perfectly insulated and no energy were lost to the outside. The reason has to do with changes in the entropy of the molecules involved.

To simplify the question, assume that the fuel is the sugar glucose. Glucose is also the product of photosynthesis. When glucose burns it produces molecules of carbon dioxide and water. A complex, orderly molecule of glucose is converted into a disordered set of more numerous, less complex molecules. Photosynthesis does just the reverse. If the photosynthesis pro-

A

B

Figure 2-13 A desk when it is (a) neat and (b) messy. Unless work is performed to clean the desk, it will tend to stay messy, because the entropy of the messy desk is higher than that of the desk when it is orderly.

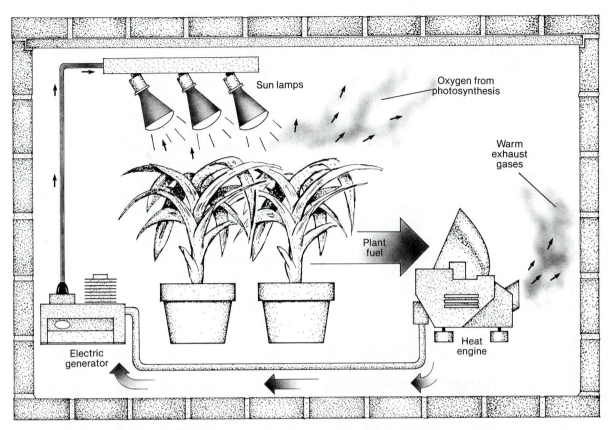

Figure 2–14 *An impossible attempt to recycle all useful energy in an isolated system. Plant matter is fuel for a heat engine that drives a generator to make electricity for sunlamps to promote photosynthesis. But because the energy in the exhaust gases of the heat engine is in the form of random motion of the gas molecules, it cannot promote photosynthesis. The cycle must die down as its entropy increases.*

duced as much glucose as was burned in the engine, the entropy of the entire system would remain constant. Such an event *never happens.* Why could it not happen in this case? The exhaust gases from the heat engine are hot. But their heat energy is in the form of the random motion of carbon dioxide and water molecules. Their energy does not have the quality of sunlight; it will not cause plants to grow and synthesize complex molecules like glucose. Thus, as this system continues to operate, even though no heat energy is lost to the outside, there will be less plant growth, hence less fuel to run the engine. The sunlamps will grow dimmer; eventually the plants will die, the engine will stop, and everything will be warm but disorderly. Entropy will have reached its maximum.

In any heat engine, some of the fuel is used to do work, and some is lost as waste heat to the environment. A small gasoline-powered engine is only about 15 percent efficient. This means that 85 percent of the potential energy in the fuel merely warms the surrounding air. (You can feel this heat if you place your hand near the exhaust muffler of the engine.) Only 15 percent is harnessed to do work. The most efficient heat engines commonly used today are those that produce electricity in large steam-fired generating stations. They operate at about 40 percent efficiency; 40 percent of the fuel energy produces work, and 60 percent is lost to the environment. Engineers have worked hard to improve this efficiency, but they cannot design an engine that escapes from the constraints of the Second Law.

(Equations for calculating energy efficiency are given in Appendix A.)

There are three important practical consequences of the Second Law:

1. The energy in a fuel can never be completely converted to work.
2. Useful energy cannot be recycled. If a lump of coal is burned, some of its capacity to do work is lost forever. It can never be reused.
3. Highly ordered systems are more difficult to maintain than less ordered systems. This is an important point. Imagine that you have two brand new cars with identical engines. One car has automatic transmission and air conditioning. It also has electrically powered windows, door locks, radio antenna, and seat adjustments—the works. The other car, built with the same quality craftsmanship, has manually operated transmission and accessories. Which car will break down more often? Of course there are more things to go wrong with the fully equipped model—it is more or-

derly. More effort must be spent maintaining that order. All other things being equal, it will spend more time in the repair shop. The same situation arises in many other instances as well. Consider agricultural systems. In a natural prairie, grains, weeds, and scrub bushes all grow together and intermingle with each other. These plants are consumed by insects, rodents, birds, large grazers, and other animals. In a modern agricultural system, on the other hand, the farmer separates weeds from the grain and cultivates only the grain (Fig. 2–15). Likewise, cattle are separated from rodents and insects, and the unwanted species are removed. This type of system produces more food for human consumption than a natural prairie. But it is more orderly and hence more vulnerable. Work must be constantly expended to maintain the order. This subject will be discussed further in Chapters 14, 15, and 16. For now the important point is this. A technical response to a problem—such as the need to grow more food—often results in the creation of a more ordered system, but additional energy inputs are needed to maintain these ordered systems. And remember, ordered systems are always more susceptible to breakdown than less ordered ones. It is a law of nature.

Figure 2–15 Many modern systems, like this vegetable farm, are highly ordered. Large inputs of energy are required to maintain this order.

Summary

The costs that are not accounted for by the manufacturer of a product but are borne by some other sector of society are external costs, or **externalities.** In recent years, the cost of pollution control in the United States has been more than offset by the savings in external costs. **Cost/benefit analysis** is an attempt to balance the cost of pollution control against the cost of externalities in order to minimize the total impact of the cost on individuals.

Three actions used in environmental law are nuisance, trespass, and negligence. Legal approaches to pollution control can assume the form of (1) restrictions, (2) penalties, (3) subsidies, and (4) pollution tax. If a factory doesn't comply with the law it may be legally susceptible to paying damages or to an injunction preventing its illegal action.

Technical solutions to environmental problems are limited by the physical laws that govern the behavior of all matter. Among the most important of these are the First and Second Laws of Thermodynamics. According to these laws, energy cannot be created or destroyed. To make matters worse, when energy is transformed from one form to another, some energy is converted to a form that cannot be used to do work. As a consequence of the Laws of Thermodynamics, useful energy cannot be recycled because some of it is lost forever whenever it is used to do work. As a result, the supply of certain fuels is inexorably running out.

Questions

Legal, Social, and Economic Aspects

1. On February 26, 1972, heavy rains destroyed a dam across Buffalo Creek in Logan County, West Virginia. The resulting flood killed 75 people and rendered 5000 homeless. The dam was built from unstable coal-mine refuse or "spoil." A United States Geological Survey had warned of the instability of many dams in Logan County, especially the one at Buffalo Creek. Soil stabilization and reclamation programs directed at mine spoil dams would have added to the cost of producing coal. After the flood, the Bureau of Mines denied responsibility for corrective action, contending that although mine spoil is within the Bureau's jurisdiction, dams are not. (1) How would you relate the concept of economic externalities to this tragedy? (2) Suggest legislation designed to prevent future Appalachian dam disasters. (3) How will your legislation affect the price of coal and its competitive position in the fuel market?

2. Explain why the control of pollution may result in a net increase in the number of jobs. Give examples.

3. If you were the mayor of a small town, and a prosperous factory, the largest single employer in the town, were illegally dumping untreated wastes into a stream, what action would you recommend? What if the factory were barely profitable?

4. If the burden of environmental control falls heavily on some particular segments of society, do you think it would be fair to provide compensation? Would you favor granting such compensation to workers who lose their jobs? To companies whose costs for environmental clean-up are high? To stockholders whose equities are reduced in value? For each of these segments, what kinds of abuses are possible? What benefits to the environment would such compensations provide?

5. Suppose you were a reporter assigned to compare the cost of stopping the pollution of the local river with the economic cost of the pollution. What sources of information would you seek out? Which category of costs would be more difficult to estimate? Defend your answer.

6. Argue for and against the validity of cost/benefit analysis.

7. Identify a source of pollution or environmental degradation in your community. Discuss various types of government action that might alleviate the problem. Argue the various merits and problems with each potential solution.

8. Discuss the following alternative proposals for reducing water consumption. (1) Tax all water usage at a constant rate per gallon. (2) Tax water usage at a progressive rate, that is, allow the tax to rise with increasing use. (3) Shut off all water from 2 to 4 P.M. (4) Shut off all water from 2 to 4 A.M. (5) Ration water at some reasonable level. What is the

purpose of the legislation? What are the side effects? Can you propose better legislation?

9. Argue for or against the proposition that noise can be a form of trespass.

The Laws of Nature

10. Referring to a discussion of the difficulty in recovering dispersed ores from the ocean, Professor Peter Frank of the Department of Biology of the University of Oregon wrote, "The Second Law of Thermodynamics comes in with a vengeance. . . ." Explain what he meant.

11. Energy from the sun, wind, and tides is considered to be "renewable" whereas energy from fossil fuels is not. Is this evaluation true in a theoretical sense or in a practical one? Discuss.

12. In 1981, homeowners in many areas of the country found that it was cheaper to heat their homes with firewood than with electricity generated by nuclear power. Some authors have claimed that this economic structure is a direct result of the Second Law. Explain what is meant by this.

Suggested Readings

A good introduction to environmental economics is given by:
Paul Burrows: *The Economic Theory of Pollution Control.* Cambridge, Mass, MIT Press, 1980. 192 pp.
Gerald Garvey: *Energy, Ecology, Economy.* New York, W. W. Norton & Co., 1972. 235 pp.
Donald T. Savage, Melvin Burke, John D. Coupe, Thomas D. Duchesneau, David F. Wihry, and James A. Wilson: *The Economics of Environmental Improvement.* Boston, Houghton Mifflin Co., 1974. 210 pp.

Several periodical articles outlining cost/benefit analysis and the cost of pollution controls include:
Gordon L. Brady and Blair T. Bower: Benefit-cost analysis in air quality management. *Environmental Science and Technology, 15,* (3): 256 ff, March 1981.

Luther J. Carter: Costs of environmental regulations draw criticism, formal assessment. *Science, 201:* 140 ff, July 1978.
Conservation Foundation: Cost-benefit analysis: A tricky game. *Conservation Foundation Letter,* December 1980.

For an emphasis on natural resources, refer to:
Ferdinand E. Banks: *The Economics of Natural Resources.* New York, Plenum Press, 1976. 267 pp.

A treatment of the Reserve Mining episode is:
Robert V. Bartlett: *The Reserve Mining Controversy: Science, Technology and Environmental Quality.* Indiana University Press, 1982. 293 pp.

Refutations of the idea that environmental improvement costs jobs are given in:
Richard Kazis and Richard L. Grossman: *Fear at Work. Environmentalists for Full Employment.* Washington, D.C., 1982.
Henry M. Peshkin et al. (eds.): *Environmental Regulation and the US Economy.* Johns Hopkins University Press, 1981. 163 pp.

United States environmental law is summarized in:
Government Institutes: *Environmental Statutes.* Washington, D.C., Government Institutes, 1982. 538 pp.
D. R. Greenwood et al.: *A Handbook of Key Federal Regulations and Criteria for Multimedia Environmental Control.* Environmental Protection Agency, EPA-600/7-79-175. 1979. 277 pp.

Citations of legal cases appear in:
Jerome G. Rose (ed.): *Legal Foundations of Environmental Planning.* New Brunswick, N.J., Center for Urban Policy Research (Rutgers University), 1974. 318 pp.

A handbook on the legal rights of citizens is provided by:
National Resources Defense Council (Elaine Moss, ed.): *Land Use Controls in the United States.* New York, Dial Press, 1977. 362 pp.

A comprehensive book that presents legal issues specifically related to energy is:
Donald N. Zillman and Laurence H. Lattman: *Energy Law.* Mineola, N.Y., Foundation Press, 1983. 841 pp.

You will find further discussion of heat engines and the Second Law of Thermodynamics from page 71 on of:

Jonathan Turk and Amos Turk: *Physical Science.* 2nd ed. Philadelphia, Saunders College Publishing, 1981. 642 pp.

An interesting book that relates thermodynamics to world planning is:

J. Rifkin: *Entropy: A New World View.* New York, Viking Press, 1980.

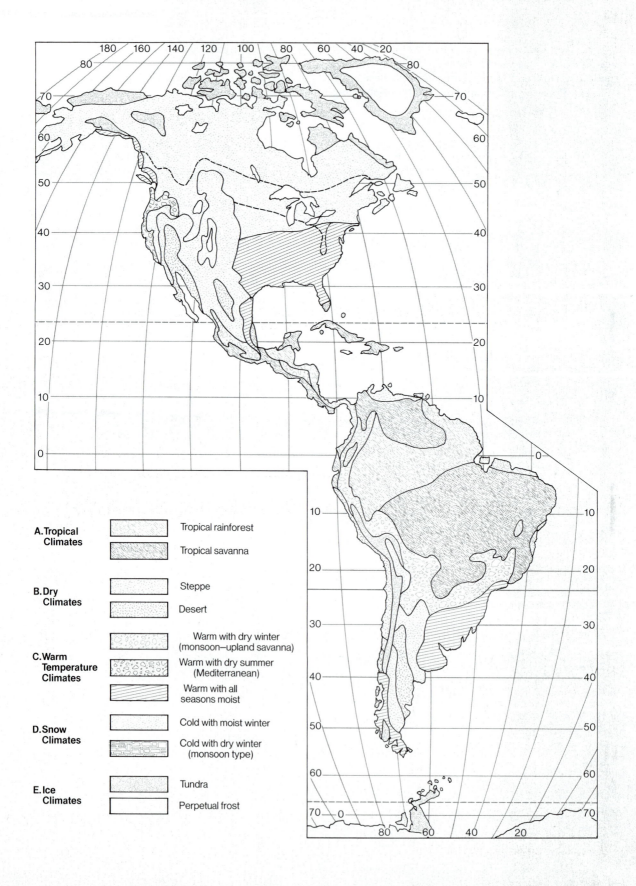

180 160 140 120 100 80 60 40 20		

A. Tropical Climates
- Tropical rainforest
- Tropical savanna

B. Dry Climates
- Steppe
- Desert

C. Warm Temperature Climates
- Warm with dry winter (monsoon–upland savanna)
- Warm with dry summer (Mediterranean)
- Warm with all seasons moist

D. Snow Climates
- Cold with moist winter
- Cold with dry winter (monsoon type)

E. Ice Climates
- Tundra
- Perpetual frost

The Ecological Background

Biomes and climates of the Western Hemisphere.

Ecosystems

If you look around a forest, a city park, or your own back yard, you can think of dozens of questions about the way animals and plants live. Grass grows in the open but not under that tree. What does grass need in the way of soil, sunshine, water, or minerals that it finds in one place but not in another? What does this bluejay feed on? That butterfly laid 500 eggs this spring; what happened to her offspring? If they are dead, why isn't the ground littered with corpses? If this forest is cut down, could crops be grown on the land or would the soil and climate be unsuitable in some way?

All these questions are about **ecology,** the branch of biology that examines the interactions between living organisms and their environments. All living things, including people, can survive only in appropriate environments. What constitutes an appropriate environment, and what is meant when people say that human beings are making the Earth less fit for human habitation? Ecologists have discovered that there are certain rules about organisms and their environments that can be used to answer these and other questions. In this and the next three chapters these general principles of ecology will be considered with an emphasis on systems that do not include people. Human affairs and the impact of civilization on natural systems are the topics of the remaining chapters.

3.1 What Is an Ecosystem?

Although the natural environment consists of the entire Earth and the atmosphere that surrounds it, this complete system is much too complex to study at one time. Even a single forest may cover thousands of square kilometers and contain thousands of different types of plants and animals living together. In order to separate these large systems into smaller areas that can be studied more precisely, ecologists work with manageable units such as a hillside, a forested valley, a lake, or a field. In 1887, Stephen Forbes, biologist for the Illinois Natural History survey, wrote:

A lake . . . forms a little world within itself— a microcosm within which all the elemental forces are at work and the play of life goes on in full, but on so small a scale as to bring it easily within the mental grasp . . . If one wishes to become acquainted with the black bass, for example, he will learn but little if he limits himself to that species. He must evi-

36

dently study also the species upon which it depends for its existence and the various conditions upon which these depend.

Nowadays we would call Forbes's lake, or any other manageably small unit with more or less distinct boundaries, an **ecosystem** (Fig. 3–1). Forbes's comments point out the other characteristics of an ecosystem: It consists of all the different organisms living in an area, as well as their physical environment. Because these all interact with each other and change one another (see Special Topic A) the study of an ecosystem is a complex undertaking. We usually regard an ecosystem as an isolated unit, but this is merely a useful simplification: In reality, ecosystems always interact with one another. For instance, plants, animals, and materials move from one ecosystem to another, as when soil and leaves wash from a forest into a lake, or birds migrate between their summer and winter homes.

Not all ecosystems are natural. A farm may be considered an ecosystem because, in order to manage the farm effectively, all the interactions between crop plants, fertilizers, soil, climate,

and the natural animal and plant life of the area must be recognized and controlled. Similarly, space stations, aquaria, and pots of houseplants are all artificial ecosystems.

3.2 The Kingdoms of Organisms

Living things are not exempt from the Second Law of Thermodynamics (Section 2.4). They need a constant supply of energy to synthesize the molecules they require, to move substances around, and to combat the universal tendency toward increasing disorganization. If an ecosystem is to survive indefinitely it must contain certain kinds of living organisms as well as supplies of energy to support the entire system. A green plant can't get its energy from steak and a cat can't live on grass: Different organisms need different types of food to supply them with energy. Indeed, all the living organisms on Earth are

Figure 3–1 A lake forms a partly isolated ecosystem enclosed by its banks.

SPECIAL TOPIC A
The Balance of Nature

Many processes go on continuously in an ecosystem. Various sets of these processes directly oppose each other; that is, the effect of one undoes the effect of another. Examples are birth and death, immigration and emigration, and gain and loss of water. An ecosystem is disrupted when some of its processes are speeded up or slowed down (for example, the increase of a death rate or a decrease in rainfall). It is a *natural tendency* that, when one process is affected, its opposite is also affected in the same direction (when there is less rainfall, there is also less evaporation). Therefore, in general, when an ecosystem is disrupted, its processes change in such a way that the effect of the disruption is partly undone. For example, drought in a grassland inhibits the growth of plants. The meadow mice become malnourished, their fertility rate drops, and their birth rate decreases. In addition, they have a protective behavioral response to lack of food; they retreat to their burrows and sleep. Thus their death rate also decreases because they are less exposed to predation. Their behavior protects their own population balance as well as that of the grasses, which are not consumed by hibernating mice. This self-regulating property of an ecosystem is called the "**balance of nature**" or, more formally, **ecosystem homeostasis**.

usually classified, largely on the basis of how they feed, into one of five different **kingdoms.**

(a) Plants. Plants are fundamental to all life on Earth because they trap energy from the Sun and use that energy to build living tissue (Fig. 3–2). Since plants do not need to feed on other organisms, they are called **autotrophs,** a word meaning self-nourishers. Animals cannot use the Sun in this manner, and they are therefore dependent on plants as their fundamental source of food.

A plant uses the process of **photosynthesis** to convert light energy from sunlight into chemical energy. The plant uses light energy to convert the energy-poor molecules carbon dioxide and water into energy-rich food molecules such as sugars. Thus, inorganic molecules (carbon dioxide and water) are converted into organic molecules (sugars, fats, and proteins). ("Organic" and "inorganic" are defined in Box 3.1.) The overall equation for photosynthesis is:

$$\text{Carbon dioxide} + \text{water} \xrightarrow[\text{sunlight}]{\substack{\text{gives in the} \\ \text{presence of}}} \text{sugar} + \text{oxygen}$$

$$6CO_2 + 6H_2O \longrightarrow C_6H_{12}O_6 + 6O_2$$

To understand photosynthesis, imagine that you take a seed, say a pumpkin seed, and plant it in a pot of dry soil. You weigh the soil and the seed together. Then you water the pot and let the pumpkin seed sprout and grow. Huge vines spread out, and large orange pumpkins appear at the ends of the stems, but the soil does not disappear. If you dry the grown plant and its soil and weigh them together, you will find that they are much heavier than the original dry soil and seed. Where did the mass of the plant come from? The green leaves of the pumpkin trapped carbon dioxide from the air and water from the soil and combined them to form sugar and other compounds.

The energy trapped by the plant during photosynthesis is not lost. If a pumpkin is burned, it will produce light and heat, which is a way of releasing the stored chemical energy (Fig. 3–3). Carbon dioxide and water are low-energy compounds; they cannot be burned to produce heat. Sugar, the product of photosynthesis, contains

Figure 3–2 Plants, like this trailing arbutus *(Epigaea repens)*, obtain the energy they need to make their food from sunlight. (Photograph by Alvin E. Stefan, National Audubon Society.)

stored chemical energy and can be burned to produce heat. Carbon dioxide and water are released as byproducts—

$$\text{sugar} + \text{oxygen} \xrightarrow[\substack{\text{energy} \\ \text{and} \\ \text{produces}}]{\text{releases}} \text{carbon dioxide} + \text{water}$$

$$C_2H_{12}O_6 + 6O_2 \longrightarrow 6CO_2 + 6H_2O$$

A fire is a rapid form of oxidation. Sugar also combines with oxygen inside living cells to produce the same products (carbon dioxide, water, and energy) in the same proportions. The difference is that in a living organism oxidation proceeds at a slower and more controlled rate. This process, called **respiration,** releases the energy stored in complex molecules for use in maintaining cell functions.

Oxygen is a waste product of photosynthesis. Some of this oxygen is used in respiration, but plants produce more oxygen during photosynthesis than all organisms on Earth use

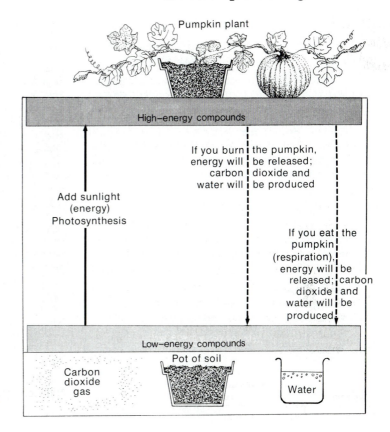

Figure 3–3 Some energy relationships of a pumpkin plant.

in respiration. Oxygen, therefore, slowly accumulates as a gas in the atmosphere. Burning consumes oxygen, and so natural fires, such as forest fires caused by lightning, use some of the leftover oxygen. In recent years, however, human activities, particularly the burning of vast amounts of fossil fuels, may be using up oxygen faster than it is replaced by photosynthesis.

A plant cannot survive on just the carbon dioxide, water, and light necessary for photosynthesis. These ingredients allow it to produce only sugars and other carbohydrates, and a plant must also produce proteins, fats, and various other substances. Plants build these other substances from carbohydrates, but they need some inorganic nutrients for this process. Proteins, for instance, contain nitrogen and sulfur, and plants must obtain these elements from the soil. This is why fertile soil is necessary if plants are to flourish.

An ecosystem must supply adequate amounts of water, carbon dioxide, oxygen, soil miner-

als (defined in Box 3.1), and sunlight if plants are to grow in it.

(b) Animals. Members of the animal kingdom are **heterotrophs** (other-nourished), which means that they cannot make their own food but must eat other organisms to obtain organic molecules. Like plants, animals respire to release energy from the food they eat. As a result, if it is to support animal life, an ecosystem must contain plants or animals for food and a supply of oxygen (Fig. 3–4). Most animals also need drinking water, although some get all the water they need from their food.

(c) Fungi. Because plants can manufacture their own food, it would seem at first glance that an ecosystem could survive indefinitely if it contained only plants. But this is not the case. Plants remove minerals from the soil, and soil minerals are replaced only very slowly from the rock beneath (see Chapter 14). In an agricultural system, nutrients are often replaced artificially

BOX 3.1 MINERALS AND NUTRIENTS: SOME VOCABULARY

An **element** is a substance that consists of only one kind of atom. Carbon consists only of carbon atoms, oxygen of oxygen atoms, and hydrogen of hydrogen atoms. Therefore, carbon, oxygen, and hydrogen are elements. Water is not an element because it consists of groups of hydrogen and oxygen atoms bonded together (H_2O). Groups of bonded atoms are **molecules,** and a substance consisting of molecules containing more than one element is a **compound.** Under terrestrial conditions, even most elements exist as molecules rather than as single, isolated atoms. For instance, nitrogen gas consists of molecules containing two atoms each—N_2. Oxygen is O_2; hydrogen is H_2. Solid elements, like carbon and iron, consist of networks of many atoms bonded together.

One particular group of substances has always been considered unique and very important. This group consists of substances found mainly in living organisms. Many of these substances, such as proteins and cellulose (in plant cell walls), are very complex and cannot be represented by simple formulas. For many years, chemists thought that these substances could be produced only by living organisms. Hence they came to be called **organic** compounds. All of these compounds contain carbon atoms bonded to hydrogen atoms or to other carbon atoms. After it was shown in a series of discoveries from 1828 to 1845 that some components of these substances could be synthesized from nonliving sources, the word "organic" came to refer to compounds containing C–C or C–H bonds, with or without other atoms. (Marble, $CaCO_3$, is not considered organic because, although it contains carbon, the carbon atom is not bonded to carbon or hydrogen.)

By this definition, substances such as proteins, carbohydrates, and fats are organic, but synthetic plastics qualify as well. Their molecules contain long chains of carbon atoms. Plastics are generally synthesized from oil or coal, organic molecules produced by living organisms millions of years ago. However, plastics could be made from inorganic carbon dioxide and water and they would still be organic compounds.

Having defined organic, we can define "inorganic" simply by exclusion. An **inorganic** substance is any substance that is not organic. Or, any substance that does not contain carbon atoms bonded to other carbon atoms or to hydrogen atoms is inorganic.

Nutrients are substances that organisms must take in from outside themselves, parts of the diet. They may be organic or inorganic. In the human diet, fats, proteins, carbohydrates, and vitamins are organic. Iron and calcium are inorganic nutrients.

Minerals are inorganic solids that occur naturally in the Earth. Each usually has a specific composition and characteristic structure. Thus the mineral quartz is composed of silicon dioxide (SiO_2) arranged in a particular crystalline structure.

The nutrients that organisms need to survive can be obtained from eating other organisms, from gases in the atmosphere, or from minerals in the ground. (**Water,** which is also vital to all living organisms, is inorganic but is not a mineral.)

by fertilization. In a natural system, if plants just went on living and growing and dying, eventually all the minerals in the ecosystem would be absorbed into the bodies of plants. Then there would be huge piles of dead plants containing minerals but no more minerals in the soil, and plant growth would stop. Every ecosystem needs, and contains, **decomposers.** These are organisms that digest the food locked up in dead organisms and return their minerals to the soil in inorganic forms that plants can absorb and use. The fungi are an important kingdom of heterotrophic decomposers, although there are de-

Figure 3–4 A reindeer in the tundra. All animals must obtain their food by eating plants or other animals. (Photograph by Eric Hosking, National Audubon Society.)

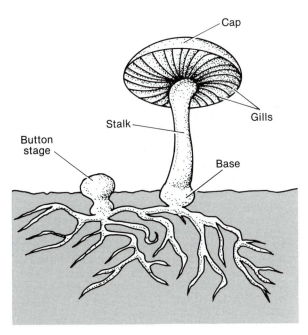

Figure 3–5 A mushroom. As with all fungi, the food-absorbing part of the organism lives buried in its food and all we see on the surface is the reproductive fruiting structure.

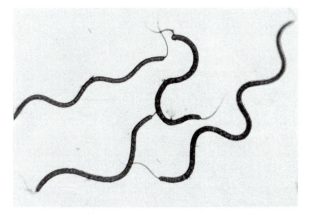

Figure 3–6 *Spirillum volutans*, a large bacterium. (Courtesy of Carolina Biological Supply Company.)

composers such as bacteria and insects in other kingdoms as well. Some fungi are not decomposers but **parasites,** meaning that they feed on living organisms rather than on dead ones.

Fungi are **saprobes,** organisms that absorb food across their body surfaces, unlike animals, which ingest food by way of a mouth. The food-absorbing part of a fungus (Fig. 3–5) usually lies buried in its food, and only the reproductive structures are exposed—black, green, white, or pink fuzz on moldy foods, or mushrooms. To obtain their own energy, fungi break down organic molecules from other organisms, excreting inorganic molecules that plants can use.

Fungi are powerful decomposers, and some cause inconvenience when they break down such substances as food, wood, leather, hair, wax, and cork. Large quantities of valuable food resources are lost every year to fungal decay. During their long years of global supremacy, the British and Spanish navies lost more ships to wood-rotting fungi than to enemy action. It was quite common for the bottom of a ship to fall out in midocean. Despite these problems, however, life on Earth would be impossible without fungal decomposers.

Any ecosystem that contains moisture, oxygen, and the dead bodies of other organisms is an appropriate habitat for fungi. Fungal decomposers, then, will be found almost everywhere except in places such as the mud at the bottom of a lake where there is no oxygen.

(d) Bacteria. The kingdom of bacteria contains one-celled organisms whose cells are different from those of other forms of life. The genetic material (DNA) in a bacterium is not surrounded by a membrane as it is in all other organisms.

Some bacteria are capable of photosynthesis, which means that they are autotrophic producers, providing food for heterotrophs. Others are saprobic decomposers like the fungi (Fig. 3–6). The main difference between fungal and bacterial decomposers is that many bacteria can survive without oxygen, whereas most fungi cannot. Deep layers of soil or mud where little oxygen exists are populated by bacteria rather than by fungi. There is almost no natural organic substance that some bacterium cannot use for food, which makes bacteria very versatile decomposers.

(e) Protista. The kingdom Protista is made up of all one-celled organisms that are not bacteria. It includes *Amoeba, Paramecium,* and most of the one-celled **algae** that float near the surfaces of oceans and lakes (Fig. 3–7). Some protists are heterotrophs, feeding on smaller protists or on bacteria, or they exist as parasites feeding on animals. Others, such as the one-celled floating al-

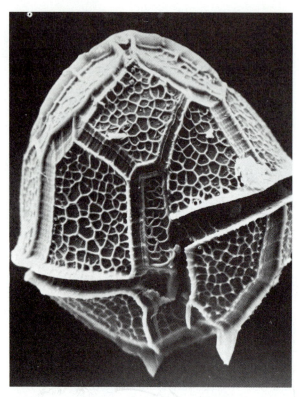

Figure 3–7 *Photograph of a dinoflagellate, a protist found in the sea, taken at a magnification of more than 2000 times. (Courtesy of Biophoto Associates.)*

Some ecosystems contain no sunlight and therefore have no producers. Consider the depths of the ocean or a stream in a dense forest where little light reaches the ground. These ecosystems depend on producers growing in sunnier areas to supply their food. Animal life in a forest stream feeds on leaves and twigs falling from the trees above; likewise, many organisms in the deep ocean depend for nourishment on dead bodies sinking from the upper layers of water where photosynthesis occurs. Ultimately, the source of energy for these ecosystems is still the Sun. A few species of bacteria can use energy from chemical reactions rather than from sunlight to produce their food. These producers, however, account for an insignificant fraction of the total energy trapped by living organisms on Earth. If the Sun were snuffed out tomorrow, life as we know it would very rapidly grind to a halt.

gae, are photosynthetic. These protists are important ecologically as the main producers of food in aquatic ecosystems such as lakes, rivers, and oceans.

3.3 Food Webs

From the previous discussion, it is clear that if an ecosystem is to be self-sustaining, it must at the very least contain some source of food, usually produced by autotrophs, and decomposers to recycle nutrients. The producers may be photosynthetic plants, bacteria, or protists. The decomposers may be bacteria or fungi and possibly some animals as well. In addition, most ecosystems need minerals, carbon dioxide, oxygen, water, and sunlight to keep the producers producing. (Special Topic B describes some un-

usual ecosystems that exist without producers or sunlight.)

An ecosystem is usually much more complicated than the basic chain of life and death of producers and decomposers. The energy and nutrients in the bodies of producers are resources that may be exploited by a variety of heterotrophic **consumers.** Consumers include **herbivores** (animals such as grasshoppers and cows that eat plants), and **carnivores** (animals such as praying mantises and owls that eat animals), as well as parasites of both. A few insect-catching plants are both producers and consumers, and many animals, such as pigs, bears, rats, and people, are **omnivores,** animals that eat both plants and animals.

The **trophic level** to which an organism belongs describes how far the organism is removed from plants in its level of nourishment. Green plants make up the first trophic level, herbivores make up the second trophic level, and the higher trophic levels are composed of the carnivores. For example, a grasshopper that eats grass belongs to the second trophic level, a shrew that eats the grasshopper belongs to the third, an owl that eats the shrew belongs to the fourth, and so on (Figs. 3–8 and 3–9). In another commonly used set of terms, plants are said to be **producers,** herbivores are **primary**

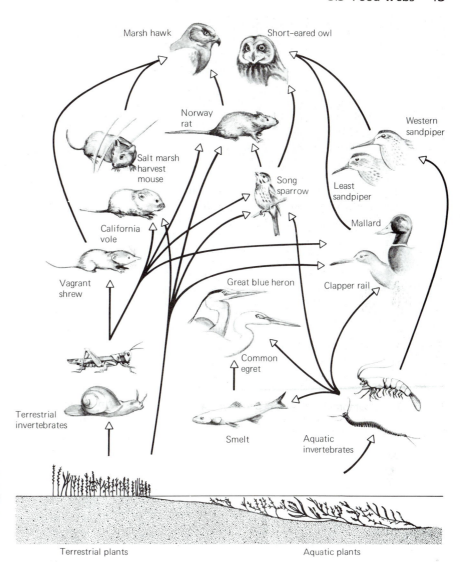

Figure 3–8 A simplified diagram of a food web in a salt marsh. (Modified from Robert Leo Smith: *Ecology and Field Biology.* 2nd ed. Copyright 1974 Robert Leo Smith. Reprinted by permission of Harper & Row, Publishers, Inc.)

consumers, and carnivores are **secondary, tertiary,** and **quaternary consumers,** depending on what they eat. Omnivores, which eat both consumers and producers, may belong to many trophic levels (Fig. 3–10 and Table 3–1). The trophic levels of an ecosystem can be represented by a diagram showing the total dry weight of living material present at each level (Fig. 3–11).

The flow of food energy in an ecosystem progresses through a **food chain** in which one step follows another: Primary consumers eat producers, secondary consumers eat primary producers, and so on. In nearly all natural ecosystems, the patterns of consumption are so

complicated that the term **food web** is more descriptive because there are many cross-links connecting the various organisms. (Figures 3–8 and 3–9 are greatly simplified food webs because they show none of the hundreds of decomposers in the systems.)

Food webs are so complex that human interference, particularly with systems that are incompletely understood, can have unforeseen results. Ecologist Lamont Cole investigated one such situation in the 1950s. The World Health Organization tried to eliminate malaria from Borneo by spraying the environment with the insecticide DDT. The spray killed the mosquitoes that carry malaria, but there was a snag. Quantities

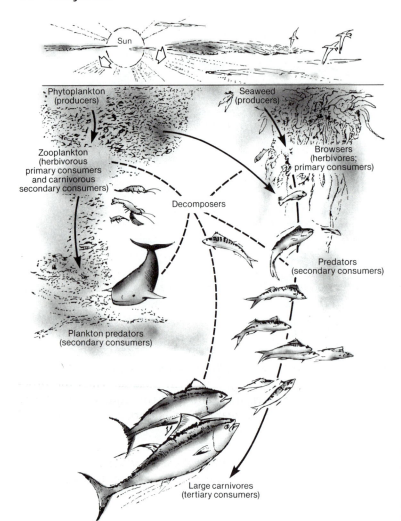

Figure 3–9 A food web in the sea showing various trophic levels. *(Plankton* is the collection of organisms that float in the surface layers. Photosynthetic members belong to the *phytoplankton,* heterotrophs to the *zooplankton.)*

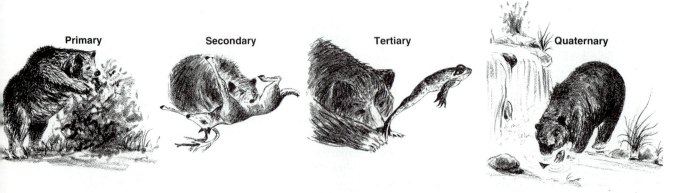

Figure 3–10 The bear, an omnivore, consumes organisms from various trophic levels in the food chain, acting as a primary, secondary, tertiary, and quaternary consumer.

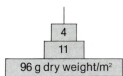

Trophic level
4. Tertiary consumers: carnivores, e.g., hawks, tigers
3. Secondary consumers: carnivores, e.g., rats, fish
2. Primary consumers: herbivores, e.g., aphids, squirrels
1. Producers: plants, protists, bacteria

Figure 3–11 *A diagram showing the biomass of organisms at various trophic levels in an ecosystem. The figures are the weight of organisms per square meter in each trophic level measured in a freshwater ecosystem. (Dry weight is used because different organisms contain different proportions of water.)*

TABLE 3–1 Ecological Classification of Organisms Based on What They Eat

Ecological Classification	Trophic Level	Level of Consumption	Examples
Autotroph	First	Nonconsumer	Trees, grass
Heterotroph	Second	Primary-consumer	Grasshopper, field mouse, cow
	Third	Secondary-consumer	Praying mantis, owl, wolf
	Fourth	Tertiary-consumer	Shrew, owl

of DDT were consumed by cockroaches. These insects, which are larger than mosquitoes and more resistant to DDT, did not die immediately. Geckoes (insect-eating lizards, Fig. 3–12) that ate the cockroaches ingested the insecticide in turn and suffered nerve damage from it. Their reflexes became slower, and many more of them than usual were caught and eaten by cats. Because most of their gecko predators were now dead, caterpillars eating the thatched roofs of local houses multiplied unchecked, and the roofs started to collapse. In addition, because the cats were soon dying of DDT poisoning from eating the geckoes, rats moved in from the forest, and with them came rat fleas carrying the bacteria that cause plague. If untreated, plague is more immediately fatal than malaria. Thus the problem escalated, and as a consequence the World Health Organization stopped spraying DDT and, in an attempt to remedy the damage already done, parachuted a large number of cats into the jungle. The whole experience was an expensive lesson on the importance of understanding a food web before trying to alter or improve a complex system.

Figure 3–12 *A gecko. The padlike toes enable this lizard to walk up walls or, completely upsidedown, on the ceiling. (Photograph by David Campbell.)*

3.4 Productivity

The **gross primary productivity** of an ecosystem is the rate at which organic matter is fixed during photosynthesis. Only about half of this productivity accumulates as new plant matter because the rest of it is metabolized in the plant's own respiration and released to the environment as heat. The material not used by the plant in respiration is its **net primary productivity**. The net primary productivity appears as plant growth and is available for consumption by heterotrophs.

Gross primary productivity =
    net primary productivity + respiration

1 m² of
field for 1 yr.

d. Net carnivore productivity: 0.4 g (0.15% of a; 6.7% of c)

c. Net herbivore productivity: 6 g (2.2% of a; 20.7% of b)

b. Every year herbivores eat: 29 g (10.7% of a)

a. Net primary productivity (plants): 270 g (dry weight)

Figure 3–13 Productivities at producer, herbivore, and carnivore trophic levels measured in a Tennessee field. The figures are the productivities for one square meter of field in one year. Considerably less than 10 percent of the food (net productivity) available at one trophic level is converted into net productivity at the next highest level. Two and two-tenths percent of plant net productivity is converted into herbivore productivity; 6.7 percent of herbivore productivity is converted into carnivore productivity. Thus, more than 90 percent of potential food is lost at each transfer from one trophic level to the next in this ecosystem.

TABLE 3–2   Net Primary Productivity of Major Ecosystems*

Type of Ecosystem	Net Primary Productivity, g/(m² year)	
	Normal Range	Mean
Tropical rain forest	1000–3500	2200
Tropical seasonal forest	1000–2500	1600
Temperate evergreen forest	600–2500	1300
Temperate deciduous forest	600–2500	1200
Boreal forest (taiga)	400–2000	800
Woodland and shrubland	250–1200	700
Savanna	200–2000	900
Temperate grassland	200–1500	600
Tundra and alpine	10–400	140
Desert and semidesert scrub	10–250	90
Extreme desert, rock, sand, and ice	0–10	3
Cultivated land	100–3500	650
Swamp and marsh	800–3500	2000
Lake and stream	100–1500	250
Open ocean	2–400	125
Upwelling zones	400–1000	500
Continental shelf	200–600	360
Algal beds and reefs	500–4000	2500
Estuaries	200–3500	1500

*From R. H. Whittaker: *Communities and Ecosystems*, 2nd ed. New York, Macmillan, 1975.

Productivity is usually expressed either in terms of energy stored (calories/square meter of land per year) or in terms of the mass of living material, called the **biomass,** gained per unit area per unit time (Fig. 3–13).

Different types of ecosystems have characteristically different productivities (Table 3–2). Because photosynthesis proceeds more rapidly when sunlight is abundant and temperatures are high, productivity generally increases from a low in the arctic regions to a high in the tropics, although there are many exceptions to this generalization. For example, if water is scarce, nutrients are lacking, or cloud cover is intense, productivity is low no matter what the latitude. In addition, many temperate regions have been converted to agricultural systems. Intensive agriculture, using special crop varieties, irrigation, and fertilizers, as well as the practice of planting two or more crops per year on the same land, can achieve net productivities as high as those of any naturally occurring vegetation. Because crop plants are maintained in peak condition, with no dying plant material, respiration losses may be as low as 20 percent of gross productivity, compared with an average of 50 percent in a natural system.

Productivity in the open ocean is generally low because nutrients from the remains of plants and animals sink to the bottom where there is not enough light to support photosynthesis. Productivity is higher in regions where nutrients are brought up to the surface layers by ocean currents and in the relatively shallow waters of the continental shelf (Fig. 3–14). Coral reefs, in contrast to the oceans surrounding them, are among the most productive ecosystems. They are very efficient at extracting nutrients from the water and at recycling these nutrients.

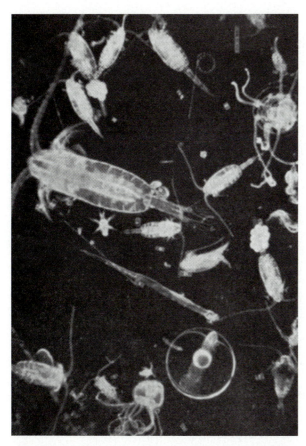

Figure 3–14 Living zooplankton magnified 16 times. Various shrimplike crustaceans and two tiny jellyfish are visible. These types of animals are the main consumers in the oceans. (From A. Hardy: *The Open Sea.* London, William Collins Sons, 1966.)

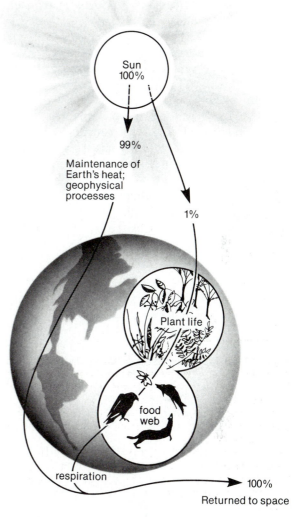

Figure 3–15 The flow of energy on Earth. Approximately 1% of the incident radiant energy is absorbed by plants, but only 0.3% of that radiation is converted to net primary productivity.

The annual net primary productivity of the whole Earth is about 170 billion metric tonnes (dry weight) of organic matter. (See the appendix for the definition of a tonne.) Of this total, about 115 billion tonnes are produced on land and 55 billion tonnes are produced in the oceans (despite the fact that the oceans occupy about 70 percent of the Earth's surface). Of this productivity, humans harvest about 1.2 billion tonnes per year as plant food.

These are enormous quantities but, for several reasons, they actually represent the use of only a tiny fraction of the incoming solar radiation. First, much of the solar energy reaching the Earth is reflected at once back into space (causing the Earth to "shine" like other planets). Solar energy is absorbed and scattered in the upper atmosphere so that even at noon on a clear sum-

mer day only about 67 percent of the incoming light reaches the Earth's surface. Second, of the radiation reaching the ground on a clear day, only about 45 percent is visible light that can be used in photosynthesis. Third, plants can use only a fraction of this visible light. Under some conditions, a barley plant may convert as much as 14 percent of the energy it receives in the form of visible light into the energy of biomass. However, the average net primary productivity over the globe represents the use of only about 0.3 percent of the radiant energy reaching the Earth's surface as visible light (Fig. 3–15).

Secondary productivity is defined as the rate of formation of new organic matter by heterotrophs. Of the net primary productivity available in a northern forest, herbivores (e.g., caterpillars, insects, deer) eat only about 1 to 3 percent. In other communities on land, as much as 15 percent of the vegetation may be eaten. In the oceans, this figure is much higher; for example, in some systems 80 percent of the net primary productivity is consumed by herbivores. Much of the plant matter that is consumed is lost, however. For example, only about half of what is eaten is actually digested and absorbed into the body. Of the food absorbed, about two thirds is used in respiration, so that only about 15 percent of ingested food appears as secondary productivity. An average figure is about 10 percent for both herbivores and carnivores, but there are considerable variations in different systems.

Even if the organisms at each trophic level were able to find, capture, and eat all of the net productivity from the previous trophic level, the tertiary consumers would receive only about one tenth times one tenth times one tenth (or one thousandth) of the energy present in the original producers in their food web. It is clear that in times of food scarcity, omnivores, including human beings, can avoid the energy losses at one trophic level and use the Earth's resources more efficiently by adopting a vegetarian diet. For example, as shown in Figure 3–16, the amount of soybeans and corn needed to support one person who eats only meat would be enough to feed 22 vegetarians (see Chapter 16).

3.5 Energy Flow

The flow of energy through an ecosystem can be represented by a pyramid that shows the total amount of incoming energy for successive trophic levels (Fig. 3–17). Energy and nutrients enter a food web together during photosynthesis as plants use the Sun's energy to convert inorganic nutrients into food. When the plant is eaten, the food it contains is transferred to a herbivore. The herbivore uses most of the food energy it takes in to move, feed, and respire. Therefore, relatively little energy, in the form of its body tissues, is left for the use of a predator that eats it. The process continues so that less and less energy is available at each successive trophic level. As a consequence, there are seldom more than five trophic levels in an ecosystem. A deer, for example, which is a primary consumer, travels only a few kilometers a day as it grazes, whereas a wolf, which is a secondary or tertiary consumer, may have to travel 30 kilometers a day to find enough to eat, and a tiger requires a home range of up to 250 square kilometers. An animal that fed on wolves and tigers would have to be extremely large and would have to cover a wide hunting area to find enough of its widely scattered prey. It is not en-

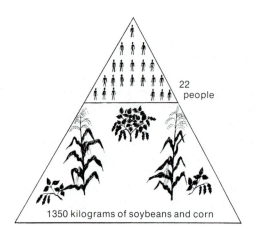

22 people

1350 kilograms of soybeans and corn

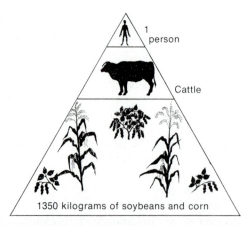

1 person

Cattle

1350 kilograms of soybeans and corn

Figure 3–16 The relative efficiencies of vegetarian and carnivorous diets for human beings. (A) In a vegetarian diet, 1350 kilograms of plant matter can support 22 people for the same length of time that (B) 1350 kilograms of plant matter can support only one person who eats only meat.

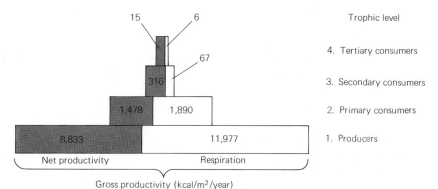

	Trophic level
15 6	4. Tertiary consumers
67	
316	3. Secondary consumers
1,478 1,890	2. Primary consumers
8,833 11,977	1. Producers

Net productivity Respiration

Gross productivity (kcal/m²/year)

Figure 3–17 A pyramid of energy for a river ecosystem, showing the total energy input at each trophic level and how much energy goes into net productivity and how much into respiration at each level.

ergetically feasible to try to harvest the small amount of food energy available in the highest trophic levels. This is why human beings eat herbivores such as chickens, cows, and pigs rather than carnivores such as hawks and lions.

3.6 The Ecological Niche

There are roughly one and a half million different species of animals and one-half million species of plants on Earth. Each species performs unique functions and occupies specific habitats. The combination of function and habitat is called an **ecological niche.** An organism's **habitat** is its address, the place where it lives; its niche is analogous to a human profession, the way in which it makes a living. An organism's niche includes all its interactions with the physical environment and with other organisms that share its habitat. To describe a niche, one would first have to describe all the physical characteristics of a species' home. One might start with specifying the gross location (for example, the Rocky Mountains or the central floor of the Atlantic Ocean) and the type of living quarters (for example, a burrow under the roots of trees). For plants and the less mobile animals, one would describe the preferred microenvironment, such as the water salinity for species living at the interfaces of rivers and oceans, the soil acidity for plants, or the necessary turbulence for stream dwellers. An animal's trophic level, its exact diet within the trophic structure, and its major predators are also important in the description of its niche. Mobile animals generally have a more or

less clearly defined food-gathering territory, or **home range,** which is another factor in establishing the physical niche.

A niche is not an inherent property of a species, because it is governed by factors other than genetic ones. Social and environmental factors also play a part in establishing the niche. For example, a certain population of tropical jellyfish, *Aurelia aurita,* swims fastest in water that is at a temperature of 29°C. Of course, it would be unreasonable to expect all the jellyfish of this population to be living in waters of exactly 29°C all the time. The weather changes, cold and warm spells occur, yet organisms survive. Imagine a situation in which a warm, sunny, sheltered bay provided an optimal physical environment for jellyfish but exposed them to abnormally high predatory pressures. Since the conditions in the bay would not be optimal, many individuals might migrate to a less favorable physical environment to find more favorable biotic surroundings. Thus, the observed niches of a bay-dwelling jellyfish and of a migrating jellyfish are different from each other and from the theoretical optimal niche. The niche of a given species in a given ecosystem is not a set of conditions that would be best suited to the genetic makeup of the organism but rather the set of conditions in which it can actually survive.

3.7 Cycling of Mineral Nutrients

Although the productivity of an ecosystem may be limited by the supply of sunlight or availability of water, in many cases productivity is limited instead by the availability of inorganic nutrients that plants need.

Living organisms require six elements in relatively large quantities—carbon, hydrogen, oxygen, nitrogen, phosphorus, and sulfur. Some of these elements are present in the form of compounds in rocks. They are released by erosion and weathering into soil, rivers, lakes, and oceans. Nitrogen (as N_2), carbon (as CO_2), and oxygen (as O_2) are also present in the atmosphere. The movements of nutrient elements through an ecosystem are called **biogeochemical cycles.** They are called cycles because nutrient elements, unlike solar energy, may be used over and over again by living systems. On the average, every breath you inhale contains several million atoms once inhaled by Plato—or any other person in history you care to choose.

Nutrients are sometimes recycled rapidly through ecosystems, as in grasslands, where the above-ground vegetation dies back each year and its nutrients are made available again the following season. In other cases nutrients may spend many years away from the activities of the biological world. For example, remains of

marine organisms may sink to the bottom of the ocean and be incorporated into rocks that are uplifted and exposed to the erosion that releases nutrients only after millions of years. Every nutrient element has a somewhat different fate, depending on its physical and chemical properties and its role in living organisms. The concept of nutrient cycling will be illustrated with a few simplified examples.

The Carbon Cycle (Fig. 3–18).   In terrestrial ecosystems gaseous carbon dioxide is absorbed by plants during photosynthesis and incorporated into organic molecules. In time the carbon in plant or animal tissue is released back into the environment as carbon dioxide during respiration. In aquatic ecosystems, the corresponding exchange is between living organisms and carbon dioxide (CO_2) or bicarbonate ion (HCO_3^-) dissolved in the surrounding water. In many situations the cycling of carbon can occur quite rapidly, completing itself in a matter of minutes, hours, days, or weeks. In part, the rapid movement of carbon in ecosystems is made possible because the large reservoir of carbon dioxide is

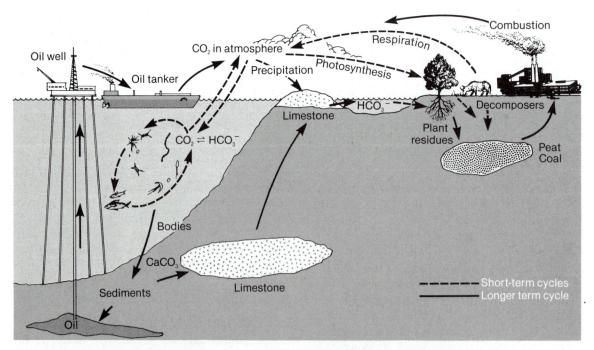

Figure 3–18 A simplified carbon cycle. Carbon passes through the processes indicated by dashed arrows much more rapidly than through those indicated by solid arrows. (Figure 3–19 shows a limestone deposit which has traveled from the bottom of the sea to the surface of the Earth as shown here.)

Figure 3–19 A chalk cliff in Sussex, England. The cliff is made up of the calcium carbonate skeletons of protists that sank to the bottom of the sea millions of years ago, forming sediment rich in carbon. This sediment was compressed into rock, which was eventually heaved to the surface of the earth where it erodes, contributing calcium and carbon to modern ecosystems. (Courtesy of British Tourist Authority.)

Figure 3–20 A portion of a hot water pipe showing "scale," which is composed of accumulated calcium carbonate.

directly available to nearly all organisms as a gas in the atmosphere or dissolved in the ocean water.

In some ecosystems, however, carbon may accumulate as undecomposed organic matter, such as that in the peaty layers of bogs and moorlands. Such carbon has entered a cycle of longer duration, and it may be millions of years before it is released. Deposits of coal, oil, and natural gas are part of the primary productivity of a bygone era. These fossil fuels are organic compounds that were buried before they could decompose and were subsequently transformed by time and geological processes. When they are burned, their carbon is finally released back into the atmosphere, largely as carbon dioxide.

The carbon cycle is complicated by the fact that carbon can also be bound into mineral formations, thus entering sedimentary cycling, which is usually much slower than the cycling of atmospheric gases. Certain aquatic animals ab-

sorb carbon dioxide and convert it to insoluble calcium carbonate ($CaCO_3$), which is then used to construct hard protective shells. When the animals die, the seashells accumulate as bottom sediments, which may eventually turn into sedimentary rocks such as limestone and dolomite. After millions of years, these rocks may be lifted above sea level and exposed to erosion, releasing their carbon as carbonate and bicarbonate (Fig. 3–19). "Hard" water has usually flowed through limestone, picking up carbonate, which may accumulate as a hard scale in kettles or pipes when the water is boiled (Fig. 3–20).

The carbon balance of the Earth as a whole is moderated by the exchange of carbon dioxide between the atmosphere and the oceans. The carbon dioxide content of the atmosphere is about 0.03 percent, but it is steadily increasing owing both to the burning of fossil fuels and to the cutting down of forests. The felled trees are rapidly digested by decomposers, which produce both CO_2 and methane, CH_4. The rate of carbon dioxide increase in the atmosphere, however, is less than half what would be expected from human activities. The remainder is dissolved in the oceans, which act as a global carbon dioxide sink.

The Nitrogen Cycle. Nitrogen is one of the most plentiful elements in the immediate environment, because the atmosphere is composed of 78 percent gaseous nitrogen. But paradoxi-

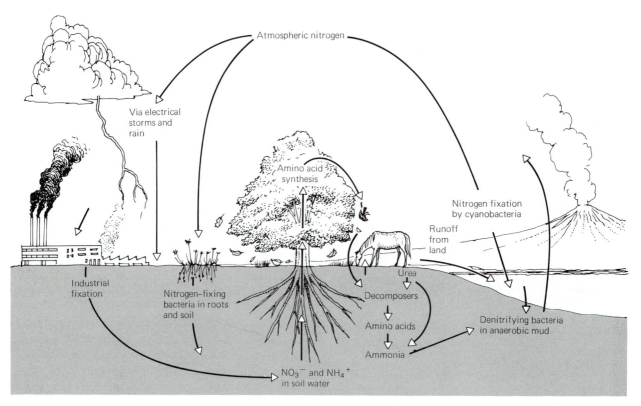

Figure 3–21 A simplified nitrogen cycle.

cally, the growth of many organisms is limited by a shortage of nitrogen. This shortage occurs because very few organisms can utilize atmospheric nitrogen (N_2) directly. Most plants can absorb this vital nutrient only when it is in the form of nitrites (NO_2^-), nitrates (NO_3^-), or ammonium ions (NH_4^+). The nitrogen cycle must therefore provide various bridges between the atmospheric reservoirs and the biological communities. In natural systems, atmospheric nitrogen is fixed (converted to usable form) by various bacteria. Plants can absorb available nitrogen compounds either directly from nitrogen-fixing bacteria or indirectly from the pool of inorganic compounds that the bacteria excrete into the soil, water, rivers, lakes, and oceans (Fig. 3–21). Available nitrogen is also added to an ecosystem by various other means. One of these is lightning, which makes atmospheric nitrogen react with oxygen and converts it, in a series of steps, into nitrate. Other means include the erosion of rocks rich in nitrates and the decomposition of organic matter. When an organism dies, the nitrogen that it contains in organic

molecules is converted back into inorganic forms via steps carried out by a series of different decay organisms. Decomposer bacteria and fungi use the proteins in dead organisms for their own dietary nitrogen, releasing the excess as ammonia or nitrates, and they convert some of the ammonia to nitrites or nitrates, which can then be absorbed by plants.

Nitrogen is returned to the atmosphere when it is once more released as a gas by **denitrifying** bacteria that break up organic molecules for their own food, releasing any nitrogen they contain as a gas. All these bacteria live in anaerobic (without oxygen) conditions in the mud at the bottom of some lakes, in bogs and estuaries, and in parts of the sea floor.

All terrestrial ecosystems continuously lose some nitrogen when nitrates and organic matter are washed, or **leached,** out of the soil into groundwater and streams (Fig. 3–22). In most cases, the nutrients are then eventually deposited in the oceans. In a few places on Earth, leached nutrients have collected in natural geological deposits. For example, certain streams

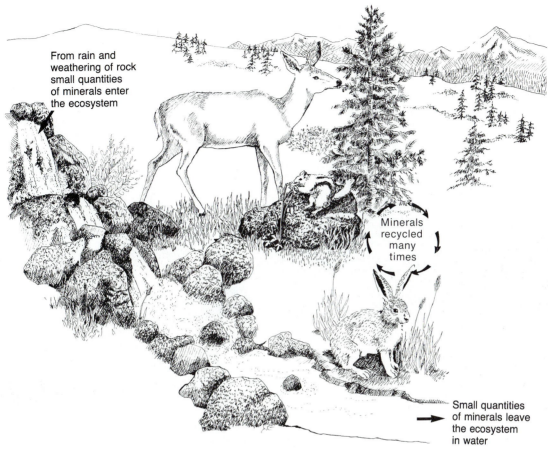

Figure 3–22 *The general pattern by which minerals in rain, snow, and groundwater pass through an ecosystem.*

From rain and weathering of rock small quantities of minerals enter the ecosystem

Minerals recycled many times

Small quantities of minerals leave the ecosystem in water

flowing out of the Andes Mountains in Chile travel across one of the driest deserts in the world. Much of this water evaporates before it completes its journey to the sea, leaving behind deposits of nitrate rocks.

Today, large quantities of fixed nitrogen are artificially added to agricultural soils in the form of fertilizers. The production of nitrogen fertilizers is accomplished by combining nitrogen and hydrogen in the presence of heat and pressure. This process, discussed in more detail in Chapter 14, has increased the fertility of agricultural systems but consumes large quantities of energy.

The Phosphorus Cycle. Phosphorus is a major constituent of biological membranes. Many animals also need large quantities of this element to make shells, bones, and teeth. Since phosphorus does not occur naturally as a gas, its cycle, unlike those of carbon and nitrogen, is a **sedimentary cycle** (Fig. 3–23). Many rocks contain phosphorus, usually in the form of phosphate (PO_4^{3-}) that is bound into their mineral structure. When rocks are eroded, minute amounts of these phosphates dissolve and become available to plants. Animals then absorb this element when they eat plants or other animals.

Much of the phosphorus excreted by animals is also in the form of phosphate, which plants can reuse immediately. Thus, phosphate cycles round and round on land from plants to animals and back again. Land ecosystems preserve phosphorus efficiently, since both organic and inorganic soil particles absorb phosphate, providing a local reservoir of this element.

In an undisturbed ecosystem, the intake and loss of phosphorus are small compared with the

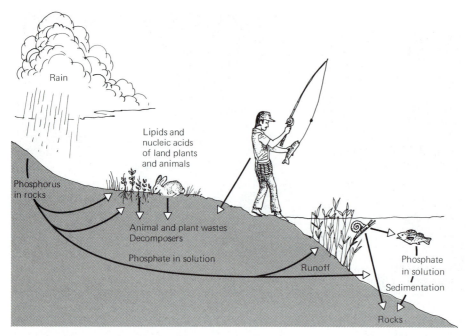

Figure 3–23 *The phosphorus cycle.*

amounts of phosphorus that are internally recycled in the day-to-day exchange among plants and animals. Some phosphorus is inevitably lost by leaching and erosion of the soil into streams and rivers. When an ecosystem is disturbed, for instance by mining or farming, erosion can become so significant that large quantities of phosphorus and other minerals are washed away. When phosphate reaches the ocean, it encounters other minerals in the water and reacts with some of them. Since most phosphates are not very soluble in water, they form insoluble compounds and eventually settle to the ocean floor. One reason for the infertility of open oceans is their low phosphorus content.

In the short term, the movement of phosphorus from ecosystems on land to the bottom of the ocean is a one-way flow. Inconsequential amounts of phosphorus are deposited on land from the droppings of ocean birds or from fish when caught by animals and brought ashore. The only other way that phosphorus can return to land is by way of extremely slow geological processes in which sea floor sediments rise from the sea to form islands or continents. Soil erosion robs an ecosystem of its phosphorus, and it may take thousands or tens of thousands of

years to recoup this loss through the weathering of rocks. Phosphate is much in demand as a fertilizer, but the richest and most easily mined deposits of phosphate are being depleted rapidly. The environmental problems associated with the depletion of mineral resources such as phosphorus are discussed in Chapter 18.

3.8 An Experimental Ecosystem

Nutrient cycles in natural ecosystems and the effect of human activities on such cycles are now being studied experimentally. One experimental ecosystem is the Hubbard Brook Experimental Forest in the White Mountains of central New Hampshire, which has been studied continuously since 1963. This system is ideal for such a study because it consists of a series of small valleys each with a creek running down the middle. The soil and rock structure prevents seepage of water from one system to another, so each valley is isolated from the others.

The first project at Hubbard Brook was to measure the amounts of water and nutrients that entered and left an undisturbed forest. To

Figure 3–24 V-shaped dam at the lower end of a valley in the Hubbard Brook experimental forest. The rate of water runoff from the watershed is measured by the height of water in the V (very low when this photograph was taken). (Photograph by Peter Marks.)

Figure 3–25 An aerial view of part of Hubbard Brook in winter, showing the deforested valley (unbroken white snow) and parts of two adjacent valleys. On the left is a watershed in which trees have been cut down in horizontal strips. (Photograph by Robert Pierce, U. S. Forest Service.)

do this, researchers built concrete dams across the creeks at the bottom of each valley. The dams were anchored in bedrock so that all the water leaving the forest, except the amount that evaporated into the atmosphere, had to cross the dams (Fig. 3–24). Here its flow was measured and its nutrient content analysed. Precipitation gauges throughout the valleys were used to measure the amount of water that fell as rain or snow and the nutrients that it contained. The information from this study showed that the forest is extremely efficient at retaining nutrients. Nutrients that entered the system as solutions in water or snow approximately balanced the quantity of nutrients that left the ecosystem in the creeks. The quantities that entered and left the ecosystem were very small compared with the total amounts of nutrients present.

The next experiment was to see what happened to nutrient balance when a forest was cut down. One winter, the investigators cut down all the trees and shrubs in one valley, leaving them where they fell, and preventing regrowth by spraying with herbicides (Fig. 3–25). With no trees to absorb nutrients and water from the soil, most of the rainwater ran through the soil into the stream, carrying soil nutrients with it. In the cut forest, water runoff in the stream passing

over the dam increased by 40 percent. More important, the loss of mineral nutrients in the cut forest was six to eight times greater than the loss in the undisturbed forest (Fig. 3–26). Later experiments showed that nutrient losses are reduced if the forest is cut in horizontal strips, leaving strips of standing trees, rather than being clearcut.

Summary

An **ecosystem** is a group of plants and animals occurring together plus that part of the physical environment with which they interact. An ecosystem is defined as a nearly self-contained system, so that the matter that flows into and out of it is small compared with that which is internally recycled in a continuous exchange of the essentials of life.

All organisms are members of one of the five kingdoms—plants, animals, fungi, protists, or bacteria. Plants are the main autotrophic producers of an ecosystem because they can perform photosynthesis, using the energy of sunlight to convert inorganic carbon dioxide and water into organic food molecules. (Some protists and bacteria are also producers.) If an ecosystem is to survive indefinitely, it must contain at least some autotrophs and some decomposers and the water, air, light, and minerals that these or-

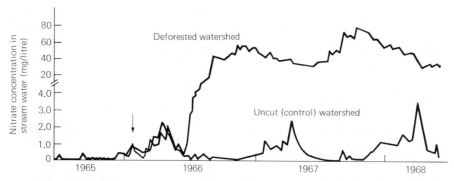

Figure 3–26 *Accelerated loss of nitrate in stream water in one of the watersheds at Hubbard Brook after it was deforested. The date the forest was clearcut is shown by the arrow. (Note the change of scale on the y axis, needed to keep the lines for control and clearcut watersheds on the same graph.)*

ganisms need to survive. Most ecosystems also contain heterotrophic consumers, which feed on producers or on each other.

Two of the most important factors determining the makeup of an ecosystem and the biomass it can support are its productivity and its nutrient cycles. The **productivity** of an ecosystem is determined by the availability of light, water, and minerals for photosynthesis. Energy passes from one trophic level to the next. Approximately 90 percent of the energy is lost at each transfer. In consequence, the biomass that an ecosystem can support at each trophic level declines rapidly, and energy flow through the system is, to all intents and purposes, one way.

The function and habitat of an organism define its **ecological niche.**

Nutrients cycle continuously through an ecosystem. They are taken in by organisms as inorganic substances and may remain as minerals or be incorporated into organic molecules. Nutrients may pass along the food chain for a time, but eventually they are once again released into the environment as inorganic substances. An ecosystem may be very efficient at conserving and recycling nutrients.

Questions

Ecosystems

1. Certain essentials of life are abundant in some ecosystems but rare in others. Give examples of situations in which each of the following is abundant and in which each is

rare: (1) water, (2) oxygen, (3) light, (4) space, and (5) nitrogen.

Food Webs

2. Is there one or more than one food web in any ecosystem?

3. It is desired to establish a large but isolated area with an adequate supply of plant food, equal numbers of lion and antelope, and no other large animals. The antelope eat only plant matter, the lions, only antelope. Is it possible for the population of the two species to remain approximately equal if we start with equal numbers of each and then leave the system alone? Would you expect the final population ratio to be any different if we started with twice as many antelope? Twice as many lions?

4. Name two organisms that occupy the first trophic level; two that occupy the second; two that occupy the third.

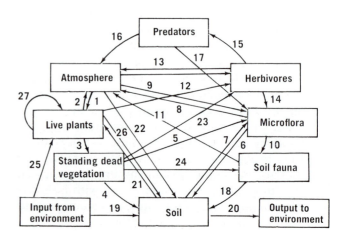

5. Consider the food web in the diagram on page 56. Arrow number 15 can be explained by the relationship: mountain lion eats deer. Write similar statements for each of the other 27 arrows.

6. An experiment has shown that the total weight of the consumers in the English Channel is five times the total weight of the plants. Does this information agree with the food pyramid shown in Figure 3–11? Can you offer some reasonable explanations for the findings?

Nutrient Cycles

7. Give three examples supporting the observation that nutrient cycling hasn't been 100 percent effective over geological time.

8. Trace an oxygen atom through a cycle that takes (1) days, (2) weeks, (3) years.

9. Why don't farmers need to buy carbon at the fertilizer store? Why do they need to buy nitrogen?

10. Describe three pathways whereby atmospheric nitrogen is converted to fixed forms that are usable by plants, and three pathways whereby fixed nitrogen is returned to the atmosphere.

11. Loggers can harvest timber either by clear-cutting (removing all the trees from an area) or by selective cutting (removing only the most desirable trees). Explain why clearcutting is an ecologically unsound practice in most woodlands.

Energy

12. Is more energy lost from an ecosystem when an herbivore eats a plant or when a carnivore eats an animal? Why?

13. How is the flow of energy in an ecosystem linked to the flow of nutrients? How do energy and nutrient flow differ?

14. More energy from the Sun is used to nourish a human being who eats meat than a human being who eats vegetables. Is this fact, by itself, an argument for or against vegetarianism? Take into account the following questions: Does the choice of human diet affect the total energy flow through the food web? Does it affect the total biomass of plant matter on Earth? Does the choice between the alternatives of a large human population living largely on vegetable food or a smaller human population living largely on meat and fish significantly affect the total biomass of animal matter on Earth?

15. Assume that a plant converts 1 percent of the light energy it receives from the Sun into plant material, and that an animal stores 10 percent of the food energy that it eats in its own body. Starting with 10,000 Calories of light energy, how much energy is available to a person who eats (1) corn, (2) beef, (3) frogs that eat insects that eat leaves.

Suggested Readings

Three basic textbooks on ecology are:
Charles J. Krebs: *Ecology.* 2d ed. New York, Harper & Row, 1978.
Eugene P. Odum: *Ecology.* 2d ed. Philadelphia, Holt, Rinehart and Winston, 1975.
S. J. McNaughton and Larry L. Wolf: *General Ecology.* 2nd ed. New York, Holt, Rinehart and Winston, 1979. 702 pp.

A periodical issue devoted in its entirety to ''The Biosphere'' is:
Scientific American. September, 1970. 267 pp.

Three books dealing with specific areas of natural ecology are:
R. Platt: *The Great American Forest.* Englewood Cliffs, N. J., Prentice-Hall, 1965. 271 pp.
B. Stonehouse: *Animals of the Arctic: the Ecology of The Far North.* New York, Holt, Rinehart and Winston, 1971. 172 pp.
G. M. Van Dyne, ed.: *The Ecosystem Concept in Natural Resource Management.* New York, Academic Press, 1969. 383 pp.

A classic study of ecology and conservation as seen through the eyes of a naturalist is:
A. Leopold: *A Sand County Almanac.* New York, Sierra Club/Ballantine Books, 1966. 296 pp.

A short account of the Hubbard Brook watershed study of nutrient cycles is given in:
F. H. Bormann, and G. E. Likens. Nutrient cycles of an ecosystem. *Scientific American,* October 1970.

A description of the nonphotosynthetic ecosystem discovered recently in the ocean depths is found in:
J. B. Corliss, and R. D. Ballard. Oases of life in the cold abyss. *National Geographic,* October 1977.

4

Populations of Organisms

The size of a population of mice in a field or of violets in a woodlot seems at first sight to vary little from year to year. Is this really the case? A violet produces several dozen seeds each year, and a mouse may give birth to 10 offspring every few weeks. Surely these organisms produce so many offspring that their populations could increase greatly from one year to the next. Do these populations in fact increase and then decline in size? What determines how many violets grow in the woodlot at any one time? How does the population of a rare species become so reduced that it is in danger of extinction? How many people will there be on Earth in the year 2000? If we introduce an animal that eats Japanese beetles, will it exterminate this pest or merely reduce the beetle population, or will it have no effect?

The answer to these and other related questions constitute the subject of population ecology. Since people are an animal species living on this planet, many of the basics of population ecology apply to us as well. On the other hand, the dynamics of human population are different enough from those of any other species to warrant a separate chapter (Chapter 9).

4.1 Population Growth

Individuals are added to a population by birth and by immigration. They leave a population by death and by emigration. The size of a population at any given time, therefore, de-

Figure 4–1 Earth, a planet that many believe is becoming dangerously overpopulated with human beings. (Courtesy of NASA)

pends upon the balance between these processes. If the number of individuals added to the population by birth and immigration in any period of time exceeds the number removed by death and emigration, the population will grow (Fig. 4–1). If the number is less, the population will decrease, and if they are the same, the population will remain the same size.

Even a cursory study shows that the populations in a natural ecosystem do not grow as fast as they might. Some individuals die when they are still juveniles so that they do not grow up to reproduce, and some adults do not reproduce even though they are able to. How would a population grow if nothing impeded its growth? The Russian ecologist G. F. Gause examined this question by studying the growth of experimental populations of the freshwater protist *Paramecium caudatum*. Every few hours, a well-nourished paramecium divides to form two new individuals. Gause set up tubes containing plenty of bacteria for food and introduced one paramecium into each tube. He then followed the growth of the *Paramecium* populations. If nothing checks its growth, each population grows *exponentially*. That is, as time goes by the number of individuals added in each time period keeps increasing (Box 4.1). If exponential growth is plotted on a graph, it forms a curve that grows steeper and steeper, sometimes, rather inaccurately, called a J curve (Fig. 4–2).

Gause found that no matter how much food, space, light, and warmth he gave them, there was a limit to the rate at which the *Paramecium* populations grew. This maximum rate, called the **innate capacity for increase,** is determined by the reproductive capacity of the organism. The innate capacity for increase of *Paramecium* is quite high, one generation every few hours. In contrast, polar bears give birth to one or two infants once every three or four years, so their potential growth rate is considerably lower.

Populations can grow exponentially even when they are not growing at this maximum rate. For instance, the human population of many areas started to grow exponentially in the mideighteenth century, although women were not bearing as many infants as frequently as is biologically possible.

Populations rarely grow exponentially in natural ecosystems because if they did they would rapidly overpopulate the Earth. Yet such exponential growth does occur in some instances when resources are abundant. For example, if a species is imported into a new environment where food, water, and sunlight are

---

BOX 4.1 EXPONENTIAL GROWTH

An exponential (also called geometric) increase is an increase whose rate is always changing (Fig. 4–2). The change can be an increase or a decrease, but exponential growth commonly refers to an increase. Think of it this way: Assume there is one bacterium at time zero, that each individual splits into two offspring in 20 minutes, and that no bacteria die. Then there will be two bacteria in 20 minutes, and 20 minutes after that each will have split again, so there will be four. After another 20 minutes there will be eight, and so on, as shown below. (The general equations are given in Appendix E.)

TIME	GENERATION	POPULATION
0	0	$1 \times 2^0 = 1$
20 min	1	$1 \times 2^1 = 2$
40 min	2	$1 \times 2^2 = 4$
60 min (1 hr)	3	$1 \times 2^3 = 8$
.	.	.
.	.	.
.	.	.
1½ days	108	$1 \times 2^{108} = 10^{33}$

At the end of 1½ days, the single bacterium would have multiplied into 10^{33} individuals, enough to cover the entire surface of the Earth to a uniform depth of 30 cm!

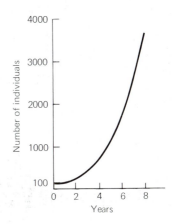

Figure 4–2 This is how exponential growth of a population looks when plotted on a graph.

plentiful and natural enemies or competitors are absent, these rapid growth rates are observed for a short period of time. Dandelions, starlings, and house sparrows that were introduced into the United States underwent dramatic population explosions. Similar exponential growth occurs when bacteria invade the intestinal tract of a newborn animal or when decomposers invade a freshly dead animal or plant.

4.2 Factors Affecting Innate Capacity for Increase

The rate at which a population can grow differs widely from species to species. Many factors affect the innate capacity for increase of a population; the factors with the most direct influence are (1) the age at which the organism first reproduces; (2) the frequency of the reproductive cycle; (3) the total number of times each organism reproduces in its lifespan; and (4) the number of offspring produced each time an organism reproduces.

All of these are important and are related to each other, but the influence of the first factor (the age of first reproduction) is especially noteworthy (Fig. 4–3). For example, a paramecium does not produce many offspring each time it reproduces nor does it have a long reproductive life. On the other hand, an oak tree produces hundreds of seeds every year and lives for many years. Yet a paramecium can reproduce within an hour of being formed, whereas an oak tree must be more than 10 years old before it can reproduce for the first time. As a result, *Paramecium* populations can grow faster than populations of oak trees.

The influence of these factors on human populations yields results that are different from those one would expect. Consider, for example, the following two human populations: In the first one, all women bear three children, one each at ages 13, 14, and 15. In the second, all women bear five children, one each year, but starting at age 30. The first population will grow at a much faster rate than the second!

Such calculations have important implications for attempts to curb the human population explosion. One effective way to curb population growth is to encourage women to have their first child as late in life as possible. If a woman reproduces for the first time at a relatively late age, she will make a smaller contribution to population growth than a woman who has the same number of children but starts earlier. (Human populations will be discussed in more detail in Chapter 9.)

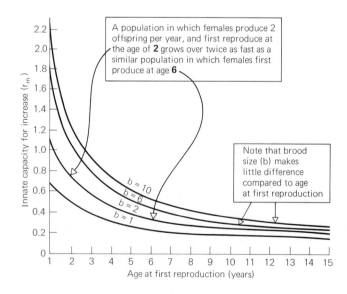

Figure 4–3 The effect of females' age at first reproduction on the innate capacity for increase of a population. This graph is for a population of long-lived organisms in which young are produced in one brood (litter) once a year. b = brood size, the number of young produced each year. (After Cole, 1954.)

4.3 Factors Affecting Population Size

No population can grow exponentially for long because if it did, vital resources such as food or water would become exhausted. Therefore, some combination of environmental pressures must act to inhibit the rate of population growth of every species.

The sum of all the environmental interactions that collectively inhibit the growth of a population is known as the **environmental resistance.** Imagine a population that is initially very small. The very fact that it is small places it in danger of extinction because it may not be able to recover from such setbacks as epidemics, famine, or poor breeding. Even if such factors do not totally destroy the population, they will limit its growth rate, and therefore the population will increase only slowly at first. However, once the population is established, its size will grow more rapidly as long as there are adequate food sources, relatively few predators, and favorable living conditions. When the population becomes very large with respect to its food supply, availability of shelter, and vulnerability to predators, and becomes so dense that disease spreads rapidly, the environmental resistance increases and the growth rate decreases. The entire curve of growth looks like an S and is said to be **sigmoid** or S-like, as shown in Figure 4–4. The rate of growth shown by the upper right portion of the sigmoid curve is very nearly zero. A zero growth rate does not mean that there are no births and no deaths. It simply means that the total number of births plus immigrations equals the total number of deaths plus emigrations. When this equilibrium is reached, the innate capacity for increase is balanced by the environmental resistance. The magnitude of this upper population level is called the **carrying capacity,** the maximum number of individuals of a given species that can be supported indefinitely by a particular environment.

When the carrying capacity has been reached, the system cannot continue to support any more individuals of that species. The carrying capacity is not constant from region to region; for example, a wheat field has an inherent

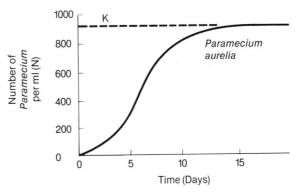

Figure 4–4 Carrying capacity (K) for one of Gause's *Paramecium* populations. Carrying capacity is the maximum population that the environment can sustain indefinitely. A single individual is introduced into a tube. The experimenter supplies food at some constant rate. At first the population grows exponentially, but when it becomes sufficiently dense, competition for food sets in. The population finally stabilizes at the carrying capacity, which in this case is determined by the rate at which food is supplied.

ability to support more locusts than a short-grass prairie can.

Smooth growth curves are not always observed in natural ecosystems. For example, some populations oscillate sharply. North of the Canadian forest lies the arctic tundra, and one common rodent of the tundra is the lemming. Lemmings exhibit predictable three- or four-year population cycles. One summer the population is extremely high; the next year the population rapidly declines, or crashes. For another year or two, the population recovers slowly; then it skyrockets for a season, and the cyclic pattern repeats itself.

Lemming abundance is associated with forage cycles (Fig. 4–5). During an abundant lemming year, the tundra plants are plentiful and healthy. Most of the available nutrients exist in plant tissue, and there is little stored in the soil reservoir. In the spring following the overabundance of lemmings, the heavily overgrazed plants become scarce, and most of the nutrients in the ecosystem are locked in the dead and dying lemmings. The process of decay and return to the soil requires another year or two, during which time the population and health of both plants and rodents recover.

Unfortunately, recognition of the relationship between lemmings and forage cycles does

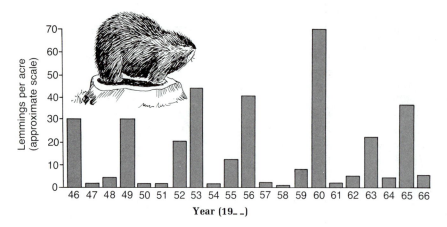

Figure 4–5 Lemming population cycles at Point Barrow, Alaska, from 1946 to 1966.

not prove either that forage cycles cause rodent cycles or vice versa.[1] Some researchers argue that perhaps internal genetic factors may be responsible for the cycles in some rodent populations.

Erratic **population blooms** occur even in old, seemingly stable systems. Since recorded history, a species of red phytoplankton has often been observed to enter a period of rapid growth in coastal areas and turn the ocean red. There seems to be no predictable cyclic pattern associated with this phenomenon; instead, the plankton grow very rapidly at unexpected times—seemingly in response to a new influx of nutrients. Traditionally, these "red tides" were often observed after flood waters washed large quantities of soil nutrients into the sea. This influx created a plentiful food source. The red-colored plants were best able to take advantage of it. Unfortunately, these organisms discharge toxic substances into the sea, killing fish and aquatic mammals, and are therefore considered a great hazard. Recently, a series of red tides apparently unrelated to natural phenomena have been tentatively attributed to the nutrient content of pollutants discharged into the sea.

In contrast to the large fluctuations in population that occur from time to time, a noteworthy feature of most large organisms is that the size of their average population changes relatively little over the years, certainly less than one might

expect from their innate capacities for increase (Fig. 4–6).

The factors that kill members of a population (mortality factors) may be classified as either density-dependent or density-independent (Fig. 4–7). A **density-dependent mortality factor** is one that kills a larger proportion of individuals as the population density increases. A **density-independent mortality factor** kills a constant proportion of the population, regardless of its density.

The two most important density-dependent factors are competition for resources and the action of predators and parasites. When the population is dense, a higher proportion of individuals will perish if they lose out in competition for food or nest spaces than in a population that is less dense. In an extreme case, if there is one nest site per square kilometer and one breeding pair in the area, no individual will die or be prevented from breeding by lack of a nest site. But if several individuals must compete for the same nest site, some must be displaced.

A frequently cited density-independent mortality factor is bad weather. A hurricane, a severe winter, or a drought may kill all of the individuals in a population, no matter what the population density (Fig. 4–8). The action of weather is not always independent of density, however, since some individuals may be able to shelter from bad weather. If the number of shelters is limited, all of the members of a small population can find shelter, whereas only a fraction of a more dense population will be protected.

---

[1] Judith H. Myers and Charles J. Krebs: Population cycles in rodents. *Scientific American*, June, 1974, p. 38.

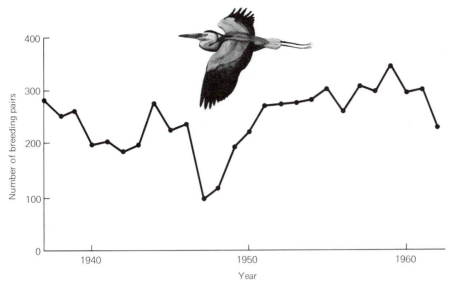

Number of breeding pairs

Year

Figure 4–6 Changes in size of a natural population. The number of breeding pairs of gray herons in part of northwest England fluctuated little over 25 years. The population recovered rapidly from the decline caused by the severe winter of 1947. We deduce that the carrying capacity of the environment is about 250 breeding pairs. Fluctuations around the carrying capacity are small compared with the heron's innate capacity for reproduction, which is about three new birds per pair per year. (After Lack, 1966.)

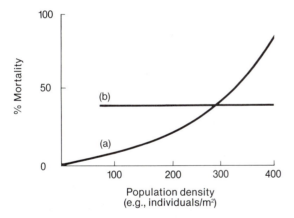

% Mortality

Population density
(e.g., individuals/m²)

Figure 4–7 Hypothetical curves describing the actions of (a) density-dependent mortality factors—an increasing fraction of the population is killed as the population density increases; (b) density-independent mortality—the same fraction of the population is killed, regardless of the population density (although higher *numbers* of individuals are killed at higher densities).

Figure 4–8 A weather-beaten tree growing on a mountain top. The state of the tree and the absence of other trees suggests that the size of the tree population here is controlled by density-independent factors such as the very high winds, which few trees can withstand.

4.4 Reproduction and Survival

In any natural ecosystem the innate capacity for increase of all organisms must ultimately be counterbalanced by mortality factors. There are two extremes of reproductive strategies. At one extreme, parents may produce many small offspring and provide each with a minimal amount of food and parental care. At the other extreme, an organism may produce very few offspring but invest a great deal of energy in each one before it becomes independent of the parents. In the first case, in which there are many small offspring that receive no care, there is extremely high mortality among the young. Because there are so many, however, there is a good chance that a few of them will survive to become parents of the next generation. In the second situation, parental care is much greater and the rate of survival among the young tends to be considerably higher, but because there were only a few infants to start with, only a limited number of individuals will survive to the next generation.

Internal regulation of populations may be triggered in various ways when an individual, a family, or a group spontaneously restricts its birth rate. A human female may choose to bear two children even though she has enough money to feed, clothe, and shelter many more and is biologically capable of bearing a larger family. Birth control is observed in many other animal species as well. Many birds in desert areas suppress their breeding, often for years, in periods of drought. After prolonged rainfall, however, such birds may breed more than once, producing several broods of young. The availability of food may also trigger breeding. Grass warblers (birds) of the genus *Cisticola* in Nyasaland normally breed in the wet season when the grass is growing. In areas where humans have planted banana plantations, however, the birds breed whenever plantation owners flood the bananas, causing grass to grow.

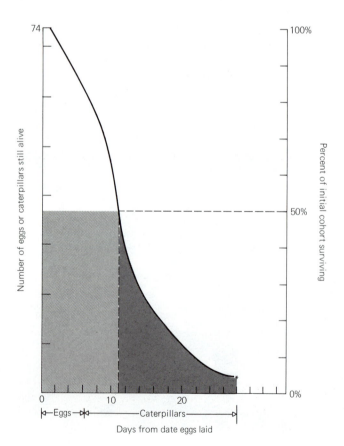

Figure 4–9 Survivorship curve for eggs and caterpillars of the black swallowtail butterfly (*Papilio glaucus*) on wild carrot plants in a New York hayfield. The initial cohort of 74 eggs was found by following butterflies and marking where they laid their eggs. These were then visited daily as they hatched into caterpillars. Fifty percent of the insects had died by day 11; that is, an egg laid on day 1 had only a 50 percent chance of surviving to day 11. Most of the deaths are due to predation by spiders and mites. (Data courtesy of Paul Feeny.)

Many social animals, such as hippopotamuses, rats, honey bees, and wolves control their reproduction. For example, a wolf pack typically consists of a dominant male (the leader), his mate (the dominant female), and a number of subordinate males and females. The dominant pair mate every year, but other members of the pack generally do not copulate even though they are sexually mature. In one study of a wolf pack kept in a large fenced enclosure, the dominant male became aroused and tried to mate with a subordinate female, but she would not allow it, and with tail between her legs, she cowered and avoided his advances. This birth control is beneficial to a family of wolves in the wild, for otherwise the wolves, who are not themselves subject to predation, would overpopulate their range and face starvation.

The patterns of mortality of each species can be shown by a characteristic **survivorship curve**. A survivorship curve is constructed by following the fate of young individuals throughout their lives, in order to describe the mortality at different ages. To construct a survivorship curve, ecologists start with a **cohort** of newborn individuals. A cohort (or **birth cohort**) is a group of individuals born in a given period of time, such as in a particular year. The members of the cohort are then followed to determine the age at which each of them dies. If the number of survivors is plotted against ages (Fig. 4–9), the curve can be used to predict the probability that a newborn individual will reach any particular age. In this way the life expectancy can be determined for any population.

Survivorship curves take various forms, depending on the characteristics of the particular population (Fig. 4–10). In a Type I survivorship curve, most individuals survive for a long time and die as a result of the diseases of old age. This type of survivorship curve is usually associated with a reproductive strategy in which the parents devote considerable care and energy to each offspring. A "perfect" Type I curve never occurs because some individuals always die during the first few months of life. Embryonic development is a precarious process; many things can go wrong with it. Genetic and developmental defects and birth accidents produce a higher death rate for the embryo and the newborn of Type I organisms than they experience at any

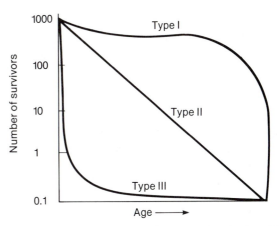

Figure 4–10 The main types of hypothetical survivorship curves.

other time until the onset of old age. The Type I survivorship curve is characteristic of prosperous human populations and of other animals such as some large birds and mammals, for example, eagles, polar bears, elephants, and whales, which produce few but large offspring (Fig. 4–11).

In a Type III survivorship curve, most individuals die at an early age. This type of survivorship is associated with a reproductive strategy in which large numbers of offspring are produced but receive little parental care. It is characteristic of many species of invertebrates, fish, plants, and fungi.

The Type II curve falls between those of Types I and III. There is again an initial period

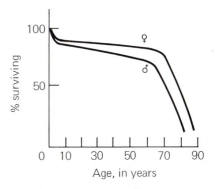

Figure 4–11 Survivorship curves for Americans based on 1970 census data. Note the low infant mortality. Human mortality in developed countries approximates a Type I curve.

of high mortality due to genetic defects or accidents during development, birth, or hatching. Once past this critical period, however, a smaller but constant number of individuals die at any subsequent age. This curve is typical of several bird species and of human beings exposed to poor nutrition and hygiene. With this curve, reproductive strategy lies somewhere between the other two. Parents usually produce more offspring than do members of Type I populations, a situation selected for by the higher mortality rate of offspring.

4.5 Interactions Between Organisms

An encounter between two individuals can be beneficial, harmful, or neutral to either or both of them. Likewise, the sum of *many* encounters between individuals can be beneficial, harmful, or neutral to the populations to which the individuals belong. The effects on the individuals are not necessarily the same as they are on the populations: If a coyote eats a mouse, it is the end for that mouse but not for mice. Ecology therefore deals with the effects of the sum of individual encounters on the total community.

The important encounters between any two individuals, A and B, are those that are potentially beneficial to at least one of them. These interactions are discussed in the ensuing paragraphs.

Competition

Competition is an interaction in which two or more individuals try to gain control of the same resource. **Intraspecific competition,** competition between members of the same species, is very common. Members of the same species generally need the same resources and thus are bound to compete for them except when they are colonizing a new habitat or when the population is very small in relation to the available resources. A few examples illustrate the point.

In one experiment, seeds of white clover, *Trifolium repens,* were planted at two different densities. Half the plants at each density were watered throughout the experiment, but the other half were watered only for the first 18 days. Among the seedlings that were watered, mortality was low, regardless of their density. Among the seedlings deprived of water, however, 300 percent more seedlings died in the high density than in the low density plot, dramatic evidence of density-dependent mortality, which was presumably caused by competition for the limited supply of water (Fig. 4–12).

In the kind of competition that occurred between these seedlings, sometimes called **scramble competition,** each individual attempts to use the resource without regard to the other individuals competing for it. Such competition is common in nature mainly in species that are incapable of complex behavior.

There is another kind of intraspecific competition. Most animals that are capable of complex behavior have conventions that permit them to avoid competition with their neighbors most of the time. Instead of competing directly for a limited resource, these animals compete for social status or for territories. This kind of competition is called **contest competition.** Territories are areas that an individual will defend against invaders. The possession of a territory may be of no value in itself, but it guarantees access to certain resources, such as food, space, or a mate. Similarly, most social animals have "pecking orders." These are social hierarchies in which low-

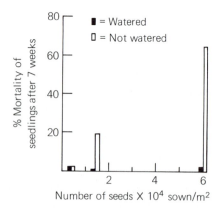

Figure 4–12 The effect of seed density on the survival of white clover seedlings. Unshaded bars represent the mortality of seedlings that were not watered after the eighteenth day. (After J. L. Harper: *Society for Experimental Biology Symposium,* 15:1, 1961, Cambridge University Press.)

ranking members defer to those of higher rank. Low-status individuals get out of the way when a higher ranking member of the group demands access to a resource such as food or a mate. Human beings are among those animals that compete for social status and so, indirectly, for the scarce resources that such status brings.

Competition for territories that guarantee their owners access to shelter and food is common among higher animals. Ecologists studied the muskrat population of an Iowa marsh for 25 years. During this time, the marsh always contained almost exactly 400 muskrats. The researchers found that the remarkable stability of the muskrat population depended on the density-dependent action of competition. Males compete for territories where they find food, hide from predators, and raise their young. The marsh always contained the same number of territories and therefore the same number of muskrat families. Males who failed to secure territories were forced to breed in less-than-ideal conditions outside the marsh where they and their offspring suffered enormous mortality. Even a flood that killed off most of the population of the marsh in one year did not affect the population size for long. Muskrats have a high innate capacity for increase and rapidly repopulated the marsh to its previous density. This kind of situation is common in the many animal species that hold territories. The population size is very constant because only those animals that possess territories breed successfully.

Interspecific competition is competition between members of two or more different species for the same resource. Mathematical models developed in the 1920s predicted that two different species cannot occupy the same niche at the same time. In other words, if individuals of two species compete intensely for the same resource, one species will always be more successful and will eliminate the other species from the area. This prediction has been borne out by studies of interspecific competition in nature.

It often looks as though several species coexist while competing strongly for the same resource. The trees in a forest, for instance, compete for light, water, and soil minerals. Several species of warblers (birds) often feed on the same species of insects in the same part of a forest. Whenever such cases have been examined in detail, however, it has invariably turned out that the different species partition the resource between them to reduce competition. Robert MacArthur found that different species of wood warblers in northeastern forests forage for insects in different parts of the trees, thereby reducing competition between them (Fig. 4–13).

It is reasonable to ask whether or not two species of plants living side by side with their

BLACKBURNIAN WARBLER   BAY-BREASTED WARBLER   MYRTLE WARBLER

Figure 4–13 *Several species of warbler in the genus* Dendroica *hunt for insects in coniferous trees in the same New England forest. Each usually forages in a different part of the tree (shaded), thus reducing the competition for food.*

roots in the same soil and their tops touching are not in fact occupying the same niche. When these situations have been studied, investigators have found that if there is no difference in nutrient requirements or root depth there usually is a difference in timing of life cycles. Of two species of clover that coexist, one was observed to grow faster and spread its leaves sooner than the other. The second species ultimately grows taller than the first. Thus, each has its period of peak sunlight, and they are both able to survive.

Interspecific competition can also restrict the abundance and distribution of natural populations. An example is competition between two species of barnacles, *Balanus* and *Chthamalus* (pronounced thamalus) on the rocky coast of western Scotland (Fig. 4–14). *Chthamalus* occupies the upper part of the intertidal zone and *Balanus* the lower, with little overlap between them. The planktonic larvae of both species settle on rocks in both zones. However, *Balanus* cannot survive in the upper zone because it dries out and dies during low tide. What goes on in the lower zone? Joseph Connell removed *Balanus* larvae from the lower zone and found that *Chthamalus* soon filled the lower zone. Obviously, only competition from *Balanus* usually keeps *Chthamalus* out of the lower zone. Because *Balanus* grows faster than *Chthamalus*, *Balanus*

grows over *Chthamalus* and crowds it out when the two occur together in the lower zone.

Populations limited by this sort of competition are very common in nature. If a species did not face competition, it could survive in many areas where it is not in fact found. A species is *actually* found only in areas where it is capable of fighting off the competition. Gardens are full of examples. Exotic flowers and vegetables may grow successfully in gardens where they are protected from competition for light, water, and space, but they would rapidly succumb to competition from native plants if they were planted in nearby fields or woods.

In almost all cases, competition between two species reduces the population of both competitors, and therefore it is tempting to say that one species would be "better off" if its competitors were eliminated. This is not always the case, however. In one section of the shoreline in the state of Washington, barnacles and sea anemones competed for suitable living space in the intertidal regions. When all the barnacles were removed during an experiment, the anemone population grew rapidly, as one might have expected, but this bloom was followed by a rapid decline, as shown in Figure 4–15. The decline did not result from a food shortage; rather, it was found that the anemones died of desicca-

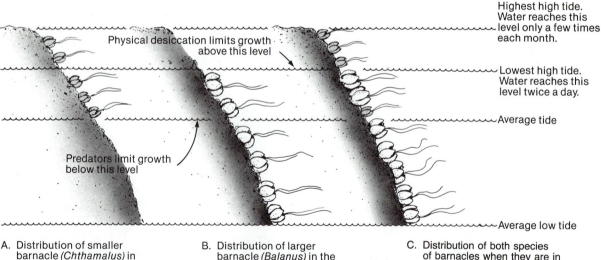

Figure 4–14 Competition between two species of barnacles on the Scottish coast.

A. Distribution of smaller barnacle *(Chthamalus)* in the absence of competition from larger species

B. Distribution of larger barnacle *(Balanus)* in the absence of competition from smaller species.

C. Distribution of both species of barnacles when they are in competition.

Physical desiccation limits growth above this level

Predators limit growth below this level

Highest high tide. Water reaches this level only a few times each month.

Lowest high tide. Water reaches this level twice a day.

Average tide

Average low tide

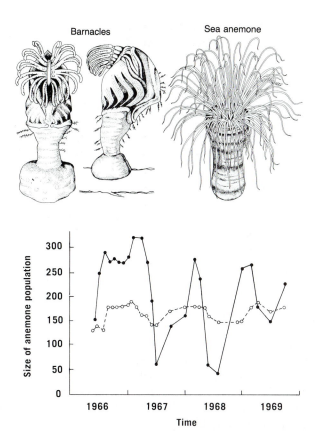

Barnacles   Sea anemone

Figure 4–15 *Effect of removal of barnacles from intertidal region. Solid circles indicate anemone population in an area where barnacles were removed. Open circles indicate control; the anemone population is in an undisturbed area. (From P. K. Dayton:* Ecological Monographs, 41(4): 373, 1972. *Reprinted with permission of author.)*

tion. In an undisturbed system anemones grow close to their competitors, the barnacles, thereby finding shelter from the hot summer wind.

Predation

Predation is an interaction in which certain individuals eat others. Since all heterotrophs must eat to survive, predation is an integral part of any ecosystem. Examples include a mountain lion eating a deer, a frog eating a fly, a paramecium eating a bacterium, and a grasshopper eating grass.

Predation is a major cause of mortality to most natural populations. However, predation exerts different effects on population size and growth, depending on the particular species in-

volved and on the ecological niches they occupy. In some situations, predation seriously depletes prey populations. An attractive, yellow-flowered perennial herb called perforated St. John's wort or Klamath weed *(Hypericum perforatum)* was introduced into the United States from its native habitat in Europe and Asia in 1793. By 1900 it had appeared along the Klamath River in northern California. Not only is this plant toxic to cattle, but it is also an aggressive competitor on grazed range land, eliminating desirable grasses and herbs. By the late 1940s it occupied over 2 million acres in California, Oregon, and nearby states, covering 80 percent of the ground in many areas and making it useless for ranching. In Europe, Klamath weed is attacked by several insects that specialize in eating only this plant, including two species of beetles of the genus *Chrysolina*. Supplies of beetles were obtained and let loose in California in 1945. They flourished. By 1959 Klamath weed had been reduced to less than 1 percent of its former abundance, thanks to the enormous damage inflicted on it by the beetles. Neither the beetles nor *Hypericum* became extinct. Today, both persist at low but stable densities in the western United States. The plant survives best in shade areas where the beetle is less effective, but it no longer represents a major threat to agriculture in the region.

In other systems it has been shown that predatory pressures either improve the health and vitality of a community of prey animals or increase the number of species of prey in a system. Studies of moose killed by wolves have shown that the old, the crippled, the sick, and the very young are killed in disproportionately high numbers (Fig. 4–16 *B*). A healthy adult moose is killed only very rarely by a wolf or wolf pack. Predation helps to select for a healthy breeding population of prey by eliminating individuals weakened by genetic disorders or susceptibility to disease.

It has already been mentioned that two species cannot occupy exactly the same niche because one always dominates or displaces the other. It has been shown that in the absence of predation, two species with different but similar niches often are unable to coexist. For example, destruction of the predator population in northwest Wyoming has led to a rapid rise in the number of elk. The abundance of elk has put so

A

B

Figure 4–16 Predators. *(A)* An archer fish knocks a lady-bug off a leaf with a well-aimed squirt of water. (Roy Pinney, Globe Photos, Inc.) *(B)* A wolf pack circling moose on Isle Royale in Lake Superior. A healthy adult moose, using its hooves or horns, is a match for up to several wolves. The wolves will attempt to stampede the moose so that aged, young, or sick individuals that cannot defend themselves will be separated from the herd and so identified. (Photograph by L. David Mech.)

much competitive pressure on the deer for access to food that they are threatened with elimination from the area, whereas when predation was more intense, both species flourished.

Other studies in other systems have led to similar conclusions. In a series of experiments performed in intertidal communities on the Pacific coast, a typical section of shoreline harbored 15 resident species. There were several predators in the community, one of which was a type of starfish. When the starfish were removed from an experimental area, the community was altered drastically. Interspecific competition for space and food became intense. Of the 15 original species, 7 were eliminated. Apparently, the niches of many of the organisms were quite similar. In spite of this similarity, however, competitive displacement did not occur in the undisturbed system because starfish are general rather than specific predators. They put the greatest hunting pressure on the most common species, thereby controlling overpopulation by any one type of organism (Fig. 4–17).

In yet another ecosystem, ecologists removed all the grazers from a section of a western range land. In the early spring the grasses grew faster and denser than the grasses in nearby areas that were subject to grazing. (Recall that grazing is a form of predation.) When the summer drought arrived, the ungrazed grasses lost so much moisture through their extensive leaf systems that many plants yellowed, and by the end of the season the ungrazed plot was less healthy than the natural one.

Parasitism

Parasitism is a special case of predation in which the predator is much smaller than the victim and obtains nourishment by consuming the tissue or food supply of a living organism known as a **host**. Just as predator–prey interactions are balanced in healthy ecosystems, parasite–host relationships have also become part of the balance of nature. It must be stressed that this type of balance observed in old systems

Figure 4–17 Several general predators. A snake, a fish, a dragonfly, and a bird capture their prey.

does not imply that a new parasite (or predator or grazer), artificially imported from another continent, will immediately establish itself as part of a stable system. On the contrary, a new species may find a new niche for itself and increase unchecked until, perhaps many years later, food supplies decline or another species migrates, is introduced, or evolves to control the rampant one.

Commensalism

Commensalism is a relationship in which one species benefits from an unaffected host. Several species of fish, clams, worms, and crabs live in the burrows of large sea worms and shrimp. They gain shelter and often eat their host's excess food or waste products but do not seem to affect their benefactors.

Mutualism or Cooperation

Mutualism or **cooperation** is a relationship that is favorable and sometimes necessary to both species. For example, crabs often carry coelenterates on their backs and move them from one rich feeding ground to another. In turn, the crabs benefit from the camouflage and protective stingers of their guests. (Not all crabs and coelenterates are mutually cooperative.)

Lichen, which grows on bare rock, resembles a thin layer of vegetation. Actually, the lichen is a mixture of a fungus and an alga. The fungus, which does not contain chlorophyll and thus cannot produce its own food by photosynthesis, obtains all of its food energy from the alga. In turn, the alga cannot retain water and, in some harsh environments, would dehydrate and die if it were not surrounded by fungus. Here the dependence is direct because the organisms must grow together in order to survive.

Another example of a mutualistic interaction can be found within our own bodies. Millions of bacteria live in the digestive tracts of every person. These organisms depend on their host for food but, in return, aid in the digestive process and are essential to our survival.

Summary

Given ideal environmental conditions, the number of individuals in a population increases exponentially at the population's **innate capacity for increase.** This capacity is determined by various factors, including the age of the female parent at first reproduction, the number of offspring produced at each reproduction, the frequency of reproduction, and the parents' reproductive lifespan. A population seldom if ever reproduces as fast as its innate capacity would permit, even when the population is growing exponentially.

The size of a population is limited by the **carrying capacity** of its environment. Some populations oscillate sharply, whereas others, especially those of large organisms, show less variation. Factors that limit population growth may be density-dependent, as for example predation or competition for resources, or they may be density-independent, as for example bad weather or birth control.

Survivorship curves for members of a population reflect the population's reproductive strategy. At the two extremes, the members of a population may produce many small offspring to which they give no parental assistance toward their subsequent survival, or they may produce a few large offspring that are nourished and trained by the parents.

Encounters between two individuals can be classified on the basis of their potential benefit or harm to one or both of them. **Competition,** in which two or more individuals try to gain control of the same resource, benefits the victor. **Predation,** in which some individuals eat others, also benefits only the predator, but it can help to select for a healthy breeding population of prey by eliminating weak individuals. **Parasitism** is a special form of predation by a small organism against a much larger host. Other interactions include **commensalism,** in which one individual benefits from an unaffected host, and **mutualism,** or **cooperation,** in which both organisms benefit.

Questions

Growth of Populations

1. Imagine that the population of a given species quadrupled (increased by multiples of four) every 10 years. If there were 10 individuals in 1950, how many would there be in 1960, 1970, 1980, 2000? Draw a graph showing the number of individuals as a function of time. (Plot time on the horizontal, or X, axis, and population on the vertical, or Y, axis).

2. The worldwide human population has been increasing continuously for the past few hundred years. Can this trend continue indefinitely? Are human populations subject to the constraints of a worldwide carrying capacity? Explain.

3. It is mentioned in the text that lemming populations vary on a three- to four-year cycle. Do you think that the populations of other species in the ecosystem would also cycle? Discuss.

4. Describe a real or imaginary organism whose innate capacity for increase depends mostly on (1) the number of offspring at each reproduction or (2) the age at first reproduction and the frequency of the reproductive cycle or (3) the total number of times it reproduces in its lifetime.

Factors Affecting Population Size

5. Select a nondomestic plant or animal with which you are familiar and discuss the primary components of the environmental resistance that affect it.

6. Does the carrying capacity of an environment for a given species change (1) if the environmental resistance changes; (2) if there is an accident, such as an earthquake, that kills off half the population of the species? Defend your answers.

7. The physical components of the environmental resistance (shortage of inorganic nutrients; adverse climate) are sometimes closely associated with the biological components (predation; competition). Discuss this statement and give examples to support it.

Reproduction and Survival

8. Describe the two extremes of reproductive strategy with respect to the rate at which offspring are produced and the care that is provided for them. What conditions would favor a strategy of producing many offspring at a fast rate? What conditions would favor producing and caring for fewer offspring?

9. Describe the kinds of organisms that exhibit (1) a Type I survival curve; (2) a Type II survival curve; (3) a Type III survival curve.

Species Interactions

10. Classify each of the following mortality factors as density-dependent or density-independent: (1) competition for food; (2) competition for nest sites; (3) precipitation of hot ashes from a major volcanic eruption. Defend your answers.

11. One ecologist stated that predators live on capital while parasites live on interest. Explain. Is this true from an individual or from a community standpoint? Defend your answer.

12. Describe the mechanisms that reduce the harm that (1) interspecific competitors and (2) intraspecific competitors can do to each other.

13. Is there any way in which predation can benefit (1) the individual prey organism; (2) the community of prey organisms? Explain your answers.

Suggested Readings

Refer to references in Chapter 3. In addition, specific references relating to this chapter include:
H. G. Andrewartha: *Introduction to the Study of Animal Populations.* 2nd ed. Chicago, University of Chicago Press, 1971.

J. G. Blower, et al.: *Estimating the Size of Animal Populations*. Winchester, Mass., George Allen & Unwin, 1980.

Gerald Elseth and Candy Baumgardner: *Population Biology*. New York, Van Nostrand, 1980.

Evelyn Hutchinson: *An Introduction to Population Ecology*. New Haven, Yale University Press, 1978.

George A. Seber: *The Estimation of Animal Abundance and Related Parameters*. New York, Macmillan, 1980.

Lawrence B. Slobodkin: *Growth and Regulation of Animal Populations*. New York, Dover Press, 1980.

James T. Tanner: *Guide to the Study of Animal Populations*. Knoxville, University of Tennessee Press, 1978.

Changes in Ecosystems

A superficial glance may suggest that ecosystems remain essentially unchanged for long periods of time. The forest that covers much of Canada, for instance, has been there for thousands of years. But is it really constant? If the forest system is examined carefully, many changes can be observed. Here an aged white pine has died and crashed to the ground, creating a gap in the tree tops. Sunlight comes through the gap and shines brightly on this part of the forest floor for the first time in 200 years. This sunlight causes seeds that have been lying on the ground for many years to germinate, and soon a collection of herbs and shrubs, and a grove of spindly pin cherries, fills the space where the giant pine stood. The animals also change: Insects, mice, and birds that feed on pin cherry move into the area, displacing the animals that lived on the pine tree. The death of this pine tree has caused a radical change in the species of organisms that exist in this part of the ecosystem. This chapter will discuss the nature of a species, why particular species occur in an area at any given time, and the forces that may cause the species in an area to change with time.

5.1 What Is a Species?

A **species** is usually defined as a group of organisms that breed with one another and not with members of other species. Groups of organisms that make up a species may be reproductively isolated from other species in many different ways. For example, closely related species often have different mating seasons or spawning grounds so that males and females of the two species never meet when both are in breeding condition (Fig. 5–1). Members of one species will not recognize or respond to the courtship behavior of another species. In some cases, mating between different species is physically impossible because the reproductive organs are incompatible. If mating between species does occur, the egg may not be fertilized, or, if it is fertilized, it may die before many cell

Figure 5–1 Five closely related species of birds of paradise from New Guinea. The members of different species breed at different times and in different places, reducing the chance that they will attempt to breed with mates from a different species.

divisions take place. An egg may even develop into an adult organism, but one with greatly reduced fertility, as is the case with the mule (a cross between a horse and a donkey). All these mechanisms operate in nature in various groups, singly or in combination, to ensure the identity of separate species.

The boundaries between species are not as clear-cut as this description suggests, however. For instance, a lion and a tiger in a zoo may mate and produce a healthy offspring. Does this mean that lions and tigers belong to the same species? As another example, many plants never reproduce sexually. Scientists cannot say that they breed with each other and not with members of other species because they never breed at all.

Organisms may breed together in the artificial conditions of captivity (a zoo or a greenhouse), although they would not interbreed in the wild. This is the case with lions and tigers, which never breed together in the wild. But it is often impossible to test whether or not two organisms would breed in nature, particularly when they live in different areas or habitats. For all these reasons, it, is sometimes difficult to know whether or not two similar organisms belong to the same species. According to common convention, however, organisms are described as belonging to particular species because it is necessary for scientists to name organisms, in order to communicate with each other about them. Box 5.1 describes how organisms are named.

BOX 5.1 NAMING AND CLASSIFYING ORGANISMS

Living organisms are grouped, or classified, on the basis of kinship. The **species**—a group of organisms closely enough related to be able to interbreed with one another—is the basic unit of classification. Each species is placed in a **genus**, which may contain other, similar species. Each genus is placed in a **family,** each family in an **order,** and so forth; in most cases, each successively higher group includes a larger number of more distantly related species.

Below are the classifications for one plant and one animal:

	HUMAN BEING	BLACK-EYED SUSAN
Kingdom	Animalia	Plantae
Phylum	Chordata	Tracheophyta
Class	Mammalia	Angiospermae
Order	Primates	Asterales
Family	Hominidae	Compositae
Genus	*Homo*	*Rudbeckia*
Species	*sapiens*	*hirta*

Note that to name a species you have to give the name of both the genus and the species: *Homo sapiens, Rudbeckia hirta*. This is because the species name is often trivial *(hirta* merely means hairy), and many organisms may have the same species names: Consider *Hepatica americana* (a spring flower), *Erythronium americanum* (trout lily), *Coccyzus americanus* (cuckoo), and *Veronica americana* (brooklime) (*americana* means American).

International commissions—one for plants, one for animals and one for bacteria—now regulate the scientific names for members of these groups. The snag? The three commissions never get together, so that *Pieris* is a genus of common garden shrubs and is also a genus of butterflies (including the cabbage white butterfly, *Pieris rapae;* see Fig. 5–B).

Scientific names are Latin because, for hundreds of years, Latin was the language of Western scholars. Many of the Latin names are full of information and folklore. A gardener, for instance, finds it useful to know that *Gypsophila* (baby's breath) means "to love chalk," a reference to this plant's preference for lime. Some organisms are named after people. The black-eyed susan, *Rudbeckia,* is named after Olaf Rudbeck, a professor of botany in eighteenth century Sweden (see

Figure 5–A *Rudbeckia hirta.* (Courtesy of Biophoto Associates.)

Fig. 5–A). Medieval folklore taught that, as an early herb book puts it, "God . . . maketh grass to grow upon the mountains, and herbs for the use of man, and hath . . . given them particular signatures, whereby a man may read, even in legible character, the use of them." Thus *Hepatica,* a spring flower, was believed to cure diseases of the liver because its leaf is liver-shaped (Fig. 5–C).

Our system of naming and classifying organisms is probably inadequate to the task. Biologists believe that there are some 10 million different species of organisms in the world, of which only about 15 percent have been described. (Today we are destroying natural habitats so rapidly by pollution, by our population explosion, and by the destruction of forests, rivers and fields, that most of the remaining species will be extinct before they are ever described!) Until better systems of classification become widespread, however, we must make do with what we have. We need scientific names for organisms so that we can communicate with one another and know when we are talking about the same species, and we need some sort of classification scheme to make sense of the vast diversity of organisms all around us.

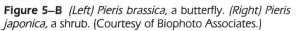

Figure 5–B *(Left) Pieris brassica,* a butterfly. *(Right) Pieris japonica,* a shrub. (Courtesy of Biophoto Associates.)

Figure 5–C *Hepatica acutiloba.* (Courtesy of Pamela Camp.)

Despite the theoretical difficulties of assigning certain organisms to species, in most cases the distinctions are relatively clear-cut. In practice, organisms are often classified by their physical features—their appearance, biochemistry, and behavior. Those organisms that are sufficiently similar physically are treated as members of the same species.

Populations

Although any member of a species can, in theory, breed with any other member, studies have shown that an organism usually finds a mate from within a very much smaller group. A **population** is the breeding group to which an organism belongs in practice. Many populations are surprisingly small, consisting of fewer than 100 individuals. Organisms usually breed only within their own populations for purely practical reasons. The mice on one island off the coast of Maine or in one particular wheatfield never meet the mice from another island or wheatfield and so cannot breed with them. When organisms interbreed, genes from both parents are passed to the offspring, so there are genetic similarities between the members of a population just as there are between members of a family. (Genes are defined in Box 5.2.) For example, skin color tells us whether a person comes from a human population that originated in Africa, Asia, or Europe, even though all people belong to the same species. As people travel and interbreed more than they used to, the genetic differences between human populations are steadily disappearing. The more interbreeding there is between two populations, the more similar their members will be.

5.2 Changes in Species

Species do not remain unchanged indefinitely. Scientists observing the animal inhabitants of a local wheatfield over a period of time may find that the members of the population of mice that live in the field have become steadily smaller and darker over the years. Similarly, many species, such as the dodo, woolly mammoth, and passenger pigeon, have become ex-

> **BOX 5.2 GENES**
>
> **Genes** are the physical entities passed from parents to offspring that determine many, but not all, of the characteristics of an organism. They determine various aspects of the biochemistry, structure, and behavior of organisms. Genes are composed of DNA (deoxyribonucleic acid), which exists in enormously long molecules in almost every cell.

tinct and disappeared forever (Fig. 5–2). Or a species may form one or two brand new species that differ from the original one. All these changes are results of evolution.

Evolution by Means of Natural Selection

According to the theory of evolution, organisms arise by modification from ancestral forms of life. Evolution can also be defined as a *change in the genes of a population from one generation to the next*. If half of a human population contains the genes for blue eyes in one generation, and in the next generation only a quarter of the population contains the blue-eye gene, that population has changed genetically, or evolved. **Natu-**

Figure 5–2 Extinct dinosaurs. Top, *Triceratops*. Bottom, *Stegosaurus*.

ral selection is the evolutionary process whereby the bearers of some genes reproduce more successfully than the bearers of other genes.

The theory of evolution is based on three observations. *First,* as can readily be seen by comparing one cat or one human being with another, the members of a species differ from one another. *Second,* some, though not all, of the differences between individuals are inherited, which means they are genetically determined. (Other differences between members of a species are not inherited but are caused by environmental factors.) *Third,* more organisms are born than live to grow up and reproduce; many organisms die as embryos, seedlings, nestlings, larvae, or infants.

The logical conclusion from these three observations is that some of the inherited, genetic characteristics of an organism must increase its chances of living to grow up and reproduce compared with the chances of organisms with other characteristics. To take an extreme example, a person who has inherited a severe genetic disease of the liver will have a much smaller chance of living to grow up and reproduce than someone without this disease.

Inherited characteristics that improve an organism's chance of living and reproducing will be more common in the next generation than characteristics that decrease the chance of reproducing. Various genes will be naturally selected for or against, from one generation to the next, depending on how they affect reproductive potential. There are a great many different ways in which an organism's genes can be selected for. A predator that is strong and powerful or a prey organism that is quick and alert may both be more likely to reproduce (Fig. 5–3). But strength and speed are only two characteristics that improve the chance of successful reproduction. A tree that produces a lot of seeds, a mosquito that is resistant to pesticides, a desert lizard that preserves water efficiently, or a moth with coloration that camouflages it from predators are all more likely to pass on their genes to the next generation than are those that are not so endowed. Anything that determines that some genes are passed to the next generation more frequently than other genes constitutes a natural selective pressure.

For natural selection to cause a genetic change in a population from one generation to the next (that is, for it to cause evolution), it is not necessary that all the involved genes affect survival and reproduction. The same result will occur if there are just a few genes that make an individual more likely to reproduce.

To summarize—

1. Individuals in a species vary in each generation.
2. Some of these variations are genetic.
3. More individuals are produced than live to grow up and reproduce.
4. Some genes make the individuals that carry them more fit to survive and reproduce than individuals with other genes. Therefore, the characteristics of these fitter individuals are more likely to be passed on to the next generation.

From these four premises it follows that genetic traits that make their owners more likely to grow up and reproduce will become increasingly common in the species, and therefore a species will change genetically, or evolve, with time.

Figure 5–3 Impalas in Africa. Like most ungulates (deer and their relatives), impalas escape potential predators by their sharp senses and great speed. These are some of the important adaptations that permit ungulates to survive for a long enough time to reproduce. (From T. A. Vaughan: *Mammology.* Philadelphia, W. B. Saunders Co., 1972. Courtesy of W. Leslie Robinson.)

Figure 5–4 The two forms of the peppered moth. *(A)* Moths on a lichen-covered tree trunk in an unpolluted area. The gray form is so well camouflaged that it is almost invisible below and to the right of the black moth. *(B)* Moths on a soot-covered tree trunk. The gray form is much more obvious than the black. (Courtesy of Bernard Kettlewell.)

A

B

Evolution in Action—An Example

In nineteenth century England, many people collected moths as eagerly as people now collect stamps. Collectors avidly sought a rare black form of the normally gray British peppered moth *(Biston betularia)*. Black and gray are two genetic forms of this species just as blue eyes and brown eyes distinguish two genetically different forms of human beings. The peppered moth flies and feeds at night and rests during the day, usually camouflaged on tree trunks covered with gray lichens (Fig. 5–4).

By looking at collections made from about 1850 to 1950, biologists found that the black form of the moth became more and more common during the century, and the gray form became scarcer, particularly near industrial cities. Why had this change occurred? Someone suggested that the industrial revolution was the cause. The large-scale burning of coal produces soot, which killed lichens and blackened tree trunks. Against this smooth, dark background, the gray form of the moth became more visible to birds that eat the moths. Birds caught more of the gray moths, whereas the black moths were better camouflaged and survived in larger numbers.

Bernard Kettlewell, an Oxford ecologist, recognized that this was an opportunity to study natural selection experimentally. He raised large numbers of both black and gray forms of the moth in the laboratory, marked them, and released them in two places: one an unpolluted rural area where the black form was more visible to a human observer, the other a polluted industrial area where the gray form was easier to see against the blackened tree trunks. Kettlewell then recaptured as many of the marked moths as he could. The percentage of black moths recovered was twice that of gray moths in the industrial area but only half that of the gray moths in the unpolluted countryside (Table 5–1). These results agreed with the prediction that the black moths were more likely to survive near the cities, whereas the opposite was true in rural areas.

Kettlewell also watched birds feeding on moths on tree trunks. On one occasion, when equal numbers of gray and black moths were released in an unpolluted area, birds picked 164 of the black moths off tree trunks but only 26 gray ones.

It is clear that, in a polluted area, many more of the black moths than of the gray ones will live long enough to reproduce. Since the color of these moths is inherited, the next generation will contain proportionately more black moths than the last generation. In other words, in a polluted area, the proportion of genes for black color increases in the population with time—and that is evolution.

TABLE 5–1   Numbers of Gray and Black Peppered Moths Recaptured After The Release of Marked Individuals In Two Areas

Location	Percent recaptured	
	Gray	Black
Dorset (unpolluted) 1953	13.2%	6.3%
Birmingham (polluted) 1953	13.1%	27.5%
Birmingham (polluted) 1955	25.0%	53.5%

The natural selective force that brings about this evolution is clear: In polluted areas birds kill a higher percentage of moths with the gene for gray color than of moths with the gene for black color. Natural selection has produced populations of moths that are well **adapted** to survive in their environments, populations whose characteristics change as the environment changes.

Following this logic, one would predict that if pollution were reduced, black moths would become rarer and gray forms more common. In fact, the Clean Air Act of 1952 has reduced air pollution in England. Collections of the peppered moth from industrial Manchester in the years since 1952 reveal a dramatic increase in the ratio of gray to black individuals in the moth population. This finding is impressive evidence that the populations of moths have indeed evolved under the influence of natural selection as Kettlewell suggested.

Origin of Species

The most dramatic example of evolution in action is the formation of new species from old **(speciation).** This usually happens because two separate populations of a single species evolve in different directions until they become two separate species, unable to breed with one another any longer.

On St. Kilda, an island off the Scottish coast, there is a species of bird found nowhere else in the world—the St. Kilda wren. This wren closely resembles the species of wren found on the British mainland and has almost certainly evolved from a population of the mainland wren that emigrated to St. Kilda (Fig. 5–5). It is easy to

imagine that a few wrens were blown out to St. Kilda in a storm and then stayed on the island and bred there. Since contact between the mainland wrens and the island wrens was practically nonexistent, the two groups could no longer breed with each other. If birds from the two populations had interbred, neither group could have become genetically distinct or evolved genetic adaptations to its own local conditions. If two populations are to become separate species, breeding between the two must stop so that they do not exchange genes (Fig. 5–6).

Populations can form new species even if they are not physically separated from one another. Such speciation usually results from a drastic genetic change that produces reproductive isolation in one big step. Many cases are known in flowering plants in which the number of genes in a plant doubles in one step. Plants with double the normal number of genes cannot breed with their normal relatives but must breed with each other to form a new species that is reproductively isolated from the parent species (Fig. 5–7). New species of animals may occasionally arise through a doubling of the normal number of genes. (Several grasshopper species are thought to have arisen in this way.) Recent research has shown that there are other rapid changes that can produce new animal species in one genetic step. For instance, adults of the apple maggot *Rhagoletis pomonella* find the apple trees they feed on by visual and chemical cues. In about 1960, one group of apple maggots colonized cherry trees in Wisconsin, probably because a slight genetic change affected the chemistry by which these insects find their food. Now distinct populations, which hatch and mate at

A

B

C

Figure 5–5 *(A) The deserted village on St. Kilda. (B) A St. Kilda wren with an insect in her beak. She is more speckled on head, wing, and tail than the mainland wren. (C) A wren from the Scottish mainland bringing insects to her nest. (Courtesy of Biophoto Associates.)*

different times, breed on cherry and apple trees. Interbreeding between the members of the two populations is becoming less and less frequent, and they are clearly on their way to becoming separate species.

5.3 Ecological Succession

Each individual species represents only one component of an ecosystem. Recall from the pre-

vious two chapters that ecosystems are characterized both by the physical environment and by the interactions among organisms that live in the area. The community that forms when the land is left undisturbed and that perpetuates itself as long as no disturbance occurs, is called the **climax community.** The climax community in New England, most of Canada, Europe, and large parts of the USSR is temperate forest. It consists of the trees and shrubs that characterize such forest as well as the animals, fungi, protists, and bacteria that depend on these plants (Fig. 5–8).

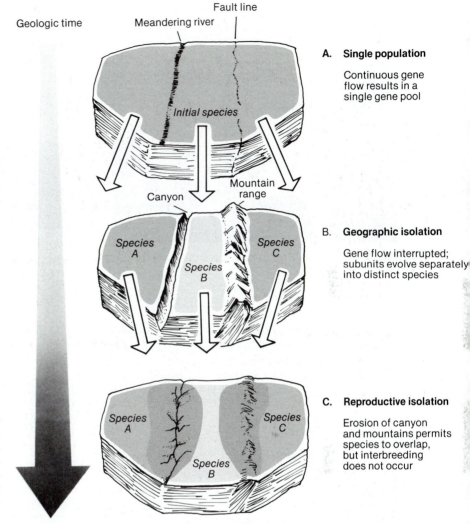

Geologic time

Fault line

Meandering river

Initial species

Canyon

Mountain range

A. Single population

Continuous gene flow results in a single gene pool

Species A

Species B

Species C

B. Geographic isolation

Gene flow interrupted; subunits evolve separately into distinct species

Species A

Species B

Species C

C. Reproductive isolation

Erosion of canyon and mountains permits species to overlap, but interbreeding does not occur

Figure 5–6 Stages by which three new species might evolve from one species. In *B* geographical boundaries form that split the initial species into three populations. A new species evolves from each. (From Mary E. Clark: *Contemporary Biology.* Philadelphia, W. B. Saunders Co., 1973.)

Figure 5–7 Many domesticated species of plants are derived from their wild ancestors by gene multiplication. The right flower is that of *Primula vulgaris,* a wild primrose. At the left is a garden primrose that has twice as many genes as the wild primrose. (Courtesy of Biophoto Associates.)

There are only a dozen or so major types of climax community in the world and these will be considered in more detail in Chapter 7. Many areas do not contain the climax community expected from their climates. They have been disturbed either by natural processes such as floods and volcanic eruptions or by human activities. When a climax community is disturbed, either by human activities or by natural means, it begins to return to its original state by a process known as succession. **Succession** is a progressive series of changes that ultimately produces a climax community. Succession can be seen even on a city street. Mosses and weeds establish themselves in cracks in the sidewalk, quite large

Figure 5–8 Temperate forest surrounding a marsh. (From *North American Reference Encyclopedia*.)

Figure 5–9 Primary succession. This photo shows the volcanic birth of Surtsey, an island off the south coast of Iceland in 1963. The bare rock shown here is still undergoing primary succession as organisms slowly colonize it. (Courtesy of US Navy.)

plants may grow in a corner where leaf litter and dirt have been deposited by a rain gutter, and moss invades a roof that needs repair. If people suddenly left the area, even the center of Manhattan would turn into a rather rock-filled woodland within our lifetimes.

Primary Succession

Primary succession occurs when the terrain is initially lifeless or almost so—when a new island rises out of the sea, when a volcano erupts, when a glacier retreats, or when a mountainside caves in leaving a pile of rocks (Fig. 5–9). Primary succession is usually quite slow because it starts without any soil. Consider an area of barren rock created by a landslide. Water seeping into cracks in the rock may freeze, expanding and breaking the rock into chunks. The surface of the rock is also weathered by the beating of wind and rain. The acid solution formed when atmospheric carbon dioxide dissolves in water helps to dissolve minerals in the rock fragments, providing nutrients. Lichens, which are easily adapted to exposed conditions, may spread over the rock surface. They produce organic acids, which further dissolve the rock. Dead lichens also contribute organic matter to the forming soil, and mosses may gain a hold in even a thin layer of lichen remains and rock dust. As the mosses break up the rock even more, adding their own dead bodies to the pile, the seeds of

---

SPECIAL TOPIC A
Succession on Mt. St. Helens

On May 18, 1980, the volcanic core of Mt. St. Helens in western Washington erupted, spewing millions of tonnes of ash into the air, and destroying many square kilometers of once lush vegetation. In the most highly devastated regions, the ash completely blanketed the existing soil, and the heat vaporized soil particles and nutrients. Four months after the eruption, biologist Roger del Moral found an active ant nest (*Formica subnuda*) high on the mountain and far from any visible vegetation. Presumably these ants were at the top of a subterranean food chain. The lower levels of this micro food chain probably survived on decomposing organic matter buried by the ash. The following spring, a mushroom was found in the original blast zone, growing in the ash and the slowly decomposing organic matter present. Within a year after the eruption, many different types of pioneer plants were colonizing the less severely disturbed areas. Migrating deer and elk passed over the ash, and their footprints formed small indentations that provided shelter and pockets of moisture favorable for the germination of seeds blown in by the wind or transported by birds. In border areas between the devastated zone and places where vegetation has survived, pocket gophers are burrowing into the ash, breaking up the impermeable layers. Full succession to a stable climax system will take many years, but the process has already begun.

small rooted plants can germinate and grow. The process continues along similar lines, with progressively more rock broken up into soil and progressively larger plants moving in, until the climax community becomes established. These processes are slow; it may take up to thousands of years for the soil and the climax vegetation to develop fully. Primary succession occurs also when a lake or pond fills up with silt and fallen leaves and the shoreline creeps toward the center of the lake (Fig. 5–10). Gradually the lake turns into a marsh and then into dry land, eventually colonized by plants of climax species from surrounding ecosystems. As succession proceeds, the species of organisms in the area change. In general, very few different species are present during the early stages of succession. As time goes on, the number of species slowly increases, but then the trend may reverse itself, and the number of species may fall again as the climax community is established.

The number of species present depends on the number of different niches available. Theoretically, two species with identical niches cannot survive together in the same area. One species will inevitably be better at capturing some resource than the other, and the poorer competitor will die for lack of this resource. During succession, the number of different niches available changes with time, so the number of differ-

ent species also changes. Bare rock provides only a few simple niches and will support only a few different species. When the rock has been colonized by lichens, mosses, and a few larger plants, however, niches exist for insects that eat each of the plants, for parasites and predators of these insects, and for animals and plants that may live in the developing soil. As grasses, bushes, and in some cases trees start to grow, even more niches become available.

Secondary Succession

Secondary succession is the series of community changes that takes place in disturbed areas that have not been totally stripped of their soil and vegetation. Although it may take a hundred years or more for the climax stage to return during secondary succession, the process is nevertheless much faster than primary succession because soil already exists.

A familiar example of secondary succession in the New England area is "old field succession," by which abandoned farms return to the climax forest (Fig. 5–11). When a farmer stops cultivating the land, grasses and weeds quickly move in and clothe the earth with a carpet of plants including wild carrot, black mustard, and dandelions. These species are the pioneers of newly available habitats (Fig. 5–12). They grow

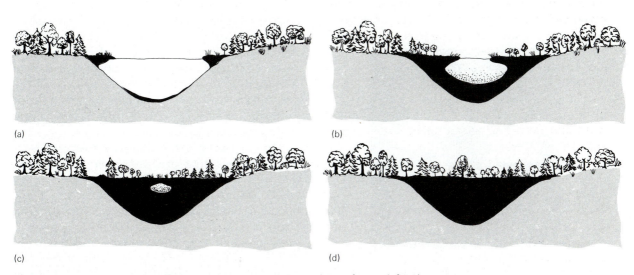

(a)
(b)
(c)
(d)

Figure 5–10 Succession as a lake turns into a bog and then into a forest. A floating mat of vegetation extends progressively out from the shore, eventually covering the open water. After thousands of years, the former lake will be covered with forest.

Figure 5–11 Old field (secondary) succession. Shrubs are starting to grow in the field on the right. Small trees, such as the aspens and pin cherries on the left, will soon invade.

rapidly and produce seeds that can spread over appreciable distances, carried by wind or by animals. In addition, the seeds of many of these pioneer plants are adapted to live for long periods in a dormant state, germinating when a disturbance provides the proper conditions, such as increased light. Soon plants that are taller than the pioneers, such as goldenrod and perennial grasses, move in. Because these newcomers shade the ground and their long root systems monopolize the soil water, it is difficult for seedlings of the pioneer species to survive. But even as these tall weeds choke out the sun-loving pioneer species, they are in turn shaded and deprived of water by the seedlings of pioneer

Figure 5–12 Weedy pioneer species, such as this black-eyed susan, produce hundreds of air-borne seeds that are well adapted to reach new habitats. (Courtesy of Pamela Camp.)

trees, such as pin cherries, dogwoods, sumac, and birch, which take longer (two to five years) to become established but command most of the resources once they reach a respectable size. Succession is still not complete, because the pioneer trees are not members of the species that make up the mature climax forest. After 5 to 30 years, slower-growing oak, maple, beech, elm, and hickory trees will move in and take over, shading out the saplings of the pioneer tree species. After a century or two, the land is covered with mature climax forest.

One of the reasons succession occurs is that many species, especially those characteristic of early successional systems, change the environment in which they live in ways that make the environment less favorable for themselves and more favorable for others. To take a simple example, a corn plant removes the nitrogen it needs from the soil, leaving soil in which corn cannot grow again (unless fertilizer is added) but in which alfalfa, which needs no nitrogen from the soil, can flourish. Similarly, in the Rocky Mountain region, when the leaves and branches of aspen trees decompose, the soil chemistry is changed, making the land less fit for aspen but quite fertile for pine, spruce, and fir.

Secondary succession can be considered as the gradual return of a climax community to equilibrium following disturbance. If the disturbance has resulted in major damage to the soil, as by erosion or removal of soil nutrients, the successional end-point may be an impoverished climax. In the eastern United States, so much topsoil has been lost because of poor farming practices in the nineteenth century that it may take thousands of years for nutrient levels in the present forests to build up again by slow release from the parent rock. Climax forest cannot grow again until the soil is replaced.

In any tract of land, there are always at least small patches that are undergoing succession following disturbance. There may be a spot where a large tree has fallen, leaving a light-filled gap where pioneer weeds can move in and begin a miniature old field succession; a small pond filled in with dead leaves and gradually developing the species composition of the surrounding forest; a slope where a landslide has occurred; or a burned forest. Because every area always contains various patches undergoing

TABLE 5–2   Some Comparisons Between Early Successional Stages and Climax Systems

Characteristic	Early Successional Stage	Climax System
Total organic matter	Small	Large
Inorganic nutrients	Mostly found in physical environment	Large amounts locked in organic matter
Diversity of species	Low	Considerably higher
Food chains	Predominantly linear	More complex, weblike
Net productivity* of organic matter	High	Low

*The *net* productivity is the total productivity minus the total consumption, or the *gain* in biomass of an ecosystem. A climax system has reached a steady state in which productivity and consumption are balanced, so the net gain approaches zero. Therefore it is not profitable to try to farm a climax system. On the other hand, an early successional stage has a high net productivity. In fact, farms are systems that are artificially maintained in an early successional stage. Such maintenance requires energy, as will be discussed in Chapter 14.

succession, there is always a supply of the pioneer plants and animals that thrive in these patches.

Some of our agricultural pest problems stem from the fact that many crop plants originated as pioneer species. These plants depend on their sparse distribution and their nomadic habits (never in the same place for many seasons in a row) to protect them from their insect predators. By planting fields exclusively with one crop year after year, farmers create a paradise for animals that eat these species, such as cabbage worms and cucumber beetles, which no longer have to spend energy to find food and have nothing to do but eat and multiply.

Repeated disturbances will prevent an area from returning to the climax state. Suburban homeowners spend considerable time, energy, and money creating a continuous series of disturbances that maintains a lawn of a few species of short grasses. They constantly interrupt the old field succession of tall weeds, shrubs, and light-loving tree seedlings that will take over if vigilance is relaxed. Many areas in the Midwest were prevented from reverting to climax forest by native Americans who deliberately set fires after they found that herds of buffalo, which live on the prairie, could be hunted more efficiently than the solitary white-tailed deer of young forests (Fig. 5–13).

Figure 5–13 Buffalo, *Bison bison,* which roamed the North American prairies in millions until the end of the nineteenth century. (From E. P. Odum: *Fundamentals of Ecology.* 3rd ed. Philadelphia, Saunders College Publishing, 1971.)

Although ecologists use the term "climax system" to refer to stable, "unchanging" communities, there is actually no such thing as a final, constant situation in a natural ecosystem. When a tree dies and falls, when a moose drops dung, when a stream floods its banks, an area is disturbed and part of the ecosystem is off again on the continuous cycle of change.

Figure 5–14 A forest fire. (Courtesy of Forest Service, U.S.D.A.)

Fire-Maintained Communities

Fires, set by lightning storms or human activities, occasionally sweep through large areas of forest, destroying whole communities of animals and plants (Fig. 5–14). Burned areas undergo secondary succession. In the spruce–fir forests of the Rocky Mountains, for example, burned areas are rapidly colonized by windborne seeds of fireweed, which grows and clothes the slopes with its purple flowers in summer, until it is displaced by aspen and then eventually by the spruce–fir system.

In many communities, fires do not often recur in the same place and thus do not prevent the system from reaching the normal climax pattern. In others, fire is sufficiently frequent to determine the nature of the climax vegetation. For example, before white people colonized the area, parts of central Wisconsin were characterized by a climax grassland system which traditionally caught fire and burned on a regular basis. When farms were established in the area,

and range fires were controlled, untilled land started to succeed into forest. This growth showed that the climate and the soil in the region was favorable for the growth of trees. In the earlier system, however, the saplings were repeatedly destroyed by frequent fires, and the fast-growing grasses dominated.

Many pines are adapted to survive and even to exploit fires. The seedlings of longleaf pine (*Pinus palustris*) in the southern United States, for example, remain as grassy tufts for the first six years or so of their lives. Meanwhile, they accumulate large food reserves in their roots, so that during about their seventh year they can grow extremely rapidly. Fires in southern forests are light ground fires, very different from the intense crown fires of northern forests, and longleaf pine saplings can survive ground fires both when they are in their young "grass" stage and when they are more than about 6 m tall (Fig. 5–15). Their peculiar growth habit permits them to go through the middle stage, when they are most vulnerable to fire, in a very short time. In

Figure 5–15 Longleaf pine forest in southern United States. When a ground fire sweeps through such an area, most of the low shrubs are killed, but these tall pines survive because their growing points are above the fire. (U.S. Forest Service photo.)

5.4 Why Are Organisms Where They Are?

At least two factors determine why organisms are where they are—the distribution of suitable habitats, and the organisms' ability to disperse to an appropriate environment.

Because most species cannot travel between continents, there are different species in similar areas of the different continents. Species in similar areas on different continents, however, may resemble each other closely. This similarity results from **convergent evolution**—the evolution of similar adaptations to similar environments. For instance, it is valuable to be able to conserve water in desert habitats. As a result, in deserts all over the world plants have evolved with thick, water-storing stems as well as spiny leaves that deter animals from using the stems as the source of their own water. Yet the plants in each continent are different. The results of independent evolution are also illustrated by many differences among representative grazing animals on various continents, as shown in Figure 5–16.

5.5 Diversity of Species

In areas that receive plenty of rainfall, tropical ecosystems contain many more species of plants and animals than do similar ecosystems to the north and south of them. There may be more than 100 different tree species in a hectare of tropical forest. The same area of temperate forest hardly ever contains more than 10 tree species. Similarly, a large number of different species of grazing animals populate the African plains. In contrast, only a few different species live in the arctic tundra. Despite much study, no one really knows why these differences exist.

One obvious answer might be that the productivity of tropical areas is higher. (Productivity is defined in Section 3.4.) The water, heat, and sunlight necessary for plant growth are all plentiful in a tropical forest, and these ecosystems support a large concentration of biomass. In addition, the stability of tropical climates over a long period of time has favored the evolution of

the Rocky Mountains and other regions, the pinecones of lodgepole pines *(Pinus contorta)* are adapted to germinate rapidly after they have been heated by a fire.

If fires are prevented in a fire-adapted pine forest, deciduous trees may become established. In addition, dead wood and litter build up on the ground, so when fire eventually does occur, it is more severe than usual and destroys not only any deciduous trees but also the original pines. Odd though it may seem at first, frequent burning is essential for the preservation of many natural communities. Some pine barrens, such as those of New Jersey and around Albany, New York, are slowly dying because human interference prevents fires in those areas. The pitch pine *(Pinus rigida)* of the pine barrens live on poor sandy soil. They can survive only in open areas where they do not have to compete with shrubs. When fires are prevented, shrubs grow under the pine trees, competing for moisture and eventually killing the pines.

Figure 5–16 *Some important grazers that have evolved on different continents.*

a considerable diversity of species that can efficiently exploit the abundant tropical resources.

Productivity is not the whole answer, however. Redwood forests, for instance, are among the most productive ecosystems in temperate regions, yet they contain relatively few species of plants and animals. Some have argued that tropical regions contain more species because they are more mature. Long periods of time, relatively free from major disturbances, have allowed more opportunities for new species to evolve in the tropics. Temperate regions, by contrast, have been subject to major disturbances such as the glaciers that covered much of the land during the Ice Ages (Fig. 5–17). According to this theory, many species became extinct during the glaciations, and not enough time has passed to permit the evolution of new species.

Ecologists have shown that the number of insect species that feed on any one species of European tree is correlated with the evolutionary age of the tree. Apparently, the longer the tree species has been around, the more insect species have adapted to feeding on it.

There is some evidence, then, that temperate regions are not yet at equilibrium. They could support more species than they in fact do and more species will steadily evolve with time. Maturity, however, is almost certainly not the whole explanation for the great species diversity of the tropics. The most obvious difference between temperate and tropical regions is that in the tropics the climate changes little from one day to another during the course of a year, whereas in temperate regions enormous differences in temperature and rainfall occur from

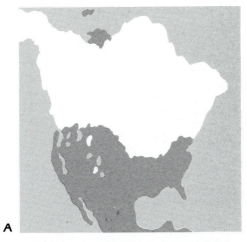

Figure 5–17 *(A)* Glaciation. The white areas on this map show the extent of the ice sheet in North America that pushed south during the Ice Age, about 11,000 years ago. *(B)* A glacier. This is Taylor Glacier in Antarctica, which flows slowly into Lake Bonney, whose frozen waters are visible at the bottom of the picture. (Courtesy of US Navy.)

month to month. It seems likely that, in some unknown way, the greater climatic stability of the tropics permits more different habitats, each supporting a particular species, to develop in these areas.

Species Turnover

Although some areas appear very stable in that some of the same species have lived there for thousands of years, there are inevitably other species in the same area that change with time. The species in an ecosystem may change quite rapidly. In 1968 ecologists did a survey of the species of birds breeding on each of the nine Channel Islands off the coast of southern California and compared their survey with one taken 50 years earlier (Fig. 5–18). They found that the total number of bird species breeding on each island had not changed in the 50 years but that the particular species had changed. On San Nicolas Island, for example, there were 11 species of birds in 1917 and 11 in 1968, but only 5 of these species were the same. Six species had become extinct on this island, and six different species had colonized it.

Species turnover is more rapid on islands than on continents because populations on islands tend to be smaller and become extinct

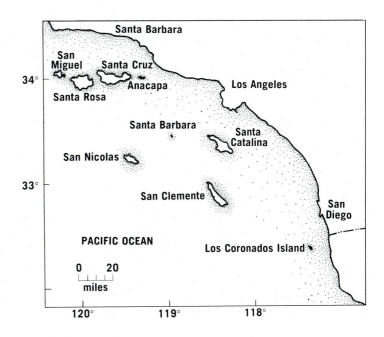

Figure 5–18 The Channel Islands off the coast of southern California.

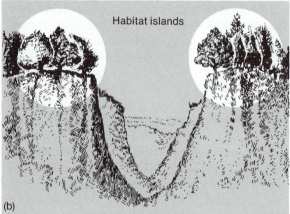

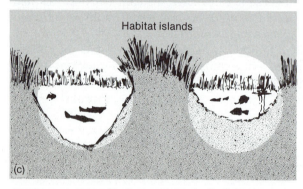

Figure 5–19 *Ecological islands. (A) Real islands. (B) and (C) Isolated islands of the same habitat on a continent.*

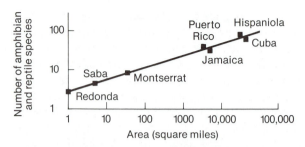

Figure 5–20 *Larger islands have more species. This graph shows the correlation between the number of different species of amphibians and reptiles on Caribbean islands and the size of the island. (From R. H. MacArthur and E. O. Wilson:* The Theory of Island Biogeography. *Princeton, Princeton University Press, 1967.)*

more easily. On a continent, unless large-scale habitat destruction has occurred, if a population becomes locally extinct in a particular ecosystem, there may be similar ecosystems nearby that contain other populations of the same species. Individuals from these ecosystems will usually recolonize the ecosystem from which the species has disappeared. Small, isolated ecosystems such as those of islands, mountaintops, bogs, or lakes, however, are not so readily recolonized, and the species in such ecosystems may change quite rapidly (Fig. 5–19).

If topographically similar islands are compared, the larger island nearly always houses more species than the smaller island (Fig. 5–20). Ecologists think this is because larger islands have a greater diversity of habitats. Additionally, any particular habitat, such as an oak grove, is more likely to be found in multiple copies on a large island than on a small one. Thus, the number of species on the largest islands approaches the number on a continent.

The number of species on an island also correlates with the island's distance from the nearest continent, which is where most of the organisms that colonize the island originate. Obviously, an island connected to the mainland at low tide would be expected to house most of the species found on the mainland, because movement from one to the other would be easy. On a small island far from land, the rate of immigration will be much slower. If a species on an isolated island becomes extinct, it is not likely to be replaced by immigration in the near future.

Studies of islands have illuminated two important points— (1) Any particular isolated area, such as an island, carries a particular number of species. This number remains relatively constant because the number of species that colonize the island is approximately balanced by the number of species that become extinct on the island. (2) The number of species on an island depends on the climate, the terrain, the size of the island, and the distance between the island and the nearest large land mass.

An understanding of species turnover in isolated ecosystems such as islands is vital to the proper design of parks and nature reserves. Such reserves are designed to preserve populations of particular species of plants or animals. They are artificial "islands" surrounded by other systems, which may be roads, towns, or farms. As a result, if the species one wishes to preserve becomes extinct within the reserve, it will not be replaced by immigration from similar nearby ecosystems. The larger the reserve, the more likely a particular species is to survive there.

Summary

The **theory of evolution** states that species are not unchangeable but arise by descent and modification from pre-existing species. Individuals in any population differ from one another, and some of these differences are inherited. **Natural selection** is the process whereby individuals with some genes produce more offspring than individuals with other genes. It leads to evolution, a change in the proportions of different genes in a population from one generation to the next.

The evolution of natural populations through natural selection was not convincingly shown until the twentieth century, when predation by birds was shown to be a selective force in the evolution of predominantly black populations of the peppered moth in polluted parts of England.

A **species** is a group of interbreeding organisms that do not breed with members of other such groups. New species may be formed after two populations of the same species become so isolated from each other that genes no longer pass from one to the other. Each then evolves under selection pressures that produce adaptations to local conditions. The two populations may become so different that they are considered separate species.

The species composition of an ecosystem changes in various ways with the passage of time. The distribution of species depends on the availability of suitable habitats and the ability of organisms to disperse to these places. Although the climate of an area mainly determines the **climax community** of an area, parts of the area are always in various stages of **succession** as a result of disturbances of the climax community.

Some ecosystems contain more different species than others. For reasons that are not entirely understood, tropical ecosystems almost invariably contain more species than similar ecosystems in temperate areas. Larger islands, and those close to large land masses, contain more species of organisms than do small islands a long way from land. Although the number of species on a particular island remains remarkably constant, the identity of some of the species may change quite rapidly with time. Isolated ecosystems on continents behave like islands in these respects.

Questions

Speciation

1. Although intermarriage between ethnic groups is relatively rare in human societies, many such unions do occur. Are specific ethnic groups separate species? If your answer is no, are they likely to become separate species in the future? Explain.

2. Many zoos are finding it advantageous to breed specimens in captivity. If this practice continues, do you feel that there is a possibility that separate "zoo species" will evolve?

3. Suggest why there are more species per unit area in the tropics than in polar regions.

4. Many arctic animals migrate southward during the winter. Discuss how these seasonal migrations might affect the number of species in the forests south of the tundra.

Fitness

5. In primitive societies, where physical strength and stamina may have counted for more than they do today, physical fitness may have endowed its possessors with Darwinian fitness. Can you imagine a type of social organization in which the best physical specimens were actually at a reproductive *disadvantage* with respect to others? Describe such a hypothetical situation.

6. A disease that infects individuals before they are old enough to reproduce or during the reproductive years will necessarily have a selective effect on a population. Suggest a way in which a disease that occurs principally in the *post*reproductive period could have a selective effect on a population.

7. Some anthropologists feel that the dark skin of the black races is adaptive for the warm climates and hot sun in which these people have spent thousands of years. What simple experiment might you design to test the hypothesis that blacks tolerate warm climates better than whites?

Habitats

8. Mountain tops are habitat islands for many resident species but not for all. Explain, and give examples of the latter category.

9. State whether each of these areas is or is not a habitat island, and defend your answer. (1) the Great Plains, (2) a small glacial lake, (3) an oasis in the desert, (4) the Mediterranean Sea, (5) the Amazon Basin, (6) an island in the center of the Mississippi River.

Succession

10 During the past 100 years, the large glaciers in Glacier Bay, Alaska, have receded approximately 65 km. Describe the types of vegetation you would expect to find in this 65-km zone.

11. Explain why a farmer's field must be maintained as an early successional system. Why is energy needed to maintain it?

12. Imagine that a new volcanic island were formed in the middle of the ocean out of hardened lava. Outline a possible successional sequence for the island, and give the time periods for each stage of development.

Suggested Reading

A good summary of modern genetics is given in:
Charlotte G. Avers: *Genetics.* New York, D. Van Nostrand, 1980.

Several excellent works dealing with various aspects of genetics and evolution are:
Otto and Dorothy Solbrig: *Introduction to Population Biology and Evolution.* Reading, Mass., Addison-Wesley Publishing Co., 1979.
J. M. Van Brink and N. N. Vorontsov, eds.: *Animal Genetics and Evolution, Selected Papers.* Boston, Kluwer, 1980.
Michael J. White: *Modes of Speciation.* San Francisco, W. H. Freeman, 1978.

A summary of the research on survival of moths in England is given in:
H. B. D. Kettlewell: The phenomenon of industrial melanism in the Lepidoptera. *Annual Review of Entomology* 6:245, 1961.

Two comprehensive books dealing with the distribution of species on islands and continents are:
Robert H. MacArthur: *Geographical Ecology—Patterns in the Distribution of Species.* New York, Harper & Row, 1972. 269 pp.
Robert H. MacArthur, and E. O. Wilson: *The Theory of Island Biogeography.* Princeton, N.J., Princeton University Press, 1967.

An excellent book that discusses mathematical models of stability in natural ecosystems is:
Robert M. May: *Stability and Complexity in Model Ecosystems.* Princeton, N.J., Princeton University Press, 1973. 235 pp.

A comprehensive discussion of the role of fire from natural as well as human sources is given in:
Stephen J. Pyne: *Fire in America. A Cultural History of Wildland and Rural Fire.* Princeton, N.J., Princeton University Press, 1982. 654 pp.

Extinction and Genetic Resources

Since life first appeared on Earth, species of organisms have originated and existed for a time, and then many have become **extinct,** which means that all members of the species have died. This is a natural process. Today, however, many people are worried that human activities are causing extinctions at an unacceptably high rate and are wondering what can be done about the situation.

To provide some idea of the magnitude of the problem, the *Red Data Book* of the International Union for the Conservation of Nature lists about 280 species of mammals and 350 species of birds in danger of extinction. These figures are dwarfed by the estimate that as many as 20,000 species of plants are endangered. Although they are less visible, the plants may be of more practical importance to the future of human life on Earth. Animals are much less important than plants as human food. All our food crop plants originate from wild plants. Plant breeding programs, which often involve crossing wild with domesticated plants, are vital to modern efforts to increase agricultural yields.

6.1 General Causes of Extinction

Probably 99 percent of all species that have ever existed are now extinct. Examination of the fossil record provides evidence of species that existed for various periods of time and then disappeared forever. The demise of hundreds of species of dinosaurs some 65 million years ago is probably the best known example of extinction. (Scientists do not really know why the dinosaurs and other ruling reptiles became extinct. Ideas range from speculation that changes in climate and competition from mammals were important factors to a theory that dust raised by a huge meteor cut off the sunlight needed by the food plants eaten by the dinosaurs.)

As discussed in previous chapters, organisms live in populations that are more or less isolated from other populations of the same species. Local populations become extinct fairly frequently and are then replaced by immigration from surrounding populations. Hard work and

95

gallons of insecticides may exterminate a local population of Japanese beetles *(Popillia japonica)*, gypsy moths *(Porthetria dispar)*, or Mediterranean fruit flies *(Ceratitis capitata)* from our yards, woodlots, or orchards. But we are gloomily aware that these species as a whole are not extinct and that the local populations will be replaced by immigration before another year rolls around. A species does not become extinct until all of its local populations have died out. Obviously, a species with a great many local populations is less likely to become extinct than one with only a few.

Extinction results when the death rate of a species exceeds the rate at which it produces new individuals. Various pressures, such as predation or competition, may reduce the numbers of individuals in a species. But these pressures in themselves rarely cause extinction. What factors may lead to the final extinction of a species containing few individuals? Or, in other words, how does an endangered species become an extinct species?

Consider the case of the North American passenger pigeon *(Ectopistes migratorius),* which became extinct at the end of the nineteenth century (Fig. 6–1). In 1871 an estimated 136 million birds nested together in central Wisconsin. Millions, but not all, of these birds were shot for food and sport. After the last "great nesting" in 1878, conservationists realized that the pigeons were endangered, and laws were passed to protect them. Even though thousands of the birds remained, it became obvious that the populations were reproducing too slowly to replace the birds that died. In this situation, the pigeon populations are said to have fallen below the **critical minimum size** necessary for survival. Twenty years after the last great nesting, the species was extinct. Social animals that breed in groups tend to have particularly large critical minimum sizes. Such species may be headed for extinction even when quite large populations exist.

The main causes of extinction may be classified as competition with other organisms, inability to adapt to new diseases or predators, commercial exploitation, and the destruction of habitat. Each of these causes will be considered in turn.

6.2 Extinction Caused by Competition

When two organisms compete for a resource that both of them need to survive, be it food, light, or shelter, the organism that is more effective at capturing the resource will survive and the other organism must usually either emigrate or die. For instance, conservationists discovered that populations of bluebirds *(Sialia sialis)* in parts of the United States were small because there were not enough of the right sort of nest-holes available for many of the birds to breed in. They put out nestboxes so that more birds could breed, and the number of birds increased. Then house sparrows *(Passer domesticus)* started to find and use the nestboxes. The sparrows were superior competitors for the boxes. They were able to evict the bluebirds, and once again the bluebird population declined.

The fossil record shows that particular groups of organisms such as fish, reptiles, or birds have replaced each other over long periods of time. Figure 6–2 shows that some 300 million years ago there were very few species of reptiles. (Mammals, reptiles, and birds are defined in Box 6.1.) The number of species of reptiles then increased, peaking about 110 million years ago, when tens of thousands of species of reptiles existed. After this, the number of reptile species decreased steadily, until today only about 6000 species remain. The reptiles were re-

Figure 6–1 *The passenger pigeon.*

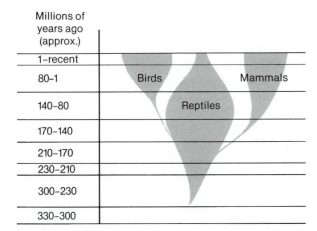

Figure 6–2 Diagram to show when birds and mammals replaced reptiles. The width of each area shows the number of species of each group that lived at any one time.

BOX 6.1 THE CLASSES OF VERTEBRATES

The vertebrates are all animals with backbones, from fish to humans. Vertebrates are divided into six different classes:

1. *Cartilaginous Fish (Chondrichthyes).* These are marine fish with skeletons made of cartilage instead of bone; they include the sharks, rays, and skates.
2. *Bony Fish (Osteichthyes).* Bony fish, found in the sea and fresh water, including tuna, cod, trout, salmon, and pike.
3. *Amphibians (Amphibia).* Adults breathe air through their moist skin and can survive on land. The eggs must have water to develop. This class includes toads, frogs, salamanders, and newts.
4. *Reptile (Reptilia).* Animals with dry skin that breathe air with internal lungs. The eggs are protected from dehydration and are laid on land. Examples include lizards, snakes, turtles, and crocodiles.
5. *Mammals (Mammalia).* These are like reptiles except that they have furry skin and only one bone in the lower jaw (so that it is stronger than the reptile jaw). In most, the eggs develop inside the mother's body and the offspring are nourished after birth by milk from mammary glands. This class includes the primates, bats, whales, rats, dogs, seals, cattle, deer, and horses.

placed by mammals and birds. A small number of birds and mammals had existed for about 100 million years. When the reptiles started to decline, mammals and birds spread widely, forming thousands of new species.

Why this constant replacement of one group by another? Organisms are constantly evolving new adaptations to changing conditions, and one might expect that once any major new habitats were filled, the existing species would be able to evolve fast enough to keep up with changing physical conditions and to outcompete any new species that came along. Clearly, this does not occur. Existing species are not necessarily more effective at making a living from the habitats they occupy than new species that may invade the area. Mammals, for example, are simply more effective at exploiting the resources of most terrestrial habitats than were the ancient reptiles, although there are exceptions to this generalization.

It is a curious fact that species invading from a continent or large island usually outcompete the species present on small islands. Sometimes these invading species themselves are later displaced in turn by new immigrants from an even larger land mass. An example of displacement by invasion can be seen in the bird species of the Caribbean islands. Adelaide's warbler (*Dendroica adelaidae*) colonized essentially all of the islands from the parent population in South America. It has now been displaced and is extinct on all ex-

cept the larger islands of Puerto Rico, Barbuda, and St. Lucia. The lesser Antillean bullfinch (*Loxigilla noctis*) has also colonized the whole island chain, forming visibly distinct populations on many of the islands. It has now become extinct on some of the smaller islands, notably the Grenadines. The bullfinch may well have been displaced from the Grenadines by the glossy cowbird (*Molothrus bonariensis*), which since 1900 has colonized these islands steadily northward. It has now reached about halfway up the island chain (Fig. 6–3).

In many cases, immigrants to islands from continents gradually lose their competitive edge, becoming less and less able to deal with competition from later immigrant species. The reason for this is not entirely clear, but it is obvious that species evolving on continents have an advantage. The number of species on an island is invariably smaller than the number on an adjacent continent. As a result, species evolving on a continent have to cope with more competition and, if they survive this process, are likely to be superior competitors. In general, the larger the

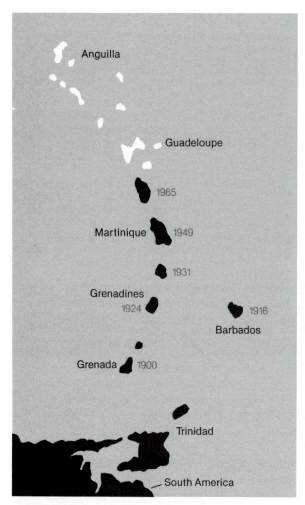

Figure 6–3 Map of the southernmost Antilles islands in the Caribbean. The islands shown in black have been colonized by the glossy cowbird, which originated in South America.

land mass, the more successful the species. For instance, most dominant vertebrate groups, including human beings, have arisen in the tropics of the Old World (Africa and Asia). Members of these groups spread from this center of origin and migrated successfully in all directions, outcompeting other animals they encountered and forming many new species on the way.

Many of the successful competitors that cause the extinction of species nowadays are introduced into an area by human beings. For instance, a unique species of giant tortoise (a land turtle) lived on Abingdon Island in the Galapagos Islands off the west coast of South America. In 1962, an expedition was sent to the island to check on the status of this historic species, which was known to Charles Darwin. A search of the island showed that the tortoise was extinct, although the expedition found the remains of tortoises that could not have been dead for more than a year or two. The reason for the extinction was clear. The tortoises' food plants had been completely consumed by goats, introduced onto the island by a party of fishermen in 1957. The goats were alive and well, which points up another lesson: Animals like tortoises, which feed on only one or a few species of plants, are much more susceptible to extinction than animals like goats, which can switch to feeding on other species of plants if they completely consume one food plant.

6.3 Diseases and Predators

The goats that wiped out the tortoises on Abingdon Island were new to the island. Their rapid destruction of the food previously eaten by the tortoises illustrates another general fact about extinction: A competitor, disease, or predator that invades or is introduced into an area may have a much more devastating effect on local populations than organisms that have coexisted for long periods of time. When predator and prey, or parasite and host, exist together for any length of time, they reach an equilibrium. An efficient predator does not wipe out its prey because then it has no food. Selection ensures that those predators and parasites that survive are those that permit the host species to survive as well. Similarly, there will be selection for the prey or host to evolve defenses that protect it against indigenous enemies. When organisms are introduced into an area for the first time, however, the mechanisms that lead to equilibrium between predator and prey do not exist. For example, the American chestnut tree (*Catanea dentata*) used to be common in the United States, but most of the trees were killed by a parasitic fungus imported from China at the beginning of the twentieth century (Fig. 6–4). Both Chinese and American trees had evolved resistance to the parasites that grew in their own habitats but not to foreign parasites. All the large chestnuts in the United States are now

Figure 6–4 American chestnuts infected with chestnut blight. (From E. P. Odum: *Fundamentals of Ecology.* 3rd ed. Philadelphia: W. B. Saunders Co., 1971.)

gone. Many chestnuts, however, survive because their roots are not destroyed by the fungus. They put up new shoots from the roots and some of these shoots form fruit before they are killed by the fungus. The fact that some chestnuts reproduce offers hope that the species may slowly evolve resistance to the fungus and reappear in our forests.

An example of the effect of introduced predators is evident in the case of domestic cats *(Felis catus)*, which were introduced into New Zealand several hundred years ago. Many of the cats escaped from domesticity and established feral (wild) populations. Predation by the cats has caused the extinction of at least five species of birds that were found in New Zealand and nowhere else. (Species found only in one place are said to be **endemic** to that place.)

Although predators, parasites, and diseases have been introduced into new habitats in various ways since life began, human activities have sharply accelerated the process. Humans are mobile animals and transport organisms all over the world in their trucks, boats, airplanes, and clothing.

Whatever their depredations, introduced parasites and predators do not usually cause the extinction of a species except on small islands where the prey population is small to begin with. Even in the case of large populations on continents, however, predators and disease may reduce the numbers of a species below its critical minimum size, or to the point where something else, such as a severe winter storm, will kill off all the remaining members. For instance, the few remaining members of a bird species called the Laysan honeycreeper were blown off Laysan Island (Hawaii) and killed in a cyclone in 1923.

Even when they do not cause extinction, newly arrived predators and parasites can cause economic problems. In many cases, invaders quickly become pests and are responsible for large financial losses. The sea lamprey, a predator–parasite that attacks various species of fish, migrated into the upper Great Lakes when a ship canal was built to bypass Niagara Falls. Shortly after gaining access to Lake Huron and Lake Michigan, lampreys virtually eliminated a lake trout industry of 3.8 million kg per year. Many such examples abound: The Japanese beetle, imported from the Orient, feeds on many crops, such as soybeans, clover, apples, and peaches; the American vine aphid, imported to France from the United States, was responsible for destroying 1.2 million hectares of French vineyards; and the Mediterranean fruit fly (medfly) originating in West Africa, caused extensive damage to fruits and vegetables in California in the early 1980s. In fact, over half of the major insect pests in many areas are imports.

6.4 Commercial Destruction

Human activities that kill organisms directly are to blame for some extinctions. Such killing takes place for various reasons. People may want the species in question for food, clothing, or decoration, or they may try to eliminate animals and plants that compete with them for food or space (Fig. 6–5).

Before the advent of modern technology, direct human destruction of animals and plants functioned much like natural predation. When an animal was common, people could kill it ef-

Figure 6–5 *The sable antelope, an African ungulate that has been hunted almost to extinction by trophy hunters avid for its beautiful horns. (Courtesy of Richard D. Estes.)*

Figure 6–7 *Bison, or buffalo, around a water hole.*

Figure 6–6 *Trailing arbutus, once on the endangered list, now spreading, since laws to protect it were enacted. (Photo by Alvin E. Steffan from National Audubon Society.)*

ficiently for food or clothing. As their depredations reduced the population size, however, it was no longer economical to trap or shoot that particular animal. When beaver became rare in an area, the local beaver trapper picked up his traps and moved to another area. The population in the first area rapidly recovered to its original size. In recent years, the situation has changed. For example, many modern beaver trappers hold regular jobs in small towns and trap for recreation and additional income. Thus they can continue to trap an area even when the beaver population is low.

In addition, people today are armed with methods of catching, transporting, and processing animals that make it economical to harvest seriously depleted populations. In the early 1900s blue whales were partially protected both by their size and by the stormy waters of the Antarctic Ocean, which is their home. Today, sophisticated whaling vessels and harpoons that carry explosives have overcome these traditional protective measures (see Section 6.9).

There are many different types of commercial exploitation. Plants such as trailing arbutus (*Epigaea repens*) and club mosses (species of *Lycopodium*) were formerly endangered in some areas of the United States partly because people picked them to decorate their houses (Fig. 6–6). These plants are now protected, and their numbers are recovering. Rhinoceroses in Africa are endangered because powdered rhinoceros horn is widely believed to be an aphrodisiac. Animals

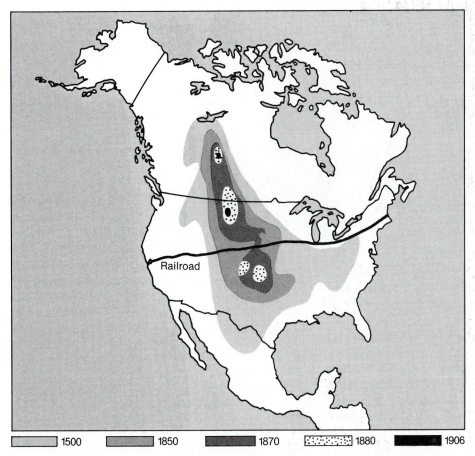

| | 1500 | | 1850 | | 1870 | | 1880 | | 1906 |

Figure 6–8 *This map shows the range of the bison at various dates during the period when the United States was developed. The railroad, whose officials did such damage, runs through the middle of the range. Today, bison exist only on farms and in a few wildlife preserves.*

may also be destroyed for "sport," which usually endangers the larger and more decorative mammals. The Union Pacific Railway, constructed in the 1860s, ran through the plains where the American bison (buffalo) roamed for thousands of years, hunted but not depleted by native Americans (Fig. 6–7). The Union Pacific encouraged passengers to shoot buffalo from the trains. Tens of thousands of animals were shot and left to rot beside the tracks (Fig. 6–8). Even today, although it is illegal to do so, people who have the money shoot polar bears from helicopters or small airplanes and bring back the stuffed heads as testimony to their courage in stalking this dangerous animal.

Perhaps the most persistent conflict between humans and other animals has been with other predators, particularly the large carnivores.

From earliest times, humans have competed for food with other predators such as wolves, lions, tigers, and bears (Fig. 6–9). In addition, some of the large carnivores are capable of killing human beings and have been feared and killed for this reason. In general, as civilization has covered the face of the globe, the large carnivores have been pushed into the remaining uninhabited areas, which grow smaller and smaller as time goes by. Farmers almost invariably believe that the benefits of killing off predators such as coyotes, wolves, cougars, and bears outweigh the disadvantages (Fig. 6–10). They see that these predators kill sheep or young cattle and may not realize that they also kill many herbivores, such as rabbits, which compete with sheep and cattle for grass. In the western United States, farmers and the government spend millions of dollars a

Figure 6–9 A mountain lion, or cougar *(Felis concolor),* cornered by bounty hunters in Colorado. (Photo by Carl Iwasaki, *Life* Magazine)

year killing coyotes and other carnivores. As the populations of the predators are reduced, the populations of many rodents, especially rabbits, have multiplied, and many more millions of dollars are spent trying to eliminate this problem.

Herbivores that compete with domesticated animals for grazing land may also be killed. The European farmers who settled in South Africa believed that the wild ungulates in the area competed with their cattle, sheep, and goats. The settlers therefore undertook a program of systematic destruction. The blue buck *(Hippotragus leucophaeus)* was brought to extinction by 1800, and the last of the quagga *(Equus quagga),* once a common zebra, died in 1878. A major disadvantage of destroying native ungulates is that these animals are often better adapted to the area than are introduced cattle and sheep. They may use the feed more efficiently and be more resistant to disease. For instance, breeding imported cattle with bison in the United States has produced animals that are now being marketed in selected areas in the western regions. Such breeding programs are impossible if farmers destroy all the native ungulates in an area.

6.5 Destruction of Habitat

With the increase in human populations, the amount of land used for agriculture, timbering,

Figure 6–10 A coyote chasing a mouse. When predators are killed by humans, their prey, such as mice and rabbits, multiply. (Jen and Des Bartlett, National Audubon Society, Photo Researchers, Inc.)

mining, roads, and towns has steadily increased. This development has destroyed the habitats that many animals and plants need to survive. Although some animals, such as pigeons, house sparrows, rodents (particularly rats and mice), and deer flourish in the modified habitats provided by human activity, others do not.

More than half the species that are in danger of extinction today are on the list of endangered species because their forest habitats have been destroyed. A majority of these species are from the tropics, where human population growth has been most explosive and habitats have been destroyed most rapidly. Two thirds (more than 13,000) of the plant species known to be endangered are in the tropics. (This staggering figure compares with about 4500 plant species known to be endangered in temperate regions.) The destruction of habitat may be quite subtle. Small changes in the environment may have surprisingly deleterious effects. For instance, even though millions of hectares of forest remain in the northeastern United States, the character of the ecosystems there has changed dramatically. Almost every hectare of forest in North America has been lumbered. Most of us, although we do not realize it, will never see the incredible size to which a white oak (*Quercus macrocarpa*) or a white pine (*Pinus strobus*) can grow. These trees, which take hundreds of years to reach maturity, are invariably cut when they are less than a hundred years old. Most American forest is **second-growth** forest, and one major difference between climax forest and second-growth forest is that in the latter there are fewer trees dying of old age. As a result, animals that depend upon the soft wood, hollow limbs, and many holes in old trees have nowhere to nest. Birds such as the wood duck (*Aix sponsa*) cannot nest unless they find holes of the right size and shape. The wood duck has been saved only because many people now put out nestboxes that can be used by these birds. A similar situation exists in many parts of the world. The Puerto Rican parrot (*Amazona vittata*) is also close to extinction because few nestholes remain in its native forest. It took researchers many years to discover precisely what sort of hole these fastidious birds required. By that time the population of *Amazona* was down to about 20 individuals, and it is not clear whether or not the species can survive.

6.6 Characteristics of Endangered Species

As discussed in the previous sections, many different factors, operating independently of each other or in concert, are responsible for the destruction of species in modern times. These pressures act on entire ecosystems, yet certain species are more susceptible than others. For example, hunting and habitat destruction have all but exterminated the bison from North America, but many species of deer have flourished, and their populations have actually increased over the past century. The general characteristics of a species that make it especially vulnerable to extinction pressures are summarized below.

Island Species. The fastest way to make it onto the endangered species list is to be a species endemic to one island. More than half of the approximately 2000 plant species endemic to the Hawaiian islands are endangered. Islands constitute less than 7 percent of the Earth's surface, yet over half of endangered species are island forms. Island species tend to have small populations that cannot be replaced by immigration if they are destroyed. They have also usually evolved in isolation from many potential predators and competitors, against which they have few defenses. Seals and birds on the Galapagos and other uninhabited islands in the Pacific are so tame that predators can just stroll up and kill them. The most devastating effects have occurred where human beings have introduced mammals to islands that evolved without them. Goats that destroy vegetation, rats that eat eggs, cats that kill birds, and mongooses that eat reptiles have all wreaked widespread destruction.

Species with Limited Habitats. Some organisms, although they live on large continents, are so specialized in their requirements that they are adapted only to a limited habitat. These are ecologically akin to island species and are likewise vulnerable. The Tiburon mariposa lily (*Calochortus tiburonensis*) is so specialized that it occurs only on one hilltop near San Francisco. This plant was not discovered until 1970 and the population is so small that it is automatically endangered. Specialized species that rely on one form of food, nesthole, or any other resource are much more vulnerable than more adaptable animals. The Everglades kite (*Rostrhamus sociabilis*) feeds largely on a snail (species of *Pomacea*) that lives in the Florida Everglades. Drainage of the Everglades for housing and airports has killed off the snails and endangered the kite. Similarly, the ivory-billed woodpecker (*Campephilus principalis*) can survive only on the beetle larvae that live in freshly killed trees. In modern forests trees are harvested and removed for lumber, thereby destroying the food source for the ivory-billed woodpecker.

At the other extreme are species whose needs are so unspecialized that they may exist for many millions of years. The cockroach (*Blattus* sp.) can eat almost anything organic and survives in temperate regions as well as the tropics. Some blue-green bacteria species, found in almost all parts of the world, appear to have survived essentially unchanged for more than a billion years. Their fossils are among the oldest that have ever been found.

Species That Require Large Territories to Survive.   Animals with large home ranges or that need large hunting territories to find sufficient food are also more liable to extinction than animals that need less space (Fig. 6–11). One example is the California condor, which is being threatened because ranchers and agriculture are usurping its territory.

Species with Low Reproductive Rates.   Many large animals have a low reproductive rate because traditionally the mortality of their offspring was low and high birth rates were not necessary for survival. Today, however, low reproductive rates make the organisms susceptible to modern hunting pressures. Examples include the California condor, the polar bear, and the blue whale.

Species That Are Economically Valuable or Hunted for Sport.   Alligators are extensively hunted for their hides, snow leopards for their fur, elephants and walrus for their tusks, whales for oil and meat, and various parrots, monkeys, and apes for sale as pets or for medical research.

Predators.   Wolves, tigers, lions, and certain species of bear all prey on domestic cattle and are therefore feared and killed.

Species That Are Susceptible to Pollution.   Bald eagles, ospreys, and peregrine falcons are all especially susceptible to certain pesticides (Fig. 6–12) and are therefore in danger of extinction. In

Figure 6–11 A California condor, *Gymnogyps californianus*. These ugly but spectacular carnivores are endangered because they need a large area if they are to find enough food. Human activities have destroyed much of their habitat. (Photo by Carl Koford, from National Audubon Society.)

Figure 6–12 A peregrine falcon's nest. The egg shells were weak because the mother had ingested DDT. The eggs are cracked, and the chicks, which were near to hatching, are dead. (Courtesy of Peregrine Fund.)

addition, many plant species are threatened with extinction because they are destroyed by herbicides.

Incompatibility with Civilization. In many cases species are endangered because their habits are simply incompatible with civilization. The red-headed woodpecker (*Melanerpes erythrocephalus*) has the death-wish habit of flying in front of cars, which most bird species have learned to avoid.

6.7 The Nature of the Loss

Does extinction caused by human activities really matter? After all, new species are continually formed, even if speciation is slower than extinction in the twentieth century. What will happen to the biosphere if species continue to become extinct as rapidly as in recent years? Ecosystems and food webs may become simplified, but the pattern of primary, secondary, and tertiary consumption will continue. If deer replace bison as the largest grazers and the coyote replaces bears and mountain lions as the largest nonhuman predators, grazing and predation will continue, so why worry about extinction?

There are several reasons for worry. In some ways the most compelling is the aesthetic and religious argument. Different individuals may express their feelings in different ways. To some, species and the wilderness should be preserved simply because they exist. Others might reformulate this by saying that humans have no right to exterminate what God has created. To still others, an unobtrusive passage through an untouched wilderness area is a source of enormous aesthetic gratification, as valid and moving an experience as a great play or a string quartet. As many of the greatest and noblest creatures of the Earth fall prey to the thoughtless acts of people, so too do the richness, variety, and fascination of life on this planet diminish with their passing.

A second reason is concerned with developments in medicine and the biological sciences, which have always been dependent on various plants and animals as experimental subjects. The range of species used in different experiments has traditionally been very wide. For example,

when Thomas Hunt Morgan initiated his famous studies of genetics, he needed an animal that was easy to breed in large numbers and had few, large, and accessible chromosomes. He chose the fruit fly, *Drosophila*, not because the life style of these insects was of particular interest in itself, but because the animals were easy to study and the lessons learned from them could be generalized to studies of other creatures. Another serious scientific loss in this category is that of wild populations of plants and animals that have traditionally been used in breeding hybrid species for agriculture. For example, strains of domestic corn are often susceptible to various diseases. Geneticists have periodically crossed high-yielding strains of susceptible corn with hardy wild maize in an effort to develop high-yielding resistant corn. However, natural maize growing along fence lines and roadways is often considered to be a weed and has been combated across North America with herbicides. The loss of these species would be a severe blow to modern agriculture.

The final reason, and perhaps the most compelling of all, concerns the richness and variety of life on the planet. The Earth is constantly changing. Climates change, environments change, products and foodstuffs change. Traditonally, species of plants and animals have always evolved along with their changing environment. If a mountain range uplifted, certain organisms evolved to live at a higher altitude; if a species of small, slow-moving horse developed into large, fast-moving horses, then quicker, smarter, stronger predators evolved to pursue these horses. Our world is now developing and changing rapidly, but evolution is slow. If species are allowed to become extinct now or in a few short years, the total global potential for adaptation and change will be greatly diminished, and the diversity of future ecosytems may be endangered.

> **BOX 6.2 T. H. MORGAN**
>
> Thomas Hunt Morgan (1866–1945) was an American geneticist whose experiments formed the basis for his theory that paired elements or "genes" within the chromosome are responsible for the inherited characteristics of the individual. He won the Nobel prize in medicine in 1933.

6.8 Actions to Save Endangered Species

Several types of actions are needed to preserve endangered species.

1. **Research and Documentation.** As a first step, lists of endangered species are established by various national and international agencies. Another important action used to save an endangered species is the compilation of information about it. What is the critical minimum size of the species and what is the current size of the population? What does the species need in the way of breeding area, food, and climate in order to survive? Why has the population been reduced to the danger point? Various governments and international agencies provide the money for this sort of research.

2. **Habitat Preservation and Development of Wildlife Refuges.** Almost every country in the world has established a system of national parks and wildlife refuges (see Chapter 17). Sometimes a small preserve can be extremely effective. For example, preservation of a few wetlands along the routes of migratory birds can ensure the completion of the migration and survival of the species. In other cases, large land areas are needed. Before a government can be induced to set aside an area as a wildlife refuge, thereby preserving habitats that endangered species need to survive, it must be convinced that the effort is worthwhile. Among the rewards for creating such refuges is the fact that such areas attract tourists who may be valuable to the economy. The refuge also encourages biological research and the exchange of information with other countries. Many of the world's largest and most impressive mammals are indigenous to parts of Africa, which boasts a number of magnificent game parks. The first such park was the Kruger Park, 12,000 square kilometers of scrubland, established in South Africa in 1895. It was followed by the Etosha Game Preserve in Southwest Africa—the largest wildlife sanctuary in the world—and the Serengeti Park in Tanzania (Fig. 6–13). All these large parks have existed long enough to prove that they serve their function well. The Kruger Park preserves nearly 500 spe-

Figure 6–13 Giraffes flourish and breed in the Serengeti Park in Tanzania where tourists can photograph them with ease. The giraffes browse on these small scrubby acacias and larger trees.

cies of birds and 114 species of mammals. Not a single species has been lost from this park since it was established.

In other situations, habitat preservation does not mandate special parks but instead requires control of pesticides or pollution. These topics are discussed in Chapters 15, 20, and 21.

3. **Providing Critical Resources.** Another way to improve the habitat of an endangered species is to determine which resource is limiting the population size and to provide more of that resource. As has already been mentioned, construction of nestboxes for wood ducks has increased the wood duck population in parts of America. The white-naped crane (*Grus vipio*) winters in Japan, but the one area where these large, beautiful birds used to feed has now been built over. In 1958 only 45 birds appeared in all of Japan. The Japanese therefore started feeding the birds at just one location in Kyushu, and more than 700 birds now winter in the area each year.

4. **Legal Actions for the Preservation of Species.** Several legal approaches have been used to preserve species. One is to enact laws regulating the killing of members of certain species, with severe penalties for breaking the law. These laws may be very effective. For instance, in the United States, the hunting of deer is controlled by issuing tags to each person who applies for a hunting license. Heavy fines await those caught with a dead deer that does not carry a tag. This system permits wildlife agencies to control deer populations very precisely by issuing as many tags as they consider proper. In many countries it is illegal to kill any members of an endangered species.

Even strict penalties do not eliminate poaching if the products can be sold for a high price (Fig. 6–14). An alternative approach is to ban trade or transport of plants or animals that may have been killed illegally. In 1969 the hides of 113,069 ocelots, 7934 leopards, and 1885 cheetahs were imported into the United States. Many more hides were shipped to Paris, London, and Tokyo. Conservationists realized that legal action was necessary. In 1973 delegates from every major nation in the world except the People's Republic of China agreed to outlaw all trade of any part of any plant or animal that was endangered. But a law on paper must be enforced if it is to save animals' lives. At John F. Kennedy Airport in New York City custom officials confiscate about $300,000 worth of products of endangered species every year. This collection includes leopard skin coats, hunting trophies, crocodile shoes, and many other items. No one knows how many more products are smuggled in. As another example, certain species of whales are currently endangered. Yet Japanese and Russian whalers continue to hunt these animals (see the Case History, Section 6.9).

In the United States two important pieces of legislation, the Marine Mammal Protection Act of 1972 and the Endangered Species Act of 1973, have been passed to protect species in the marine and the terrestrial environments. The Endangered Species Act states that it is illegal to import or transport endangered species or products of endangered species. Furthermore, it is illegal to destroy the habitat of any endangered species. Canada, Great Britain, Taiwan, Kenya, and several other nations have also passed similar legislation.

The power and controversy of the Endangered Species Act is illustrated by the following example. In 1967, the construction of a large dam, called the Tellico Dam, was started in Tennessee (Fig. 6–15). Six years later, when the En-

Figure 6–14 Leopard skins and other hides on sale in a street market in Srinagar, India. This author (Jon) asked the head of game management in Srinagar if it was legal to trap leopards and sell their skins. He informed me that the animals are endangered and carefully protected. I then asked about the hides for sale in the street. He told me that sometimes the law is difficult to enforce.

Figure 6–15 The Tellico dam when it was partially completed. (Courtesy of the Tennessee Valley Authority.)

dangered Species Act was passed, the dam was nearly complete. Construction had already cost $103 million. The dam was to provide enough electrical power to heat about 20,000 homes in the region. In August, 1973, a zoologist studying the rivers in the area discovered a small species of snail-eating fish that he called the snail darter (Fig. 6–16). Early reports claimed that the snail darter was found only in the water behind the Tellico Dam, and that if the construction were completed the habitat of the darter would be destroyed and the species would become extinct. A United States District Court ruled that since the Endangered Species Act states that it is illegal to destroy the habitat of an endangered species, the completion of the dam must be halted. The Tennessee Valley Authority (TVA), builders of the dam, took the case to the Supreme Court. In June, 1978, the Supreme Court upheld the decision of the lower court. Further construction of the dam was forbidden. The majority opinion of the court stated that ''The plain intent of Congress in enacting this statute was to halt and reverse the trend toward species extinction, whatever the cost.'' But the case was not closed.

In 1978, Congress amended the act to allow exceptions in federal disaster areas, in cases involving national defense, and in cases permitted by a special review committee. After a continued legal and legislative struggle, snail darters were relocated in nearby streams, and in 1979 permission was granted to complete the dam. Later, another natural population of snail darters was found some miles from the dam, so continued existence of the species seems more secure.

After the lengthy battle over the Tellico Dam and the snail darter, many people claimed that the Endangered Species Act was excessively strict, and some even suggested that it be scuttled entirely. However, in June 1982, the Act was renewed for three years with some changes.

On one hand, the revised law contains procedures that would allow future projects such as the Tellico Dam to proceed with only minor delays. At the same time, the general operating procedures were streamlined to strengthen the protection of many biologically important species of plants and animals.

5. **Breeding in Captivity.** When a species has become very rare, biologists sometimes capture some of the few remaining animals and attempt to breed them in captivity. The idea is that if they can increase the population size, the animals can be released again into the wild. Breeding in captivity may involve artificial insemination, farming out juveniles or eggs to females of other species to act as foster mothers, and various other devices for increasing the number of offspring produced. A recent program of this kind in the United States is devoted to the peregrine falcon *(Falco peregrinus)*. Whole populations of this bird were wiped out by chemicals such as the pesticide DDT. Carnivorous birds such as falcons and pelicans ingest large amounts of pesticides with their food. DDT causes the birds to lay soft-shelled eggs that are crushed when the parent sits on them. In 1973 the sale and use of DDT was banned in the United States. Therefore, many habitats have once again become favorable for the falcons. The object of the peregrine program is to breed these birds and reintroduce them into areas of the country where they used to exist but have recently become extinct. This program has been relatively successful, but it has also pointed up a number of reasons why captive breeding is considered by many biologists to be a poor way of preserving an endangered species. First, it is very difficult and expensive to induce many animals to breed in captivity. Second, even if breeding is successful, animals bred in captivity often cannot cope with life in the wild when

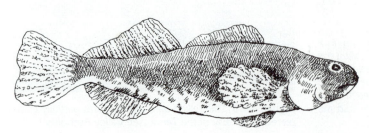

Figure 6–16 The snail darter.

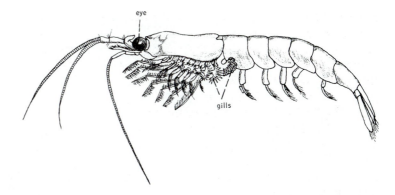

Figure 6–17 Krill, *Euphausia superba*.

they are released. Third, it is doubtful whether any except the most experienced biologists should be encouraged to attempt to breed endangered animals. Many amateur breeders exist, paying enormous sums for animals captured in the wild. The often illegal trade in wild animals that this hobby encourages has killed millions of endangered animals.

6.9 Case History: The Blue Whale

Antarctica is a cold, frozen continent, covered with ice and nearly devoid of vegetation. By contrast, the southern oceans that surround this continent are rich, fertile, and teeming with life. In fact, life is more abundant in the Antarctic Ocean than in many central tropical oceans, for deep-sea currents carrying valuable nutrients rise in the lower latitudes and fertilize the surface layers of ocean. The nutrient-rich waters support large populations of phytoplankton, the primary producers in the ocean. In turn, there are a great many zooplankton that feed on the tiny plant species. The zooplankton population consists in part of various species of small, shrimplike crustacea, known collectively as **krill** (Fig. 6–17). An individual krill organism is only about a centimeter or two long, but these animals congregate in schools that can be 10 meters thick and cover a surface area of several square kilometers. Schools of krill this size naturally represent a significant food supply; in fact, the schools are large and plentiful enough to feed the largest animal that has ever lived on this planet—the blue whale, *Sibbaldus musculus* (Fig. 6–18). A blue whale is about 30 m (100 ft) long and weighs 90,000 kg (90 tonnes). It has no true teeth, but rather a set of elastic, horny plates called **baleen** that forms a sievelike region in the whale's mouth (Fig. 6–19). A blue whale feeds by swimming through a school of krill with its mouth open, allowing the krill to pass through the baleen but excluding any larger objects.

Little is actually known about the life and habits of the blue whale. We know that they are mammals; they are born alive and breastfeed for some time after birth. Scientists believe that a young animal becomes sexually mature when it is about six years old and that a mature cow gives birth to a single infant every two or three years. Other species of whales that have been

Figure 6–18 A blue whale.

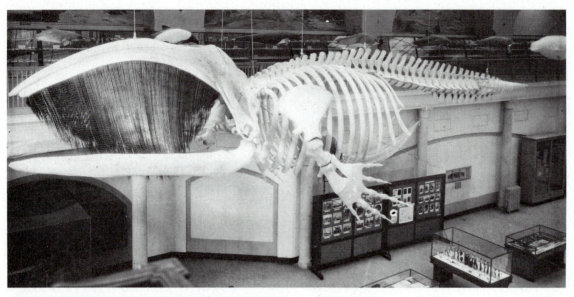

Figure 6–19 Skeleton of an Atlantic right whale, *Eubalaena*. Note the numerous sievelike baleen in the whale's mouth. Blue whales have the same feeding mechanism. (Courtesy of American Museum of Natural History)

studied more intensively have proved to be highly intelligent.

Blue whales appear to be well adapted to life in their natural environment. Their bodies are streamlined and their skin, smooth and nearly hairless, offers little resistance to water flow. Body heat is conserved with the aid of a thick insulating layer of fat, called blubber. A mature blue whale has no significant nonhuman predators. Thus, the low reproductive rate of these animals has ensured that they do not overpopulate their range.

The Antarctic is one of the most dangerous and tempestuous oceans in the world. Early whalers, therefore, did not venture south after these giants. But gradually, as whaling grounds in the more central latitudes became depleted, and as ships were built to be stronger and more seaworthy, people ventured south in search of the blue whale. At about the turn of the century, it was estimated that the blue whale population numbered about 200,000 individuals. In 1920, over 29,000 whales were killed. The blubber was boiled down to oil to be used directly as a lighting fluid or as a raw material for the manufacture of cosmetics, medicinals, lubricants, shoe polish, paint, and other products. Some of the meat was eaten by humans, some thrown away,

and in later years some was ground into food for dogs and domestic mink. Whaling was a dangerous but profitable business. The whales, having evolved in a predator-free environment, could not adapt to the new predatory pressure. The population started to decline precipitously. By the middle 1950s the species had been decimated. Marine biologists warned the whaling communities that not only would blue whales be threatened with extinction if current practices continued but also that the indiscriminate slaughter of whales was economically unsound in the long run. If whalers could practice reasonable conservation methods and allow a large breeding population to survive, they would be able to harvest several thousand individuals annually on a prolonged and sustained basis. On the other hand, if the wholesale slaughter were continued, and the breeding population were destroyed, then in a very short time there would be no more whales. Both whales and whalers would suffer. Unfortunately, the open ocean is subject to little national or international control and is a type of commons. Owners of whaling ships must meet interest and loan payments and cover maintenance and docking fees whether whales are caught or not. Since few companies are willing to accept immediate losses in exchange for long-term stability, especially when others continue to exploit the resource, the slaughter has continued.

The International Whaling Committee (IWC) was established in 1946 to regulate whaling practices. The Committee ignored the recommendations of marine biologists and set limits that greatly exceeded the sustained catch levels. The whalers ignored even these generous guidelines and killed more whales than the IWC recommended. Predictably, the whale population declined so that the total catch decreased steadily. In 1964–65, only 20 whales were harvested, and in 1974–75 the catch dropped still further. With the blue whale population nearly extinct, whalers turned to hunting other smaller species—fin, sei, and brydes whales.

At the present time, the blue whale population is estimated at a few thousand. Some experts feel that the critical level may have been reached already and that the blue whale, largest of all animals ever to live on this planet, is destined for extinction. At the same time, the total yield, whether we count that yield in number of animals killed, or in kilograms of food harvested, or in dollars, rubles, or yen, has declined well below the theoretical productivity of the southern ocean. As a result, the world community has directly suffered from the slaughter of the whales.

Even though it is now illegal to import products of the blue whale into many nations, such as the United States, Great Britain, and Canada, a few nations, including Japan and Russia, continue to hunt Antarctic whales. If current practices continue, and species of whales become extinct, not only will we suffer a great aesthetic and scientific loss but also a hitherto productive region of the Earth's surface will produce less protein and fewer raw materials for the human population.

In July 1982, after several lengthy battles and a few stopgap measures, the International Whaling Commission (IWC) voted to outlaw commercial whaling entirely, beginning in 1986. Until then, Spanish whalers are allowed to harvest 400 fin whales and the Japanese are allowed a quota of 850 sperm whales. These concessions were made to secure the necessary votes and to discourage key nations from withdrawing from the commission entirely. The effectiveness of the

ruling is still based on voluntary participation, for no nation or group owns the seas and no guidelines have the strength of law. However, political and economic pressures may be an effective deterrent against breaking the IWC ruling. For example, the United States passed legislation stating that any nation that ignores the IWC directives will not be allowed to fish in U.S. territorial waters or to sell fish products within the country. It is hoped that these measures, taken together, will save many species of whales from extinction.

Summary

During evolutionary history, organisms continually become extinct, usually because they are replaced by other species that capture a limited resource more efficiently. Often the later-evolved species is more efficient in many ways than the species it replaces. In addition, organisms usually cannot compete effectively with species that have similar needs but have evolved on larger land masses.

Diseases and predators, particularly those that have recently invaded an area, may drastically reduce the populations of their prey. Such factors do not usually completely wipe out a species but may reduce its numbers below the critical minimum, or to a point where an accident may kill all the remaining members of the species.

Human activities that kill organisms directly endanger many species. Hunting and fishing by modern methods are so efficient that they may cause extinctions, particularly of large predators and herbivores.

The most important way in which human activities cause extinction is by destroying or altering habitats. Today, organisms that live in tropical forests are particularly endangered by habitat destruction.

Species that are most susceptible to extinction are those that live on islands, have limited habitats, are highly specialized, require large territories to survive, have low reproductive rates, are economically valuable or hunted for sport, are predators, are susceptible to pollution, or are generally incompatible with civilization.

When a species becomes extinct everyone suffers an aesthetic loss. In addition, genetic resources are lost that are needed to breed animals and plants for human food and other commercial purposes.

Actions taken to preserve endangered species include research and documentation, habitat preservation and the establishment of wildlife refuges, providing resources needed by endangered species, legal protection of endangered species, and captive breeding.

Questions

Preservation of Endangered Species

1. What arguments would you use if your task were to persuade the residents of the county where you live that it would be a good idea to convert part of the county into a wildlife refuge?

2. Would you advocate a ban on importing all new plant and animal species to the United States? Why has such a policy never been developed and enforced?

3. What criteria should be used to decide whether or not a new species should be introduced into an area?

4. Do you think that money spent to preserve endangered species is well-spent? Why? Consider the other things upon which you might spend the money instead.

Causes of Extinction

5. List the major factors that lead to habitat destruction in the modern world. What factors would you list as being most disruptive to wild animals? Explain.

6. Elk normally feed in the high mountains in summer and travel to lower valleys during the winter months. Imagine that a developer was planning to build a housing complex in a mountain valley in Montana. The developer claims that since only 10 percent of the elk's annual range is being preempted, the herd will not be seriously affected. Do you agree or disagree with this argument? Defend your position.

7. When ships pass from ocean to ocean through the Panama Canal they are raised through a series of locks, sail across a freshwater lake, and then are lowered through a second series of locks. Passage through the canal would be facilitated if a deep trench were dug to connect the Atlantic and Pacific oceans directly. Would such a canal affect the survival of aquatic species? Discuss.

8. Give some reasons why the bison herd in North America has been virtually eliminated whereas the white-tailed deer population has actually increased during the past century.

9. Consider the following fictitious species and comment on the survival potential of each in the twentieth century: (1) A mouse-sized rodent that gives birth to 40 young per year and cares for them well. This creature burrows deeply and eats the roots of mature hickory trees as its staple food. (2) An omnivore about the size of a pinhead. Females lay 100,000 eggs per year. This animal had evolved in a certain tropical area and can survive only in air temperatures ranging from 80° to 100° F (27° to 38°C). (3) A herbivore about twice the size of a cow adapted to northern temperate climates. This animal can either graze in open fields or browse in forests. It is a powerful jumper and can clear a 15-foot fence. Females give birth to twins every spring.

Suggested Reading

Good introductions to the topic of wildlife preservation are:
M. Soule and Bruce Wilcox, eds.: *Conservation Biology.* Sunderland, Mass., Sinauer Associates, 1980.
R. F. Dasmann: *Wildlife Biology.* 2nd ed. New York, John Wiley & Sons, 1981.
D. W. Ehrenfeld: *Biological Conservation.* New York, Holt, Rinehart and Winston, 1970.
O. O. Owen: *Natural Resource Conservation: An Ecological Approach.* 3rd ed. New York, Macmillan, 1980.

Important books on extinct and endangered animals are:

C. Cadieux: *These Are the Endangered.* Washington, DC, Stone Wall Press, 1981.

R. A. Caras: *Last Chance on Earth.* New York, Schocken Books, 1972.

V. Ziswiler: *Extinct and Vanishing Animals.* The Heidelberg Science Library, Vol. 2. Revised in English by F. and P. Bunnell. New York, Springer-Verlag, 1967.

Friends of the Earth (FOE) and the Whale Coalition: *The Whaling Question.* 1982. 344 pp.

Discussions of the effect of introducing new species into an area are found in:

C. Elton: *The Ecology of Invasions by Animals and Plants.* New York, John Wiley & Sons, 1958.

C. Roots: *Animal Invaders.* New York, Universe Books, 1976.

Two of many introductions to evolutionary theory are:

Francisco Ayala and James Valentine: *Evolving: The Theory and Process of Organic Evolution.* Menlo Park, Calif., Benjamin Cummings Press, 1979.

Max Hecht, et al.: *Evolutionary Biology.* New York, Plenum Publishing Corporation, 1980.

The Biosphere

Human beings adapt to a wide range of different environments and ecosystems, from tropical jungles to arctic snows. Different ecosystems often have different susceptibilities to environmental disruption. Therefore, when human problems such as overpopulation, food production, and pollution are considered, the particular local environment must be taken into account. For example, sewage pollutants might do more damage in the Arctic than in the tropics because cold areas contain many fewer decomposer organisms to degrade waste materials. In this chapter the main types of ecosystems on Earth are outlined along with some of the features of these natural environments that affect the lives of people who live in them.

Life on Earth requires water, a source of energy (usually light from the Sun), and various nutrients found in the soil, water, and air. Suitable combinations of these essentials cannot be found high in the upper atmosphere or deep underground. They exist only in a narrow layer near the surface of the Earth. This layer is called the **biosphere** because it is, as far as we know, the only place where life can exist. The biosphere extends over most of the surface of the Earth, about 8 km into the atmosphere (where insects and the spores of bacteria and fungi are found), and as much as 8 km down into the depths of the sea. Living organisms are not distributed uniformly through the biosphere—few organisms live on polar icecaps and glaciers, whereas many live in tropical rain forests.

As Western naturalists explored the world and catalogued its life, two general patterns were observed. First, each newly discovered area contained previously unknown species of organisms. Second, in spite of the ever-increasing numbers of known species, there are only a few basic types of ecosystems. A tropical forest in South America contains tall trees with large leaves and fruits festooned with immense climbing vines, while colorful butterflies and birds flit through the gloomy shade. A tropical forest thousands of miles away in Africa would look very much the same, although the particular species of trees and vines, of butterflies and birds, would be different. Other types of ecosystems—desert, shrubland, grassland, or tundra—also look much the same wherever they occur. Why are similar communities of organisms found in very different parts of the world?

7.1 Climate and Vegetation

The front endpaper of this book contains a map of the world showing the types of different ecosystems found in different geographic locations. This figure demonstrates that ecosystems of the same type have similar climates. Climate is the main factor determining the type of plants that can grow in an area. The kinds of plants that are present in turn determine the appearance of the ecosystem and what other kinds of organisms can live there.

Climate depends basically on the amount of sunlight in an area, which influences the temperature and thus affects air motion and rainfall. Near the Equator, the Sun's rays strike the Earth almost vertically, thus giving tropical plants large amounts of the Sun's energy. Outside the tropics, the Sun's rays strike obliquely (Fig. 7–1); plants outside the tropics therefore receive less of the Sun's energy. Because of the tilt of the Earth on its axis, in nontropical areas the seasons change with the time of year, whereas in the tropics there is little seasonal difference in day length and temperature. Tropical climates, therefore, have fairly steady, high temperatures. In other areas, the temperature varies roughly with the amount and intensity of sunlight at different seasons.

Global climate systems are roughly determined by the movement of masses of air of different temperatures. At the Equator, air is heated and rises vertically upward. This vertical movement produces only intermittent wind on the surface of the Earth. At sea this region is known as the **doldrums.** The rising air cools as it mounts higher into the atmosphere and releases some of its moisture as it does so. Therefore, many areas near the Equator receive large amounts of rain. Examples include the steamy tropical jungles of the Congo River basin in Africa and of the Amazon River basin in South America.

The air that rises at the Equator moves both north and south at high altitudes. When the air masses reach latitudes of about 30 degrees north and south they have cooled enough to sink to earth again. Falling air warms up and absorbs moisture. Thus these latitudes are among the driest on Earth. The world's great deserts, such as the Sahara in North Africa and the coastal desert in Chile, are found here (Fig. 7–2). At sea these regions are called the "horse latitudes." This name was given because sailing ships were often becalmed for long periods in these latitudes, and horses transported as cargo often died of thirst and hunger. Still further north and south, in the temperate latitudes that include most of the United States and Europe, swirling winds pull masses of air sometimes from warm tropical areas and sometimes from frigid polar regions. This activity leads to varied weather patterns, sometimes wet and sometimes dry. The world's most productive agricultural land lies in the temperate zone, for reasons that will be examined later.

In 1889 C. Hart Merriam, a young naturalist surveying the biology of part of Arizona, noticed that San Francisco Mountain, near Flagstaff, showed a variation in vegetation from base to peak that was similar to that seen when traveling farther and farther north (or south) from the Equator (Fig. 7–3). Since temperature varies with altitude as well as with latitude, Merriam concluded that the type of vegetation in an area is determined by the temperature.

Later it was demonstrated that moisture plays just as important a role as temperature. Heavy rainfall is needed to support the growth of large trees, whereas progressively lighter rain-

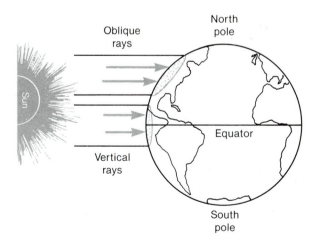

Figure 7–1 A beam of sunlight striking the Earth at high latitudes is spread over a wider area, and is therefore less intense at any one point than a similar beam near the Equator.

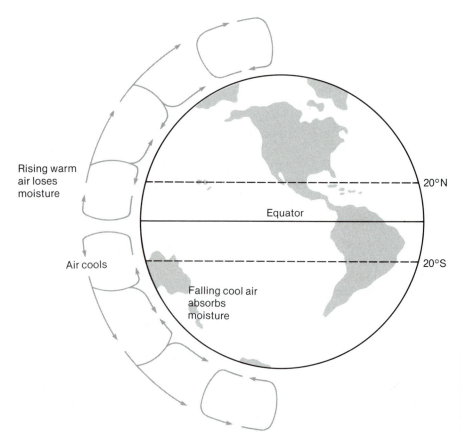

Figure 7–2 Major air movements over the surface of the Earth. (For the sake of simplicity, wind movements caused by the rotation of the Earth are not included.)

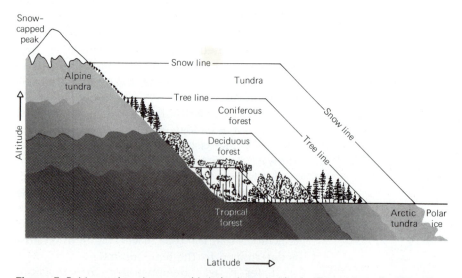

Figure 7–3 Vegetation changes with latitude and altitude. Temperature, which affects vegetation, falls as one travels up a mountain or away from the Equator, so that if there is plenty of moisture, vegetation is similar at high altitudes and at high latitudes as shown here.

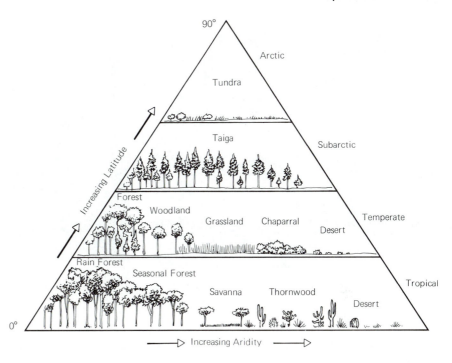

Figure 7–4 Simplified scheme of the major terrestrial biomes, arranged along gradients of increasing aridity at different latitudes, illustrating the dominant influence of moisture and temperature on the structure of plant communities.

fall supports communities dominated by small trees, shrubs, grasses, and finally scattered cactuses or other desert plants. In extreme cases, lack of rainfall results in total lack of plants.

Biomes. Major kinds of terrestrial ecosystems on the Earth are known as **biomes.** Each is recognized by the characteristic structure of its dominant vegetation. Ecologists differ as to the number of biomes that should be recognized. This confusion reflects the fact that biomes seldom have sharp boundaries. Instead, they gradually merge into one another.

The world's communities of organisms can be divided into 12 major biomes, which will be discussed in order of increasing latitude (and therefore of decreasing temperature). Within a given temperature zone, biomes can be arranged along gradients of increasing aridity, or lack of moisture (Fig. 7–4).

7.2 Tropical Biomes

Tropical rain forests occur in areas where high, fairly constant rainfall and temperatures permit plants to grow rapidly throughout the year. In such areas, a month with less than 10 cm of rain is considered dry, and annual precipitation may exceed 400 cm (Fig. 7–5).

Imagine a walk through a tropical rain forest. The forest is both familiar and strange to a visitor from a temperate area. It is familiar because so many of the plants grown in our homes and offices come from this biome. The dark, leathery leaves of familiar house plants are all around, although the *Dieffenbachia* that is less than a meter tall in your living room may soar as high as a house in the forest. The dominant plants here are tall trees with slender trunks that branch only near the top, covering the forest with a dense **canopy** of leathery evergreen leaves. The canopy blocks out most of the light. Plants that can survive in the permanent twilight under the canopy can also survive at the low light levels that exist in buildings (although tropical forests are more humid). The dense canopy permits as little as 0.1 percent of the sunlight to reach the forest floor. Since only a few types of plants can grow in this permanent twilight, the lower levels of a tropical rain forest are fairly open and easy to walk through. The kind of impenetrable "jungle" for which a person would need a machete to walk through occurs only in open areas, along river banks or in clearings cre-

Figure 7–5 Familiar house plants in tropical rain forest on a Caribbean island.

Figure 7–6 Tropical jungle along a river in the Amazon forest in Brazil. (Courtesy of the Brazilian Consulate.)

ated by fallen trees, where sunlight can reach the ground (Fig. 7–6).

The ground in a tropical forest is soggy, the tree trunks are wet, and water drips down everywhere. The canopy overhead is so dense that a person cannot tell whether it is actually raining or whether the water dripping down is just coming from the tree tops. Whatever the time of year, some of the trees are in flower and some are in fruit. There are no definite seasons in the forest.

Many plants are not rooted in the soil but rather on the surfaces of other plants in the canopy overhead. These plants are called **epiphytes.** Epiphytes include a great variety of orchids, largest of all plant families, as well as various bromeliads, and ferns. The ferns produce their own soil high in the air by acting as traps for dust and organic matter that will hold water. The organic matter trapped by epiphytes may amount to several tonnes per hectare. In addition to conserving moisture for the roots of epiphytes, this organic matter provides a habitat for many other plants and animals. Because most of their plant food is in the canopy, most of the animals live there too. Also found in the canopy are the vines, or **lianas,** which have their roots in the soil but are supported by the giant trees.

There are few animals on the forest floor, but huge colorful butterflies and beetles fly about in the canopy. The chatter of monkeys and the call of many species of birds and frogs can be heard from above. Some of these brightly colored tropical frogs are very poisonous. Colombian Indians tip their poison arrows with batrachotoxin from the skin of a tropical frog. A couple of micrograms of this potent nerve toxin will kill a human being.

Although the productivity of a tropical rain forest is high, if the land is cleared of forest most of it does not make productive farmland. As one ecologist pointed out recently, the main reason the world's developing countries are poverty-stricken is that they have inherited poor real estate. The land is unsuited to the intensive agriculture needed to feed populations that have roughly quadrupled in the twentieth century. As will be discussed in Chapter 14, high agricultural productivity is possible only in areas with adequate sunshine, rainfall, and soil. Sunshine and rainfall exist in abundance in the tropics, but much of the soil is very poor. Fertile soil is made up largely of **litter** or **humus,** the partially decomposed remains of the plants that have grown in the area for thousands of years. In a rain forest, temperature and moisture are ideal for decomposer organisms, so that organic matter falling to the forest floor is quickly decomposed. The minerals released by the decomposers are rapidly taken up again by the plants, and almost the entire nutrient pool of the forest is locked within the bodies of living organisms. So efficient are these systems at decomposing litter and absorbing and retaining nutrients that very little organic matter remains in the soil, and therefore soil quality is quite poor. When the forest is cut down, the exposed soil has little humus to retain moisture. Since the trees, which normally absorb nutrients, are gone, the minerals that remain in the soil are rapidly washed away by the heavy rainfall and run off downstream, leaving soil that may be only a few centimeters deep and that contains little humus and few nutrients. Except in some young, nutrient-rich volcanic soils (as in Indonesia and Central America), the cleared land rapidly loses its fertility. Sustained agriculture is impossible on such soils.

Sadly, most of us will never see the incredible beauty of a tropical rain forest. These forests are rapidly being cut down (see Chapter 17). If the present rate of clearing continues, nearly all of the rain forests will be gone by the year 2000. The hope is that there are a few tropical countries in which enlightened governments have realized that much of their rain forest will never make productive agricultural land. The Brazilian government has recently completed a soil survey of the entire country and pinpointed the land (about 30 percent of the country) on which agriculture may be possible. Of the remaining land, the government has set aside a remarkable 100,000 square kilometers to be preserved as rain forest.

North and south of the Equator, climates are characterized by distinct and distinguishable seasons. In these regions, precipitation is concentrated during part of the year, and there is an increasingly pronounced dry season. Tropical rain forest grades more or less sharply into **tropical seasonal forest.** Canopy heights are lower, and the proportion of **deciduous** trees (trees that lose their leaves for a season) increases as precipitation decreases and the length of the dry season increases. Tropical seasonal forests include the monsoon forests of India and Southeast Asia.

Tropical savanna consists of grassland dotted with scattered small trees and shrubs, such as acacias. It extends over large areas, often in the interiors of continents, where rainfall is insufficient to support forests or where the development of forests is prevented by recurrent fires. Some savannas are entirely grassland, while others contain many trees.

The proportion of trees in a savanna reflects competition between trees and grasses for water. Where rainfall is light, grass roots lying close to the surface absorb all the water during the dry season, and woody plants, which have deeper roots, cannot survive because there is no water for them during this time of year. Where rainfall is greater, the grasses are unable to absorb all of it, leaving water available for scattered trees. Where rainfall is sufficient to support a woodland, the canopy shade inhibits the development of grasses, and this relationship is reversed (Fig. 7–7).

Savannas are most extensive in Africa, where they support a rich fauna of grazing mammals such as zebras, wildebeest, and gazelles. The spectacular migrations of some of these species are related to shifting patterns of local rainfall that permit the growth of the young, nutritious foliage of grasses (Fig. 7–8).

7.3 Desert

Deserts generally occur in regions having a rainfall of less than about 20 cm per year. Typical hot deserts are found at latitudes of about 20 to 30 degrees north and south, where dry air from the Equator falls from the upper atmosphere, becoming warm as it is compressed at lower altitudes.

The Sahara desert, stretching across Africa from the Atlantic coast to the Red Sea, is the largest desert in the world. Desert areas with less than 2 cm of rain per year support little life of any kind, and the landscape is dominated by rocks and sand. Less extreme areas, including parts of the Sahara, have highly specialized plants, many of them annuals that grow, bloom, and set seed in the few days every year when water is available. Most desert perennials, such as the American cactuses, are **succulents,** plants that store water in their tissues, or small woody shrubs that shed their leaves during the dry season (Fig. 7–9). Desert animals have adaptations that restrict loss of water through their skins and lungs and in their urine and feces. Many are

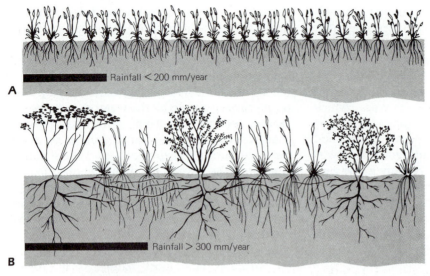

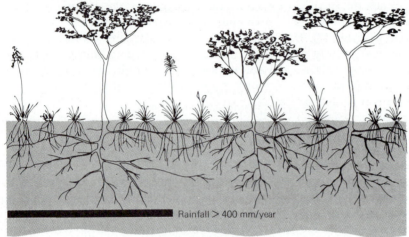

Figure 7–7 Relative proportions of trees and grasses in a savanna are influenced by rainfall (black bars). In *(A)* where annual rainfall is 200 mm or less, the roots of grasses use up all of the water available during the rainy season. If annual rainfall reaches 300 mm or more *(B)* not all of the water is used up by grasses and some is available to permit scattered trees and shrubs to survive the dry season. At annual rainfalls above 400 mm *(C)*, trees come to predominate, forming tree savanna or (at higher rainfall levels) tropical woodland. (After H. Walter, *Vegetation of the Earth.* New York, Springer-Verlag, 1973.)

Figure 7–8 A herd of impala and gazelles migrating across savanna in Kenya (From C. A. Spinage: *Animals of East Africa.* Boston, Houghton Mifflin Co., 1963.)

Figure 7–9 *Cacti and succulents in a desert in Arizona.*

nocturnal, avoiding the desiccating heat of the day by burrowing into the cooler soil. Desert soil is often fertile enough to support agriculture if enough water can be supplied. The Israelis have started agricultural projects by irrigating parts of the Negev desert. Deserts cannot be counted on to increase the amount of land devoted to agriculture in the future, however, because the supply of water for irrigation is running low in most parts of the world (see Chapter 14).

7.4 Temperate Biomes

North and south of the tropics and their adjacent deserts lie the world's temperate regions, so-called because on the average they experience moderate temperatures. (This moderation represents an average temperature over the entire year; some parts of the temperate zones are hotter in the summer than the tropics.) The **temperate forest biome** occurs in temperate regions with abundant rainfall. The composition of temperate forests, the proportions of deciduous to evergreen trees, and the spacing and height of the trees depend largely on the seasonal distribution of precipitation, the severity of the winters, the nature of the soil, and the frequency of fires. Three major categories of temperate forest can be distinguished—deciduous forest, evergreen forest, and rain forest.

Temperate deciduous forests occur in moderately humid (usually inland) climates where precipitation occurs throughout the year but where winters are cold, restricting plant growth to the warm summers. Most of the trees are deciduous. Thus they lose little water by transpiration (evaporation from the leaves) in the winter when their roots could not replace it from the frozen soil. Broad-leaved deciduous trees, such as beeches, oaks, hickories, and maples dominate this kind of forest; there is also a well-developed understory of shrubs and herbaceous plants on the forest floor (Fig. 7–10). The soil is rich in minerals and organic matter because decomposition proceeds relatively slowly, leaving a thick layer of litter.

Mammals typical of North American deciduous forests include white-tailed deer, chipmunks, squirrels, opossums, raccoons, and

Figure 7–10 *Young temperate deciduous forest in the northeastern United States. (Photograph by Paul Feeny.)*

foxes. Wolves, black bears, bobcats, and mountain lions roamed widely until they were largely eliminated by human activities. As winter draws near, up to three quarters of the birds migrate south, and many of the mammals hibernate. In the spring, herbaceous plants such as skunk cabbage, violets, Solomon's seal, and *Trillium* produce their leaves and flowers before the tree canopy has leafed out and cut off most of the light from the forest floor.

Temperate evergreen forests occur over wide areas where conditions favor needle-leaved conifers or broad-leaved evergreens over deciduous trees. These conditions include poor soils and a high frequency of droughts and forest fires. Various species of pine can grow in poor soils and have specialized adaptations for surviving fire. In some pines, for example, the cones open and their seeds germinate only when exposed to temperatures of several hundred degrees. The seeds thus germinate in areas that have just been burned. In the western United States, temperate evergreen forests include impressive stands of ponderosa pine, spruces, and firs. Extensive pine forests are also found in the southern states. These are areas in which pines are now grown extensively for timber.

Temperate rain forests occur in cool climates near the sea with abundant winter rainfall and summer cloudiness or fog. They include the forests of giant trees along the Pacific coast of North America, stretching from the mixed coniferous forest of the Olympic peninsula of Washington to the coastal redwood forests of northern California (Fig. 7–11). The heights of these forests may reach 60 to 90 m, and they include some of the tallest trees in the world. Although there is little rainfall in California in summer, the foliage of redwoods can absorb water from the frequent fogs. Other temperate rain forest types include the coastal Sitka spruce forests extending north to Alaska, forests of southern hemisphere conifers in New Zealand and Chile, and various forests at higher elevations on tropical mountains.

Temperate woodland occurs in climates too dry to support forests, yet with sufficient moisture to support trees as well as grasses. The dominant trees may be conifers, evergreen flowering trees, or deciduous trees. Pygmy conifer

Figure 7–11 Temperate rain forest—a redwood forest in California.

woodlands of piñon pine and juniper cover extensive areas of the American west, between the grassland semidesert biomes at lower elevations and the pine forests higher up. Oak woodlands are common in central California, and evergreen oak and oak-pine woodlands are extensive in the southwestern states and in Mexico.

The **temperate shrubland** biome is best represented by the **chaparral** communities that occur in all five regions of the world that have a Mediterranean climate—coastal California and Chile, the Mediterranean coast, southern Australia, and the southern tip of Africa (Fig. 7–12). These areas have moderately dry climates with little or no summer rain. Most of the shrubs have leathery leaves and range in height from about 1 to 5 m. They are often distinctly aromatic, with volatile and flammable compounds in their leaves. Fires are frequent and pose a constant threat to the residents of Santa Barbara and other cities within this biome. After fires,

Figure 7–12 Chapparal on the hills of southern California. (Photograph by Paul Feeny.)

the dominant shrubs regrow from surviving tissues near the ground.

Temperate grassland, known variously as prairie (North America), steppe (Asia), pampas (South America), or veldt (South Africa), covers extensive areas in the interiors of continents where there is not enough moisture to support forest or woodland (Fig. 7–13). When the grasses are grazed, water loss due to evaporation from the leaves ceases, leaving enough water in the soil to support woody plants such as mesquite, which has invaded overgrazed grassland in the southwestern United States.

Although grassland vegetation forms only a single layer, many plant species may be present. The persistence of many prairie wildflowers depends on periodic fires. Mammals of North American prairies include small burrowing species, such as prairie dogs and ground squirrels, and large grazing herbivores such as bison and pronghorn antelope. Because of the rich deep soil that underlies many temperate grasslands, these regions, including the midwestern United States and the Ukraine in Russia, have become prime areas for sustained and highly productive agriculture. The original prairie of North America is now represented only by a few tiny, scattered, unfarmed remnants.

Figure 7–13 Temperate grassland (prairie) in central North America. Grassland in the Red Rock Lakes National Wildlife Refuge, Montana, with a herd of pronghorn antelopes. (From E. P. Odum: *Fundamentals of Ecology,* 3rd ed. Philadelphia, W. B. Saunders Co., 1971.)

Temperate desert or semidesert occurs in regions too dry to support grassland (Fig. 7–14). Temperate semidesert occupies much of the Great Basin east of the Cascade and northern Sierra Nevada mountain ranges in the western

Figure 7–14 *Temperate desert in Utah.*

7.5 Taiga

The term "taiga" derives from a Russian word meaning "primeval forest." The **taiga** biome is dominated by subarctic needle-leaved forest, consisting mainly of conifers (spruces, pines, and firs) that can survive extreme cold in winter. Trees in the taiga tend to be further apart than the trees of a deciduous forest, and light penetrating to the forest floor is used by an extensive ground cover of shrubs, mostly of the Ericaceae (the blueberry and heath) family. The taiga, or **boreal forest,** as it is sometimes called, stretches in almost unending monotony in a giant circle through Canada and Siberia (Figs. 7–15 and 7–16). This monotony is due to the low diversity of tree species, which is occasionally interrupted by extensive areas of bog or "muskeg."

Much of the precipitation in the taiga falls as snow, and in the winter many of the resident animals grow fur or plumage that blends with

United States. In these regions, large areas are dominated by sagebrush *(Artemisia),* interspersed with perennial grasses. Typical animals include jack rabbits, sage grouse, and various pocket mice and kangaroo rats.

Figure 7–15 Taiga, or boreal forest, borders a lake in the San Juan Mountains of Colorado. This photograph also shows the timberline, the altitude above which trees cannot grow on a mountain.

Figure 7–16 *Taiga at the northern end of its range. The MacKenzie River, with some summer ice remaining, lies in the foreground. (Northwest Territories, Canada)*

Figure 7–17 *A domestic reindeer herd in flight across the arctic tundra. Photo taken on the shore of the Beaufort Sea in northern Canada.*

the white background. Animals characteristic of the North American taiga include moose, wolverine, wolves, lynx, spruce grouse, gray jays, crossbills, and (in summer) many species of warblers (birds).

7.6 Tundra

The **tundra,** a treeless biome, occurs far north in arctic regions, where winters are too cold and dry to permit the growth of trees (Fig. 7–17). In many areas the deeper layers of soil remain frozen as **permafrost** throughout the year, and only the surface layer of soil thaws during the summer. Although the number of different species of organisms is low, the tundra is often teeming with life during the warmer seasons. Another important characteristic of the tundra is that decomposition occurs very slowly because the ground is so cold for most of the year. Because of the low rate of decomposition, the shallow soil, and the slow growth rate of the plants, tundra takes a long time to recover when it is destroyed. Thus tundra is especially vulnerable to destruction by human activities. This is why conservationists are so concerned about the effects of running oil pipelines through the tundra.

Tundra vegetation is dominated by sedges, grasses, mosses, lichens, and dwarf woody shrubs. Bogs are common because the permafrost retards drainage. Large animals of the tundra include caribou, musk ox, and grizzly bears

in North America and reindeer in Europe and Asia. Lemmings, ptarmigans, waterfowl, arctic foxes, and wolverines are also typical. Hordes of mosquitoes, deerflies, and blackflies breed in the wet spots during the brief arctic summer. These insects contribute to the food available for a variety of migratory birds, including various plovers and sandpipers, snow buntings, longspurs, and horned larks, which nest in the tundra.

Neither taiga nor tundra occurs at sea level in the Southern Hemisphere because the continents do not extend far enough south. Antarctica harbors only a very scanty population of organisms around its edges. In Antarctica, as in the northern parts of the Arctic, life on the land (or, more often, on the ice) depends largely on the productivity of the sea. Many birds and mammals, though they breed on land, live on fish. These animals may die on land or leave feces that supply the main nutrients for the few land plants that survive in areas of almost constant ice.

A variety of **alpine grasslands, alpine shrublands,** and **alpine semideserts** are found on mountains, between the **timberline** (the greatest height at which trees can grow) and higher regions where few life forms survive (Fig. 7–18). Alpine tundra resembles arctic tundra in many ways, although nights are cool throughout the year in alpine areas, whereas they are warm during the brief arctic summer. In northern temperate mountains, alpine meadows cover extensive areas. These meadows are dominated by

Figure 7–18 Alpine tundra in early spring in the Rocky Mountains in Colorado.

Figure 7–19 An alpine meadow in the Andes, near LaPaz, Bolivia, with llama in the foreground.

7.7 Marine Communities

Strictly speaking, the term "biome" is used only to refer to communities on land. However, there is a great array of aquatic communities, both marine and freshwater, which, like biomes, exhibit similarities wherever in the world they occur.

Like the land, the ocean can be divided rather artificially into various zones based on the prevailing physical conditions and the different types of organisms that these conditions support (Fig. 7–20). As on land, temperature and the intensity of sunlight determine what grows where. Lack of water is not a problem in the sea, but there are large areas where a shortage of dissolved nutrients limits ocean life.

Along the seacoasts, many kinds of plants and animals thrive in the **littoral,** or **intertidal zone,** the area between the high and low water marks. This zone is submerged for part of the day and exposed to the air for the remainder. The organisms in the littoral zone must be adapted to withstand desiccation when the tide is out, whether by having a waterproof covering, by hiding or burrowing in moist places, or by retiring into a tube or shell. Many different types of organisms live in a littoral zone, depending on the type of ground surface available. Muddy regions provide a habitat for numerous types of algae and mollusks, worms, and some crustaceans (Fig. 7–21). Firmly anchored species

sedges and grasses, interspersed with dwarf willows, heaths, and other shrubs (Fig. 7–19). Alpine cushion plants increase at higher and drier sites. Many plants cultivated in rock gardens, such as gentians, saxifrages, and edelweiss, are alpine species. Alpine meadows in North America are inhabited by mountain sheep, mountain goats, grizzly bears, and marmots. Many of the larger animals migrate to lower elevations during the winter, and all organisms, like those of the tundra, are adapted to take advantage of the short growing season.

Alpine grasslands also occur above the timberline on tropical mountains. The **paramo** of the South American Andes is partly alpine grassland. Communities of similar structure but widely different evolutionary origin occur in the alpine zones of African mountains. African alpine communities also contain heaths, and these shrubs dominate the alpine shrublands of the Himalayas from which come many of our cultivated azaleas and rhododendrons.

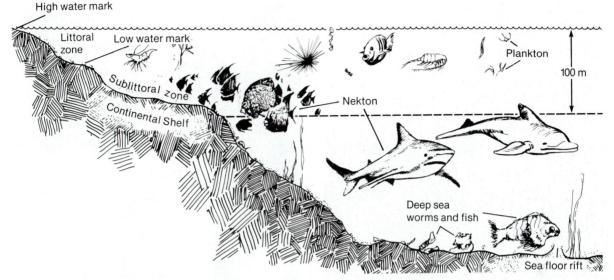

Figure 7–20 A diagram showing the major types of different habitats to be found in the sea (not to scale).

such as specialized algae and animals such as barnacles live in rocky zones exposed to wave action (Figs. 7–22 and 7–23).

Coastal bays, river mouths, and tidal marshes are all physically contiguous to the open ocean and are also close to fresh water and to land. These areas, known as **estuaries,** have (1) easy access to the deep sea, (2) less salinity than the open ocean, (3) a high concentration and retention of nutrients originating from both land and sea, (4) protective shelter, and (5) rooted or attached plants supported in shallow water. As a result of these factors, estuaries are very productive areas. They provide nurseries

Figure 7–21 Mudflats and marsh in an estuary.

Figure 7–22 Waves pound a rocky shore where a Californian sea lion *(Zalophus californianus)* surveys the view.

for many deepwater fish that cannot produce viable young in the less nutritious environment of the open sea.

The **sublittoral zone** occupies the continental shelves (the edges of the continents), extending from the low tide mark to a depth of about

Figure 7–24 Some of the inhabitants of a coral reef. A sluglike nudibranch crawling over sponges and sea anemones. (Photograph by Harold Wes Pratt.)

Figure 7–23 Barnacles.

200 meters. In the sublittoral zone, waters are deep enough to support large fish and other aquatic animals. In addition, mineral nutrients, washed from the land by rivers, are readily available. Sublittoral areas are among the most densely populated on Earth. They support many of the world's great fisheries. Even the productivity of the sublittoral zone, however, has its limits. Almost everywhere in the world, the combined effects of unlimited fishing and pollution have reduced the commercial fishing catch. (These topics will be discussed in Chapters 16 and 20.)

Coral reefs are found only in the tropics. They are restricted to warm oceans where the water temperature seldom falls below 21°C. Corals are small animals related to jellyfish that live

Figure 7–25 Deep sea fish attack a piece of bait. (Photograph by Richard A. Schwartzlose, Scripps Institution of Oceanography.)

in association with photosynthetic protists. The reef itself is made up of calcareous (calcium-containing) material, secreted by the coral animals and by green and red algae. Since photosynthetic organisms are so important to their formation, coral reefs are found only in clear, shallow water (less than 100 m deep) where there is enough light for photosynthesis. The reef acts physically like a rocky shore in providing anchorage for algae and sessile animals. Since most of the reef is usually under water, the movement of water is usually less than that on a rocky shore, and a great variety of fish and swimming invertebrates can find shelter within the crevices of the reef (Fig. 7–24). Coral reefs are among the most productive ecosystems in the world. Coral is easily destroyed, however, and because it regrows very slowly, reefs must be carefully preserved if they are to survive.

The **open ocean** can be divided into two main regions—the top 100 m or so where sufficient light penetrates for photosynthesis to occur, and the ocean depths. Small organisms in the surface waters make up the **plankton,** consisting of protists, plants, and animals not powerful enough to swim against the ocean currents. Plankton can be subdivided into autotrophs,

called **phytoplankton,** and consumers, called **zooplankton.** Zooplankton range in size from roughly 0.2 to 20 mm in diameter and consist of both permanent zooplankton and larvae of larger forms of life. Most of the production and respiration in the ocean is carried out by plankton. The larger and more familiar animals such as fish, sharks, and marine mammals such as whales and dolphins have a relatively small role in the energy balance. These animals are collectively called **nekton,** creatures that are powerful enough to swim in any direction, independent of water currents. Nektonic organisms feed mainly on plankton and on each other, so their presence in an area depends largely on whether the water contains enough mineral nutrients to support a large population of phytoplankton.

The productivity of the central oceans is limited by the facts that light is available only at the surface, whereas gravity tends to pull nutrients downward toward the sea floor. As a result, key nutrients, especially nitrates and phosphates, are scarce in the photosynthetic zone in many parts of the ocean, and productivity is so low that these regions have been likened to a great desert. Biologically productive marine environments are limited to nutrient-rich zones. These fall into three categories. (1) Coastal zones and continental shelves are fertilized by runoff from land and are shallow enough so that wave and current action causes rapid vertical mixing of nutrients. (2) Certain regions exist where rifts and volcanic action on the ocean floor lead to movement of minerals to the surface. (3) In a few areas, for example along the coast of Peru and in regions near Antarctica, deep-water ocean currents rise vertically to the surface. This phenomenon, called **upwelling,** produces nutrient-rich, highly productive deep-water ocean systems.

Seventy-five percent of the oceans' water lies more than 1000 m deep. For many years, people assumed that there was little life in the depths of the ocean because it was too dark for photosynthesis to occur. Improved diving techniques that permit sampling at depths of more than 6000 m have, however, revealed fascinating communities on the ocean floor, and many dives have turned up hitherto unknown organisms. It is now clear that the ocean floor, at all depths, supports populations of both large animals and of decomposer bacteria. The food chain for these organisms depends largely on the carcasses of dead plants and animals or on feces falling from the surface layers above them (Fig. 7–25).

Summary

Two main patterns can be seen in a world-wide survey of the distribution of organisms:

1. Different areas of the world are inhabited by different species of plants and animals.
2. Terrestrial communities in different parts of the world can be divided into a fairly small number of categories, or **biomes,** on the basis of vegetational structure. These biomes are worldwide and are not restricted to single continents.

The kind of biome found in an area depends mainly on rainfall, temperature, and soil type. Similar changes in biomes occur with increasing altitude and with increasing latitude.

The biome with the greatest number of species is a **tropical rain forest,** where high temperature and abundant rainfall permit plants to grow throughout the year. Most of the plant and animal life is found in the canopy among the broad evergreen leaves of tall trees. Decomposition is rapid, and the soil is usually poor. At any one time, most of the nutrients in the forest are locked in the bodies of organisms.

In temperate deciduous forest the soil is much richer in nutrients because the trees lose their leaves in fall, creating a litter layer that decomposes only slowly. Deciduous forest is an important biome of North America, Europe, and Asia in areas with warm, moist summers and cold winters.

Where the soil is poor or fires are frequent, **temperate evergreen forest** replaces **temperate deciduous forest.** Further north, both are replaced by **taiga,** a biome dominated by coniferous trees adapted to growing in sparse soil and to resisting extreme cold and water loss during the winter.

North of the taiga and above the timberline lie the **tundra** and **alpine grasslands,** dominated by cold-resistant woody shrubs or by sedges and grasses, depending on the soil type and the amount of moisture in the soil.

Grasslands receive more rain than deserts and less than deciduous forests. Grassland occurs in the dry interiors of continents in the Americas, Asia, and Australia. Shrubs and trees may be scattered among the tall grasses.

Deserts have hot days, cold nights, and very little rainfall. Their plant life is mainly annuals with very short growing seasons and succulent perennials adapted to the low rainfall.

The distribution of marine organisms is determined by water temperature and the availability of light and minerals. The **littoral** and **sublittoral zones** are well supplied with both light and minerals and support dense communities of life. **Coral reefs** are specialized sublittoral communities found only in tropical waters. In the open ocean, the availability of light for photosynthesis restricts plankton to the upper layers of the water, but scarcity of nutrients in these layers may limit the numbers of organisms. Larger nektonic organisms are found mainly where plankton is abundant. Dead organisms from the surface layers of the ocean supply food for a community of bacteria and other organisms that live on the deep-sea floor.

Questions

Biomes

1. What biome do you live in?
2. The 30° N latitude line runs through southern Louisiana and northern Florida as well as through desert country in Mexico and Texas. Why is the area in Louisiana and Florida not desert like the area in Mexico and Texas?
3. Why is it proving difficult to carry out largescale "agribusiness" farming in vast tracts of land cleared of their tropical rain forest vegetation?
4. Study a map of the world and answer the following questions. (1) Why does the temperate zone in the northern hemisphere produce more food for humans than the temperate zone in the southern hemisphere? (2) Compare the latitude of the Sahara desert (North Africa) with that of the desert in central Chile (north of Santiago). What conclusions can you draw? (3) Compare the latitude of New York City with that of Rome. What do you know about the climate in these two regions? Is latitude the only factor that determines the temperature and rainfall of a region?
5. Do you feel that it would be economically advantageous to fertilize the central oceans with nitrates or phosphates to improve the world's fisheries? Discuss.

Suggested Readings

A periodical issue devoted in its entirety to "The Biosphere" is:
Scientific American. September, 1970. 267 pp.

Four books dealing with specific areas of natural ecology are:
R. Platt: *The Great American Forest.* Englewood Cliffs, N.J., Prentice-Hall, 1965. 271 pp.
B. Stonehouse: *Animals of the Arctic: the Ecology of The Far North.* New York, Holt, Rinehart and Winston, 1971. 172 pp.

G. M. Van Dyne, ed.: *The Ecosystem Concept in Natural Resource Management.* New York, Academic Press, 1969, 383 pp.
John Madson: *Where the Sky Began. Land of the Tallgrass Prairie.* Boston, Houghton Mifflin, 1982. 321 pp.

A classic study of ecology and conservation as seen through the eyes of a naturalist is:
A. Leopold: *A Sand County Almanac.* New York, Sierra Club/Ballantine Books, 1966. 296 pp.

A book that specifically discusses shore erosion is:
Joseph M. Heikoff: *Politics of Shore Erosion: Westhampton Beach.* Ann Arbor, Michigan, Ann Arbor Science Publishers, 1976. 173 pp.

An excellent introduction to the physical processes that shape the Earth's climate and geography is:
A. N. Strahler: *The Earth Sciences*, 2d ed. New York, Harper & Row, 1971.

Despite its formidable title, this book is short and readable; probably the best account of world vegetation zones and the conditions that determine what biome occurs where:
H. Walter: *Vegetation of the Earth in Relation to Climate and the Eco-Physiological Conditions.* Translated from 2d German edition by Joy Wieser. London, The English Universities Press Ltd.; New York, Heidelberg, Berlin, Springer-Verlag, 1973.

Wildlife in the Earth's biomes is described in:
International Wildlife Series: This Fragile Earth.
Part I: Doomed jungles? by Peter Gwynne. July–August 1976.
Part II: The island dilemma, by Mariana Gosnell. September–October 1976.
Part III: Mountains besieged, by Edward R. Ricciuti. November–December 1976.
Part IV: Shifting sands, by Frederic Golden. January–February 1977.
Part V: Margin of life, by Robert Allen. March–April 1977.
Part VI: The living sea, by Arthur Fisher. May–June 1977.

The microbiology of the ocean is described in:
H. W. Jannasch, and C. O. Wirsen. Microbial life in the deep sea. *Scientific American*, June 1977.

A short and beautifully illustrated introduction to the tropical rain forest, by a leading authority is:
P. Richards: *The Life of the Jungle.* New York, McGraw-Hill, 1970.

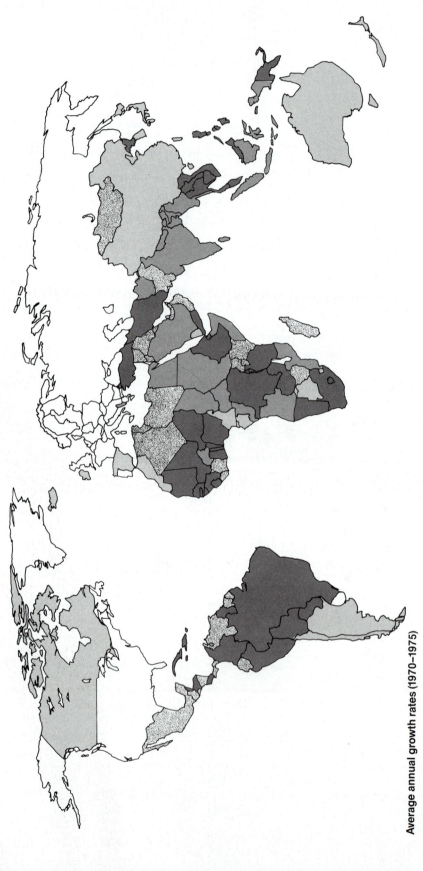

Average annual growth rates (1970–1975)

Above 3.0%

2.5% to less than 3.0%

2.0% to less than 2.5%

1.0% to less than 2.0%

Below 1%

Human Impact on the Earth

Average annual growth rates of the human
population (1970–1975). Asia, Latin America, and
Africa have the world's highest population growth
rates. Latin America averaged 2.7 percent a year
during this period; some African countries exceeded
3 percent. Overall, the world growth rate declined
from an all-time high of almost 2 percent at the
start of the decade to 1.6 percent 5 years later.
World population, however, grew in that time by
more than 300 million to a total of 3.92 billion.
(From *Our Magnificient Earth.* © 1979 by Mitchell
Beazley as *Atlas of Earth Resources.* Published in
the U.S.A. by Rand McNally & Company.)

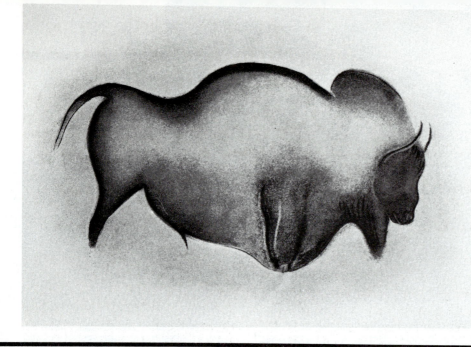

8

Human History and the Environment

This chapter traces the history of some important human interactions with the environment. Human lifestyles have changed in major ways over the years, and as a result, the impact of human activities on the Earth has changed.

Human history began at the time our ancestors came down from the trees of the African forest and started to make a living by hunting game and gathering plants on the African savanna. Prehuman primates and early humans continued to be hunters and gatherers for many generations. In this role, their effect on the environment was comparable to that of many other animals. A major change occurred when humans started to develop agriculture. In addition to its direct impact on the environment, agriculture has a profound effect on human societies. This effect, in turn, resulted in further environmental changes.

8.1 Human Origins

Few areas of research have produced as much argument and confusion as the search for the fossils of our ancestors. The fossil evidence of human ancestry is fragmentary. Seldom does the most spectacular find consist of more than a few teeth and part of a jawbone. You can imagine the difficulty of trying to deduce, say, the brain size of a fossil species from such evidence, let alone of deciding whether the individuals walked upright or on all fours. Yet such deductions are made all the time. One of the reasons for the scarcity of prehuman primate fossils is that many of the animals in question were forest-dwellers. Few fossils are preserved in forests because the acidity of the soil and conditions favorable to decomposition tend to destroy the remains quickly.

Part of the muddle, though, can only be ascribed to human vanity. Researchers inevitably hope to discover vital clues to human ancestry. As a result, dozens of fossils that in retrospect are those of monkeys (or even of modern humans!) have been hailed as the "missing link" between apes and humans. In addition, for many years almost every fossil of a human relative that was found was given a new species name or even assigned to a new genus to make it more memorable.

Human Ancestors

Modern humans are members of the species *Homo sapiens*. In turn, *Homo sapiens* belongs to the mammalian order **Primates.** Primates can be divided into the lower primates, which include tarsiers, lemurs, and lorises, and the **anthropoids,** including monkeys, apes, and humans. The modern apes consist of only four genera— *Hylobates,* the gibbons; *Pongo,* the orangutans; *Gorilla,* the gorillas; and *Pan,* the chimpanzees. All anthropoids have stereoscopic color vision and rounded heads. Many of them can learn complex behavior patterns, thanks to their relatively large brains.

Human beings belong to the hominid family of anthropoids. There are more similarities than differences between hominids and apes. In fact, some people question whether humans would have been classified as a separate family if classification had been invented by any other species. Nevertheless, the four genera of modern apes have some characteristics that are common to all of them but are not found in hominids. For instance, all anthropoids show adaptations of the arms and backbone that adapt them to **brachiation,** swinging by their arms through trees with the body held vertically. Adaptations to brachiation are more extreme in apes than in hominids. Modern apes have powerful jaws and specialized feet, features that are not shared by humans. In addition, the human brain is relatively larger than that of any ape.

Research on human origins is currently active, and the conclusions are still uncertain. Consequently, the establishment of a system of hominid and early human classification from the fossil record is by no means fixed; in fact, the subject is often controversial. One of the recent discoveries of the remains of early hominids that has received much popular attention was made in a remote region of Ethiopia in East Africa in 1974. At that time, two scientists, Donald Johanson and Tim White, unearthed nearly three fourths of a skeleton of a young female whom they named Lucy (Fig. 8–1). Dating techniques indicate that Lucy lived approximately 3.5 million years ago. Smaller fragments of several other individuals were found nearby. From the interpretation of their findings, Lucy's discoverers concluded that these hominids represented a species that was ancestral to two evolutionary

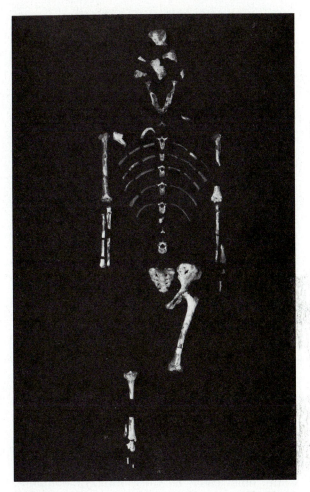

Figure 8–1 Photo of Lucy's bones. (Courtesy of Cleveland Museum of Natural History.)

lines. According to their theory, one of these lines eventually died out; the other gradually evolved into the genus *Homo* and finally into the species *Homo sapiens*. These conclusions have been challenged, and other views of human origin have been advanced.

Even though hominid and early human history is clouded in uncertainty, we do know that the emergence of *Homo sapiens* generated a potential for altering the environment in ways that no other species has ever duplicated. What was so special, however, about this new species? Other forms of life can also cause major environmental disruptions. Small herbivores like locusts or large ones like elephants (Fig. 8–2) can destroy vegetation over a wide area. A species that invades a region in which natural controls

Figure 8–2 *The vegetation in some areas in Kenya is seriously disrupted by elephants.*

against its growth are ineffective can overwhelm its competitors, or even drive them to extinction. Each species brings its unique set of attributes to the struggle: Some use their teeth and claws, some their potential for rapid reproduction, some their ability to store food or to lie dormant during periods of scarce resources. Early hominids did not rule the Earth then, any more than monkeys or apes do today. The evolving human species, however, began to acquire what proved to be a unique combination of attributes. At some time early in their history, hominids started to walk on their hind legs in an upright position. This attribute, known as **bipedalism,** was associated with a shift from the forests to the plains. No one is sure of the precise sequence of events that led to this change of both habitat and posture. Whatever the reason, the result was that the hands, no longer used for walking, were free to catch animals, throw stones, use objects as tools, and later to make tools. Piles of animal bones are found with the skeletons of early hominids, indicating that they ate more meat than their ancestors did. There is some evidence that the shift from a primarily vegetarian diet to a more carnivorous one was associated with the advent of bipedalism, although here, too, no one is sure which came first. Nor do we know when language evolved; scientists can only speculate that the advantages of group cooperation for hunting and defense may have promoted natural selection for improved communication among members of the group. Improvements in communication in turn led to the development of language.

Early Technology

The first fossils that were similar enough to modern humans to be classified in the genus *Homo* date from the African Pleistocene, about 2 million years ago. These fossils, such as those of *Homo erectus,* are those of a fully bipedal, omnivorous, tool-using hominid. Some *H. erectus* bones are found in caves, suggesting that these hominids used more or less permanent home bases. Besides animal bones and stone tools, some of the caves contain heaps of charcoal and charred bones, showing that fire had been domesticated and brought indoors.

The discovery of fire undoubtedly had far-reaching effects. First, it opened up a wide range of new foods for hominids. Many wild plants are poisonous if eaten raw but are nutritious if cooked, partly because boiling water extracts many poisonous chemicals, reducing the toxicity of certain plants. Second, fire can be used to thaw out frozen food and to provide warmth in winter, permitting hominids to colonize areas where the climate is colder than it is in Africa. The domestication of fire is correlated with the colonization of central Europe and China. Their anatomical and physiological adaptations alone would not have permitted hominids to survive the cold winters in these areas; behavioral adaptations and technological expertise were necessary in addition. Plainly, the early human brain had developed to the point where *Homo erectus* could produce social and technological solutions—such as fire, clothing, stored food, and communal living in caves—to the problems of surviving cold winters.

The First Humans

The technological solutions for surviving cold weather probably led to the further development of *Homo erectus* and the eventual evolution of *Homo sapiens,* first identified as a separate species from fossils that are thought to be about 300,000 years old (Fig. 8–3).

Scientists are still debating the reasons why natural selection favored the evolution of the human brain, which is exceptionally large in rela-

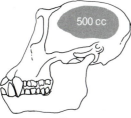

500 cc

Female Champanzee

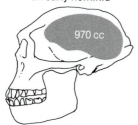

550 cc

Reconstruction of
Australopithecus Africanus ,
an early hominid

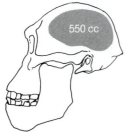

970 cc

Reconstruction of
Homo Erectus

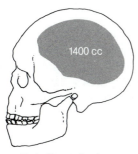

1400 cc

Early *Homo Sapiens*

Figure 8–3 *Changes in proportions of the skull from great ape to human. Note the increase in size of the brain case, the change in position of the connection of head to neck (arrows) as hominids became more bipedal, and relative reduction in size of the teeth and jaws as hominids changed from a purely herbivorous to an omnivorous diet.*

is clear that the more precise its control over the hand muscles and the more complex its learned behavior, the more brain cells we would expect to find in a hominid. However, large areas of the human brain can be destroyed without significant alteration of behavior in any way. There seem to be many reserve cells in the brain that we can normally do without. In effect, scientists do not know enough about the organization and function of our own brains, let alone about the brains of ancestral hominids, to speak in any but the most speculative terms about when and how particular changes occurred.

8.2 Hunter-Gatherer Societies

Early humans were **hunter-gatherers** (Fig. 8–4); they obtained their food by collecting it from the wild rather than by cultivating plants or domesticating animals. Studies of the few hunter-gatherer tribes that have survived to modern times (Fig. 8–5) make it clear that the early human diet contained a much greater variety of plant species than it does in agricultural societies. These plant materials can be divided into two main groups. The first includes fruits, nuts, berries, seeds, and pods. These are mainly the seeds and fruits of trees, bushes, vines, and annual grasses. Today these species comprise more than half of the human diet of plant food because they include the cereals, which are grass seeds. The stems, leaves, and roots of such plants are protected against herbivores mainly by physical defenses such as toughness and spines and by possessing lesser nutritional value for an animal.

The second group of plant foods is quite different ecologically. These include lettuce, spinach, carrots, and plants of the cabbage family—today's vegetables. In nature, these are "pioneer" plants that escape their predators by being short-lived and widely dispersed in patches that take time and effort to find. They contain chemicals that are toxic except to herbivores that have evolved the specific enzymes needed to feed on them. Some of these plants have probably always been eaten by mammals, whose livers contain a number of detoxifying enzymes. Others, such as many members of the

tion to body size. One of the difficulties is that there is no consensus as to how much brain is necessary for intelligent behavior. In general, it

Figure 8-4 Early hominids hunting.

cabbage family, have been bred until they contain lower levels of toxic substances than their wild cousins, and still others are edible only when they are cooked. Even today some of these plants are toxic except for certain edible parts. For example, potatoes, tomatoes, and rhubarb are all good foods, but the leaves of these plants are poisonous.

Many early hunter-gatherer societies were nomads who were constantly on the move following the seasons and the migration of animals. Population growth in nomadic tribes is often quite slow. Very young children are a severe liability because they must be carried when the

Figure 8-5 A Kalahari bushman beside a partly constructed hut in the family camp. These people are one of the last surviving groups of hunter-gatherers. Today profound changes are taking place in their society. (Courtesy of Biophoto Associates, N.H.P.A.)

band moves from one camp site to another. The ideal interval between births is therefore long enough so that a mother has to carry no more than one child. Hunter-gatherer societies typically have relatively small family sizes; the total population of a such a society is usually well below the biological carrying capacity of the area it occupies. Hunter-gatherers control their populations by infanticide, abortion, and primitive forms of contraception.

The first permanent settlements, in Europe and Asia, may well have been a response to the difficulty of a nomadic life during cold winters. However, the early settlers, many of them cave dwellers, were still hunter-gatherers. They ate plant matter during the warm spring and summer months and hunted animals whenever they could. There is also evidence that these early humans trapped game animals that migrated through passes in the mountains between summer and winter feeding grounds in spring and fall.

A settled hunter-gatherer society of this sort must have been the precursor of the development of agriculture. Such a society would demand social organization and communication between individuals. Such social interactions formed strong selective pressures for the development of the language, social rites, laws, and customs that make society more efficient. These traits are reflected in the decorated tools, pots, and dwellings that began to appear in Europe and Asia about 20,000 years ago (Fig. 8–6). Another important change was that infant care be-

came much easier when families settled. Furthermore, children are valuable on a farm, and this factor also encouraged population growth.

8.3 The Agricultural Revolution

Agriculture, the process of breeding and caring for animals and plants that are used for food and clothing, is a relatively modern development in the 300,000-year history of human beings. As recently as 2000 years ago, farmers had still occupied only about half the land suitable for agriculture. After that, farming societies steadily pushed hunter-gatherers off the face of the Earth. Today, very few hunter-gatherer populations remain.

It is not clear what pressures induced early human populations to abandon their wandering hunter-gatherer existence for a hard life on a primitive farm. For instance, hunter-gatherers do not face the constant battle with pests, drought, and famines that beset all agricultural communities. Studies in Southern Africa during a drought showed that farmers starved while the population of hunter-gatherer bushmen in the Kalahari desert remained stable in size and the people were well-fed (Fig. 8–5). This probably occurs because most hunter-gatherer populations stay well below the carrying capacity of their territories. Furthermore, hunter-gatherers have a more balanced diet than most farmers, and their incidence of chronic and disabling diseases is no higher. Their life expectancies are similar to those of agricultural peoples in most parts of the world. It has been suggested that the bushman diet contains too few calories for perfect health, but this is hard to reconcile with the observation that these people are healthy and spend very few of their waking hours finding food. Even though they live in inhospitable deserts, Kalahari bushmen and Australian aborigines spend only about 15 hours a week collecting and preparing food. Children do not have to work until they are married, and the aged are revered and cared for. In contrast, even in modern times, many people in agricultural societies work for at least 60 hours a week and spend around 70 percent of their pay on food (Fig. 8–7). Thus, about 42 hours a week are de-

Figure 8–6 Cave painting of a bison. (Courtesy of American Museum of Natural History.)

Figure 8–7 People in primitive agricultural societies must work harder to survive than those who live in hunter-gatherer societies. However, agricultural systems can support more people in a given area of land. This photograph shows a wheat harvest in the Andes Mountains in Peru.

voted to acquiring enough to eat. Even in the affluent West, with the world's most advanced and mechanized forms of agriculture, people still devote about a third of their incomes, or 13 hours a week, merely to buying food, and this does not include food preparation time.

Clearly, then, hunter-gatherers in many ways have an easy life compared with even the most affluent members of agricultural societies. Following this logic, it seems reasonable to assume that farming societies began of necessity rather than choice. An agricultural system can usually feed more people in a given area than can hunting and gathering. Thus domestication of plants and animals probably arose from a response to pressure caused by overpopulation. One theory states that gradually, as hunting techniques became more efficient, the traditional balances that kept human population levels low were altered, and a slow increase in population size occurred. In one study of aboriginal people in North America, archeologists learned that just before the start of agriculture, the density of nomadic camp sites was high, hunters frequented

more marginal habitats, and people consumed a broader range of foods. These indicators imply that population levels were rising in the area and that food was becoming harder to find. Furthermore, the onset of agriculture took place in a period of hardship, not plenty. Studies of skeletal remains of the people in early farming communities in North America show evidence of a variety of diseases and a general lowering of life span. Presumably, people would accept such hardships only after they had overpopulated their ecosystem to the point where there was no longer enough game to hunt or wild food to gather.

The adoption of agriculture by human societies can be traced back about 12,000 years. It has long been thought that agriculture originated east of the Mediterranean and spread from this area, with a separate center of origin in America. Fossils of domesticated dogs dating from 11,000 years ago have been found in Iraq, and cultivated plants date back at least 9000 years in the same area. Now that archeological excavation has become more widespread, however, evidence is accumulating that agriculture originated, probably independently, in many different places at about the same time (Fig. 8–8). This is not surprising, because agriculture seems to have been merely an extension of what people already knew. Studies show that modern hunter-gatherers know enough to settle down as farmers, and there is evidence that this knowledge is many centuries old.

Once people began to plant seeds deliberately, cultivated strains of plants soon came to differ from their wild counterparts. For instance, the seed heads of wild grasses tend to burst open easily and shed their seeds so that the plants can propagate readily. Scattered seeds are not easy for humans to gather, however. Therefore, a farmer selects plants whose seed heads do not burst open easily because these seeds are easier to collect for replanting. As a result, plants with nonbursting seed heads rapidly increased in the population of cultivated plants. Today's cereal plants evolved in just this way.

Farmers also select seeds that contain large quantities of stored nutrients. The benefits are twofold: These seeds contain the most food for human consumption and can also compete most successfully in a crowded seed bed where rapid germination is necessary for survival.

Similarly, docile animals with small horns and animals with woolly coats were most likely to become the domesticated breeding stock. It is easy to imagine that an early agricultural society would rapidly have pushed a large ram with dangerous horns into the wild again or, more likely, have slaughtered him for a tribal feast. Thus, by a combination of conscious and unconscious selection, early farmers rapidly produced animals and plants that were considerably different from their wild forebears.

The change to an agricultural way of life has had such profound effects on human history that it is often called the **agricultural revolution.** One of the most important effects was that it permitted the accumulation of material goods.

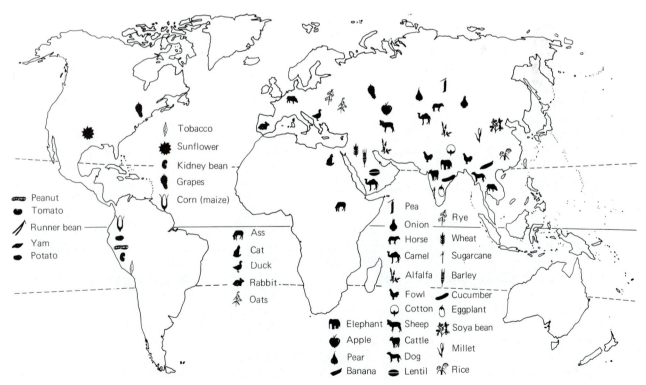

Figure 8–8 Areas where plants and animals were probably first domesticated. Some species that were probably domesticated independently in two or more areas are shown in both places.

Nomadic hunters travel with few possessions. Farmers, living in one place, can accumulate as much as they can afford. Even land can be owned and passed on by inheritance or sale.

A striking consequence of agriculture is that the division of labor is more pronounced than that seen in hunter-gatherer societies. As population and civilization grow, some workers remain on the farms, while others are freed to become builders, bakers, and merchants. In successful societies, the population may even be able to afford the luxury of poets, scholars, and students, who make no immediate contribution to the group's physical well-being but are the basis of its cultural life and its future development.

The population explosion, which is such a problem today, is also a direct result of agriculture. Population control is usually abandoned by agricultural communities, partly because chil-

dren, who are generally a liability in a hunter-gatherer society, are useful as laborers on farms (Fig. 8–9). In addition, the inheritance of land and goods becomes more important. The desire to have children who will inherit the property and care for their aging parents is a recurrent theme in the mythology and literature of agricultural societies. These feelings persist in many of today's agricultural societies as well.

8.4 Civilization and the Environmental Perspective

The story of the expansion of agriculture and the rise of civilization is a topic for another book. This section will deal with people's attitudes toward the natural world. Of course, different civ-

Figure 8–9 Even young children in agricultural societies can help support the family. These children in Southeast Asia are protecting the rice by shooing away a flock of ducks. (Photo courtesy of Ken Heyman.)

ilizations have had their own unique perspectives, religions, and styles of life. Our attention will focus on several of the most influential of human societies whose attitudes have had a broad effect on ecological history.

Mesopotamia

Some of the earliest civilizations of the Western world are believed to have risen in the broad plains between the Tigris and the Euphrates rivers of western Asia (Fig. 8–10) in the region called Mesopotamia (now Iraq). Highly sophisticated systems of letters, mathematics, religion, architecture, law, and astronomy originated in this area. Obviously, then, people had the time and the energy to educate themselves and to philosophize, and therefore it seems reasonable to deduce that the food supply must have been adequate. Today much of this region is barren semidesert, badly eroded and desolate. Archeologists dig up ancient irrigation canals, old hoes, and grinding stones in the middle of the desert. What must have happened?

In many ways the environment of Southwestern Asia is harsh. The climate is very dry, with an average rainfall of 15 to 20 cm (about half the rainfall that collects in the wheat-growing "breadbasket" of North America). Summer temperatures are among the hottest recorded in

Figure 8–10 The Fertile Crescent.

the world, only a few degrees cooler than temperatures in the central portions of the Sahara Desert. Yet this region was known in antiquity as the Fertile Crescent. In ancient times the fertility of much of this region was a direct consequence of an extensive irrigation system, one of the great wonders of early civilization (Fig. 8–11). Thousands of kilometers of sophisticated aqueducts were built to support the agriculture that in turn supported ancient cities such as Babylon and Ur.

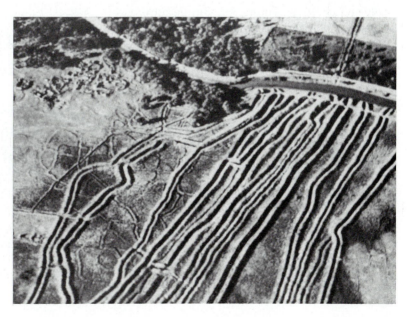

Figure 8–11 Aerial photograph of old irrigation canals along the Tigris River. (Photo courtesy of Iraq Tourist Authority.)

The attitudes of the people of the region are well preserved in many remarkable writings. According to the religion of the area, nature was represented as a monstrously chaotic being. This chaos was tamed and controlled, and order was maintained, by the constant labor of people aided by their patron gods. Thus, one legend tells about the hero-god Enlil, who slayed the primeval monster of chaos. Such legends mirrored the labor of the people themselves, who tamed a hostile desert and built a civilization.

From the beginning of recorded history the Fertile Crescent was plagued by recurrent outbreaks of war. In addition to warfare, a true ecological disaster influenced the fall of these great civilizations. Part of the story starts at the source of the great rivers in the Armenian highlands. The forests were cleared to make way for pastures, vineyards, and wheat fields. But croplands, especially if poorly managed, cannot hold the soil and the moisture year after year as well as natural forests or grasslands can. As a result, large water runoffs such as those from the spring rains or melting mountain snows tended to flow down the hillsides rather than soaking into the ground. These uncontrolled waters became spring floods. The flood waters carried large quantities of clayey silt into the irrigation canals of Mesopotamia. At first these floods caused no major problems. Every spring the canals were dredged and the recovered soil was spread onto the fields. Since flood-silts are generally fertile, crops flourished. But because clay soils are relatively impermeable to water, the natural drainage pattern of the fields was gradually disrupted.

All river water is slightly salty. A common property of all irrigated farmlands is that the continued evaporation of river water may eventually lead to a gradual increase in the concentrations of these small quantities of salt in the soil. This problem was enhanced in Mesopotamia by the hot dry climate, which promotes rapid evaporation, and by the lack of drainage caused by the silt. Over a period of a few centuries, salination became so severe that eventually agriculture had to be abandoned in many areas. Today some of these regions are barren deserts, and in others, agricultural yields are less than they were 4000 years ago.

Ancient Greece

Today, a traveler in Athens can see bare rocky hillsides lying beyond the ruins of the Acropolis (Fig. 8–12), the center of the ancient city-state. The Mediterranean climate is also dry, although it is considerably more moist than that of the Fertile Crescent. In many regions there is enough water to support open forest systems. In fact, the hills outside of Athens were once forested and cooled by springs and small streams. In ancient times, the trees were cut to supply the city with firewood, building material, masts for ships, and various other amenities of civilization. After the forests were destroyed, the grasslands were overgrazed by sheep and goats. These animals ate any new tree seedlings that started to sprout, and their hoofs cut into the remaining sod. As a result, the rich loamy soils were eroded to form barren rocky hillsides. Rocky surfaces do not retain water nearly as well as soil does, so the sparse rainfall flowed rapidly downhill. Reservoirs of underground water became depleted, and many natural springs dried up. This ecological disaster is documented by the writings of some of the great naturalists and philosophers of the era. Plato, who lived during

Figure 8–12 The rocky plateau of the ancient Acropolis overlooks modern Athens. The Parthenon (center), temple of the goddess Athena, dates from the fifth century B.C. The bare rocky hills lying beyond the city were once forested. (Greek National Tourist Office Photo.)

the height of the period of deforestation, wrote his observations, as quoted below.

> What now remains compared with what then existed is like the skeleton of a sick man, all the fat and soft earth having wasted away, and only the bare framework of the land being left. But at that epoch the country was unimpaired, and for its mountains it had high arable hills, and in place of the "moorlands," as they are now called, it contained plains full of rich soil; and it had much forest-land in its mountains, of which there are visible signs even to this day; for there are some mountains which now have nothing but food for bees, but they had trees not very long ago, and the rafters from those felled there to roof the largest buildings are still sound. And besides, there were many lofty trees of cultivated species; and it produced boundless pasturage for flocks. Moreover, it was enriched by the yearly rains from Zeus, which were not lost to it, as now, by flowing from the bare land into the sea; but the soil it had was deep, and therein it received the water, storing it up in the retentive loamy soil; and by drawing off into the hollows from the heights the water that was there absorbed, it provided all the various districts with abundant supplies of springwaters and streams, whereof the shrines which still remain even now, at the spots where the fountains formerly existed, are signs which testify that our present description of the land is true.

The Roman Empire

The Romans, too, were beset by ecological problems. In the second century A.D. the population of the city of Rome is believed to have been between 1 and 1¼ million people. Cooking and heating were done mainly over uncontrolled fires, and lighting was accomplished through the use of smoky oil-burning lamps. The air pollution must have been severe. Writers commented that people travelling from the countryside to the city would lose their tan within a few days under the smoky pallor that blocked out the rays of the Sun.

Other environmental problems of the Roman Empire included deforestation, erosion of hillsides, salinization of irrigated soils, and overgrazing. One extreme example is the destruction of the grasslands of North Africa. As the population in Italy expanded, the Romans began to import grain from distant provinces, much of it from the North African lands captured during the Punic wars. These dry but fertile lands were overplowed and overfarmed, leading to a dramatic reduction of the available groundwater. Eventually, the productivity of the land was lost in many places, and today the sands of the Sahara Desert blow over the ruins of ancient farmhouses (see Box 8.1).

The Dust Bowl in the United States

Ecological disasters have not been limited to the ancient world. The early European settlers found millions of acres of virgin land in America. The eastern coast, where they first arrived, was so heavily forested that even by the mid-eighteenth century, a mariner approaching the shore could detect the fragrance of the pine trees over 300 kilometers from land. The task of clearing land, pulling stumps, and planting crops was arduous. Especially in New England, long winters and rocky hillsides contributed to the difficulty of farming. It was natural that pioneers should be lured by the West, for here, beyond the Mississippi, lay expanses of prairie farther than the eye could see. Deep, rich topsoil and rockless, treeless expanses promised easy plowing, sowing, and reaping. In 1889 the Oklahoma

---

BOX 8.1 OZYMANDIAS

I met a traveller from an antique land
Who said: Two vast and trunkless legs of stone
Stand in the desert . . . Near them, on the sand,
Half sunk, a shattered visage lies, whose frown,
And wrinkled lip, and sneer of cold command,
Tell that its sculptor well those passions read
Which yet survive, stamped on these lifeless things,
The hand that mocked them, and the heart that fed:
And on the pedestal these words appear:
"My name is Ozymandias, king of kings:
Look on my works, ye Mighty, and despair!"
Nothing beside remains. Round the decay
Of that colossal wreck, boundless and bare
The lone and level sands stretch far away.

P. B. Shelley (1817)

The desert to which Shelley referred was the Sahara.

Territory was opened for homesteading. A few weeks later the non-Indian population there rose from almost nil to close to 60,000. By 1900, the population was 390,000—a people living off the wealth of the soil. In 1924 a thick cloud of dust blew over the East Coast and into the Atlantic Ocean. This dust had been the topsoil of Oklahoma (Fig. 8–13).

In each of the earlier examples one might contend that the destruction of the land was really caused by changes in climate rather than by mismanagement of the land. Indirect evidence, however, strongly implicates farmers. For example, geological evidence indicates that the climate in Southwest Asia and the Mediterranean has been relatively constant for the past 6 thousand years. Also, the land in those regions that have not been disturbed remains fertile. In the case of the Oklahoma Dust Bowl, however, the evidence is direct, not indirect. Scientists *know* that agricultural practices, not climate, destroyed the land. The southwestern plains of the United States are at best a near-arid ecosystem. In addition, periodic droughts have occurred every 20 to 25 years in the region for centuries.

The natural prairies on which the bison grazed were resistant to these droughts. This resistance arose out of the diversity of the grassland system. Both annual and perennial plants grow in the prairie. The perennial grasses and bushes live continuously from year to year and grow deep roots to absorb moisture from lower levels of the subsoil. Annuals, on the other hand, grow from seed in the springtime, flower, and then die during the hot summer months. During dry years there is so little water that many annuals die. However, the perennials, which use water deep underground, are able to live; and in doing so, they hold the soil and protect it from blowing away with the dry summer winds. In years of high rainfall, the annuals sprout quickly, fill in bare spots, and, with their extensive surface root systems, prevent soil erosion from water runoff (Fig. 8–14). Survival of both types of grasses is ensured by minimal root competition because the different plants have root systems that reach different depths. In addition, not all species flower at the same time of the year, so the seasons of maximum growth and consequent maximum water consumption differ.

Practices of the new settlers upset this naturally balanced and resistant system. They killed the bison to make room for cattle, then killed the wolves and coyotes to prevent predation of the herds. Moreover, cattle generally crop grasses lower than wild herbivores do, and the settlers

Figure 8–13 During the 1920s and 1930s, windblown dust had a devastating effect on agriculture in the western United States. This photograph, entitled "Buried machinery in barn," was taken in Dallas, South Dakota, on May 13, 1936. (Reprinted with permission of the National Archives.)

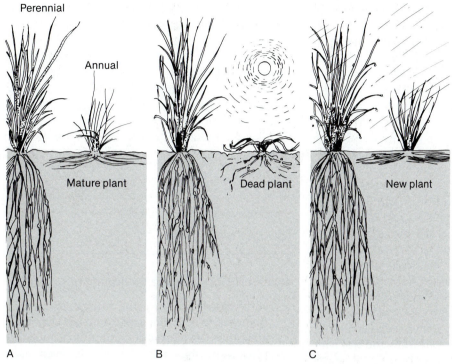

Figure 8–14 The diversity of a natural prairie system promotes stability in the event of either drought or heavy spring rains. *A,* Perennials have deep roots; annuals have shallower ones. *B,* In times of drought, annuals die but perennials survive by using deeper water; their roots hold soil. *C,* In a wet year, annuals sprout quickly, fill in bare spots, and prevent erosion of surface soil. Both perennials and annuals thrive.

often did nothing to prevent the resulting overgrazing. In overgrazed land, the plants, especially the annuals, become so sparse that they cannot reseed themselves. The land itself, therefore, becomes very susceptible to soil erosion during heavy rains. In addition, the water runs off the land instead of seeping in, resulting in a lower water table. Because the perennial plants depend upon the underground water levels, depletion of the water table means death for all prairie grasses. The whole process was further accelerated as the grazing cattle packed the earth down with their hooves, blocking the natural seepage of air and water through the soil.

The introduction of the plow to the prairie had an even more severe effect because the first step in turning a prairie into a farm was to plow the soil in preparation for seeding. At this point, of course, the soil was vulnerable to drought and wind erosion, because the perennial grasses, which normally held the soil, had already been killed.

On the other hand, if, after plowing, the spring rains were too heavy, the soil was easily washed away before the seeds had an opportunity to grow. Even after seeds had sprouted, the practice of pulling weeds between the rows left some soil susceptible to erosion by heavy rains.

As a result of all these factors, the soil fertility in Oklahoma slowly declined over a period of 20 to 35 years. Incomplete refertilization and loss of soil from wind and water erosion took their toll. Finally, when a prolonged drought struck, the seeds failed to sprout, and a summer wind blew large quantities of topsoil over 1500 km eastward into the Atlantic Ocean.

The droughts that wrecked the Oklahoma farms had no lasting effect on those prairies left untouched by humans. In fact, these virgin lands are still fertile. In a few thousand years, perhaps, the wind-scarred Dust Bowl will regain its full fertility.

In spite of these and other calamities that have occurred over millennia, civilization has grown steadily. There have been regional setbacks, but on the other hand, both the agricultural productivity and the human population of our planet have increased sharply for many years.* The average health of a large number of the people living on Earth has improved as well. Today, many people live longer and with greater freedom from disease than ever before in our long history. So why sound the alarm of ecological doom? The answer is that a great many people feel that we are threatened by, and moving toward, an unprecedented ecological catastrophe. They claim that the historical perspective reminds us that people should not battle against the Earth but instead must learn to work with it. The remainder of this book deals with human interactions with the environment and explores the relationship between civilization and the natural world. We will also examine the practical steps that can be or are being taken to avoid repetition of past environmental mistakes.

Summary

Because the fossil evidence of human ancestry is so fragmentary, the classification of hominid and early human species is uncertain. During its evolution, *Homo sapiens* acquired a potential for altering the environment in ways that no other species has ever duplicated. This set of attributes included **bipedalism** (walking on hind legs); free use of the hands for throwing and tool-making; the shift to a more carnivorous diet; and the improvement of communication, leading to the development of language.

Early humans were nomadic hunter-gatherers. Populations of these societies remained small, partly because very young children are a liability. People living in settled societies prefer to raise larger families. Scientists believe that agriculture arose in response to pressures of overpopulation.

During the 12,000 years that people have practiced agriculture, populations have expanded and civilization has developed. Ecological disasters have occurred throughout history. Before modern industry, most such environmental disruptions were the result of improper agricultural practices.

Questions

Human History and the Environment

1. What attributes gave the human species its potential to alter the environment? Did these attributes develop independently? If not, explain how they may have influenced each other.

2. Choose a primitive society (agricultural or hunter-gatherer) and read an article on it in a magazine or a good encyclopedia. Prepare a short report outlining the general lifestyle, health, population size in relation to the environmental carrying capacity, social and religious customs, amount of free time, and the collective attitude toward the environment as evidenced by the religion, literature, or customs.

3. It has been argued that advances in human civilization can be traced to exploitation of new sources of energy. Major steps forward (or backward if you are a pessimist) included the addition of meat to the herbivorous diet of our hominid ancestors, the taming of fire, and the use of fossil fuels. Other advances along the way have been domestication of beasts of burden and harnessing of wind and water power. Suggest how each of these has made it possible for human beings to alter the environment.

4. The human race has, in effect, enlarged for itself the carrying capacity of the Earth. How has this been accomplished? Why is it difficult to predict the stable population that can be supported in the future?

---

* Agricultural productivity, however, now appears to be leveling off.

Agricultural Disruptions

5. In 1973 a series of devastating floods ravaged sections of Bangladesh. Many observers have attributed these floods to unsound logging practices in the mountains of Nepal, at the headwaters of the rivers that flow through Bangladesh. Explain how logging in Nepal could affect farming in Bangladesh.

6. Drought recurs in North America approximately every 20 to 22 years. Discuss the relationship between these periodic droughts and the Dust Bowl disaster of the 1930s. Did the drought cause the land destruction? Defend your answer.

Suggested Readings

A readable book that discusses the origin of the human race is:
Donald Johanson, and Maitland Edey: *Lucy: The Beginnings of Humankind.* New York, Simon and Schuster, 1981. 409 pp.

Several shorter articles related to the same general issue are:
C. Owen Lovejoy: The origin of man. *Science, 211:*341, January 23, 1981.
John Pfeiffer: Current research casts new light on human origins. *Smithsonian,* June, 1980, p. 91.
Elwyn L. Simons: Ramapithecus. *Scientific American,* May, 1977, p. 28.

Two references on early civilization are:
Jean-Francois Jarrige, and Richard H. Meadow: The antecedents of civilization in the Indus Valley. *Scientific American,* August, 1980, p. 122.

Roger Lewin: Disease clue to dawn of agriculture. *Science, 211:*41, January 2, 1981.

An excellent short book about environmental problems in Mediterranean civilizations is:
J. Donald Hughes: *Ecology in Ancient Civilization.* Albuquerque, New Mexico, University of New Mexico Press, 1975. 181 pp.

For those interested in reading further about civilizations and ecological disruptions, the following books are of interest. A text on early civilization in China is:
Edward H. Schafer: *The Vermilion Bird.* Berkeley, University of California Press, 1967. 380 pp.

A text containing a chapter on ecology and Mayan civilization is:
John Harte, and Robert Socolow: *The Patient Earth.* New York, Holt, Rinehart and Winston, 1971. 364 pp.

Two interesting books about hunter-gatherer societies are:
Elizabeth M. Thomas: *The Harmless People.* New York, A. A. Knopf, 1959.
Robert S. O. Harding and Geza Teleki: *Gathering and Hunting in Human Evolution.* New York, Columbia University Press, 1981. 673 pp.

A more theoretical approach to the same subject is found in:
Bruce Winterhalder, and Eric A. Smith, eds.: *Hunter-Gatherer Foraging Strategies.* Chicago, University of Chicago Press, 1982. 268 pp.

On adaptive interactions between humans and their environments, see:
Emilio F. Moran: *Human Adaptability.* Boulder, CO, Westview Press, 1982. 404 pp.

The Human Population

It is estimated that the species *Homo sapiens* has been living on this planet for over 300,000 years. For some 99 percent of this time span, most humans have lived either as hunter-gatherers or in small communities of subsistence farmers. Within the last few centuries, however, tremendous changes have occurred. Many of these changes have happened so fast that they can often be chronicled in the memory of an immediate family. Think for a moment of the place where you grew up. Picture in your mind how it looked when you were young and what it looks like now. Most probably some of your favorite haunts such as woodlands, swamps, open fields, or empty lots have since been developed and are now sites for shopping malls, buildings, or parking lots. Now ask your parents or grandparents to compare the childhood appearance of the place in which they grew up to its appearance now. Their memories will probably provide sharper contrasts than your own (Fig. 9–1). If your family is from New York City, they may remember farms in the Bronx. Californians will remember vast expanses of uninhabited land. Southerners will recall small cities and very rural areas. Looking back a mere 100 years, when our

great grandparents were alive, the northwest plains of the United States were still dotted by herds of bison, and Indian children were riding their ponies across the prairies.

The past is relatively easy to study. But what will happen in the future? What kind of world will our children or grandchildren inherit? Of course, the future is always shrouded in some mystery and uncertainty. This chapter will examine past changes in human populations and study the methods for projecting the changes that may occur during the next generation or two.

9.1 Extrapolation of Population Growth Curves

The most obvious method for predicting population growth is to construct a graph that plots past population size against time and then to guess how the curve will continue. Guessing points on a curve outside the range of observation is called **extrapolation.** Extrapolation is a

A   B

Figure 9–1 Washington, D.C., and the Capitol building as photographed (A) in 1882, and (B) in the 1960s. A, Reprinted with permission of the National Archives; B, Wide World Photos.

subtle art. Figure 9–2 shows the growth of world population size from the emergence of *Homo sapiens* to 1980. A glance at the curve shows that world population growth is becoming more and more rapid. Indeed, Figure 9–2 may well cause the reader to panic. If the population continues to grow ever more rapidly, or even if it continues to grow at its current rate, very soon there will be too many people for the Earth to support.

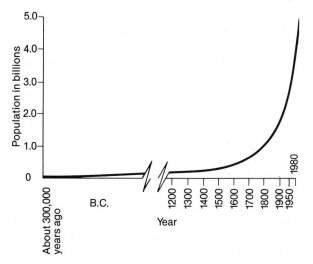

Figure 9–2 Growth of world population size from emergence of *Homo sapiens* to 1980.

For most of its 300,000 years on this planet, the human population increased very slowly indeed. Then the growth rate started to accelerate rapidly. At the time of the discovery of America, there were about one quarter of a billion people alive on Earth. In 1650, about a century and a half later, world population had doubled to one-half billion. In another 300 years, world population multiplied fivefold to 2.5 billion persons. During the 1950s, the population increased almost another one-half billion. By 1980, world population was approximately 4.5 billion persons. In other words, the *increase* in world population from 1950 to 1980 was about three times the *size* of world population in 1650. Today people are being added to the population at a rate of 2½ new children every second, which amounts to 215,000 per day, or 78 million people per year. If poverty and starvation exist now, how can economic and agricultural development be expected to keep pace with this exploding population?

Look back at Figure 9–2. What reasonable extrapolation can be made? One approach is to model the human population curve after population growth rates for animals. But all population curves follow different patterns under different conditions. In Chapter 4, we encountered a sigmoid, or S-shaped, curve, as shown in Figure 9–3B. Populations may oscillate in size (Fig. 9–3C) or even become extinct (Fig. 9–3D).

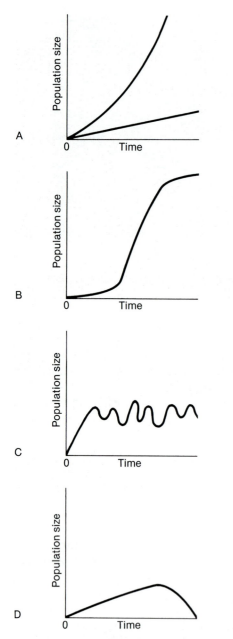

Figure 9–3 Schematic growth curves. *A,* Arithmetic (straight line) and geometric (curved line) patterns of growth. *B,* Sigmoid curve of growth. *C,* Oscillating population curve. *D,* Growth curve of a population that becomes extinct.

tend to grow faster than food supplies, which can grow only arithmetically. Therefore, he predicted, any uncontrolled population would eventually outstrip its food supply. Malthus concluded that when food supplies become exhausted, the human population must stop growing. However, he gloomily predicted that the leveling process would be accompanied by misery and vice. He wrote:

> By that law of our nature which makes food necessary to the life of man, the effects of these two unequal powers must be kept equal.
>
> This implies a strong and constantly operating check on population from the difficulty of subsistence. This difficulty must fall somewhere and must necessarily be severely felt by a large portion of mankind.
>
> . . . The race of plants and the race of animals shrink under this great restrictive law. And the race of man cannot, by any efforts of reason, escape from it. Among plants and animals its effects are waste of seed, sickness, and premature death. Among mankind, misery and vice. The former, misery, is an absolutely necessary consequence of it. Vice is a highly probable consequence, and we therefore see it abundantly prevail, but it ought not, perhaps, to be called an absolutely necessary consequence. The ordeal of virtue is to resist all temptation to evil.

Nearly 200 years have passed since Malthus wrote his famous essay. World population has indeed risen rapidly. In many places, conditions have been getting harsher, and people are suffering on a worldwide basis; however, Malthus's pessimistic prediction that shortages of food would put an end to the growth of human population has not come true up to now. Instead, the development of new agricultural techniques during the nineteenth and twentieth centuries has helped to improve economic conditions in much of the world.

No one can predict with certainty the shape of the curve of human population for the years to come. Perhaps humans will act with foresight and will gradually control their population to coincide with the carrying capacity of the biosphere. However, the core of the Malthusian argument cannot be ignored, for there are limits to the number of persons Earth can support, and unless growth is checked rationally, something

One famous prediction for human population growth was set forth in 1798 by the Reverend Thomas Robert Malthus, who claimed that populations, which can grow exponentially,

akin to the Malthusian's "misery and vice" *will* afflict humankind.

9.2 An Introduction to Demography

Demography is that branch of sociology or anthropology that deals with the statistical characteristics of human populations, with reference to total size, density, number of deaths, diseases, migrations, and so forth. The demographer attempts to construct a numerical profile of the population viewed as groups of people, not as individuals. For this purpose, the demographer needs to know facts about the size and composition of populations, such as the number of females alive at a given time or the number of infants born in a given year. This information is then used to predict population change with time.

The simplest way to measure growth is by subtracting the population at an earlier date from that at a later time. The following example does this for the populations of India and the United States during the 30-year period from 1950 to 1980 (Fig. 9–4).

Figure 9–4 *Between 1950 and 1980 the population increase in India was four times as great as that in the United States. Large increases in population often lead to poverty. Here, homeless people sleep on the street in New Delhi, India.*

Populations (in millions)

	INDIA	U.S.
1980	694	226
1950	360	152
Increase	334	74

These differences show that the United States, with many times the wealth of India and about three times the land area, had less than one quarter the population increase.

Unfortunately, these population differences give little indication about how fast a country is growing in relation to its population. More useful are **rates of growth.** For India, the average annual rate of growth from 1950 through 1980 was approximately

$$\frac{[694\,(10)^6 \text{ persons } - \, 360\,(10)^6 \text{ persons}] \times 100\%}{360 \times 10^6 \text{ persons } \times \, 30 \text{ years}}$$
$$= 3.1\% \text{ per year}$$

For the United States in the same period the average growth rate was

$$\frac{[226(10)^6 \text{ persons } - \, 152(10)^6 \text{ persons}]}{152(10)^6 \text{ persons } \times \, 30 \text{ years}} \times 100\%$$
$$= 1.6\% \text{ per year}$$

Average rates of growth represent only the crudest form of demographic data. They tell us how fast a population is growing at present but provide few clues that can be used to base predictions about the future. A more thorough demographic analysis must be based on a great deal more data.

A demographer is interested in the number of **vital events**—births, deaths, marriages, and migrations—that occur in a given period of time. Two very basic measures of population growth are the **crude birth rate*** and the **crude death rate.** The difference between these two rates is the **rate of natural increase.** For any geographical area or ethnic group being studied, these rates are computed as follows:

$$\text{Crude birth rate in year X} = \frac{\left(\begin{array}{c}\text{Number of live children}\\\text{born in year X}\end{array}\right)}{\begin{array}{c}\text{Midyear population}\\\text{in year X}\end{array}} \times 1000$$

$$\text{Crude death rate in year X} = \frac{\begin{array}{c}\text{Number of deaths}\\\text{in year X}\end{array}}{\begin{array}{c}\text{Midyear population}\\\text{in year X}\end{array}} \times 1000$$

$$\text{Rate of natural increase in year X} = \begin{array}{c}\text{Crude birth rate}\\ - \text{ Crude death rate}\end{array}$$

Stated in words, the crude birth and death rates are the number of births and deaths per 1000 people in a population (counting the population size at the midpoint of the year). The rate of natural increase represents the change in population per 1000 people (assuming that migration is negligible). For long-term assessment of historical trends, these three rates are concise, useful, and graphic. However, even more information is needed to predict future population trends accurately. To understand how populations grow, it will be helpful first to review the principles of geometric or exponential growth (see Box 4.1 and Appendix E) and then to see how the growth of populations differs from, say, the growth of money in a bank.

Let us start with money. Imagine you decided to deposit $100 in a bank that offered 5 percent interest per year. Suppose you started walking to the bank carrying a collection of change and bills of various denominations for a total of $100. If you deposited the entire $100, you would expect to have $105 at the end of the year. But on the way to the bank you bought an irresistible ice cream sundae for $1. Thus, you

had only $99 to deposit when you arrived at the bank. No matter how you paid for your sundae, whether you used coins, or a dollar bill, or a bill of higher denomination and received change, your $99 would grow at a rate of 5 percent. In one year you would have $99 + $99 × 0.05 = $103.95.

How different a population is! Imagine a population of 100 people—3 infants, 7 children, 50 adults under 65, and 40 people at least 65 years old. Suppose that during an entire year, no one moved in or out of the population, seven of the women under 65 had babies, and two of the people over 65 died. These were the only vital events. At the end of the year, the population would be 100 + 7 − 2 = 105, for an annual rate of growth of 5 percent.

Now suppose the population had contained only 99 people at the beginning of the year. What would the rate of increase have been? If the population grew in the same way that money in the bank grows, the rate of growth would be 5 percent no matter which person in the original population were no longer there. People, however, are not interchangeable like dollar bills. If the population had been missing an infant, there still would have been seven births and two deaths. There would have been 99 + 7 − 2 = 104 persons at the end of the year. The annual rate of growth would have been

$$\frac{104 - 99}{99} \times 100\% = 5.05$$

On the other hand, if the population had been missing one of the women who had a child, only six births would have occurred, and the rate of growth would have been

$$\frac{103 - 99}{99} \times 100\% = 4.04\%$$

If one of the elderly persons who died had been missing from the population, the population would be 99 + 7 − 1 = 105 persons at the end of a year, for an annual growth rate of

$$\frac{105 - 99}{99} \times 100\% = 6.06\%$$

---

* "Crude" refers to the fact that the rate is not specifically adjusted for variables such as age or sex. The midyear population is chosen as the base because that is most likely to be closest to the average population for the year.

This very simple example has pointed out some of the difficulties confronting the student of population size, but it also leads to an important insight that is necessary for an effective approach to the investigation of growth. Since the probability of dying or of giving birth within any given year varies with age and sex, the **age–sex composition,** or **distribution,** of the population has a profound effect upon a country's birth rate, its death rate, and hence its growth rate.

Figure 9–5 shows a hypothetical age–sex distribution for an imaginary population. In this situation, each age group has the same number of males as females. In particular, there are 500 boys and 500 girls under 10, and 50 men and 50 women between 90 and 100 years of age. Furthermore, there are exactly 50 fewer men and 50 fewer women at each succeeding age decade. Thus, there are 450 males and females between the ages of 10 and 20, 400 between 20 and 30, and so on. This graph can be used to predict future population growth. It shows that there are more babies than teen-agers, more teen-agers than young adults, and more young adults than older people. Women in the age group 17 to 40 are most likely to bear children. In five years the large population of young teen-agers will become adults, and bear children. In ten years the

even larger population of children will grow up and bear more children. Therefore, the graph in Figure 9–5 indicates that the population will increase rapidly in the near future.

Most human age–sex distributions don't look at all like Figure 9–5. Figure 9–5 would represent a population in which (1) boys and girls were born with equal frequency; (2) the same number of persons were born every year for over a century; (3) everyone died by the age of 100; and (4) any person, at birth, had an equal chance of dying throughout each year of his life span. However, in real human populations, on the average about 106 boys are born for every 100 girls. Nor is the probability of dying constant throughout one's life span. Instead, a relatively large proportion of people die when they are very young, comparatively few die between the ages of 10 and 50, and the proportion of people dying each year after 50 increases rapidly. In addition, there are marked sex differences in **mortality,** the number of deaths occurring in a given period. Women have a higher probability of surviving from one year to the next throughout the life span except during the childbearing years in areas without modern medical care.

Consider the effects of realistic patterns of vital events on a group of people born in the

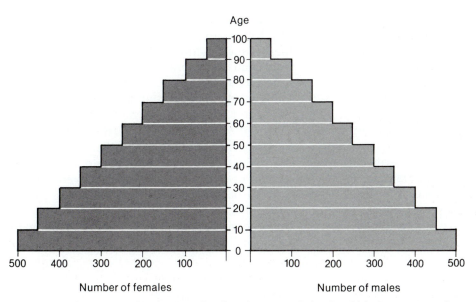

Figure 9–5 Age and sex distribution of an imaginary population in which the numbers of males and females are always equal.

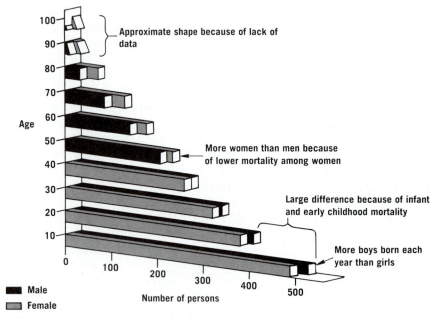

Figure 9–6 Typical age–sex distribution.

same five-year period. The greater survivorship of women over men means that even though more boys are born than girls, the ratio of women to men increases as the group grows older. By the time the group is elderly, there are considerably more women than men. Also, data are usually collected in such a way that we know only the total population of each sex over 70, 80, or 85. Therefore, the graphs can be only approximate for the very old age groups. Figure 9–6 presents an age–sex distribution that more nearly reflects these demographic characteristics.

In addition to these reasonably predictable phenomena, many changes that are less predictable can occur. Population growth is affected by such events as war, migration, famine, medical advances, and changes in social customs. Thus, each nation, and in some instances even each region, has its own individual distribution. Figure 9–7 shows the age–sex distribution for three nations—India, Norway, and West Germany. Note that the curve for India has a very broad base. There are a great many young children in the

population. Therefore, if no catastrophes occur, the population can be expected to expand rapidly during the next generation. In Norway, there are approximately an equal number of people in each age group between 0 and 30. Ten years from now there will be about the same number of men and women in the reproductive age group as there are now. Most probably, the population will remain fairly constant in the near future. In West Germany, there are *fewer* infants and children than young adults. Ten years from now there will be fewer people in the reproductive age group than there are at present. Therefore, if current trends continue, the population will eventually decline.

Another statistic that is useful in conjunction with age–sex distribution is the **total fertility rate** (TFR). The total fertility rate is the total number of children a woman in a given population can be expected to bear during the course of her life if birth rates remain constant for at least one generation. There is considerable variability in TFRs. In the less developed countries

Figure 9–7 Age–sex distributions for three nations. *A,* India—a rapidly expanding population. *B,* Norway—a stable population. *C,* West Germany—a declining population. (From *U.N. Demographic Yearbook,* 1979.)

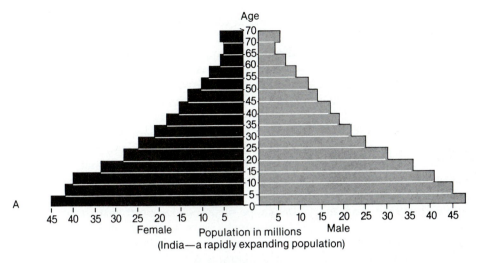

Age

A

45 40 35 30 25 20 15 10 5   5 10 15 20 25 30 35 40 45
Female   Population in millions   Male
(India—a rapidly expanding population)

Age

B

160 140 120 100 80 60 40 20 0   0 20 40 60 80 100 120 140 160 180
Female   Population in thousands   Male
(Norway—a stable population)

Age

C

700 600 500 400 300 200 100 0   0 100 200 300 400 500 600 700
Female   Population in thousands   Male
(West Germany—a declining population)

the average TFR was 4.4 in 1980, whereas it was 2.0 in the developed regions of the world. If every woman in the developed world continues to bear only 2.0 children, the population will eventually decline, for some infants or children invariably die before they reach reproductive age. The **replacement level** is the value of the TFR that corresponds to a population exactly replacing itself. In the developed countries, where hygiene and medical attention are superior, the replacement level is 2.1, whereas in the less developed world the insufficiency of medical services has led to high infant mortality, and the replacement level is about 2.7.

If a demographer knows how many females of reproductive age there are in a population from the age–sex distribution and the average number of births per female from the TFR, and *if birth rates do not change,* then it is relatively easy to predict the population in the next generation.

In reality, birth rates do change through changing patterns of social behavior or as a result of external factors such as war or migration. If the TFR changes unpredictably, the demographic prediction will be wrong.

The change in population distribution over time has been likened to the digestion of a mouse by a snake. A snake swallows a mouse whole and thereby gets a big lump in its throat. The body of the mouse then moves slowly through the digestive system of the snake. The bump in the snake gets smaller and smaller as the mouse is slowly digested. The movement from head to tail can be considered analogous to the aging of a generation, and digestion to its gradual dying. The problem with guessing what a snake will look like tomorrow is that the observer doesn't know when, or what, the snake will eat next. Once a mouse is in the snake's body, it is easy to predict what will happen to the shape of the snake. So it is with population distributions. Once a generation has been born, demographers can predict quite accurately how that generation will change. However, accurate prediction of the size of the coming generation is extremely difficult. The size of the generation of childbearing age gives clues but not definitive information. For example, Figure 9–8 shows that the distribution for Sweden looked like a triangle in 1910. Not knowing anything about changes in vital rates, one would predict a triangle in 1930. However, in 1930 the base of the age distribution was pinched. This pinch occurred as a result of the social and economic effects of World War I. With fewer eligible males in the population, there were fewer marriages and thus fewer births. A demographer would have had to pre-

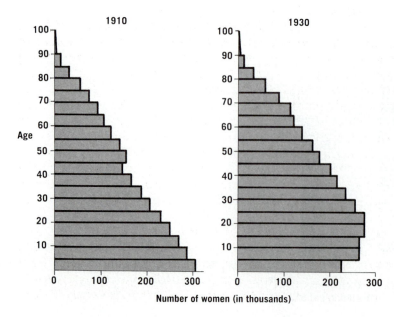

Figure 9–8 Female age distribution in Sweden in 1910 and 1930.

dict World War I as well as its profound economic and social effects on all of Europe to have projected accurately the population of Sweden from 1910 to 1930.

9.3 Population Changes in the Developed World—Example: The United States

The demographic history of the United States is largely the history of rapidly changing patterns of fertility; by contrast, mortality has remained relatively constant (Fig. 9–9). Fertility reflects social and economic events: In the 1920s, a period of economic boom and the Great Gatsby, fertility was quite high. The TFR was then about 3.2. The TFR fell sharply throughout the years of the Great Depression. It reached a low of about 2.2 in 1936. During World War II, in spite of the fact that many men were overseas, the fertility rate increased steadily. In 1946, the year after the war ended, the TFR was about 2.5. At that time, demographers and social scientists believed that the trend was due to the return of war veterans. They predicted that fertility would soon decrease. Instead, it rose rapidly to a peak of 3.7 in the late 1950s. Analysts then attributed the rise to economic and social well-being. Businesses that specialized in baby products flourished; the middle class began in earnest its flight from the cities to provide their children with "fresh air" and places to play. Schools became overcrowded; communities couldn't build schools fast enough to keep pace with the growing numbers of children. Experts predicted a continued pattern of high births. Then, suddenly and inexplicably, the TFR began to drop quite rapidly. In 1971, it fell below the replacement level of 2.1 and continued to fall. By 1976 it bottomed out at 1.75.

Population experts were so surprised by the continued decline in TFR that in 1975 the Census Bureau drastically revised its assumptions concerning U.S. population growth. In the 1970s communities were forced to close many of the new schools they had built less than a decade before because the number of school-age children had dropped so precipitously. Businesses that had specialized in baby goods were forced

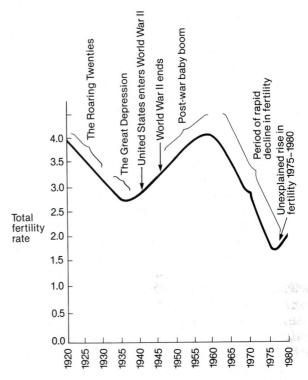

Figure 9–9 Total fertility rate for the United States 1925–1980. (From U.S. Bureau of the Census.)

to diversify. Older teachers were fired and many young teachers were unable to find jobs.

During the early years of the 1980s the TFR has begun to creep upward again. It reached 1.8 in 1980 and continues to rise. But birth rates are still below the replacement level.

When the TFR is below replacement level, the population may continue to grow for a while if there is a disproportionate number of young persons. To understand this concept, look at the age–sex distribution for the United States in 1980 (Fig. 9–10B). Note the bulge formed by the age group between 15 and 25. These adults were the baby boom infants born in the late 1950s and early 1960s. Today, this large group of people is in the reproductive age group. Even if all the women in this group bear children at less than the replacement level, the population will increase.

When will this increase level off? The Census Bureau has attempted to predict the U.S. population into the early part of the twenty-first century. There are two major uncertainties in formulating this prediction: (1) How will the TFR change in the future, and (2) what will the level of immigration be? Since these factors cannot be predicted with certainty, three different scenarios are proposed.

U.S. Population as of April 1, 1980, was 226.5 Million People

IF	THEN THE POPULATION WILL	AND WILL
1. The TFR rises to 2.1 and immigration (legal and illegal) is 800,000 per year,	be 337 million by the year 2025	continue to increase indefinitely.
2. The TFR stabilizes at 1.8 and immigration stabilizes at 400,000 per year,	reach 270 million by the year 2035	stabilize at that level.
3. The TFR decreases to 1.6 by 1986 and immigration decreases to 150,000 per year,	reach 241 million by the year 2007	stabilize at that level.

Look back at Figure 9–10*A* and *B*. Note that between 1970 and 1980 the baby boom bulge in the population distribution moved steadily upward. (Remember the snake digesting the mouse.) This trend will continue, so that by the year 2000 the average age of the population will be significantly older than it is today. Also contributing to the same effect is the fact that people can expect to live longer now than they used to.

As a result, there will be a higher proportion of old people in the future population than in the present one. This change in the age distribution has profound effects on the economy. Old people are often retired, and are therefore not direct wage earners. Many must be supported by pensions or Social Security. The Social Security system was established on a pay-as-you-go basis. Thus the money an individual pays in this year is not saved for his or her pension but is used directly for paying the pension of an older person alive today. The system was established under the premise that the ratio of workers to retired persons would remain high.

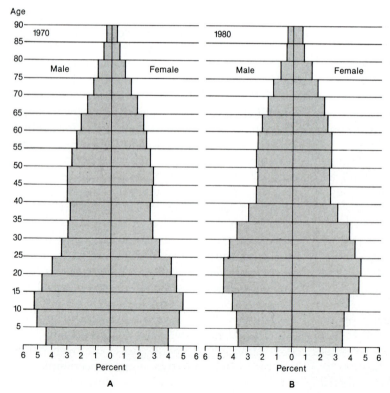

Figure 9–10 The age distribution in the United States, 1970 and 1980. Note that in 1970 the postwar baby-boom children were between 5 and 15 years old. A decade later this group was between 15 and 25 years old and had entered the reproductive age group. Therefore, we can expect an increase in population even if women bear children at the replacement level.

In 1950 there were 16 workers for every retiree. This ratio dropped to 3 to 1 by 1980 and is expected to drop further to 2 to 1 by the year 2020 (Fig. 9–11). If this change does in fact occur, then the Social Security tax will have to be increased drastically or benefits will have to be reduced.

9.4 Populations in the Less Developed Countries—The Demographic Transition

For many thousands of years, human population grew slowly. Suddenly, in the fifteenth century, population began to rise rapidly. What happened? Have women been bearing more and more children during the past 500 years? The answer is no. In fact, the average number of births per family has become smaller. The change has occurred because people live longer, on the average, than they used to.

> **BOX 9.1 "DEVELOPED" AND "LESS DEVELOPED" NATIONS**
>
> The developed nations of the world are those that are most heavily industrialized. The United States, Canada, all of Europe, U.S.S.R. and Japan are among the developed nations. The less developed nations are those that are less heavily industrialized. India, all of Africa, Southeast Asia, and most of South and Central America are less developed. One third of the world's population lives in the developed nations but consumes 85 percent of the global resources.

A simplified explanation for the patterns of change in the vital rates of a population proceeds as follows: When nutrition is poor, water unclean, and infectious disease common, relatively few people live to adulthood. Many children are born, but many die. In some societies, half of all live-born infants do not reach their fifth birthday. As modern principles of health care are introduced into a society, death rates start to fall.

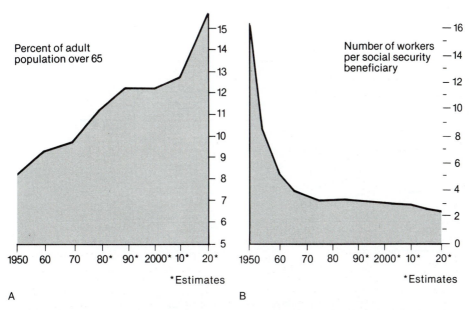

Figure 9–11 Changes in the age distribution in the United States will place a growing burden on the active workers to support an increasing number of retired people. (A from U.S. Bureau of the Census; B from U.S. Social Security Administration.)

In the developed countries, the death rates have been dropping steadily for over a century. During the past one hundred years, clean water has become increasingly available, milk has been pasteurized, and new drugs have reduced death from disease. In addition, improved agriculture and more humane distribution of food have continued to aid in improving children's health. Therefore, death rates have declined gradually. During this time, people have come to understand that it is possible to raise a family without having many babies. Consequently, the birth rate has been declining for almost a century. Since birth rates and death rates have *both* declined, the populations have remained relatively stable (Fig. 9–12).

Poor countries of the world have not experienced such gradual change. For centuries, women bore large families, and some half of the children died. Then in the twentieth century modern medicine arrived suddenly from the rich nations. Health care improved dramatically in a relatively few years. Death rates dropped very quickly. People's patterns of behavior in the less developed nations have not had time to adjust. Birth rates are still high. With high birth rates and lower death rates, population has increased rapidly.

Demographers summarize the population change of a country in the following way. Societies with primitive medical care are characterized by both high birth and high death rates. Since the difference is small, there is little or no growth. When modern medicine is introduced death rates among children drop. Birth rates, however, remain relatively constant. The combined effect of these two trends—falling death rates and constant birth rates—causes the population to grow rapidly. After some time, people get used to the fact that fewer of their children will die young, and birth rates drop. Therefore, the developed countries have low growth because birth and death rates are both low. This series of changes from high birth and death rate to low birth and death rate is known as the **demographic transition**. Table 9–1 lists some countries at various demographic stages.

Although this standard explanation for the demographic transition is still widely accepted, many researchers believe that it may be oversimplified. In several European countries, for instance, birth and death rates declined simultaneously and at nearly the same rates in the nineteenth century. In France, examination of records of births and deaths indicates that in many parishes the decline in birth rate actually

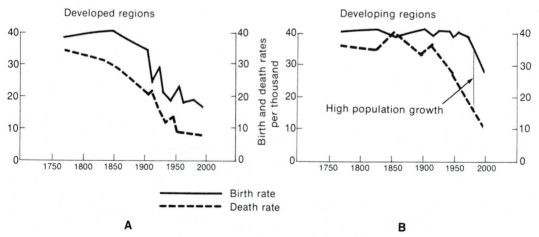

Figure 9–12 Past and estimated birth and death rates in the developed and less developed countries. Note that the high population growth in the less developed countries is due to the fact that death rates have declined faster than birth rates. Rapid oscillations in the curves have been caused by wars, epidemics, and periodic "baby booms."

TABLE 9–1 Some Typical Birth and Death Rates Before, During, and After Demographic Transition*

	Birth Rate/1000	Death Rate/1000	Rate of Natural Growth/1000
Very high birth and death rates:			
Afghanistan	45.2	21.1	24.1
Niger	51.4	22.4	29.0
High birth rate, moderate death rate:			
India	33.2	14.1	19.1
Peru	39.7	12.2	27.7
Moderate birth rate, low death rate:			
Spain	15.1	7.7	7.4
United States	15.8	8.7	7.1
Low birth rate, low death rate:			
Switzerland	11.6	9.3	2.3
Germany, Federal Republic of	10.0	11.5	− 1.5

* *U.N. Population and Vital Statistics Report,* January 1, 1982.

preceded the decline in death rates. Further study of human behavior and the demographic transition is a valuable area for research.

9.5 Current Worldwide Population Trends

Predictions of future populations are always uncertain. Dramatic upsets such as major wars, cataclysmic geological changes, and global climatic shifts are by their very nature unforeseeable. But even in a relatively stable global environment, fertility and mortality rates may change quite rapidly in response to social and political trends.

If the Earth were infinitely large, the human population might possibly continue to increase indefinitely. But our planet and our resources are certainly finite, and therefore the continued increase in population must eventually stop; that is, **zero population growth** must be achieved. Perhaps the two most important questions in demography are (1) when will zero population growth be realized, and (2) what will the population be at that time?

As one limit, imagine what would happen if by some miracle, birth rates were reduced to the replacement level today. As already explained, even if this nearly impossible event occurred, the human population would continue to increase because in most countries, especially the less developed ones, there is a disproportionately large number of women in the reproductive and prereproductive age groups. For example, in 1980 approximately 40 percent of the females in India were in the prereproductive age groups, and an even larger number of women were between 15 and 39. Even if women began to bear children at the replacement level immediately, the population would rise until it is 1.6 times as great as it is now.

Of course this discouraging analysis is only a beginning, for no one expects birth rates to drop to the replacement level overnight. In order to extrapolate trends accurately, it is necessary to return to an analysis of birth and death rates. A summary of vital rates for the developed and the less developed countries is given in Table 9–2.

(a) Less Developed Countries. As shown in Table 9–2, the death rates in the less developed countries have been more than halved during

TABLE 9–2 Vital Rates* for the Less Developed and the Developed Nations 1900–1970

	Less Developed Nations			Developed Nations		
Period	Birth Rate	Death Rate	Rate of Natural Increase	Birth Rate	Death Rate	Rate of Natural Increase
1900–1950	41	32	9	26	18	8
1950–1960	43	22	21	22	10	12
1960–1970	41	17	24	20	9	11
1970–1980	37	15	22	19	9	10

* Per 1000 population.

this century. Since 1960, birth rates have been dropping as well, but at a much slower rate. However, as the death rates in the less developed countries approach the death rate in the developed areas, the decline will necessarily level off. If birth rates continue to decline as expected, the rate of natural increase will diminish. However, population growth rates are still extremely high, and there is a long way to go before zero population growth is reached.

There are marked differences in birth rates within the less developed countries (Fig. 9–13). Birth rates are declining rapidly in China, parts of Southeast Asia, and a few countries in South America. Significant but less dramatic changes have occurred in India, Egypt, Mexico, Chile,

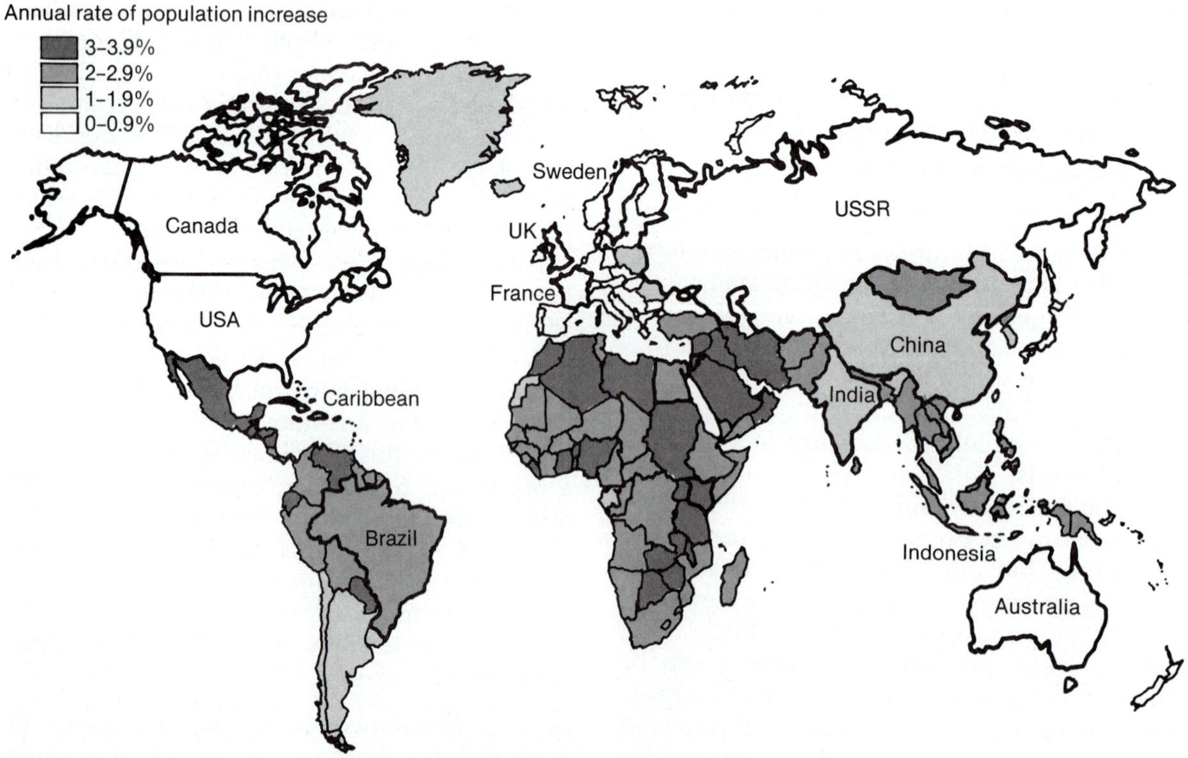

Figure 9–13 Annual rate of population increase for the countries of the world.

Argentina, and Paraguay. However, birth rates have hardly declined at all throughout most of Africa, Southwest Asia, and other parts of South America, although many of the people in these regions are already experiencing debilitating poverty, inadequate diets, and abysmal living conditions. It is difficult to predict when birth rates in these regions will drop.

(b) Developed Countries. The TFR in many developed countries is already well below the replacement level. In a few regions, population growth has all but stopped, and in many more, rates of growth are expected to stabilize by the turn of the century. Figure 9–14 shows a comparison of the doubling time (the time required for a population to double in size if current trends continue) between representative developed and less developed countries.

(c) *Population Predictions.* In 1981 the world population was about 4.5 billion people. Although the rates of growth are expected to decrease in all areas of the world except Africa, the actual population almost certainly will continue to increase. The present population of the developed countries is expected to increase by a mod-

erate 12 percent between 1980 and the year 2000. However, population in the less developed countries is growing much faster and is expected to increase by an extraordinary 50 percent during the same time period. The largest increase is expected in Africa (76 percent) and the next largest in Latin America (65 percent).

Demographers predict that if fertility levels can be reduced to the replacement levels throughout the world by the year 2000 to 2005, the world population will increase to 5.9 billion by the year 2000 and eventually stabilize at 8.5 billion. But this scenario may be too optimistic. If fertility rates drop more slowly, world population may increase to 10 or even 13.5 billion (see Fig. 9–15). Since the population in the developed countries is stabilizing fairly rapidly, most of these added people will be born into the less developed regions where poverty is already all too common.

9.6 **Consequences of Population Density**

What do these numbers mean in terms of human values and lifestyles? What would life on our planet be like if there were two or even three times as many people in the world as there are today? As the population density in a given area increases, each person's proportionate share of the available supplies of land, water, fuels, wood, metals, and other resources must necessarily decrease. (Since the distribution is never proportional, the poor become very much worse off.) In the past, people in many parts of the world have raised their standard of living despite a rising population by developing newly discovered concentrations of resources or by using available resources with increased efficiency (Fig. 9–16). But millions of others have recently fallen into dismal poverty leading to hunger and starvation.

No one knows what the ultimate carrying capacity of the Earth really is. Some suggest that, given a stable political climate (in particular, a sharp drop in today's excessive global military expenditures) and some key advances in technology, the planet could support a human population of 13 or 14 billion without Malthu-

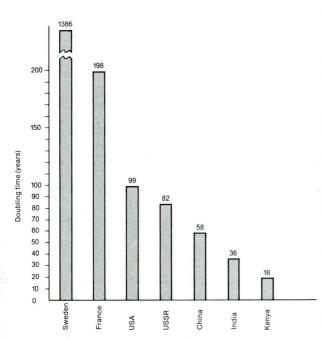

Figure 9–14 *The time (in years) required for the populations of representative countries to double.*

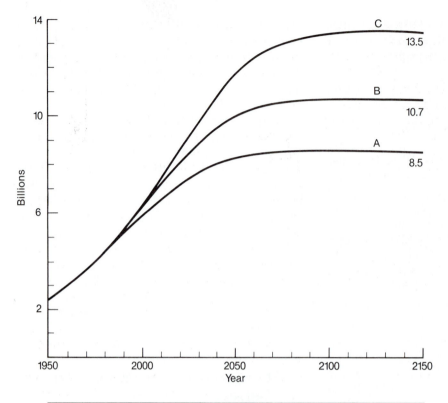

Figure 9–15 Three different predictions for world population growth.

If the world attained replacement fertility in	World population would then stand at	And world population could be expected eventually to stabilize at
(A) 2000–2005	5.9 billion (2000)	8.5 billion
(B) 2020–2025	8.0 billion (2020)	10.7 billion
(C) 2040–2045	10.8 billion (2040)	13.5 billion

sian misery and vice. Others claim that we have already exceeded the carrying capacity of global ecosystems and are sustaining ourselves artificially on nonrenewable resources. This argument predicts that the human race is rushing headlong toward an unprecedented calamity that may lead to the deaths of billions of people.

One question to ask is what happens to animal societies as populations approach or exceed the carrying capacity of an ecosystem. Unfortunately, studies of animal societies provide no consistent models that can be used as bases for confident predictions about human populations. In one experiment, 4 male and 21 female reindeer were introduced on St. Paul's Island, near the coast of Alaska. The environment was favor-

able, vegetation was lush, and there were no predators. At first the population increased exponentially. Even after the sustained carrying capacity was surpassed, vegetation was still abundant, and the reindeer continued to multiply until the population numbered 2000 (Fig. 9–17). Then, quite suddenly, almost all the food was gone. The island became barren, and mass starvation and death occurred, until only eight animals remained alive. Before too many conclusions are drawn from this experiment alone, consider what happened when reindeer were introduced onto St. George Island, which is geographically close to, and environmentally similar to, St. Paul's Island. The reindeer population on St. George increased moderately and reached a

Figure 9–16 *Children working in a grinding shop in New Delhi, India, to add to the family income.*

provides a warning that in some instances over-population followed by famine is possible, but the results of all the experiments taken together provide no clue for predicting what will actually happen either to animal or to human populations.

In a sense, much of the rest of this book is an examination of the limits of human population growth. Appropriate chapters will address such key questions as: Will useful energy be exhausted in the near future? What is the potential productivity of global agricultural systems? How much metal, fertilizer, and water will be available? How does pollution affect our lives?

If the human future is to be a gloomy one, then the quality of life on Earth will steadily deteriorate as the population increases. Many forests and wild places will disappear and be replaced by cities and indoor environments. Under ideal conditions, a city can be a pleasant environment for many. Given money, mobility, and freedom from oppressive pollution, a wealthy urbanite can enjoy excellent career opportunities, the best medical attention, theaters, art, fine restaurants, and an intellectually stimulating society. But this reality is a privilege only for a fortunate minority. In less developed countries, unskilled workers often cannot find jobs in the countryside. As a result, millions of jobless poor have been moving into the cities. Yet the cities are often unable to provide relief for these people. Consider, for example, the situation in Mexico City. In 1975 Mexico City had 11 million

stable, healthy herd of about a dozen animals. Other introductions of reindeer onto islands in Alaska have exhibited population growth patterns intermediate between these two extremes. The starvation that occurred on St. Paul's Island

Figure 9–17 *Growth function for reindeer on two similar islands. (From C. J. Krebs: Ecology. New York, Harper & Row Publishers, 1972.)*

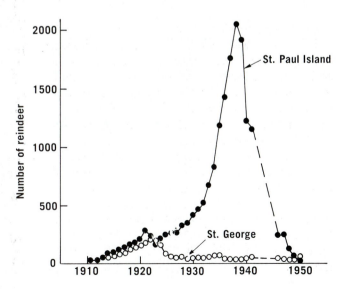

people. That total rose to 13 million by 1979 and 15 million by 1981, and is expected to multiply to 32 million by the year 2000. That is almost four times the present population of New York City. No one knows how an impoverished city of 32 million can be maintained. Certainly there cannot be enough jobs and houses for all these people. In all probability, the government won't have enough money to provide fundamental necessities such as fresh water, sewage disposal, and medical attention. Furthermore, pollution problems, already serious, are almost certain to become acute. Some highlights of the problem follow:

• Independent observers estimate that the unemployment rate in Mexico City ranges anywhere from 20 percent up to about 40 percent. At least 4 million poverty-stricken people have no homes. Instead, they live in illegal squatter communities, in tents and shacks, without running water or toilets (Fig. 9–18). Dysentery, typhoid, and hepatitis are common.

• The city is constantly building new public works projects, but the construction of the new

Figure 9–18 Dark clouds form over Mexican residents of Iztacalco slum in Mexico City, where police and residents have clashed over the shortage of housing that has created disputes over land ownership and housing. The sign in the background expresses support for the inhabitants and calls for an immediate solution to the problems. The ditch in the middle of the photograph is an open sewer. (World Wide Photos.)

facilities is slower than the increase in population. According to an engineer working on a new water project, "If we install new services to provide water for an additional 500,000 people, the good news spreads and 1,000,000 new immigrants move to the city."

• Mexico City is located in a high mountain valley ringed on all sides by higher peaks. As a result, the natural movement of air is often obstructed, and pollutants tend to be concentrated over the region. The problem is aggravated by the fact that industrial air pollution control laws are not particularly strict. In addition, many automobiles are old, and their engines are worn and in a poor state of repair. As a result, these automobiles release relatively high concentrations of air pollutants. Average carbon monoxide concentrations in certain neighborhoods in Mexico City have been measured to be approximately 20 parts per million (ppm), and rush-hour peaks of 35 to 40 ppm have been recorded. According to the United States Environmental Protection Agency, an 8-hour exposure to 20-ppm carbon monoxide is described as "very unhealthful."

Social Consequences of Population Density. Some people have claimed that a high population density naturally leads to violence, disunity, and political upheavals. In a series of experiments with strains of rats, John Calhoun studied the effects of extreme crowding. He constructed cells supplied with enough food and water for many more rats than the space would normally hold. A few animals were placed in each cell and allowed to breed. The population and the density grew quickly, and the animals began to act bizarrely. The females lost their ability to build proper nests or to care for their infants. Some of the males became sexually aggressive; most retreated from communication with others. In short, the normal processes of socialization were destroyed. Other literature presents evidence of decreased fertility and strange behavior in many animal species under conditions of overcrowding.

However, people do not necessarily respond in the same way that rats do. Research into the relationship between human population density and social problems has yielded a morass of conflicting conclusions. No consistent patterns emerge when national population densities are defined as population size divided by total national area. Such densities are grossly misleading because they do not measure population densities in populated areas. For example, the Netherlands, one of the most densely populated areas of the world, had 340 persons per square kilometer in 1976. By contrast, the population density in India was only 180/km² and in Algeria, only 7/km². Since, however, nearly all of the Netherlands is inhabitable, whereas much of India is jungle and most of Algeria is desert, the fact that the density of the Netherlands is so high and its society so stable does not by itself disprove a hypothesis that high population densities are socially detrimental. These examples are not isolated instances because the 62 percent of the Earth's surface that is semi-arid, taiga, tropical jungle, arctic, tundra, or desert holds only 1 percent of its population.

The relation between population density on arable land and social problems is also confusing. Japan, which supports 1700 persons per square kilometer of arable land, is an example of a very densely populated country that maintains a prosperous and relatively crime-free society.

Some psychologists and sociologists claim that density by itself is not relevant to the feeling of being crowded. A more important factor, they say, is the amount of space available in the individual's dwelling unit. Even here the evidence is ambiguous. Hong Kong has probably the highest residential density ever known in the world. Nearly 40 percent of residents of Hong Kong share their dwelling unit with nonrelatives; almost 30 percent sleep three or more to a bed, and 13 percent sleep four or more to a bed. Most of the population lives in a dwelling unit of a single room; most dwelling units are homes to more than nine persons; most dwelling units are homes to two or more unrelated families. Even under these conditions of extreme crowding, there is little or no proof that antisocial behavior is attributable to the crowding itself.

On the other hand, when one picks up a newspaper and reads about continued warfare in Central America, in Southwest and Southeast Asia, and in various regions of Africa, it is easy to think that Malthusian misery and vice have already overtaken us (Fig. 9–19). Again, the facts do not support any direct correlation between population density and warfare or political instability. Despite the misery of the poor

Figure 9–19 High infant mortality in Bolivia is reflected by this casket shop in La Paz.

throughout Mexico and the dismal conditions in Mexico City itself, the Mexican government has survived 50 years without a revolution. This record of stability is better than that in many countries with a much lower population density. In Central America, the population density in Costa Rica is approximately 2.5 times as great as that in neighboring Nicaragua, yet Nicaragua has experienced bloody civil warfare in the late 1970s whereas Costa Rica has been internally peaceful for decades. Similarly, India, which is large, heavily populated, and poor, has experienced relatively stable conditions for decades.

Thus, there is no conclusive proof that density itself breeds antisocial behavior. One difficulty in studying the effects of population density on humans is that spatial requirements may in part be culturally determined. Much sociological research is needed to gain an understanding of the factors that cause members of different societies to feel crowded.

9.7 Case History: Population and Public Policy in the People's Republic of China

One fifth of all the people in the world, approximately 1 billion, live in the People's Republic of China. Between 1950 and 1980, the population of that country increased by some 400 million people; this increase is nearly twice the size of the population of the United States in 1980. In 1957, one of China's most prominent economists, Ma Yinghu, published a paper claiming that the rapid population rise was interfering with economic development and threatening to nullify important gains in agriculture and industry. At that time, the official government policy stated that more people meant more production, and therefore population expansion should be encouraged. For his boldness in contradicting established policy, Mr. Yinghu was removed from his post at the Peking University and publicly denounced.

During the 1960s and 1970s continued modernization in China led to further increases in agricultural and industrial output. However, the level of living conditions of the average person remained constant during that time because the additional goods and services were needed to support the growing population. In 1978, official policy changed, and the government announced its intentions to reduce population growth. The rate of natural increase in China had been 23.4 people per thousand in 1971; it had declined by nearly half to 12.1 per thousand in 1978. The official government policy is to reduce the growth rate further to 5 per thousand by 1985 and to zero by the year 2000.

In order to realize this goal, birth control devices and medical abortions are available, free, for all citizens. In addition, a countrywide information program exhorts people to "marry late, to space their births at long intervals, and to have few children" (Fig. 9–20).

But advice and free contraceptive devices are only the first step. Couples who bear one or two children are rewarded, and those with three or more are punished. According to the new plan, all married couples will be allocated living space large enough for a family of four. Thus a family with only one child will have more than average

Figure 9–20 Billboard in Beijing, capital of China. (Photo courtesy of Jack Tackle.)

living space, whereas those with three or more will be crowded. Couples who live in rural areas will be given a garden plot for four, regardless of family size.

In addition, a strong economic incentive program has been established. A family with one child receives a direct pay raise equivalent to one month's additional salary every year. The first child will be given preference for admission to school, tuition will be waived, and all medical attention for the child will be free. In addition, any couple who has only one child during their lifetime will be rewarded with a larger pension upon retirement. A second child is neither encouraged nor discouraged by official policy. However, a couple electing to have a third child will be penalized severely. A pay cut of 10 percent is levied, and further career promotion for the parents will be refused as well.

The preliminary results of the 1982 census, which showed a mainland population of 1,008,175,288, reinforced China's birth control objectives. The State Family Planning Commission hopes to limit the population to 1.2 billion by the year 2000, but this target can be reached only if 70 percent of married couples have only one child.

No one knows what the outcome of this radical policy will be. Certainly the program would not be acceptable in many political systems. People ask why a third or fourth child, who didn't ask to be born, should be punished. The example of China therefore represents the recognition of the problem of population growth by an alert government along with the use of extremely severe measures to control it.

an age–sex distribution reflects past history and provides insights into future growth.

The problem in predicting population size is that one needs to guess the birth and death rates that will prevail in the future, and these rates depend on a myriad of social and historic forces.

Mortality has remained relatively constant in the United States but fertility has been variable. There is a wide difference in opinion concerning future growth patterns in the United States. The prediction for the year 2000 and beyond is for an increasingly elderly population.

Societies that have inadequate medical services are characterized by both high birth and high death rates and very little overall population change. Modern medicine leads to a rapid decrease in mortality among children but does not affect birth rates appreciably. Therefore, the population rises rapidly. Slowly, in response to economic trends, the birth rate falls, and the population size stabilizes once more. This pattern of changing vital rates is called the **demographic transition.**

Birth rates are falling in many countries of the world, but an increase in the present world population of 4.5 billion to 6 billion by the year 2000 is likely. A global stable population of 8.5 billion to 13.5 billion is also predicted.

Dense populations strain the available resources, and the worst possible circumstances may lead to dismal living conditions or mass starvation. However, it is very difficult to assess the social effects of human crowding. Experiments with animals suggest that crowding produces antisocial behavior, but the relevance of these findings to the human condition is unclear.

Summary

In recent years, the human population has been increasing at a geometric rate. Human population growth is strongly affected by social and economic forces; therefore, models for population dynamics applicable to other animals do not usually apply to people. Prediction of future population size should be performed on the basis of changing birth and death rates and the **total fertility rate** (TFR). In addition, the shape of

Questions

Demography

1. How would you expect each of the following to affect population growth? Consider which age groups are most likely to be affected by each event and how the event affects population change: (1) famine; (2) war; (3) lowering of marital age; (4) development of an effective method of birth control; (5) outbreak

of a cholera epidemic; (6) severe and chronic air pollution; (7) lowering of infant mortality; (8) institution of a social security system; (9) economic depression; (10) economic boom; (11) institution of child labor laws; (12) expansion of employment opportunities for women.

2. What is a demographic transition? How does it arise? Compare birth and death rates before, during, and after demographic transition.

3. Name four factors that have been responsible for reducing death rates in developed nations. How have these factors affected developing nations differently from developed nations? Discuss.

4. Explain why populations in many developing countries have increased rapidly in recent years.

5. What is the replacement level? If women in the United States bear children at the replacement level, will zero population growth be realized immediately? Why or why not? Discuss.

6. Would it be practical to stop population growth in India immediately with an active family planning program? Discuss.

7. List and discuss briefly an economic, a political, and a social factor that is important in determining birth rates.

8. Is it likely that birth rates in the developing nations will decrease to the replacement level within the next five years? Discuss.

Social Policy

9. In the United States, families are allowed an income-tax deduction for each child. In most developed countries, each family is allotted an annual grant for each child. How do you think these tax laws affect family size?

10. If you had the responsibility of discouraging population growth, would you consider reducing income tax deductions if a family has more than four children? Reducing health benefits? Whom would such policies harm?

11. How large was the family that your grandparents grew up in? Your parents? How many children are there in your family? How many children do you think an average

family in a country such as the United States should have? Compare your answers with the answers of your classmates. Discuss in terms of changes in population growth rates.

12. Discuss the effects of population growth in your community. How do these problems compare with the problems of population growth in developing nations?

13. Rapid introduction of medical care in poorer countries has led to rapid population growth. In many cases, this population growth has, in turn, led to starvation and misery. Do you feel that people should stop sending medical aid to developing nations? Or should they send more aid? Discuss.

Suggested Readings

This chapter has introduced demographic techniques for analyzing population growth. There are several valuable texts available for those interested in further study of demography. Two excellent introductory texts are:

Peter R. Cox: *Demography.* 4th ed. Cambridge, England, Cambridge University Press, 1970. 469 pp.

Donald J. Bogue: *Principles of Demography.* New York, John Wiley & Sons, 1969. 899 pp.

More mathematical introductions are:

Mortimer Spiegelman: *Introduction to Demography.* Rev. ed. Cambridge, Mass., Harvard University Press, 1968. 514 pp. (Spiegelman includes an extremely large bibliography covering a wide range of topics related to population size, control, measurement, and so forth.)

Nathan Keyfitz: *Introduction to the Mathematics of Population.* Reading, Mass., Addison-Wesley Publishing Co., 1968. 450 pp. (This highly technical and mathematical text is especially careful in its presentation of interrelationships among various measures of population composition and vital rates.)

The population data for this chapter was taken from the following sources:

Philip Hauser: The census of 1980. *Scientific American, 245*(5):53, 1981.

W. Parker Mauldin: Population trends and prospects. *Science, 209:*148, July 1980.

U.N. Statistical Office: *Demographic Yearbook 1979.* New York, 1980.

U.N. Statistical Office: *Population and Vital Statistics Report.* New York, supplement update to January, 1982.

U.N. Statistical Office: *World Population Trends and Prospects by Countries: 1950–2000, Summary Report.* New York, 1979.

Gerald O. Barney: *The Global 2000 Report to the President of the United States.* New York, Pergamon Press, 1981. 360 pp.

For a discussion of the effects of crowding on rats, see:
John B. Calhoun: *Scientific American,* 206:139, February, 1962; and Environment and society in transition, *Annals of the New York Academy of Sciences,* June, 1971.

A book that presents a simplified introduction to demography and then deals with human values and humanity's place in the universe is:
Jonas Salk and Jonathan Salk: *World Population and Human Values: A New Reality.* New York, Harper and Row, 1982. 170 pp.

The Environment and Human Health

In places where food and medical attention are readily available, the average span of human life is longer than it has ever been. Modern medical science has drastically reduced death from infectious diseases, surgeons can repair or replace many damaged organs and joints, and even those who suffer from cancer are often cured or given years of relief. Yet at the same time there is increasing concern that the environment that we live in today—the air we breathe, the water we drink, the food we eat, even the sounds that assail us—endanger our health and our general well-being. This chapter will explore some of the issues of human health in our modern world.

10.1 Chemicals in the Environment—Cancer and Birth Defects

The human body is extremely complex. It has many systems and organs such as blood, bones, brain, heart, stomach, and kidneys, all of which must operate properly if the individual is to survive. Although all the mechanisms are still not precisely understood, scientists recognize that the development of these organs in the embryo and the maintenance of their functions in an adult are controlled by genes, which in turn are composed of molecules of DNA (see Box 5.2). In a discussion of heredity and regulation of body functions, biologists distinguish between **germ cells** and **somatic cells.** Germ cells are involved in reproduction—the sperm of the male and the eggs of the female. Somatic cells comprise the other parts of the body. The complete hereditary information, carried by DNA, is found in the nuclei of cells (Fig. 10–1).

Somatic cells from many tissues and organs are being replaced all the time. When you get a sunburn, the burned skin peels and falls off, and a new layer takes its place. DNA is responsible for directing the development and the replacement of our tissues so that skin cells produce more skin cells, bone marrow produces new blood cells, and so on. However, sometimes the internal regulatory processes break down. Then cells may start to reproduce in an unorganized, uncontrolled way. They spread beyond their

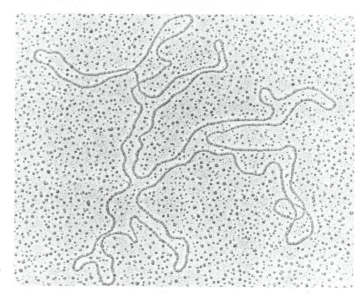

Figure 10–1 A circle of DNA from a bacterium. (Courtesy of Biophoto Associates.)

normal limits and invade other areas of the body. This abnormal growth is called **cancer,** and a clump of such cells is called a **tumor.**

No one knows precisely how cancer develops. Recent research indicates that viruses may be involved in some instances. Two types of environmental agents, chemicals and high energy radiation, have been shown to cause cancer. The link between certain chemicals and cancer was first recognized over 200 years ago in 1775 by a British scientist, Sir Percival Pott. He observed that chimney sweeps suffered an abnormally high incidence of cancer of the scrotum, and correctly deduced that chemicals in soot must be responsible.

Since that time, a great many other materials have been shown to increase the incidence of various cancers in a population. These correlations are so convincing that scientists are certain that exposure to certain chemicals leads to a higher probability that an individual will develop cancer. Substances that cause cancer are called **carcinogens.**

If chemicals can disrupt DNA in somatic cells, it should not be surprising that they can alter the DNA in germ cells as well. If the DNA molecules in germ cells are altered, the effect may be passed on to the following generation. Such changes in the genetic material in either parent are called **mutations,** and substances that cause mutation are called **mutagens.**

Although a carcinogenic chemical is not necessarily a mutagen, nor is every mutagen a carcinogen, studies have shown that some environmental chemicals do cause mutations. For example, certain chemical sprays used to control agricultural weeds contain very small concentrations of the chemical dioxin. Reports have indicated that, after these sprays have been used in certain areas, there has been an unusually high incidence of birth defects. The statistical evidence for a cause-and-effect relationship, however, is still incomplete.

Mutations occur naturally at the rate of about 1 in every 100,000 genes. Indeed, everyone alive on Earth carries some mutations. Mutations are the source of the genetic variation that, together with natural selection, accounts for the evolution of species. Why, then, are people concerned because environmental factors produce additional variation? Concern arises because some mutations are the cause of harmful genetic defects that lead to many debilitating conditions. For example, hemophilia may cause its victims to bleed to death from trivial cuts; Tay-Sachs disease in children may lead to death from nerve degeneration (see Fig. 10–2); and phenylketonuria and Down's syndrome produce mental retardation.

Radiation has also been implicated as a cause of cancer and mutations. This subject will be discussed in Section 10.6.

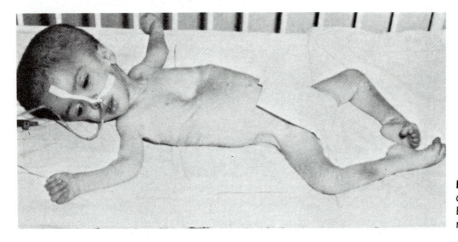

Figure 10–2 A three-year-old child with Tay-Sachs disease. Enlarged head and atrophied muscles are characteristic.

10.2 What Is a Poison? What Is a Toxic Substance?

In popular terms, a poison is a substance that can destroy life rapidly, even when taken in small amounts. However, our concern covers a wider range of health effects, not merely death. Effects that are slow or delayed, such as cancer, or mutations that are not observed until later generations, also represent serious problems. This chapter will therefore refer to the wider category of **toxic substances,** which include any substance whose physiological action is harmful to health. Adverse health effects include lung and heart disease, kidney damage, skin rashes, cancer, birth defects, reproductive malfunction (such as sterility), and nerve and behavioral disorders as well as acute poisoning. There may also be more subtle effects that are difficult to detect such as mental depression or lowering of intelligence.

Still, substances cannot be divided sharply into two categories, "toxic" and "nontoxic." The difficulty is that "toxic" is a relative term, which depends on several factors:

1. **It Depends on the Route of Entry.** There are four ways for a substance to invade your body: It can be inhaled; it can be ingested (eaten or drunk); it can be absorbed through the skin; or it can be forced into the blood stream, as with a snake bite or a puncture wound. For example,

if you swallowed some mercury from a broken thermometer, it would not be particularly harmful; the mercury would simply pass through your digestive system and be eliminated in a few days. But if you were sealed in a room in which a pot of mercury was being kept warm on a stove, the mercury vapors you inhaled would rapidly poison you. Some substances are harmful no matter how they invade the body. Benzene, for example, will poison you if you drink it or if you breathe its vapor, and if the liquid touches you it can pass right through your intact skin.

2. **It Depends on the Dose.** Obviously, the more poison you absorb or ingest, the more serious are the effects. The amount of toxic material that enters your body is called the **dose.** The relationship between the dose and the health effect can be plotted on a graph (Fig. 10–3). However, it is no simple matter to acquire the information from which such a graph can be plotted. The next section explains how such data are obtained. Here the meaning of the graph will be considered. First, note the vertical axis labeled "Effect on health." Since there can be different kinds of effects, the data can be plotted in different ways. Now look at the horizontal axis labeled "Dose." You have just learned that there are different routes of entry into the body. Therefore, the route of entry must be specified. The shape of the graph depends on all of these factors as well as on the nature of the toxic substance. For example, the effects of some poisons

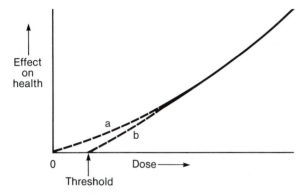

Figure 10–3 *Generalized function showing the relationship between the dose of a toxic substance and its effect on health.*

become sharply worse as the dose increases; the graph would then curve upward.

But the most uncertain (and controversial) parts of the graph are the two dashed lines *a* and *b*. Both of these lines represent the health effects *when the dose is low*. Here is where it is most difficult to get reliable information. For instance, a graph may show data on the carcinogenic effects of dimethylnitrosamine (DMN), a chemical found in salami, bacon, and frankfurters (up to 440 parts per billion [ppb]) as well as in many varieties of fish (up to 40 ppb) and dairy foods (up to 10 ppb). Experiments with animals reveal that DMN can cause liver cancer. But the experiments use high doses of DMN. Common sense and ordinary experience tell us that if we eat a frankfurter or salami sandwich today, we are not likely to get liver cancer tomorrow. Isn't it possible that the body can handle small amounts of DMN without any danger at all and that only an *excessive* amount of DMN is a threat? Yes, it is possible, and that possibility is represented by the dashed line *b*. This curve shows that below a certain dose there is no effect at all on human health. The dose at which the danger starts is called the **threshold.** Above the threshold, the hazard increases as the dose increases.

But there is another possibility, represented by the dashed line *a*. In this case, even the tiniest amount of the substance could have an adverse effect on health. What if even one molecule of DMN could change a normal cell into a

cancer cell? The chance would be small, but not zero. For the danger to be zero, the dose would have to be zero, and that would mean that there is no such thing as a threshold dose.

Scientists know that a threshold does exist for many chemicals. In fact, some substances that are essential to life in small amounts are toxic in excess. But for other chemicals, the threshold vs. no threshold question is a difficult one and is often subject to much controversy. You can see why. If a threshold exists, then a "toxic" chemical is no danger to anyone if the dose is small enough. Therefore it would be wasteful to try to remove all of it from our food, air, or water. The argument could be pressed even further to say that the best way to remove the danger of a toxic chemical is to disperse it so widely in the environment that the "dose" anywhere is below the threshold. This doctrine is summarized in the often cited short sentence, "Dilution is the solution to pollution." But, on the other hand, if there is no threshold, then this argument is invalid, and any trace of an environmental contaminant poses a threat to health.

3. **It Depends on the Susceptibility of the Individual.** Individuals are different from one another, not only in personality and appearance but also in the chemistry of their bodies. For any specific poison, some people are more resistant than others. For a given dose, some may get sick, some may die, others may not be harmed.

4. **It Depends on Other Factors.** Two different toxic substances may produce much worse effects when they are combined. For example, the toxic effects resulting from inhalation of trichloroethylene (a dry-cleaning solvent) are increased if the individual also drinks alcohol. It has been shown that alcohol interferes with the ability of the body to get rid of trichloroethylene. This enhancement is called **synergism,** which is a way of saying that the combined effect of the two chemicals is greater than the sum of the effects of the two components taken individually. Another example of synergism is the combination of smoking and exposure to asbestos fibers, which results in a greatly increased probability of cancer compared with the risk of either smoking or asbestos exposure alone.

10.3 How Are Toxic Substances Tested?

In practice, it is often quite difficult to determine the toxicity of a particular substance. Acute poisoning is relatively easy to document, but it is much more difficult to determine the effects of low doses. After all, there are hundreds of thousands of different chemical compounds present in our environment; how can the effect of a small dose of one of them be identified?

(a) Epidemiology. **Epidemiology** is the study of the distribution and determination of health and its disorders. A frequent objective of epidemiology is to establish cause-and-effect relationships. For example, one way to test a suspected toxic substance is to separate a population into two groups, one that is exposed to the chemical and one that is not. A classic example of this type of study is analysis of the effects of cigarette smoke. Since there is a reasonably clear distinction between people who smoke cigarettes and those who do not, it is relatively easy to ex-

amine the medical records (especially the death certificates) of large numbers of people in these two distinctly different groups. The knowledge that cigarette smoking is harmful comes from many studies of this kind, involving records from hundreds of thousands of subjects (Fig. 10–4).

Now suppose that the effects of a certain widely used pesticide are being studied. If this material is dispersed throughout the global environment, it will be impossible to find a group of people who have had zero exposure to it. In such a case it is impossible to compare health records of those who are exposed to the pesticide with those who are not, and therefore epidemiological studies are much more difficult to interpret and the conclusions are more tentative. One approach to this problem is to compare people with high levels of exposure with those who have lower levels of exposure. In general, workers in industry are in contact with higher concentrations of certain materials than the general public. In a classic example of an industrial study, one quarter of the male employees who worked in a coal tar dye factory in the United

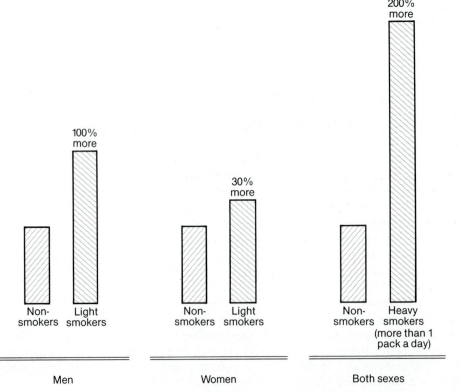

Figure 10–4 The relationship between cigarette smoking and risk of death by cancer.

States between 1912 and 1962 contracted bladder cancer. This study led to the conclusion that several of the chemicals in the coal tar dyes were carcinogenic. Unfortunately, not all companies are willing to release such data, because the information may lead to lawsuits or may generate adverse publicity. Another problem with this approach is that the number of workers affected is often small, and the results may therefore be ambiguous.

Another approach is to survey the geographical incidence of a certain disease. Figure 10–5 shows the probability of dying of cancer of the large intestine in relation to place of residence in the United States. This graph shows that a person living in the Northeast has a higher risk of dying of intestinal cancer than does someone living in the Southwest. These data suggest that regional pollutants may be associated with cancer, but the suggestion does not constitute proof. Several other factors must be considered as well. Perhaps people in the Northeast smoke more cigarettes or get less exercise than people in other parts of the country. Perhaps they eat a different diet. Moreover, the ethnic distribution

in the United States differs by location, and different ethnic groups have been shown to have different susceptibilities to cancer. Thus no unequivocal conclusions can be drawn.

Another reason why it is difficult to prove a direct relationship between an environmental substance and cancer is that cancer may not appear for 10 to 20 years after exposure to a harmful chemical. For example, a high percentage of people who are exposed to asbestos fibers eventually die of lung cancer, but acute symptoms do not usually occur until at least 20 years after contact with this substance. Yet, when a new substance is released into the environment, people want information about its toxicity immediately, not a generation later after deaths have occurred.

(b) Laboratory Studies. Because of the inherent problems of epidemiological studies, laboratory studies of toxicity are often performed as an alternative approach. In a few instances, laboratory studies using people are carried out, but only with small doses that may produce only slight effects, such as the onset of discomfort or

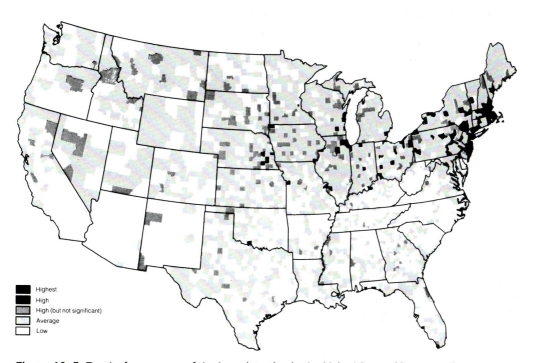

Figure 10–5 Deaths from cancer of the large intestine in the United States. (Courtesy of Dr. William Blot and *The Sciences.*)

irritation. Of course, acute toxicity studies with humans are immoral. But even if such wrongful acts were contemplated, the results would not be helpful in cancer studies, since the onset of cancer may be delayed by a decade or two.

The most expedient alternative is to experiment on animals. The usual choices are small animals (mice, rats, guinea pigs, cats, dogs, and small monkeys). Of course, a rat is not a person, but it will dine quite happily on the food that people eat and can be poisoned by chemicals that are harmful to people. Scientists can't wait 20 years to see whether a rat that eats a synthetic chemical will get cancer. For one thing, the rat doesn't live that long. So scientists do to the rat what they could never do to a human being— they feed the rat a concentrated diet of the suspected chemical. If the rat gets cancer, the chemical is considered a carcinogen for rats and a suspected carcinogen for humans.

The argument against such tests is that they are too exaggerated. No one eats concentrated amounts of synthetic chemicals, so how do the rat experiments apply to humans? The answer to this objection takes the following form: Suppose that the possible carcinogenicity of an additive used in a soft drink—perhaps a synthetic color, flavor, or sweetener—is being tested. Imagine that the additive is fed to the rats in such concentrated form that a human would have to drink 10,000 cans of soda per day to get the same equivalent dose. In this experiment, one rat out of every five develops a cancerous tumor (Fig. 10–6). Then the dose is reduced to the equivalent of 1000 cans of soda per day for a person, and on this dose one rat in every 50 develops a tumor. At a dose equivalent to 100 cans of soda per day per person, one rat in 500 gets a tumor.

Now the research is becoming more difficult. The experiment is using too many rats, and the dose is still too high—no one drinks 100 cans of soda per day. Ultimately, the important question is whether it is dangerous to drink *one* can of soda per day. The data obtained thus far are plotted in Figure 10–6. As plotted, the three points lie on a straight line. If the straight line is continued down to the one can per day equivalent dose, there would be a 1 in 50,000 chance that a rat would develop a tumor. What does this number mean? Let us make two assumptions: (1) It is valid to continue the straight line as shown. This assumption is called "linear low-dose extrapolation." (2) After correction for differences in body weight, the chances of a tumor in rats and in people are equal. Both of these

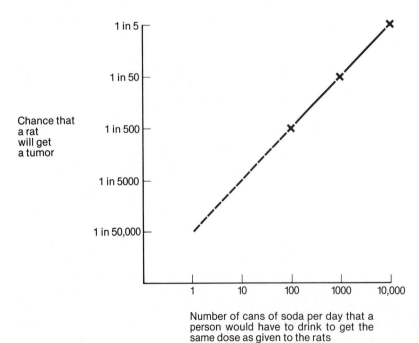

Number of cans of soda per day that a person would have to drink to get the same dose as given to the rats

Figure 10–6 An attempt to extrapolate a laboratory study on rats to the probability that a food additive will cause cancer.

assumptions are really guesses. They may or may not be true. If they are true, we can calculate the effect of this additive on a population. There are approximately 225,000,000 people in the United States. If everyone drank one can of soda per day, and the probability of developing a cancerous tumor from this dose was 1 chance in 50,000 for each person, the results would be $^{225,000,000}/_{50,000}$ or 4,500 human cancers in each generation from this exposure.

No one really knows whether such arithmetic makes sense. This uncertainty forms the basis of a difficult dilemma. Decisions must be made on many levels—international, national, and personal—yet in the absence of definitive knowledge, these decisions must be based on opinions and educated guesses.

As of 1983, there were fewer than 30 *known* human carcinogens. Many more substances, however, have been shown by laboratories to be carcinogenic to animals. All of these animal carcinogens are considered *suspected* human carcinogens. Two reasons support such suspicions: (1) Several of the known human carcinogens were predicted from animal studies; and (2) No animal carcinogen has ever been shown to be noncarcinogenic in humans.

Data on toxic substances are maintained by the following organizations:

- Registry of Toxic Effects of Chemical Substances, National Institute of Occupational Safety and Health, Cincinnati, Ohio.
- Toxicology Data Bank, National Library of Medicine, Washington, D.C.
- International Register of Potentially Toxic Chemicals, United Nations Environmental Program, Geneva, Switzerland.

10.4 Chemicals in Foods

Representatives of the chemical industry remind us that the words "chemical" and "chemical substance" refer to *any* kind of matter. "Chemicals" therefore include the chlorophyll in a leaf of spinach, the protein in a lamb chop, and the salt in the sea. Everything is a chemical. Since this is true, why worry about chemicals in food? You might answer that you are concerned

---

BOX 10.1 THE AMES TEST

A laboratory test of the possible carcinogenicity of a substance using mice or rats takes about three years and costs over $200,000. It has therefore been recognized that a faster, less costly screening test is needed. A very useful procedure was developed in the 1970s by Bruce Ames and his coworkers at the University of California (Berkeley) that can be done in three days for a very modest cost. The method utilizes a particular strain of bacteria that is very sensitive to mutation. These bacteria are exposed to the chemical to be evaluated, and the number of mutated bacterial colonies is then counted. The Ames test detects mutagens, not carcinogens, but about nine out of ten carcinogens are mutagenic. Therefore, a chemical that is shown by the Ames test to be mutagenic can be classified as a suspected carcinogen and can then be studied more thoroughly.

---

about environmental problems caused by *synthetic* chemicals. In response, the industrial chemist could point out that the properties of a chemical have nothing to do with where it came from—whether from natural or from manufactured sources. Vanillin, for example, is the chemical that imparts a vanilla flavor to food. Vanillin can be extracted from vanilla beans or it can be synthesized. It is the same substance either way. Vanillin is vanillin, so why distinguish between natural and synthetic vanillin? Furthermore, the chemist continues, not all natural substances are good and not all synthetic ones are bad. Cobra venom and the aflatoxin found on moldy peanuts are both natural substances. But the venom is a deadly poison and the aflatoxin is a potent carcinogen.

Many natural foods are believed to cause specific diseases. For example, saturated fats may contribute to hardening of the arteries and consequent heart failure, sugar leads to obesity and dental caries, and too much alcohol can cause cirrhosis of the liver. Studies have also indicated that spinach, cabbage, and charcoal-broiled meats contain measurable quantities of some carcinogenic chemicals that are also found in automobile exhaust and in tobacco smoke.

These arguments are well presented, yet they leave a sense of unease. The list of ingredients on the label of a loaf of "enriched bread" may make you wonder why all those ingredients are needed and how they affect your body.

Almost every food available in modern supermarkets contains small amounts of chemicals that are not natural to the food itself (Fig. 10–7). Fruits and vegetables contain pesticide residues. (Pesticide pollution will be discussed in Chapter 15.) Cattle are fed artificial growth compounds before they are slaughtered. Some of these chemicals remain in the meat when it is sold. Almost all pre-prepared "convenience foods" contain a variety of additives.

There are several different categories of food additivies.

1. Preservatives are used to prevent spoilage.
2. Vitamins and minerals enhance the nutritional value of foods.
3. Artificial colors, flavors, and sweeteners alter the taste or appearance of various products.
4. Other additives, such as emulsifiers, are used to blend foods. For example, the oil in natural peanut butter separates and floats to the top of the jar. Emulsifiers are added to peanut butter so that the oil will not separate out.
5. Many chemicals are added to animal feed to encourage the production of meat, milk, or eggs. Similarly, use of pesticides and herbicides leads to chemical contamination of plant products.

Figure 10–7 Most packaged foods contain a variety of additives.

TABLE 10–1   **Some Commonly Used Classes of Food Additives**

Additive	Purpose	Common Compounds Used
Preservatives	Retard spoilage caused by bacterial decay and growth of molds	Benzoic acid, calcium propionate, citric acid, potassium sorbate, salt, sodium benzoate, sodium nitrate, sodium nitrite, sorbic acid, sulfur dioxide
Antioxidants	Maintain freshness by retarding chemical oxidation	Butylated hydroxyanisole (BHA), butylated hydroxytoluene (BHT), propyl gallate
Nutritional supplements	Add nutrients or replace those lost in food processing	Minerals, vitamins, and certain amino acids
Coloring agents	Alter the color of foods to make them more appealing and also identify products for advertising purposes	Many compounds, natural and synthetic
Flavoring agents	Enhance flavor or identify a product for advertising purposes	More than 1500 different chemicals, both natural and synthetic. Two of the most common synthetic ones are monosodium glutamate and saccharin
Emulsifiers	Disperse droplets of one liquid in another liquid (used for example, to disperse oil in water)	Lecithin, polysorbates, mono- and diglycerides, propylene glycol
Stabilizers and thickeners	Change texture, produce smooth consistency	Dextrin, gelatin, seaweed extracts, vegetable gum

The use of all these chemicals has risen in response to modern ways of life (Table 10–1). Homemakers no longer bake bread every morning; instead, many shop once a week. Since bread without preservatives may go moldly in a week, chemical preservatives are added to reduce the possibility of spoilage. Similarly, many people do not have time to cook an elaborate breakfast before work. So they pour some cereal out of a box, add milk, and eat an instant meal.

The vitamins and minerals in the packaged foods add to these people's dietary intake. Hormones added to animal feed speed growth and reduce the cost of many foods. Other additives do not improve the quality of the food. For example, natural peach ice cream is a dull pink. However, with the simple addition of some coloring, a bright, almost "day-glo" color can be achieved. The color does not improve the taste, flavor, or nutritional value of the ice cream. But

BOX 10.2 SHORT NOTES ON FOUR ENVIRONMENTAL CARCINOGENS

Tobacco Smoke

The relationship between tobacco smoke and cancer has been established beyond any reasonable question. The effects of smoking are described in Section 21.4. Tobacco smoke is a mixture of gases, liquid droplets, and solid particles, all of which contain complex mixtures of a great many different chemicals. No one knows exactly which combinations of chemicals are responsible for which specific diseases. Cigarette filters do *not* remove gases, such as carbon monoxide and nitrogen dioxide, and they only partially remove droplets and solid particles. There is a relationship between tar and cancer, so that a low-tar cigarette is preferable to a high-tar one, although not enough is known to be able to say how much better it is. However, smokers of low-tar cigarettes (which are also low in nicotine) often smoke them down to a shorter butt in an unconscious effort to make up for their nicotine deprivation. Since the tar accumulates at the butt end, such smokers inhale a larger proportion of the tar content of their cigarettes.

Asbestos

About 5 to 10 percent of all workers employed in asbestos manufacturing or mining operations die of mesothelioma, an otherwise rare form of cancer of the membranes lining the chest and abdominal cavities. There are about 1100 new cases of mesothelioma each year. Of course, asbestos workers die of other forms of cancer as well, and if they are smokers, their risk of cancer is much higher than would be expected from a simple combination of the two exposures. The risks also extend to the families of these workers, presumably because of asbestos dust carried on clothes and in workers' hair.

Nitrosamines

Nitrosamines were referred to in Section 10.2 in the discussion of threshold effects. These carcinogenic chemicals appear in food as a result of a series of chemical reactions that start with common, naturally occurring substances. One reaction starts with nitrates, such

as sodium nitrate, which are common chemicals in soil and plants. Nitrates are also used for curing meats, such as those sold in a delicatessen, and fish. Nitrates are reduced to nitrites by various microorganisms; therefore, nitrites also occur naturally in many foods. A second reaction starts with amines, which are another class of nitrogen compounds related to ammonia. Amines also occur naturally in many foods, including fish, dairy products, various fruits and vegetables, cereals, wine, beer, and tea. Some amines react with nitrites to produce nitrosamines, which are carcinogenic. What is one to do? One cannot cut out all the foods that are involved in the sequences of reactions leading to nitrosamines. As noted in Section 10.2, however, cured meats such as frankfurters contain up to about 440 ppb of dimethylnitrosamine; foods not cured by nitrates contain much less. A reasonable personal choice, then, would be to eat mostly fresh or fresh-frozen fish and meat, rather than that which has been cured.

Diethylstilbesterol (DES)

DES is a synthetic hormone that, when fed to cattle, increases their rate of growth. As a result, in the 1960s DES was fed to about three fourths of all the cattle raised in the United States. DES was also prescribed for pregnant women to reduce the chances of miscarriage. However, DES was shown to be carcinogenic in mice and is also linked to vaginal and cervical cancers in women. (Women born in the years from 1950 to 1966 whose mothers took DES during their pregnancies are at increased risk of developing a vaginal growth. Any such family history should be reported to a physician.) DES was therefore banned by the FDA in 1971. However, the federal courts reversed this ban on the grounds that DES was fed to cattle, not added directly to foods. The FDA countered that measurable quantities remain in meat after the cattle are slaughtered, and the ban was reinstated. Today, the entire issue of legislation of growth additives used in meat production is still not completely resolved.

BOX 10.3 DIET AND CANCER

The Commission on Life Sciences of the National Academy of Sciences issued a 496-page report on *Diet, Nutrition, and Cancer* in June, 1982. According to the report, there is strong evidence that what we eat affects our chances of getting cancer, especially particular kinds of cancer. It is not yet possible to predict *how much* cancer is linked to diet. The report does recommend some dietary changes, however, as summarized below.

Recommendation	Expected Benefit	Recommendation	Expected Benefit
1. Reduce your consumption of both saturated and unsaturated fats. (The present average U.S. level of dietary fat is 40 percent of the total calories. A suggested decrease is one fourth of this intake, down to a level of 30 percent.)	Lowered incidence of certain cancers, particularly of the breast and the colon.	3. Limit your consumption of salt-cured, salt-pickled, and smoked foods, such as bacon, frankfurters, and smoked fish.	Reduced incidence of cancers of the stomach and esophagus.
2. Include fruits, vegetables, and whole-grain cereal products in your daily diet. Especially include citrus fruits and vegetables from the family of cabbage, broccoli, and brussels sprouts.	These foods contain Vitamins A and C. They also contain some compounds that, although non-nutritive, appear to inhibit chemically induced cancers. Therefore, dietary supplements such as vitamin pills are not a substitute for these foods.	4. Try to limit your intake of food additives.	The evidence here is not firm. The data are insufficient to show that any individual food additive is a *major* contributor to the risk of cancer. The combination of these chemicals, however, may pose significant risks.
		5. Alcoholic beverages should be consumed only in moderation, if at all. The combination of alcohol and smoking should be avoided.	Reduced incidence of cancers of gastrointestinal and respiratory tracts.

advertising companies have conditioned us to expect highly colored foods and to buy their clients' products.

There are two levels of decisions to be made about food additives. The first is a political one, the second is on a personal level. One legitimate role of government is to protect the health of its citizens. As a result, most nations have established various agencies to monitor the quality of foods. In the United States, the Food and Drug Administration (FDA) was first established in 1906 to set guidelines for the safety of various foods and drugs. In the beginning, the law was not focused primarily on chemical additives. In fact, before 1958 a new chemical could be released on the market *before* it was tested for toxicity. The law was then changed to require extensive tests on each new chemical additive or drug before it is released on the market. But what about those products in use before 1958?

Are they safe or not? It costs upward of $1 million and several years' work to test a single product completely, and the cost and time required to test all the products traditionally used in food preparation seemed prohibitive. As a result, in 1958 FDA officials made up a list of all the chemicals that were at that time added to foods and sent the list to a panel of several hundred experts. Each expert was asked to choose, on the basis of professional opinion (not direct research), which chemicals were safe and which were suspect. After analyzing the reports, the FDA made up a list, called the **GRAS** list, of those compounds that were "Generally Recognized as Safe."

Another important piece of legislation relating to food additives is called the Delaney Clause, after its author, Representative James Delaney. This regulation states that the FDA is required to ban any food additive that has been

argument may be, well, it's only one glass, and besides, the risk seems too far off to worry about. Even if you do drink one glass every day, the danger is that there *might* be a 1 in 50,000 chance that you *might* get cancer *some* day. So you accept the soda. But now look at it another way. Why risk even a small chance of a premature death for a soft drink? If you don't think this advantage is worth the risk, you don't drink the soda.

10.5 Hazardous Wastes

Food additives are of particular importance because we ingest them. But a great many other compounds are present in our environment. Some of these are benign, and others are deadly poisons. The poisons are permitted because they serve useful purposes and are not intended for human consumption. Yet any product manufactured in quantity will inevitably be dispersed into the environment.

As an example of this type of problem, consider a class of chemicals known as the polychlorinated biphenyls (PCBs for short). PCBs are organochloride chemicals that are structurally similar to the pesticide DDT (see Appendix D). Their unique electrical insulating properties make them useful in the manufacture of transformers and other electrical components. They have also been used in the production of plastic food containers, epoxy resins, caulking compounds, various types of wall and upholstery coverings, and as ingredients in soap, cosmetic creams, paint, glue, self-duplicating ("no-carbon") paper, waxes, brake linings, and many other products. In the late 1960s scientists began to realize that PCBs were also serious environmental poisons. In 1968 the accidental contamination by PCBs of some cooking oil in Japan caused several thousand people to suffer from enlarged livers, disorders of the intestinal and lymphatic systems, and loss of hair. In New York City, workers in an electrical factory that used PCBs complained of similar ailments. Laboratory studies have shown that PCBs interfere with reproduction in rodents, fish, and many species of birds and monkeys. They are also suspected of being carcinogenic, but conclusive evidence is lacking. Moreover, PCBs are not easily biodegradable. Because these chemicals are en-

shown to cause cancer in animals or in humans. The Delaney Clause has been both praised and condemned. Supporters claim that it protects our health and well-being. Critics claim that a product that causes cancer in large doses in rats may be perfectly safe in small doses in people, and therefore the law is unnecessarily harsh.

Using the Delaney Clause, the FDA has banned several chemicals that were originally on the GRAS list, including cyclamate sweeteners, certain vegetable oils, and several food dyes, including the controversial Red Dye No. 2 (see Box 10.4). As of early 1982, several other additives were undergoing scrutiny. In 1978, the FDA discovered that an alternate artificial sweetener, saccharin, caused cancer in laboratory animals. However, under heavy pressure from the food industry, Congress suspended the ban on this additive pending the establishment of a more definite link between laboratory tests and human health.

When experts disagree, how can you as an individual know how to act? Well, you can't know for sure, so you have to make certain value judgments. You could think about it in the following way. Suppose you were seriously ill, near death, and your doctor prescribed a medicine to cure you. Say the medicine works 99 percent of the time, but there is a 1 percent chance that there might be some bad side effects. You would probably take the medicine because the benefit (99 percent chance of cure) is well worth the 1 percent risk. In the case of the soft drink, the risk is much less—in our example we assumed 1 in 50,000, which is only two thousandths of 1 percent. But the benefit is also much less—after all, you could drink water or orange juice. So what do you do? If you are offered a glass of soda, should you drink it? One

vironmentally persistent, small quantities that have been accidentally spilled into waterways or that have leaked onto the ground may be dispersed throughout the environment. Dispersal is facilitated by the fact that PCBs are soluble in the fat of animals and are stored in living tissue.

In the early 1970s, PCBs were found in cow's milk, many inland and deep sea fish, most meats, and in people's bodies as well. Their persistence means that the contamination will continue far into the future. Even when PCBs do decompose, problems will continue, because evidence indicates that the decomposition products are even more poisonous than the original material.

As evidence of the harmful effects of PCBs mounted, chemical manufacturers voluntarily curtailed their production and use, so that by 1972 applications of PCBs were restricted to the manufacture of electrical components. Monsanto Chemical Corporation, the last manufacturer of PCBs in the United States, stopped all production in 1977. Yet it will require decades for all these compounds to be removed from the natural environment.

PCBs are an example of a product that enters the environment mainly through accidental spillage. Another serious problem arises from the disposal of chemical wastes. Chemical wastes are produced in a great many industrial processes. A chemical change rarely produces only the desired product. In most cases there is no way to avoid some unwanted materials, or byproducts. Sometimes an entire batch goes wrong because it overheats or was not mixed correctly.

The resulting wastes may be solids, liquids, or gummy, gooey mixtures that are messy to handle. They may contain compounds of chlorine, bromine, or sulfur, which would generate air pollutants if they were burned. Many of the wastes are very toxic; some are deadly. They cannot (or should not) be mixed with ordinary domestic or commercial garbage. They certainly must not be emptied into any river or lake, nor into the ocean. They can be destroyed or converted to harmless products by incineration followed by liquid scrubbing (see Chapter 21) or by other costly chemical processes. But the easiest way to dispose of them is to pack them in 55-gallon steel drums (the "garbage cans" of the chemical industry) and store them somewhere.

BOX 10.5

Another illegal and sneaky way to dispose of toxic wastes is to mix them with home heating oil, diesel fuel, or gasoline. Such adulteration makes it more hazardous to handle the fuels and adds to air pollution when the fuels are burned.

As these drums accumulate, often in hundreds or thousands, a "toxic waste site" is born.

The story of the discovery of one of these sites is an interesting one. The Hooker Chemical Company is a manufacturer of chemicals with its main plant located in Niagara Falls, New York. Early in the 1940s Hooker purchased an old, abandoned canal called **Love Canal.** Employees loaded many of the chemical wastes into 55-gallon steel drums and dumped approximately 19,000 tonnes of these wastes into the canal. In 1953, the company covered one of the dump sites with dirt and sold the land to the Board of Education of Niagara Falls for $1. The deed of sale stated that the site was filled with "waste products resulting from the manufacturing of chemicals." The deed also specified that Hooker would no longer be responsible for the condition of the land. The site was then used for an elementary school and a playground. But steel drums, exposed to moist soil from the outside and often strong acids or bases from the inside, cannot remain intact indefinitely. They rust from the outside, corrode from the inside, and eventually leak. The chemicals then seep into the soil and may travel through groundwater systems into rivers, streams, lakes, and reservoirs.

In the spring of 1977, heavy rains raised the level of the groundwater and turned the region into a muddy swamp. But this was no ordinary swamp. Mixed with the water and dirt were thousands of poisonous chemicals. The chemical goo floated about the playgrounds, flowed into people's basements, and settled over flower gardens and lawns (Fig. 10–8). Many of the children who attended the school and many of the adults who lived nearby suffered serious illnesses. Epilepsy, liver malfunctions, miscarriages, skin sores, rectal bleeding, severe headaches, and birth defects were reported.

Investigators learned that the drums, buried decades before, were leaking. A study revealed

Figure 10–8 *A,* Steel drums full of poisonous chemicals were buried near the city of Niagara Falls, N.Y. The landfill was then covered, leveled, and the property sold to the city as a site for a school playground. However, the drums rusted, and some of the liquids leached to the surface. Many children contracted strange and serious diseases. This photograph shows a pool of rainwater polluted with industrial chemicals from the dump beneath the surface. (Wide World Photos.) *B,* A school child's opinion of the problems in Niagara Falls, N.Y. (Wide World Photos.)

that several other similar dumps were scattered about the city. Practically overnight, a new and serious environmental concern was brought to the attention of the public. Experts argued and disagreed. Some said that the Love Canal issue was being blown out of proportion because it was impossible to prove that the illnesses in the Niagara Falls area were directly caused by chem-

ical poisoning. Many others disagreed and were alarmed. A number of scientists believed that the diseases caused by the chemical spills were severe and that even greater disasters could follow. For example, an estimated 1000 kg of wastes contaminated with dioxin have been buried in the area. Dioxin is one of the most deadly poisons known; approximately 75 g (0.075 kg) dissolved in drinking water could kill 1 million people. In 1978, the state of New York spent $37 million to relocate the families who lived closest to the Love Canal dump. Two years later, President Carter declared the region a Federal disaster area and released an additional $30 million to aid families in the area. Hooker Chemical was faced with some $2 billion in lawsuits. The legal issues are complex because it is difficult to *prove* that the released chemicals *caused* illnesses. Therefore, the final outcome is uncertain.

Government experts believe that there may be 14,000 dangerous chemical dumps in the United States. No one is exactly sure how to find them all and how to remove the hazards. But concern has been great enough so that in December 1980, the United States Congress passed a law commonly known as the Superfund Legislation.* This law provides a $1.6 billion emergency fund to deal with the problem of hazardous wastes. Under old laws, if a hazardous dump site is discovered, it may take months or years to find the guilty parties and force them to remove the dangerous chemicals. Under the new law, government officials can use money from the Superfund to respond immediately to the situation and clean up the dump site. Then, if the polluters can be found, the government can sue them and retrieve the cost of the clean-up operation. The bill contains no provision for corporate liability for damages from toxic wastes. In other words, if you are poisoned by a dump site in your neighborhood, the federal law provides no assistance; you must battle out your claim in the state courts. Under the 1980 law, nearly $1.4 billion of the $1.6 billion fund would come from a tax on oil and certain chemicals. The remaining $200 million would come from the general budget.

---

* It is formally called the Comprehensive Environmental Response, Compensation, and Liability Act. Other federal laws related to toxic chemicals are listed in Section 2.2.

As with most environmental issues, the story of hazardous wastes is not complete. By the spring of 1982 critics of the Superfund complained that the "quick clean-up of hazardous waste sites" was going too slowly. The main activity of the Environmental Protection Agency (EPA) up to that time had been to prepare a list of the most hazardous sites and to develop a ranking system to establish priorities for clean-up. Only 10 actual clean-up operations had been undertaken.

10.6 Biological Effects of Radiation

Radiation is a generic term for any oscillating electromagnetic disturbance. Such radiation has been grouped into various categories, including radio waves, microwaves, infrared, visible light, ultraviolet, X-rays, and gamma rays. All electromagnetic waves travel at the speed of light. They differ from each other in frequency and wavelength (Fig. 10–9). Most important, they differ in energy—the higher the frequency, the more energetic the radiation. Furthermore, the more energetic the radiation, the more damage it can do to living organisms.

Microwaves

Microwaves are low in frequency and therefore carry little energy, far too little to destroy molecules or alter genes. However, it is known that high doses of low-energy radiation can affect biological tissue. Microwave ovens are used to cook many types of food. The microwaves cause the atoms and molecules in the food to rotate rapidly; molecular motion is heat, and, presto, dinner is ready.

But what about much lower dosages of radiation? If you stand in a busy airport, microwaves from hundreds of different radar transmitters are absorbed by your body. Do they have an effect? Soviet scientists report that low levels of microwave exposure lead to headaches, eye strain, weariness, dizziness, irritability, emotional instability, depression, diminished intellectual capacity, partial loss of memory, and loss of appetite. They believe that the electromagnetic energy may somehow interact with the normal functions of the brain. Many investigators argue that these reports cannot really be verified because symptoms such as "weariness" and "irritability" are subjective. The criticism continues that since it is impossible to *prove* that someone is irritable, the study cannot be fully substantiated. However, direct and reproducible evidence that microwave radiation can affect the

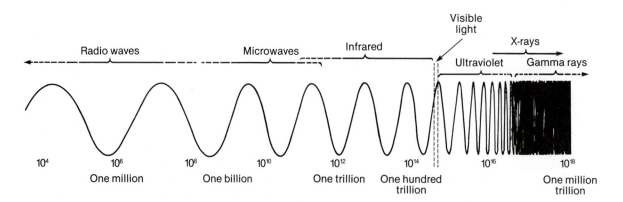

Frequency in cycles per second

Figure 10–9 The electromagnetic spectrum is a continuous range of radiation ranging from low-energy waves to high-energy waves. The names of the various sections are used for descriptive purposes only. All waves have similar properties and travel at the same speed; they differ only in frequency and wavelength.

brain is available. When electrodes were implanted in the brains of cats, and the animals exposed to low levels of microwaves, many parts of the brain, including the region where emotions are believed to originate, responded measurably. At the present time, there is no conclusive evidence that microwaves in the density normally present in the environment can affect human behavior. There is enough uncertainty about this relationship, however, to warrant caution.

High-Energy Radiation: Effects on Living Cells

Electromagnetic radiation in the high-frequency portion of the spectrum (Fig. 10–9) can damage living cells. In particular, X-rays or gamma rays are energetic enough to knock electrons away from their atoms or molecules. Ions are thereby produced. If certain key molecules in a living cell are ionized, cellular function may be disrupted and the cell may die. Other effects of radiation may fall short of producing ions but may nonetheless alter or break chemical bonds. The affected molecule may be essential to the cell. For example, DNA is a sensitive target for radiation; when a cell is irradiated, the DNA strands tend to break into fragments. If the rate of delivery of the radiation is low, the cell's repair mechanisms can seal the breaks in the strands, but above a certain dose rate, the repair process cannot keep up, and the DNA fragmentation becomes irreversible.

Early Somatic Effects: Radiation Sickness. On several occasions during the past 75 years, groups of people have been exposed to large doses of ionizing radiation over periods of time ranging from a few seconds to a few minutes. The holocausts at Hiroshima and Nagasaki, together with accidents at civilian nuclear installations, have provided much information about what radiation can do when a lot of it is administered to the entire body over a short period of time. Consider first the simplest and most drastic measure of radiation effect—death. Figure 10–10 shows the relation between the dose administered to a population of animals and the percentage of the population remaining alive three weeks or more after the exposure. Up to a dose of about 350 rads (see Table 10–2 for definition of a rad), vir-

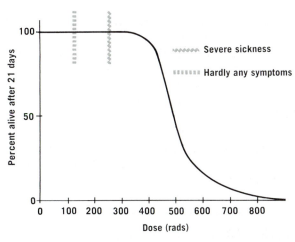

Figure 10–10 Curve showing the approximate relationship between the dose of radiation administered to a whole animal (such as a mouse or a man) and the percent of the treated population that survives three weeks afterward. Mice have, of course, been intensively studied in the laboratory; accidents in industry and the nuclear explosions in Japan have provided the approximate data for humans. (From *American Scientist* 57:206, 1969. Copyright 1969 by Sigma Xi National Science Honorary.)

tually everyone survives. When the dose is increased above this point, survival begins to drop sharply, and with a dose of more than about 800 rads, everyone dies.

Even though very few people are killed outright with doses of under 250 rads, many may become quite ill. At doses of between 100 and 250 rads most people will develop fatigue, nausea, vomiting, diarrhea, and some loss of hair within a few days of the exposure; the majority, however, will recover from the acute illness. At doses of around 400 to 500 rads, however, the outlook is not so rosy. During the first few days, the illness is similar to that of the previous group. The symptoms may then go away almost completely for a time, but beginning about three weeks after exposure, they will return. In addition, because the radiation has impaired bone marrow function, the number of white cells and platelets in the blood will decrease. This is of great significance, because without white cells the body cannot fight infection, and without platelets the blood will not clot. Figure 10–10 shows that about 50 percent of those exposed in

TABLE 10–2   Units Related to Radioactivity

Unit	Abbreviation	Definition and Application
Disintegrations per second	dps	A rate of radioactivity in which one nucleus disintegrates every second. The natural background radiation for a human body is about 2 to 3 dps. This does not include "fallout" from manmade sources such as atomic bombs.
Curie	Ci	Another measure of radioactivity. One Ci = 37 billion dps.
Microcurie	μCi	A millionth of a curie, or 37,000 dps.
Roentgen	R	A measure of the intensity of X-rays or gamma rays, in terms of the energy of such radiation absorbed by a body. (One R delivers 84 ergs of energy to 1 gram of air.) The roentgen may be considered a measure of the radioactive dose received by a body. The dose from natural radioactivity for a human being is 5 R during the first 30 years of life. A single dental X-ray gives about 1 R, a full mouth X-ray series, about 15 R.
Rad		Another measure of radiation dosage, equivalent to the absorption of 100 ergs per gram of biological tissue.
Rem		A measure of the effect on man of exposure to radiation. It takes into account both the radiation dosage and the potential for biological damage of the radiation. The damage potential is based on the following scale of factors: X-rays, gamma rays, electrons : 1 neutrons, protons, alpha-particles :10 high-speed heavy nuclei :20 The rem is then defined by the relationship: Rems = Rads × Biological damage factor. Therefore, 100 ergs per gram (X-rays) = 1 rad × 1 = 1 rem, but 100 ergs per gram (neutrons) = 1 rad × 10 = 10 rems.

this dose range will die, and of these, most will die of either infection or bleeding.

If, instead, the dose administered is about 2000 rads, the first week of the illness will again be the same as that in the previous groups, but the symptoms do not disappear for three weeks. Instead, these people become very ill in the second week, with severe diarrhea, dehydration, and infection leading to death. At these dose levels, the cells of the gastrointestinal tract are affected before the bone marrow toxicity has a chance to become severe, and these patients may die even before their blood counts have dropped to life-threatening levels.

At doses above 10,000 rads, animal experiments have shown that death, which may occur within hours of administration of the dose, is due to injury of the brain and heart.

Delayed Somatic Effects. Of the delayed somatic effects of radiation (that is, those occurring months or years following the exposure), none is better studied or of more concern than the increased incidence of cancer in those with a history of prior exposure to radiation. Although the

molecular mechanisms at work here are still largely obscure, there is evidence that relatively high doses of radiation do increase the incidence of cancer in exposed populations. Before the dangers of radiation were appreciated, early workers were careless in their handling of radioactive materials and suffered a greatly increased incidence of skin cancers. The famous case of the radium dial workers in the 1920s also deserves mention. These women were responsible for painting the dials of watches with the phosphorescent radium paint in use at the time and routinely tipped the end of the brush in their mouths before applying the paint to the dial face. In later years this group experienced a very high incidence of bone tumors. (Ingested radium, like strontium, is preferentially incorporated into bone.) There is evidence that the survivors of the atomic attacks at Hiroshima and Nagasaki developed leukemia more frequently than one would have expected from a group of this size.

Anyone who has read this far should not be surprised to learn that there is considerable disagreement about the health effects of low doses

TABLE 10–3 Exposure of the U.S. General Population (1978) to Radiation

Source	Person-Rems* per Year (in thousands)
Natural background	20,000
Medical X-rays	17,000
Technologically enhanced**	1,000
Nuclear weapons	
Fallout	1,000 to 1,600
Development, testing, and production	0.165
Nuclear energy	56
Consumer products	6

*"Person-rems" are calculated by multiplying the total number of people exposed by their average individual doses in rems.

**This category refers to the increase over natural radiation background caused by technological factors such as processing of uranium ores or living near uranium mines or processing plants. (Data are from *Science*, 204:162, 1979.)

of radiation. Even if there were no X-ray machines or nuclear power plants, people would be exposed to measurable amounts of ionizing radiation in the form of gamma rays originating from outer space and various forms of radiation from natural substances such as uranium found in common rocks and soils. Table 10–3 compares this natural background radiation with other sources of radiation to which the general population of the United States is exposed.

A ruling of the International Commission on Radiation Protection, also adopted in the United States by the Nuclear Regulatory Commission, allows a maximum exposure of 5 rem per person per year for occupational workers in nuclear industries. This guideline has been challenged by some epidemiologists and by various individuals who have sued for damages related to cancers they claim were caused by excessive exposure to radiation. Some of these claims have been made by workers at nuclear facilities, others by people who have lived near nuclear weapons plants or near sites where bombs were tested. It is always difficult, however, to prove or disprove a cause-and-effect relationship in such instances.

Genetic Effects of Radiation. In every experimental system studied in the laboratory, in organisms as diverse as viruses, bacteria, fruit flies,

and mice, radiation has been shown to induce mutations. Although both ethical and practical considerations preclude genetic experiments with living people, human cells grown in cultures have been studied; the results have confirmed other observations that high-energy radiation can disrupt the cell's genetic material.

Scientists have looked for evidence of increased mutation rates among irradiated human populations, particularly the survivors of the atomic bombs dropped on Hiroshima and Nagasaki. They have found no evidence that these people have more genetic defects than anyone else. Scientists believe that the atomic bomb survivors have in fact undergone more mutations than other people, but it has been known for some time that all humans carry many detrimental mutations that are not expressed as genetic defects. One study in a British hospital showed that 2.2 percent of apparently normal newborn babies had chromosomes so badly damaged that their abnormality was visible under the microscope.

The evidence indicates that humans are quite resistant to the effects of mutations caused by radiation. However, if an increasing number of mutations occurred, the population as a whole would carry a large number of detrimental genes. As the percentage of detrimental genes increased, the number of children born with genetic defects would grow. This problem has already been documented, particularly in industrial countries. Radiation is only one of the causative factors, and probably among the least significant. Many chemical pollutants are known to cause mutation. Possibly more important, medical technology now permits many people with genetic defects to have children. People with hemophilia and phenylketonuria used to die young. Now they survive, as do others with more subtle genetic defects. Many of these people have children who in turn pass their genetic defects on to future generations.

One other factor that deserves mention is fear and its effect on health. Fear can be a response to any perceived danger, but recent legal action focuses attention on nuclear power. On May 14, 1982, the U.S. Appeals Court of the District of Columbia released its opinion that psychological stress (anxieties that produce physical effects) among residents near the Three Mile Island nuclear plant is a form of "nuclear power pollution." Therefore the fears of local residents must be taken into account as part of the total environmental impact of nuclear energy. A dissenting opinion held that such a view would require a measure of perceived rather than real danger, that it would lead to avoidance of any risks at all, and that it could result in economic paralysis.

10.7 Noise

Sound is not a chemical; it is simply a wave motion in air. As such, it does not accumulate in the environment. Ring a bell or whisper a few words to your friend and in about a thousandth of a second the sound is gone. But sounds, especially loud ones, do affect humans, and these effects do not disappear so easily (Fig. 10–11).

The human ear is sensitive to a wide range of sound intensities. The loudest sounds that we can hear, such as a rocket taking off or the noise of battle, are billions of times more powerful than the softest sounds, such as the patter of raindrops on soft earth or a child's whisper. Sound intensity is measured on a scale of values called a **decibel** (abbreviated as dB) **scale.** The decibel scale is set up as follows: (1) The softest sound that can be heard by humans is called 0 dB. (2) Each *tenfold increase* in sound intensity is represented by an *additional* 10 dB. Thus, a 10-dB sound is 10 times as intense as the faintest audible sound. (That still isn't very much.) The sound level in a quiet library is about 1000 times as intense as the faintest audible sound. Therefore the sound level in the library is 10 + 10 + 10 or 30 dB. Typical sources of sound from 0 to 180 dB are shown in Table 10–4.

Noise can be defined simply as unwanted sound. Noise can interfere with our communication, diminish our hearing, and affect our health and our behavior. An occasional loud noise interferes with sounds that we wish to hear, but we recover when quiet is restored. However, if a person is exposed to loud noises for long periods of time, then there may be significant permanent loss of hearing.

The general level of city noise, for example, is high enough to deafen us gradually as we grow older. In the absence of such noise, hearing ability need not deteriorate with advancing age. Thus, inhabitants of quiet societies hear as well in their seventies as New Yorkers do in their twenties.

It is important to understand that most instances of loss of hearing that result from environmental noise are not immediately noticeable. Victims are often unaware that they are slowly losing their ability to hear well. Occupational noise such as that produced by bulldozers, jackhammers, diesel trucks, and aircraft is deafening many millions of workers. There has been recent concern that rock-and-roll music in night clubs is often indeed very loud. Sound levels of 125 decibels have been recorded in some discothèques. Such noise is at the edge of pain and is unquestionably deafening. Noise levels as high as 135 dB should never be experienced, even for a brief period, because the effects can be instantaneously damaging. If the noise level exceeds about 150 or 160 dB, the eardrum might be ruptured beyond repair.

Many investigators believe that loss of hearing is not the most serious consequence of excess noise. The first effects are anxiety and stress or, in extreme cases, fright. These reactions produce body changes such as increased rate of heart beat, constriction of blood vessels, digestive spasms, and dilation of the pupils of the eyes. The long-term effects of such overstimulation are difficult to assess, but we do know that in animals it damages the heart, brain, and liver and produces emotional disturbances. The emotional effects on people are difficult to measure, but psychologists have learned that work efficiency goes down when the noise level goes up.

There are three techniques that can be used to control noise.

(a) Reduce the Source. Machinery should be designed so that parts do not needlessly hit or rub against each other. It is possible to design machines that work quietly. For example, rotary saws can be used to break up street pavement.

They do the job perfectly well and are much quieter than jackhammers. Another approach is to change operating procedures. If a suburban sidewalk must be broken up by jackhammers, it would be better not to start early in the morning, when many people are asleep. Also, aircraft takeoff can be routed over less densely inhabited areas. All too often, machines are built to per-

Motorcycle noise.

Traffic noise.

Jack hammer noise.

Grinding noise.

Figure 10–11 *Various noises common in our society.*

TABLE 10–4   Sound Levels and Human Responses

Sound Intensity Factor	Sound Level, dB	Sound Sources	Effects		
			Perceived Loudness	Damage to Hearing	Community Reaction to Outdoor Noise
1,000,000,000,000,000,000	180	• Rocket engine	Painful	Traumatic injury	
100,000,000,000,000,000	170				
10,000,000,000,000,000	160				
1,000,000,000,000,000	150	• Jet plane at takeoff		Injurious range: irreversible damage	
100,000,000,000,000	140				
10,000,000,000,000	130				
		• Maximum recorded rock music			
1,000,000,000,000	120	• Thunderclap / • Textile loom / • Auto horn, 1 meter away	Uncomfortably loud	Danger zone; progressive loss of hearing	
100,000,000,000	110	• Riveter			
		• Jet fly-over at 300 meters			
10,000,000,000	100	• Newspaper press			
1,000,000,000	90	• Motorcycle, 8 meters away / • Food blender	Very loud		Vigorous action
		• Diesel truck, 80 km/hr, 15 m away			
100,000,000	80	• Garbage disposal		Damage begins after long exposure	
10,000,000	70	• Vacuum cleaner			Threats
		• Ordinary conversation	Moderately loud		Widespread complaints
1,000,000	60	• Air conditioning unit, 6 meters away			
		• Light traffic noise, 30 meters away			Occasional complaints
100,000	50				
		• Average living room	Quiet		
10,000	40	• Bedroom			No action
		• Library			
1000	30	• Soft whisper			
100	20	• Broadcasting studio	Very quiet		
10	10	• Rustling leaf			
			Barely audible		
1	0	• Threshold of hearing			

Figure 10–12 Man wearing acoustical earmuffs while using chainsaw.

Figure 10–13 Combat noise has partially deafened many soldiers. (Photo courtesy of the U.S. Army.)

form a task most efficiently without consideration of how much noise is produced. If machines were originally designed properly, noise levels could be substantially reduced.

(b) Interrupt the Path of Transmission. Sound waves travel through air. They also travel through other media, including solids such as wood. However, some materials, especially soft or porous ones, absorb sound. Such sound-absorbing media are called **acoustical materials.** Acoustical tiles and wallboard can be used in house construction to reduce noise levels. The muffler used in automobile exhaust systems is another example of a sound-absorbing device.

(c) Protect the Receiver. The final line of defense is strictly personal. We protect ourselves instinctively when we hold our hands over our ears. Alternatively, we can use ear plugs or earmuffs as shown in Figure 10–12. (Stuffing in a bit of cotton does very little good.) A combination of ear plugs and earmuffs can reduce noise

by 40 or 50 decibels, which could make a jet plane sound no louder than a vacuum cleaner. Such protection could prevent the deafness caused by combat training (Fig. 10–13) and should also be worn for recreational shooting.

Summary

Cancer is an unregulated growth of cells. A **mutation** is an abnormal change in an embryo. In some cases, environmental chemicals may cause cancer or mutations.

A **poison** is a substance that can destroy life rapidly. A **toxic substance** is any substance whose physiological action is harmful to health. Toxicity depends on the route of entry, the dose, the susceptibility of an individual, and the possible synergistic effects. Some compounds may have threshold levels below which the substance does not affect human health. The effects of toxic substances are evaluated by epidemiological studies and by laboratory tests.

The GRAS list is a compilation of food additives released before 1958 that are "generally *recognized as* safe" by a panel of experts. Occasionally a product on the GRAS list is removed after tests show it to be toxic. The Delaney Clause states that any product that has been shown to cause cancer in animals or humans may not be marketed.

Many poisonous chemical wastes are buried in dumps or landfills. In 1980, the Superfund Legislation was passed in the United States to speed the clean-up of dump sites of hazardous wastes.

The biological effects of radiation include early radiation sickness, delayed somatic effects, and genetic effects. The degree of harm or risk depends on the dose.

Noise, defined as unwanted sound, can cause loss of hearing or lead to anxiety, stress, or fright with consequent adverse physiological effects. Noise can be reduced by more careful design of machinery, by installation of acoustical materials, or by using earplugs.

Questions

Routes of Entry

1. In each of the following cases specify the route(s) of entry of the toxic substance to the body: (1) A child living in an old house in which the paint is peeling is found to have an elevated level of lead in his blood. (2) A worker in a retail dry cleaning establishment who specializes in removing stains by rubbing them with special solvents develops liver disease, diagnosed to be caused by chlorinated hydrocarbons. (3) A hiker becomes violently ill after being bitten by a rattlesnake.

Dose vs. Effect

2. If a dose-vs.-effect curve were plotted in each of the following cases, which one(s) would show a threshold level at a dose greater than zero?

DOSE	EFFECT
Electric current passed through your body	Electric shock
Number of bullets entering body	Injury from a gunshot
Concentration of odorous gas	Odor
Height of fall	Injury from a fall
Number of times you drive a car	Chance of an auto accident

3. Under what conditions is the statement "Dilution is the solution to pollution" correct? Under what conditions is it wrong? Are there any conditions under which a diluted pollutant can do more harm than a concentrated one?

Chemicals in Food

4. Which substances among the various categories of food additives can be advantageous to health? Which are added for the convenience of the consumer? Which are mainly for the benefit of the manufacturer?

5. Too much salt in a person's diet may promote hypertension, whereas some salt is an absolute necessity for survival. Discuss the concept of a threshold level for toxicity from excess salt.

6. Figure 10–14 shows a graph of cancer of the large intestine plotted as a function of meat consumption in several countries. Do these data prove that meat causes intestinal cancer? What other possible conclusions can be suggested by the data?

7. Large doses of the artificial sweetener saccharin have been shown to cause cancer in laboratory animals. Yet this additive was available on the market during the early 1980s. Prepare a class debate. Have one side argue in favor of banning saccharin and the other argue against the ban.

Hazardous Wastes

8. Is it possible to prove conclusively that the chemicals released in the dump site in Niagra Falls, N.Y., have caused diseases in the

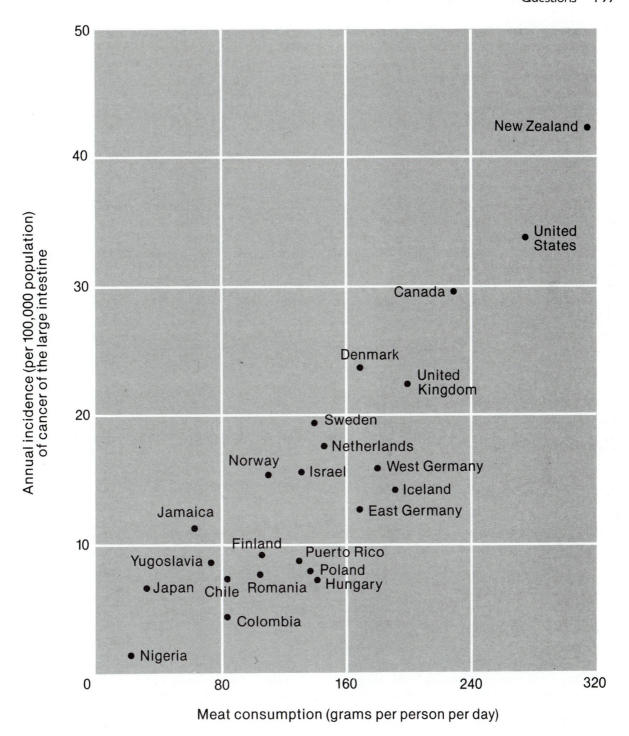

Figure 10–14 Statistical correlation between meat consumption and cancer of the large intestine in several countries.

local population? Discuss the implications of your answer.

Radiation

9. Because a six-year-old girl is too young to bear children, there is no need to shield her body from the radiation produced by dental X-rays. True or false? Explain.

Noise

10. Make a list of the five most distressing noises you hear in a specific day. How could each one be reduced or eliminated?

11. It has often been stated that some damage to hearing may be "inevitable" to people living in a developed society. Do you agree? If so, defend your answer. If not, describe some ways of living that would prevent damage to hearing over a lifetime.

Suggested Readings

Several books on chemicals and human health include:
Bill Boland, ed.: *Cancer and the Worker.* New York, New York Academy of Sciences, 1977. 77 pp.
John Cairns: *Cancer: Science and Society.* San Francisco, W. H. Freeman and Company, 1978. 199 pp.
Erik P. Eckholm: *The Picture of Health—Environmental Sources of Disease.* New York, W. W. Norton and Company, 1977. 255 pp.
John H. Gibbons, dir.: *Environmental Contaminants in Food.* Washington, D.C., U.S. Office of Technological Assessment, Government Printing Office, 1978. 227 pp.
R. W. L. Goodwin, ed.: *Chemical Additives in Food.* Boston, Little Brown & Company, 1977. 128 pp.
R. J. C. Harris: *Cancer.* Baltimore, Penguin Books, 1976. 175 pp.
Lawrence E. Hinkle, and William Loring, eds.: *The Effect of the Man-Made Environment on Health and Behavior.* Washington, D.C., U.S. Department of Health, Education and Welfare (Pub. No. CDC 77–8318), Government Printing Office, 1900. 315 pp.

A book that emphasizes chemical poisons, especially pesticides is:
M. A. Q. Kahn, and John P. Bederka, Jr., eds.: *Survival in Toxic Environments.* New York, Academic Press, 1974. 553 pp.

Books that cover the subject of environmental health include:
Edward J. Calabrese: *Nutrition and Environmental Health,* Vols. 1 and 2. Somerset, N.J., John Wiley & Sons, 1980, 1981. Vol. 1, 585 pp., Vol. 2, 468 pp.
Herman Koren: *Handbook of Environmental Health and Safety,* Vols. 1 and 2. New York, Pergamon Press, 1980. 697 pp.
William W. Lawrance: *Of Acceptable Risk—Science and the Determination of Safety.* Los Altos, Calif., William Kaufman, 1976. 180 pp.
A. R. Rees, and H. J. Purcell, eds.: *Disease and the Environment.* Somerset, N.J., John Wiley & Sons, 1982. 206 pp.
Carl E. Willgoose, et al.: *Environmental Health.* Philadelphia, W. B. Saunders Company, 1979. 476 pp.
Ruth Winter: *A Consumer's Dictionary of Food Additives.* New York, Crown Publishers, 1972. 235 pp.

Three recent books on the Love Canal disaster are:
Michael Brown: *Laying Waste: The Poisoning of America by Toxic Wastes.* New York, Pantheon, 1979.
Adeline Gordon Levine: *Love Canal: Science, Politics, and People.* Lexington, Mass., D. C. Heath, 1982. 266 pp.
Lois M. Gibbs: *The Love Canal: My Story.* Albany, N.Y., State University of New York Press, 1982.

A discussion of the health effects of radiation can be found in:
W. J. Bair, and R. C. Thompson: Plutonium: Biomedical research. *Science, 183,* 715, 1974.
D. J. Crawford, and R. W. Leggett: Assessing the risk of exposure to radioactivity. *American Scientist, 68:* 524, 1980.
Merrill Eisenbud: *Environmental Radioactivity,* 2nd ed. New York, Academic Press, 1973.
Constance Holden: Low-level radiation: A high level concern. *Science, 204:* 155, 1979.

Several books that emphasize the effects of noise are:
Robert Alex Baron: *The Tyranny of Noise.* New York, St. Martin's Press, 1970. 294 pp.
Clifford R. Bragdon: *Noise Pollution: The Unquiet Crisis.* Philadelphia, University of Pennsylvania Press, 1971. 280 pp.
Karl D. Kryter: *The Effects of Noise on Man.* New York, Academic Press, 1970. 632 pp.

William Burns: *Noise and Man*, 2nd ed. Philadelphia, J. B. Lippincott, 1973. 459 pp.

Henry Still: *In Quest of Quiet.* Harrisburg, Pa., Stackpole Books, 1970. 220 pp.

A delightfully written, nonmathematical, yet authoritative paperback book that covers the entire field very well is:

Rupert Taylor: *Noise.* Baltimore, Penguin Books, 1970. 268 pp.

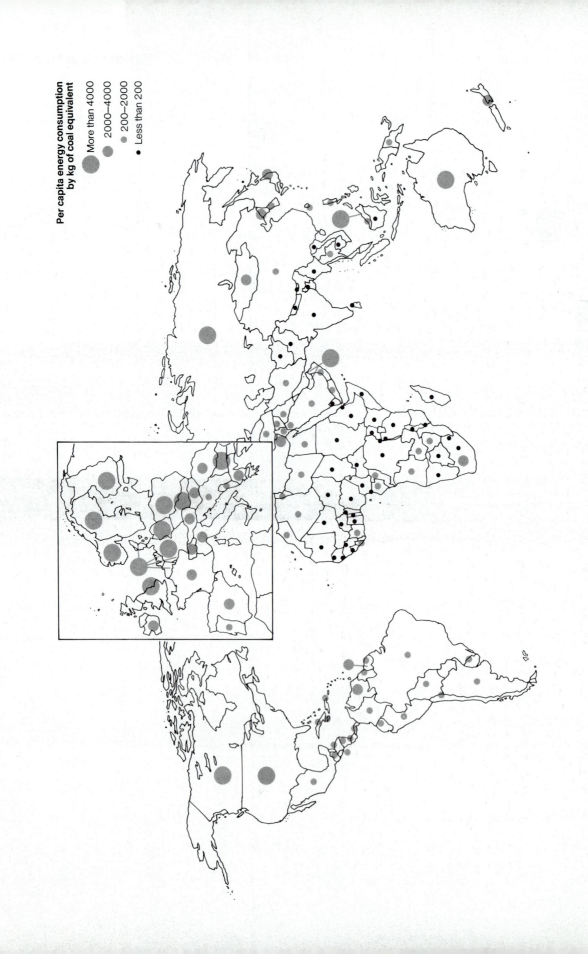

**Per capita energy consumption
by kg of coal equivalent**

More than 4000

2000–4000

200–2000

Less than 200

Energy

World per capita energy consumption per year.
Each unit is equivalent to 1kg of coal. High per
capita energy consumption is typical of countries
with cold climates that require large quantities of
heating fuel, such as Canada and the countries of
northern Europe. Other circumstances that lead to
high energy consumption are widely dispersed
populations that use much transportation or high
per capita incomes, such as in the United States.
(From *Our Magnificent Earth.* © 1979 by Mitchell
Beazley as *Atlas of Earth Resources.* Published in
the U.S.A. by Rand McNally & Company.)

Sources of Energy

11

Primitive people lived within constraints posed by natural ecosystems. They collected fruits and vegetables and hunted game. Before the use of fire, each person needed approximately 2000 kcal per day, all in the form of food energy. In later times, primitive farmers domesticated animals, raised grains and vegetables, and used fuels for cooking and heating. Energy requirements per person (per capita) rose to about 12,000 kcal per day (Fig. 11–1). The fuels burned for cooking were largely wood and animal dung, the same fuels characteristic of less-developed societies even today.

Gradually, over a period of centuries, manufacturing and trade became important in human societies. At first, trade systems were powered by animals, which were used to pull wagons and to carry loads, and by wind, which drove ships. But manufacturing was another matter. Heat was needed to convert ores into metals such as iron, copper, and tin and to fuse sand into glass. For many years, wood and charcoal (made from burning wood slowly) were the only fuels used for manufacturing. As populations expanded and manufacturing became more widespread, timber resources began to be de-

pleted. Forests were cut down to make room for more farmland. The wood was used as a building material for ships and houses and as fuel for manufacturing and home heating. By the time of the Middle Ages, trees in some regions were being cut faster than they were being replaced by natural growth. In Europe, the British Isles felt the lack of wood sooner than the countries on the continent did. During the period from 1550 to 1640, the price of wood in England skyrocketed, leading to rapid inflation and economic hardships. Deforestation became so severe that military leaders feared there would be no wood left for shipbuilding.

Then, when the situation looked particularly bleak, people discovered that coal could be substituted for wood. Mining operations flourished, and the age of fossil fuels began. Fossil fuels produce more energy for a given weight of fuel than do "biomass" fuels such as wood and dung. Therefore, the use of fossil fuels permits manufacturing processes that are not economically feasible using biomass fuel. These manufacturing processes completely changed the face of society because they provided unskilled jobs in urban factories and led to the demise of rural

202

1. **Man without fire**
    (2000 kcal/day)

2. **Primitive agriculture**
    (12,000 kcal/day)

3. **ca. 1860**
    (70,000 kcal/day)

4. **ca. 1980**
    (230,000 kcal/day)

Figure 11–1 Human energy consumption has risen from the 2000 kcal per day needed by stone age people in the form of food to 230,000 kcal per day used by North Americans.

industries. This is why the adoption of manufacturing processes powered by fossil fuels in one country after another has been dubbed the "industrial revolution."

By the mid 1800s manufacturing boomed, steam engines had been invented, and the per capita fuel consumption in London was about 70,000 kcal per day. At that time, people in western Europe were using fossil fuels faster than they were being replaced. This was possible then because large supplies of fossil fuels represented a reservoir of stored energy. However, in the United States in 1980, per capita energy consumption was approximately 230,000 kcal per day. Most of this energy was derived from the use of fossil fuels. This enormous rate is unique in the history of the world. At no other time and in no other place have people used energy faster than North Americans do today.

Rapid energy consumption is not confined to the United States. All over the world people are burning large quantities of fossil fuels. This consumption cannot continue indefinitely. Supplies of coal, oil, and gas will eventually be depleted. Furthermore, fluctuations in availability and cost of energy have led to worldwide economic instability. Note that the energy crisis today is similar in many ways to the crisis caused by the fuel shortage in England 400 years ago.

11.1 Modern Energy Use

When most North Americans think about energy consumption, they may think about automobile traffic on a busy freeway, heating systems in large apartment buildings, or mountains of coal stored near a steel mill. Use of energy in other parts of the world is quite different. Several years ago, I (Jon) was riding in a truck in the Himalayan mountain region of Ladakh in northern India (Fig. 11–2), travelling through high back-country farmland at an elevation approximately equal to that of the highest peaks in the North American Rockies. The truck was carrying a load of sticks wrapped in bundles. If that wood were used as a fuel in the United States, it could supply perhaps one rural household in the northern Rockies for a winter. But here in India it represented the main source of fuel for an entire town. Once delivered, the fuel would be far too precious to burn for heat; it would be used solely for cooking. As a substitute for a cozy fire during the cold snowy winter, people simply herded their sheep into an anteroom of the main house. The heat released from the bodies of these animals provided a small amount of warmth for the house. The entire population of the town was waiting for the arrival of the truck. There were some rapid negotiations, and the bundles of sticks were thrown onto the ground. I have never seen a pile of wood disappear so quickly; men, women, and children grabbed what they could carry and ran home. Within a few minutes every small scrap and twig had been taken, and only a few small children, too young to run very quickly, were still scurrying home with handfuls of twigs.

This story is told to personalize the concept of energy consumption in the developing countries. Statistics of global energy production and consumption are certainly useful and are pre-

Figure 11–2 In this village in the mountains in northern India, fuel is far too valuable to be used for heat and is used only for cooking.

sented in this book, but at the same time, it helps to remember that these numbers relate to individuals and to human lifestyles.

SPECIAL TOPIC A
High Technology—Low Consumption in Energy Planning

Modern technology is often associated with grandiose schemes such as nuclear power plants and gigantic hydroelectric facilities. But the benefits of much of this type of advanced technology often do not filter down to the poorest farmers of the less developed countries. Yet, villagers who don't have electricity or access to industrial machinery could benefit greatly from modern science and engineering if it were applied directly to their needs. Such direct application of modern resources is called **appropriate technology.**

As an example, around the world some 400 million draft animals—horses, oxen, donkeys, mules, cows, water buffaloes, camels, yaks, llamas, and elephants—work for people. In many cases harnesses and plow yokes for these animals are poorly designed. For example, the traditional wooden yoke weighs about 45 kg and burdens the animal unnecessarily. In addition, the strap under the neck tightens around the animal's windpipe, choking it as it pulls. Although a well-designed, anatomically efficient yoke would improve agricultural efficiency around the world, yoke design is not a popular engineering field. As N. S. Ramaswamy, Director of the Indian Institute of Management, said, "India has put a satellite in space and harnessed the atom, but our yokes are 5000 years old because professors are scared they may not get promoted if they work on finding better ones."

In 1980, the United States, with about 5 percent of the world's population, consumed 25 percent of the available commercial energy in the world. In contrast, India contains 15 percent of the global population and used only 1.5 percent of the global energy supply. Another statistic to think about: In 1981, the 228 million people in the United States used more energy for air conditioning alone than the 985 million people in China used for all their needs!

As yet another example, consider a comparison of the per capita energy use in the United States with that in two other nations, Bangladesh, one of the poorest countries in the world (Fig. 11–3A), and Sweden, one of the most developed (Fig. 11–4). The average American uses about 50 times as much energy as the average person in Bangladesh. Notice that 34 percent of the total consumption in Bangladesh is used directly in the form of food. Agricultural products, food, crop residue, and manure account for 88 percent of this consumption; fossil fuels and wood make up the remaining 12 percent. In contrast, look at the chart for energy consumption in the United States (Fig. 11–3B). Notice that Figures 11–3A and 11–3B don't even look as though they are representing the same subject. The use of agricultural goods and firewood, which composes 91 percent of the energy consumed in Bangladesh, is squeezed into 2.2 percent in the United States. Food energy is hardly even on the chart in North America because it is so small compared with other forms of consumption. Thus, on the one hand, energy is used for bare survival in Bangladesh, and on the other hand, energy in the United States is a resource exploited mainly for comfort and convenience (see Fig. 11–5).

Many people object to comparisons between the United States and Bangladesh. They argue, and rightly so, that no one would choose to live on the very edge of abject poverty. Everyone hopes for a comfortable world with an enjoyable life. Perhaps a comparison between the United States and Sweden would be more meaningful.

Both the United States and Sweden have well-developed economic systems. In both countries people are well fed. Excellent medical care is available. Infant deaths are low. Educational levels are high. Most families have a car, telephone, television, refrigerator, and vacuum

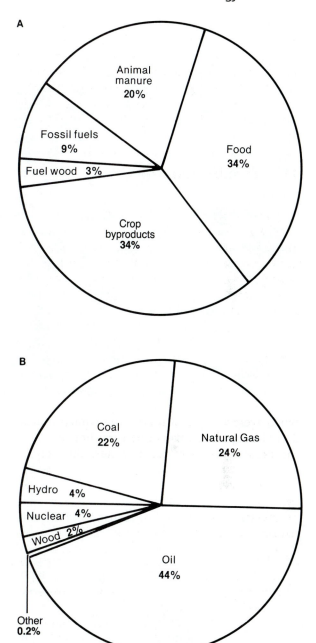

Figure 11–3 A, Sources of energy used in Bangladesh. Note that 91 percent of the energy is taken from renewable plant or animal matter. B, Sources of energy used in the United States in 1980. Note that only 6 percent of the energy supply comes from renewable sources.

cleaner. Yet the average person in Sweden uses less than two thirds the energy that an American does. North Americans may well look to people

Figure 11–4 A view of Stockholm, Sweden. People in Sweden enjoy a healthy, modern, industrial society but use only two thirds of the energy per capita used in North America. (Photo courtesy of Swedish Information Service.)

of other nations to see how consumption can be reduced.

(a) Transportation Overall, Swedes use approximately one fourth the energy for transportation that Americans do. Some of this fuel-saving results because Sweden is much smaller than the United States and the average distance between cities is considerably less. However, this difference alone does not account for the wide discrepancy in energy consumption; in addition, (1) people in Sweden frequently walk or use bicycles for short trips; (2) mass transit is used more frequently; (3) on the average, Swedish cars consume considerably less fuel per kilometer than American cars do.

(b) Space Heating Winters are considerably longer and more severe in Sweden than they are in most of the United States. Yet energy consumption for heating per person is less. Swedish houses are generally better insulated and more efficient than their American counterparts.

(c) Industrial Consumption Swedish industries are generally highly efficient because newer, innovative technology is in use. Thus, less energy is used to produce and refine a kilogram of steel, oil, paper, cement, or chemicals in Sweden than in the United States.

There are several reasons why Swedish consumption is so low. Heavy government taxes have raised fuel prices so that, for example, gasoline costs considerably more than it does in North America. But price alone is not the only factor. The government in Sweden has taken vigorous steps to promote energy conservation. Some of these steps are listed in Box 11.1. The Swedish example teaches people in the United States that it is possible to use considerably less energy without lowering the quality of life.

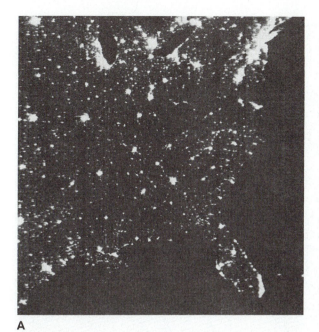

A

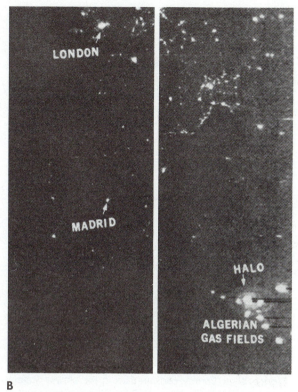

B

Figure 11–5 A dramatic example of the differences in energy consumption between the United States and Europe can be seen in these two photographs. At midnight, when most people are asleep, vast regions of the United States are lit up, as can be seen from this satellite photograph *(A).* On the other hand, most of Europe is dark, and much less electricity is needed *(B).* (A, U.S. Air Force photo; B, U.S. Defense Meteorological Satellite Program photo.)

11.2 Fossil Fuel Sources and Availability

One of the most important questions of our times, and one of the most difficult to answer with any real confidence, concerns how long our fossil fuel supplies will last. For a realistic esti-

mate of the number of years remaining before all the Earth's fossil energy reserves are gone, the quantity of fuel remaining in the ground must be estimated, human population growth forecast, and future rates of consumption predicted. All such forecasts are subject to large errors.

The most reasonable method of estimating the energy requirements from the 1980s into the early twenty-first century is to graph past energy consumption, and then try to guess how the curve will continue in the future. Figure 11–6 shows the consumption of energy in the United

---

BOX 11.1 ENERGY USE: UNITED STATES VS SWEDEN

United States

Building codes in many areas still permit construction of leaky, poorly insulated homes.

Low priority is given to energy conservation in housing loans.

Mass transit is marginal to poor . Bus service is unavailable to many suburbs of United States cities. Intercity rail system has been drastically reduced in recent years.

Sweden

It is illegal to build without adequate insulation.

High priority is given to energy conservation in housing loans.

Mass transit is highly efficient. Bus service is four minutes apart during peak hours in major cities. Intercity rail system provides direct service to most regions.

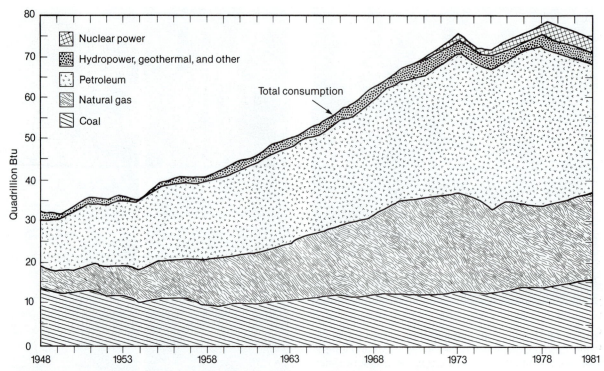

Figure 11–6 Energy consumption in the United States from 1948 to 1981. Energy demand grew steadily from 1948 to 1973 but oscillated during the following eight years. These data were taken from information issued by the Department of Energy. Note that consumption of wood is not included, even though other sources state that wood provided 2 percent of the energy requirement in 1981. (See Appendix A for definitions of energy units.)

States from 1948 to 1981. Notice that the curves are far from smooth. Total consumption increased steadily from 1948 to 1973. In fact, from 1948 to 1970, energy use doubled. It is obvious that the United States, or the world in general cannot double its consumption of a depletable resource every 22 years for very long, for each doubling corresponds to an increasingly large growth. Think of it this way: If energy consumption is doubling every 22 years, then in the past 22 years we have used as much energy as in our whole previous history. Or another way: If energy consumption is doubling every 22 years, then by the time we have used up half our total reserves, we will have only enough left for another 22 years. Perhaps still another outlook will be helpful. In 1979 the worldwide consumption of petroleum was 65 million barrels per day. If this consumption rate were doubled every 22 years, then 660 years from now the people in the

world would consume as much petroleum every year as there is water in all the world's oceans. Absurd? Of course. No one thinks there is anywhere near that much petroleum in the Earth's crust. This analysis shows that whatever is done, however much oil is found, the age of fossil fuels as we know it today cannot last long. Perhaps you will argue that a few hundred years is really a long time. But remember, people have lived on this Earth for over 300,000 years. Matched against this time scale, a few hundred years is minuscule. Our fossil fuel age is destined to be a tiny burst of time in the history of our species.

What will happen in the near future? No one really knows. Notice, in Figure 11–6, that consumption of fuel in the United States has declined in the past few years. The same trend has been observed worldwide. This decline in consumption is directly related to the perceived

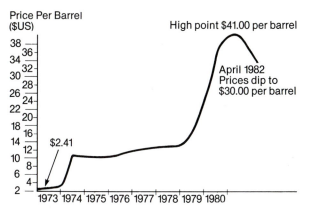

Figure 11-7 Price per barrel for OPEC crude oil.

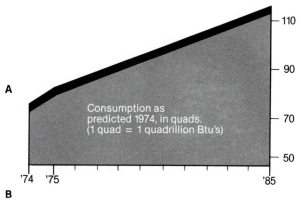

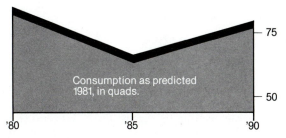

Figure 11-8 A, In 1974 the Ford Foundation predicted a steadily rising U.S. energy consumption through the year 1985. But in fact, energy consumption has declined in recent years. B, New forecasts, provided by the Department of Energy, reflect recent declines in energy consumption.

scarcity of resources. In the mid 1970s, countries that owned fossil fuel reserves began to realize just how valuable their fuels were, so they raised the prices (Fig. 11-7). As prices went up, consumers started to use less fuel. Individuals drove fewer miles and purchased smaller, more fuel-efficient automobiles. They have added additional insulation to buildings and saved in many other ways. This reduction in consumption came as a surprise to many world planners. In 1974, the Ford Foundation published its Energy Policy Predictions (Fig. 11-8A). Notice that the Ford Foundation predicted a steady rise in energy use through 1985. But this prediction was wrong. In 1981, new predictions were made, reflecting the decrease in consumption due to conservation. The results are shown in Figure 11-8B.

The sum of all this discussion is that all predictions are suspect. Since fossil fuel reserves are certainly limited, our civilization must face an energy transition in the near future. But no one really can say exactly when and how this will happen. Within the limits of our uncertainty, it is still helpful to try to make the best prediction possible. These predictions are meant to be only rough guidelines.

Oil

Petroleum is perhaps the most versatile fossil fuel. Crude oil, as it is pumped from the ground, is a heavy, gooey, viscous, dark liquid. The oil is refined to produce many different materials such as propane, gasoline, jet fuel, heating oil, motor oil, and road tar. Some of the

chemicals in the oil are extracted and used for the manufacture of plastics, medicines, and many other products. It is difficult to imagine what would happen to our civilization if the supply of liquid fuels ran out. Automobiles, airplanes, most home furnaces, and many appliances could not operate. Many industries would have to redesign their factories. Yet reliable estimates indicate that before the year 2000 there will not be enough petroleum to meet worldwide demand. Figure 11-9 shows one projected graph of global oil production and demand for the years 1975 to 2025. This graph, first published in 1975, predicted that oil production would continue to meet demand until about 1997. At that time, according to this prediction, many of the richest fields will be depleted, and production will slow down. Yet the need for oil will continue to increase. People will want more oil than is available. Therefore, a real and permanent shortage will result. As shown by the

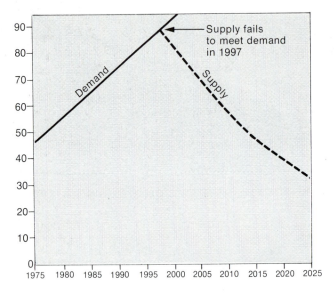

Figure 11–9 *One projected estimate of the supply and demand for petroleum through the year 2025.*

graph, oil will still be available, but there won't be enough to supply people's wants.

How accurate is Figure 11–9? For one thing, the demand for oil between 1975 and 1981 was less than was predicted. Thus, conservation is expected to provide a few more years before the oil runs out. But huge amounts of oil are still being consumed, and unless drastic reduction of energy demand is realized, conservation will extend the prediction in Figure 11–9 by a matter of years, not decades.

What about the assumptions that have been made about the amount of petroleum available? There is some belief that tremendous and as yet undiscovered oil deposits may exist in many parts of the world. For instance, an article in the magazine *Science** reports:

> Intensive exploration in Mexico is turning up oil fields so immense that they could overturn the conventional wisdom about world oil supplies and significantly alter the geopolitics of energy.
>
> For the past five years, the conventional wisdom has been that most of the world's major oil fields have already been discovered, that the United States will have to rely more and more heavily on the Middle East for future supplies of oil, and that the giant oil fields around the Persian Gulf are the result of a unique geological occurrence that is unlikely to be matched anywhere else.

Not only does the conventional wisdom appear to be wrong, it appears to be spectacularly wrong. Oil fields apparently equivalent to those in Saudi Arabia have been found only about 1000 kilometers from the U.S. border.

Other explorations are being conducted in China, the North Sea region of Europe, and elsewhere. Is it possible that new finds alone will extend our present global energy use for generations to come? The answer is no. If current consumption rates continue, a newly discovered deposit as large as the one in Saudi Arabia would extend our oil supply by merely 10 years or so. The North Sea oil is expected to produce fuel for the world for an additional one and a half years. Recent discoveries on the continental shelf on the east coast of the United States are expected to supply enough fuel to satisfy global needs for one week!

Natural Gas

Natural gas is composed mainly of methane (CH_4), the simplest organic hydrocarbon. Methane is found in underground rock layers both by itself and lying above natural deposits of petroleum. It is generally believed that natural gas is our least abundant fuel. As recently as 1975, many planners believed that our peak consumption would occur sometime in the late 1980s, and gas companies started refusing to supply many new homes and factories. In the early 1980s new discoveries increased the availability of gas, but world supplies are still limited. The scarcity of

* *Science*, 202: 1261–1265, 1978.

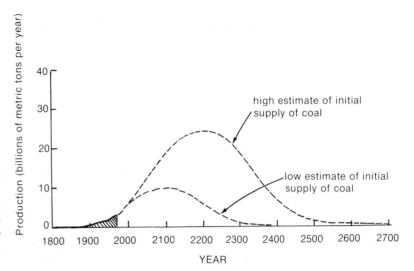

Figure 11–10 Past and predicted world coal production based on two different estimates of initial supply. (Adapted from M. King Hubbert: The energy resources of the earth. In: *Energy and Power, A Scientific American Book.* San Francisco, W. H. Freeman & Co., 1971.)

natural gas is environmentally unfortunate, since it burns more cleanly than any other widely used fuel.

Coal

Large reserves of coal exist in many parts of the world. As shown in Figure 11–10, widespread availability of this fuel can be expected at least until the year 2200. However, there are problems with coal. When coal is mined, large areas of land surface are disturbed. More air pollutants are released from burning coal than from burning oil or gas. These issues will be discussed further in Chapter 13.

Another difficulty arises because coal cannot be used directly in conventional automobiles, in most home furnaces, or in many industries. One solution to the problem is to convert coal to liquid or gaseous fuels. The theory of converting solid to liquid fuels is well understood and has been practiced both in the laboratory and in industry. In fact, conversion of coal was common in the 1920s. During World War II the Germans converted coal to gasoline for military use. By 1978, there were over fifty plants in various parts of the world that produced gaseous or liquid fuels from coal. Most of these were small industrial operations. Large scale conversion of coal is still in the developmental stage, owing mainly to economic problems. In the United States in 1982 liquid fuel from coal cost about one and one-half times as much as the most expensive imported petroleum. Undoubtedly, liquid fuels from coal will become relatively

cheaper than other fuels in future years. In fact, conversion of solid to liquid and gaseous fuel will probably be common before the end of the century.

Heavy Oil

Conventional oil wells do not tap huge underground lakes or pools of oil. There are layers of porous rock under the surface of the Earth, and the pores of these rocks are filled with petroleum, much like a sponge filled with water. An oil well is a hole drilled into a porous rock formation that is saturated with oil. The petroleum in the nearby rock drips down through the pores, collects in a small pool at the bottom of the well, and then can be pumped out to storage tanks above ground. In many places in the Earth, underground deposits of oil are available, but the petroleum is too thick to flow. This situation develops either when the oil is particularly viscous or when the pores in the rock are too small, or both. Petroleum deposits that cannot be pumped in the conventional manner are called **heavy oils**. The three main types of heavy oils are discussed below.

1. Oils from Conventional Wells. Sometimes less than one tenth, and on the average less than one half of the petroleum in a conventional underground deposit is fluid enough to flow through pumps and pipes. The remainder is left in the ground after the oil well has gone "dry." In the United States alone, there are more than 300 billion barrels of oil of this type trapped in known

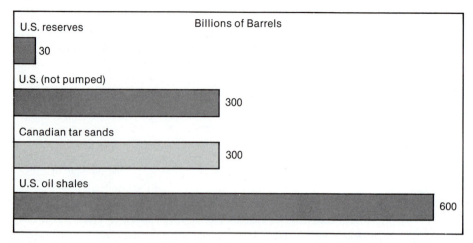

Figure 11-11 Oil reserves in North America. Note that supplies of heavy oil dwarf conventional oil deposits. However, not all heavy oils are expected to be profitable to extract. In addition, the environmental consequences of heavy oil extraction are severe (see Chapter 13).

deposits in known oil fields (Fig. 11-11). This represents approximately one hundred times as much oil as was produced in the United States in 1980, and almost as much oil as has been found to date in the Middle Eastern oil fields. Obviously, people are interested in recovering this wealth. Several different methods are being investigated today. One is to force superheated steam into old well holes at high pressure. The steam heats the remaining underground oil and makes it fluid enough to flow. Of course, a lot of energy is needed to heat the steam, so this type of extraction is not always efficient. Another process involves pumping detergent into the rock and "washing" the petroleum out. The petroleum can then be salvaged and the detergent recycled. Still other procedures involve mining the rock and then heating it to extract the oil. No oil geologists believe that all of the 300 billion barrels can be recovered, but if the price of fuels rises, a sizable portion will probably become economical.

2. Tar Sands.   Oil-saturated sandstone deposits lying near the surface of the Earth are called **tar sands,** although there is not always a clear distinction between tar sands and concentrations of heavy oil remaining after conventional drilling. Sandstone fields laden with oil have been discovered in Africa, the United States, Canada,

and the U.S.S.R. In 1980 a Canadian operation produced 45,000 barrels of synthetic crude oil per day from their tar sands deposits, and plans were under way to extract an additional 125,000 barrels per day (Fig. 11-12*A*).

3. Oil Shales.   The largest quantities of heavy oil in the world are locked into shale deposits in the western United States (Fig. 11-12*B*). If all the petroleum in all the United States oil shales could be recovered, there would be enough oil to supply the United States for 100 years at the level of petroleum consumption in 1980. However, the shale must first be mined in some fashion and then heated to extract the petroleum. In the poorest deposits, more energy would be used to mine and extract the fuel than would be gained, and some of these deposits will probably never be used. The richest deposits, containing 100 liters or more per tonne of rock, represent approximately one third of the total. Taken all together, a realistic estimate is that all the potentially recoverable heavy oil sources in North America could supply that continent for an additional 75 years or so if consumption remained at current levels. Worldwide, the situation is not so promising, because the richest heavy oil deposits are in the United States. However, these figures by themselves do not address the environmental problems of extracting and refining heavy oils. This topic will be discussed in Chapter 13.

11.3 Energy from the Sun

What will happen when the fossil fuel reserves are depleted? Will people be driving horses to town (Fig. 11–13)? Will civilization as we know it collapse? Is there any way out? Nuclear energy is at least part of the answer. Proponents of a nuclear future claim that this source can provide all the energy we will need. But in actuality, the nuclear power industry has stagnated in some countries in recent years (Fig. 11–14). In 1981 the nuclear industry in the United States did not file a single application for construction of new plants. The nuclear future is therefore uncertain. (This subject will be discussed in more detail in Chapter 12.)

Others suggest that we should use naturally renewable energy sources such as the Sun, the wind, fuels from plants, the energy of moving water, and the heat of the Earth. Could our society be supported in such a manner?

Every 29 seconds the solar energy that falls on our planet is equivalent to human energy

A

B

Figure 11–12 *A,* Mining tar sands in Alberta, Canada. (Courtesy of Suncor, Inc.) *B,* Oil shale country in western Colorado. (Courtesy of Atlantic Richfield Company.)

Figure 11–13 Will people be driving horses to town when the oil runs out? (Photograph courtesy of National Archives.)

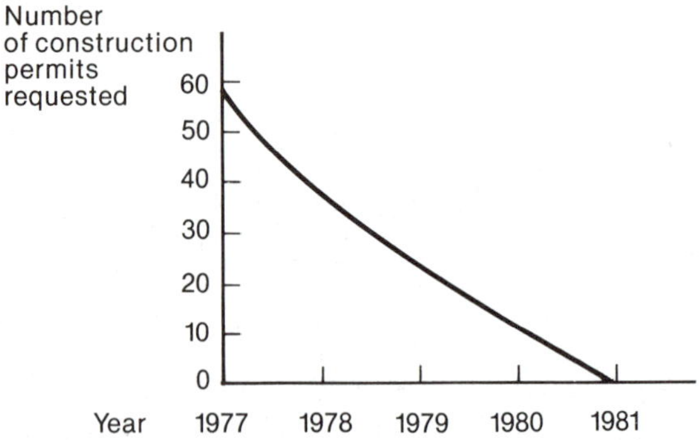

Number of construction permits requested

Figure 11–14 Number of construction permits requested by the nuclear power industry in recent years. The rapid decline reflects a smaller than predicted growth in electric consumption and the fact that nuclear power plants are not good business. They are no cheaper than coal-fired plants and present a complex set of environmental problems.

needs for a day at the 1980 consumption level. In the *least* sunny portions of the United States (excluding Alaska) an area of only 80 square meters (a square approximately 9 m, or 29 ft, on a side) receives enough sunlight to supply the total energy demands of the average American family. Can this energy be harnessed?

Passive Solar Design

The simplest way to use solar energy is to design and orient a house so that the structure itself collects and stores heat from the Sun. This concept is not a new or complicated technological trick. When the Anasazi Indians built their cliff dwellings in the American Southwest, they always chose south-facing cliffsides. As shown in Figure 11–15, the winter sun shines directly into the buildings, providing heat. In the summer, when the Sun is higher in the sky, the edge of the cliff serves as a visor, or awning, to provide cooling shade, as shown in the figure. In contrast, many people in the Southwest today live in rectangular houses oriented and constructed so that large quantities of fossil fuel are needed to heat them in winter and provide air conditioning in summer.

Many ancient civilizations used passive solar design in home construction. It was standard in ancient Greece and Rome. The Greek playwright Aeschylus (524? to 456 B.C.) believed that only uncivilized barbarian societies didn't understand the concept of solar design. In discussing people who did not use passive solar heating systems, he wrote:

> Though they had eyes to see, they saw to no avail; they had ears, but understood not. But like shapes in dreams, throughout their time, without purpose they wrought all things in confusion. They lacked knowledge of houses turned to face the sun, dwelling beneath the ground like swarming ants in sunless caves.

In a more subdued tone, the Roman architect Vitruvius (first century B.C.) wrote a treatise on building solar bathhouses:

> The site for the baths must be as warm as possible and turned away from the north . . . . They should look toward the winter sunset because when the setting sun faces us with all its splendor, it radiates heat, rendering this aspect warmer in the late afternoon.

The design of a practical passive solar home is quite straightforward. An ordinary window admits sunlight. A house without windows would be dull and dreary indeed! Windows may also allow solar heat to enter the house. When the Sun shines directly on a window facing south, the radiant energy enters the room and warms it. Because the heat is trapped within the room, the house can become much warmer than

A

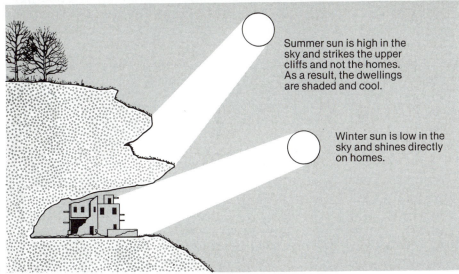

Summer sun is high in the sky and strikes the upper cliffs and not the homes. As a result, the dwellings are shaded and cool.

Winter sun is low in the sky and shines directly on homes.

B

Figure 11–15 The Anasazi cliff dwellings in Mesa Verde, Colorado. *A*, Photograph of the dwellings as they stand today. This picture was taken in the summer; note that the dwellings are comfortably shaded. In the winter, when the Sun is lower in the sky, sunshine strikes the buildings directly, thereby warming them. (Photo by Mel Davis, courtesy of U.S. Bureau of Reclamation.) *B*, Schematic view of the way in which the rock and the Sun combine to provide a warm environment in the winter and a cool one in the summer.

the outside air. But heat from inside also escapes outside through the glass. During the night, heat escapes and no sunlight enters.

Most homes existing today are poorly planned. They are heated with a furnace and then typically lose 20 per cent of their heat through their windows. This trend can be reversed through proper planning. If houses were built with large, double pane, south-facing windows and only small windows on the north side, then the Sun would warm the house during the daylight hours (Fig. 11–16*A*). An efficient passive design requires some sort of heat storage system built into the structure of the house to conserve the heat overnight. A massive masonry wall inside the building is one example

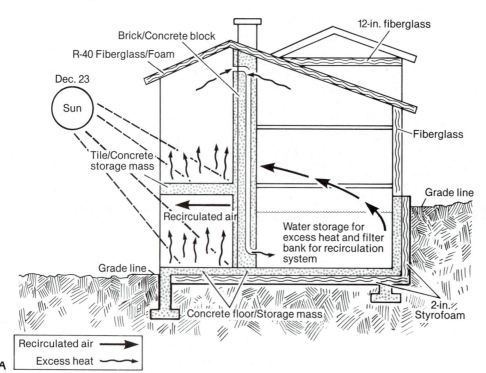

A

B

Figure 11–16 *A,* The Rush-Hampton House, cosponsored by Rush-Hampton Industries, Inc., of Longwood, Florida, and researchers at the University of Massachusetts, relies on passive solar heating. Living space is directly heated by sunlight through an expanse of south-facing glass. Masses of concrete and water are strategically located for heat absorption and storage. Air circulates through the house by natural convection and forced circulation to carry excess thermal gain to thermal storage masses, such as the central column and a water storage area in the basement.

B, The living room of the Rush-Hampton House has a floor of tile-covered concrete. The tile is dark to absorb maximum heat. Warm air rises by natural convection to the apex of the slanted ceiling at the back of the room (away from the windows) near the central concrete column, which is part of the indirect-gain storage system. The column contains ducts to transport (by forced drafts) warm air from the living room down to the lower level. A small wood stove is used to provide extra heat in winter and in cloudy weather.

of a storage system (Fig. 11–16*B*). During the day the room heats up, and some of the heat is absorbed by the masonry. In this way, heat is removed from the living space, and the house is prevented from becoming too warm. At night, when the Sun is no longer shining, the brick and concrete wall radiates the stored heat and keeps the house warm. In addition, since heat is lost and none is gained through windows during the night, heat can be conserved by installing drapes

Figure 11–17 Ontario Hydro's head office in Toronto, Canada. There is no furnace or heating plant within this building, yet internal temperatures are comfortable throughout both the harsh Canadian winter and the hot summer months. Energy conservation is realized through use of south facing, double-glazed reflective glass, and a system that stores and circulates the heat from lights, people, and machinery to supply heating needs. (Courtesy of Ontario Hydro.)

or insulated shutters for use after the Sun goes down.

Many homes, even in such cold climates as the Rocky Mountains, have been designed so that the windows alone provide 50 percent or more of the heat needed in the winter.

Considerable energy savings can be realized in commercial buildings as well. One example of a particularly innovative design is the twenty-story office complex built in Toronto, Canada, for Ontario Hydro (Fig. 11–17). This structure uses no fuel at all for space heating during the harsh Canadian winter. Instead, heat from lights, from machinery, and from people's bodies is conserved, stored, and circulated throughout the structure. In addition, double-glazed glass on the southern wall serves as a passive solar collection system. The energy saved every year is enough to supply all the power used by 2500 average homes.

Solar Collectors

Solar heat can be trapped even more efficiently by the use of various types of **solar collectors.** One type of collector consists of a coil of copper pipe bonded to a blackened metal base. The whole assembly is covered by a transparent layer of glass or plastic. The operating principle is uncomplicated. Sunlight travels through the glass and is absorbed by the blackened surface.

Metal conducts heat readily, so the water in the pipe gets hot. The glass traps the heat within the collector so it does not easily escape back into the atmosphere (Figs. 11–18 and 11–19).

Hot water produced in this fashion can be pumped through radiators to heat a building. A heating system of this type is called **active solar** heat because hot water is actively pumped and regulated by motors and thermostats. Of course, sunshine is not available at night or on cloudy days. Hot water can conveniently be stored overnight, but it is expensive to build a system large enough to heat and store enough hot water to last during several days of cloudy weather. Therefore, most active solar systems are installed together with a conventional furnace. The solar collector is used on sunny days, and the conventional heater is used when it is cloudy. Naturally, such a dual system is initially more expensive than a simple fossil fuel system, but large amounts of fuel are saved every year.

It is difficult to assess the economics of active systems accurately. A system that will save money in Colorado may be uneconomical in Michigan. Overall, active solar heating is not the cheapest heating system in many areas. Therefore, the concept of solar energy has been looked upon unfavorably in some instances. This is unfortunate, because in most instances when active systems do not pay, it is because

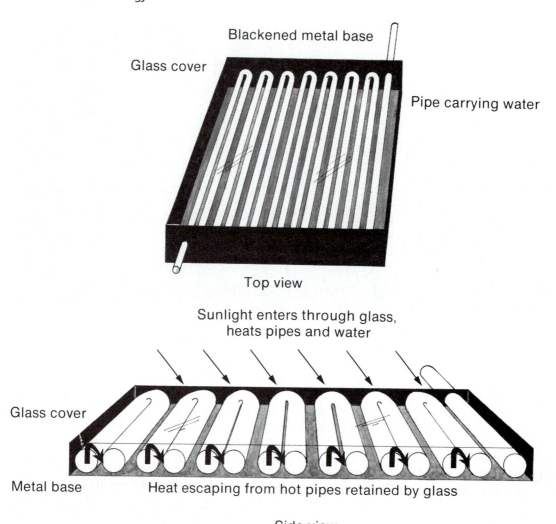

Blackened metal base

Glass cover

Pipe carrying water

Top view

Sunlight enters through glass,
heats pipes and water

Glass cover

Metal base   Heat escaping from hot pipes retained by glass

Side view

Figure 11–18 *Schematic view of a solar collector. The most efficient collectors of this type can actually boil water, but most practical units heat water to about 80°C.*

passive solar architecture is less expensive. There is no place in the world where it is economical to build an inefficient structure and make up the difference with fossil fueled heating and air conditioning.

Solar collectors are most practical for the production of hot water for washing or bathing. One reason for this is that active heating systems are used for only roughly half the year. The rest of the time the expensive capital investment lies idle. But domestic hot water is used all year, summer and winter. The first active solar water heater was patented in 1891 in the United States. It consisted simply of a bare metal tank,

painted black and tilted facing the Sun. At a time when natural gas sold for more than ten times the 1980 cost (calculated in equivalent dollars), these early heaters were a commercial success. At the present time, the use of solar collectors for the production of hot water is economical in most places in the world. Solar water heaters are required by law in all new homes in northern Australia and in San Diego, California. Millions of units have been installed in Japan and in Israel. Despite the fact that they are economically attractive, solar hot water units are not particularly popular in the United States. But the trend is changing slowly. In 1979 the av-

Figure 11–19 *Solar collectors on a residential home. (Courtesy of Energy Systems Division of the Grumman Corporation, manufacturers of Sunstream Solar Collectors.)*

erage solar hot water system was expected to pay for itself in five to seven years. After that time, it would save a homeowner money. In that year, 100,000 units were installed in this country. If passive solar design and solar hot water units were incorporated in every house in the United States, overall national energy consumption would be greatly reduced.

Solar Generation of Electricity

When a beam of light is directed onto certain materials, electrons can be energized. The energy from these moving electrons can be harnessed for production of power. Thus, light energy can be converted directly to electrical energy. A device that produces electricity directly from sunlight is called a **solar cell.** Solar cells are commonly used today to convert sunlight to electricity in spacecraft (Fig. 11–20) and have even been used in experimental aircraft (Fig. 11–21). They are quiet and trouble-free. They emit no pollution and appear to have a long life expectancy. A time can be envisioned when a homeowner would merely nail a panel of solar cells to the roof and from it would derive a significant portion of the household electrical energy without depleting the resources of the Earth. Large power plants could be built as well. A 1000-megawatt power plant (equivalent to a

large fuel-burning facility) would occupy only 10 square kilometers if it were built in the southwestern American desert.

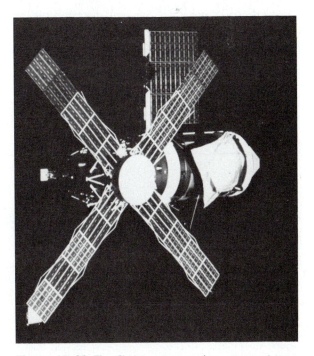

Figure 11–20 *The Skylab space station, powered by a group of photoelectric cells mounted on "windmill" arms. (Courtesy of NASA.)*

BOX 11.2 SOLAR HEATING OF HOMES

Advantages of Solar Energy

Limitless supply

Produces no air pollution

Produces no water pollution

Produces no noise

Produces no thermal pollution

Produces no harmful wastes

No possibility of a large-scale explosion or disaster

Conserves the Earth's resources

Technology available for immediate widespread use.

Disadvantages of Solar Energy

Some active systems are slightly more expensive than oil heat in some parts of the country. (But passive systems are economical in most applications.)

$10 per watt. Those numbers by themselves look discouraging, and many people believe that solar cells are an idea for the distant future, although not everyone agrees. Some inventors believe that breakthroughs for developing inexpensive solar cells are very near. Early in 1982, one research team published a report that a new process had just been developed to manufacture inexpensive cells, but at that time the units were not commercially available and the factories to produce them had not yet been built.

Of course, solar-generating stations alone could not entirely replace fossil and nuclear fuels, because the sun does not shine all the time.

There are several ways to store solar energy. Perhaps the most practical is to use solar electricity to produce hydrogen fuel according to the equation

$$\text{Water} + \frac{\text{Electrical}}{\text{Energy}} \rightarrow \text{Hydrogen} + \text{Oxygen}$$

Hydrogen is a versatile and useful fuel that can burn in air and be used as a replacement for gasoline and other liquid fuels. It is a particularly clean fuel, because water is the only by-product released when it is burned.

$$\text{Hydrogen} + \text{Oxygen} \rightarrow \text{Water} + \text{Heat energy}$$

These dreams haven't been realized, however, because solar cells are so expensive that solar generation of electricity is uneconomical in most places even though the fuel is free. It is estimated that if the price of petroleum remains constant, solar cells will become competitive if they can be produced at a cost of about 50 cents per watt of output. The cost in 1981 was $5 to

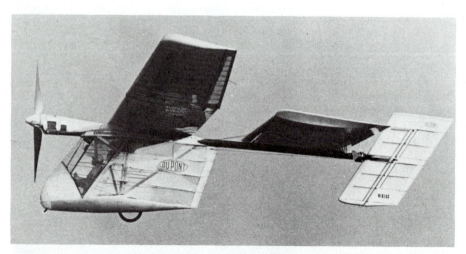

Figure 11–21 Solar Challenger was designed and built by a team headed by Dr. Paul MacCready, Pasadena, California. It is completely powered by an array of solar cells mounted on the wings and contains no batteries or other energy storage devices. (Photo courtesy of Randa Bishop, Wide World Photos.)

Production of High-Temperature Steam

If you take an ordinary magnifying glass and focus sunlight onto a piece of paper, you can easily burn a hole in the paper. The lens concentrates the solar energy from a large area to a small one so that high temperatures can be realized. Sunlight can also be concentrated through the use of specially designed mirrors. A large solar furnace of this type, completed in France in 1970, reaches temperatures up to 3500°C, hot enough to melt any metal. Such a device can be used to make hot steam to drive a turbine and produce electricity. In the United States two "solar power towers" have been built. One in Albuquerque, New Mexico, was put into operation in the fall of 1978. In this system, an array of mirrors focuses the sunlight from a large area onto a small space on the tower (Fig. 11–22). At the present time, the solar power tower is an experimental facility only; it is not economically competitive with traditional energy sources.

11.4 Energy from Plants

Most people in the world still get most of their energy from the Sun and from plants and animals. Such energy takes the form of grains, vegetables, and fruits for food, and wood and dung for fuel. In recent years, energy from plants has aroused interest in more developed societies as well. This interest has taken several different forms.

Wood

One hundred and fifty years ago, the major resource for energy in North America was wood. Today, again, there is a renewed interest in this fuel. In 1981, approximately 7 percent of the homes in the United States used wood at least partially for heating (Fig. 11–23). In some regions, such as New England, 50 percent of households used wood heat. According to one

Figure 11–22 A forest of mirrors focuses sunlight on the 200-foot-high "power tower" at the Department of Energy's Solar Thermal Test Facility at Sandia Laboratories. Each of the 222 heliostat arrays at the world's largest solar facility contains 25 four-foot square mirrors. At the present time this is a research facility only. It cannot produce power at a price competitive with fossil fuel–fired generating stations. (Courtesy of Department of Energy.)

Figure 11–23 In the first project of its type, the Burlington, Vermont, Electric Department converted two power generators at its 30-megawatt plant to burn chipped wood. (Photo courtesy of Tim Cronin, Burlington Electric Department.)

report, in 1981 more homes nationally were heated with wood than with electricity derived from nuclear energy. The reason is simple. In many cases, wood is the cheapest source of heat available. Many people are finding that a few weekends' work of cutting and splitting logs can lower their winter heating bills by 75 percent or more. Utilities, too, have discovered this economic advantage. The Electric Department of Burlington, Vermont, has constructed a 30-megawatt power plant fueled by wood chips, all obtained from a 70-mile radius around the plant (Fig. 11–23). The energy it produces is expected to cost about 20 percent less than energy produced from coal or nuclear fuel.

In some cases, this gain is balanced by various economic externalities. For one, uncontrolled wood fires in individual homes are likely to be smoky, and air pollution levels in certain regions have risen as a result. In addition, as the demand for wood fuel rises, the price of logs has risen as well. As a result, paper and lumber have become more expensive in some areas.

In Europe and the United States, trees are being planted and wood is grown as fast as it is used up. In many other places, particularly less

developed countries with large populations, this is not the case, and forest systems are being destroyed. For example, in August of 1978, the monsoon rains in India caused massive floods that ravaged the land. Entire sections of the city of New Delhi were under water, and an untold number of people were rendered homeless. Many farmers' fields were flooded and completely destroyed. The tragedy of the floods goes far beyond one season's rains. True, the monsoon downpours were unusually heavy, but in past decades heavy rains had caused far less flooding. Part of the problem stems from improper forestry and agriculture high in the Himalayas. In recent years, farmers in the mountain regions have cut nearly all the timber for fuel and have allowed their sheep and goats to overgraze the grasslands. As a result, there is little vegetation left to retain the water. When rain falls in the Himalayas now, relatively little is absorbed by the sparse plant cover. Therefore, great quantities of water run rapidly into streams and rivers. Mud slides are common, and lowland valleys become much more susceptible to floods than they were in the past. Thus the need for fuel has led not only to a shortage of wood, but also to poor soil conservation, which in turn has resulted in loss of crops, homelessness, and death.

Energy Plantations

Some experts suggest that trees grow too slowly to be used as a renewable energy source. They suggest instead that traditional crops such as sugar cane, pineapple, corn, soybeans, peanuts, or sunflowers be grown for the energy content in their sugars, stalks, or oils. Nontraditional crops have been examined as well. For example, the sap of a Brazilian tree species *Cobaifera langsdorfii* is equivalent to purified diesel oil. Tests have shown that each of these trees produces approximately 3 liters of sap every month; the sap can be placed directly in the fuel tank of a diesel-powered car.

The economic and environmental problems related to energy plantations are complex. Consider energy production from corn. If ordinary corn is cooked and then fermented, a watery solution of ethyl alcohol (ethanol) is produced. If this solution is distilled, the alcohol can be separated from the water until it is 95 percent pure. Of course, there is nothing new about this process, it is the same process as that traditionally used for the production of corn whiskey. But the next step is different: If 10 percent ethanol is mixed with 90 percent gasoline, a motor fuel called **gasohol** is produced. In the late 1970s the Carter administration subsidized gasohol manufacture and distribution in an effort to reduce the dependence of the United States on foreign oil. Some scientists objected to this program on the basis that fossil fuels are needed to plant, harvest, and transport the corn. Additional supplies of conventional fuels are required to manufacture fertilizers and pesticides. Once the corn is harvested, fuels are needed to cook the grain, warm the sugar as it ferments, and then distill the mixture to obtain the pure alcohol (Fig. 11–24). In all, about 40 MJ (megajoules) are required to produce one liter of ethanol from corn. However, the energy content of a liter of ethanol is only 21 MJ. In other words, there is a net loss of some 19 MJ per liter! Why is ethanol even considered as an economical fuel? To answer this question, one must first compare the energy balance for ethanol with that for the production of gasoline. Normally, foreign fuels are imported in

Manufacture and spread pesticides and fertilizers

Plow, plant and cultivate the crop

Cook the grain and heat the mash

Distill the mash to extract the alcohol

Add 90% gasoline

Gasohol

Figure 11–24 *Energy inputs in producing gasohol.*

the form of crude oil, a heavy, viscous mixture of many chemicals. This mixture must be refined into useful products such as gasoline and kerosene. The refining process consumes slightly more energy than that needed to produce ethanol. This implies that ethanol production

TABLE 11–1   Comparison of Energy Balance of Production of 1 Liter of Gasohol* with Production of 1 Liter of Gasoline.

Process	Gasoline (MJ/L)†	Gasohol (four different estimates by four different scientists) (MJ/L)†			
Fossil fuel input to produce fuel	42	42	39	41	38
Energy content of fuel	35	33	33	33	33
Total energy balance	−7	−9	−7	−8	−5

*Ninety percent gasoline plus 10 percent ethanol.
†Megajoules per liter.

may have value, but our energy accounting still isn't complete. A liter of ethanol contains less energy than a liter of gasoline. Table 11–1 compares the energy requirements for the production of gasoline with four different estimates of the energy requirements for the production of gasohol. Depending on which estimate of fossil fuel input for gasohol production is used, gaso-

line production may or may not be more efficient than gasohol production. To add to the confusion, even though ethanol has less energy per liter than gasoline, it burns more cleanly and efficiently in a standard automobile engine for reasons that are not well understood. Taking all factors into account, the authors of one technical article on the subject wrote, "The energy accounting can be confusing."

Well, are fossil fuel supplies conserved or not by using ethanol in fuel? Overall, small savings are probably realized. In addition, coal or agricultural wastes such as corn stalks could be used to cook and distill the grain. If this were done, low-grade solid fuel could be used to produce high-grade liquid automotive fuel, and therefore, gasohol probably provides a net gain in terms of useful energy.

On the other hand, analysis of energy balance may not be the main point anyway. According to Lester Brown (see Suggested Readings), the use of high-quality grains for automotive fuel represents a poor set of priorities in the modern world (See Table 11–2). He writes:

> The demand by motorists for fuel from energy crops represents a major new variable in the food/population equation. The stage is set for direct competition between the affluent minority, who own the world's 315 million automobiles, and the poorest segments of humanity, for whom getting enough food to stay alive is already a struggle. As the price of gasoline rises, so, too, will the profitability of energy crops. Over time, an expanding agricultural fuel market will mean that more and more farmers will have the choice of producing food for people or fuel for automobiles. They are likely to produce whichever is more profitable.

Energy from Waste

Many societies in the developed world are incredibly wasteful. Half of the household trash in the United States and Canada is paper. Huge piles of bark, wood scraps, and logging wastes rot slowly near many sawmills. If people collected these wastes and used them as fuel, considerable quantities of energy could be salvaged. In France, there are 20 generators that burn garbage for use as domestic energy sources. A large facility in Paris produces electric power for 130,000 people and 20 percent of the total steam

---

SPECIAL TOPIC B
Alcohol as an Automobile Fuel

For several years, large quantities of alcohol produced from sugar cane were used for automotive fuels in Brazil. As late as the summer of 1981, the government predicted that 20 percent of all Brazilian vehicles would be powered by alcohol by 1985; in 1980 cars designed to burn alcohol accounted for 80 percent of all new car sales in the country. A secondary program was started to convert oilseeds into diesel fuel, and biomass energy seemed to be a growing reality.

Then suddenly, the dream began to fade. In late 1981, the cost of alcohol rose to $2.15 per gallon, alcohol-burning cars started to exhibit chronic mechanical problems, and even under the best of conditions such cars lacked the power of gasoline-fueled vehicles. In addition, serious pollution problems were generated by the distilling industry, and the loss of food-producing land became serious.

As a result, people began to return to using gasoline. In the fall of 1981, alcohol-burning cars accounted for only 8 percent of new car sales, 150,000 of the new units sat idle in dealers' parking lots, and Fiat and General Motors stopped production of these models.

TABLE 11–2 Annual Per Capita Grain and Cropland Requirements for Food and for Automotive Fuel

	Grain (kilograms)	Cropland (hectares)
Subsistence diet	180	0.1
Affluent diet	730	0.4
Typical compact automobile (11,000 km/yr at 10 km/L)	2800	1.3
Typical full-size automobile (16,000 km/yr at 6.3 km/L)	6600	3.2

used for heating in the entire city. In North America a few facilities burn trash for fuel, but in most regions garbage is simply buried in a landfill.

Fuel can be produced from other waste products as well. When organic wastes such as sewage, garbage, manure, or crop residues decompose in the absence of air, methane gas is released. Methane produced in this manner is called **biogas.** Despite the difference in names, however, methane produced from wastes is identical to methane extracted from an underground gas deposit. In an ideal sequence, a farmer would collect cow manure and deposit it in a specially designed underground concrete pit with a steel cap to hold the gas released. The manure would be allowed to rot naturally, and the gas would be collected and used. The decomposed manure remaining in the concrete tank would then be removed and spread on the fields as a high-quality fertilizer. Studies in China have shown that a small biogas digester using human wastes and manure from a few cows can supply enough fuel to cook three meals and boil 15 liters of water a day for one family. Construction costs are returned in one to three years in the form of fuel savings. Many thousands of these units have been built throughout Asia.

There has been some interest in biogas in the developed world as well. For example, in Modesto, California, methane from the sewage treatment plant has been collected and used in city vehicles. The cost to the city in 1980 was equivalent to buying gasoline at 30¢ a gallon! It is incredible that in this age of fuel shortages, there are only a few places in North America where there is a commercial interest in this fuel.

11.5 Energy from Wind, Water, and the Earth

Hydroelectric Energy

Many early settlers in North America used the power of falling water to drive their mills and factories (Fig. 13–17). Today many large rivers are dammed (Fig. 11–25). The energy of wa-

Figure 11–25 A large hydroelectric dam. (Courtesy of Bureau of Reclamation; photo by E. E. Hertzog.)

Figure 11-26 A large turbine for generating electric power. The turbine spins when water falls past its blades. Note its size compared with the height of the workers in the background. (Courtesy of Tennessee Valley Authority.)

ter dropping downward through the dam is used to produce electricity. Energy produced in this way is called **hydroelectric energy.** The principle here is uncomplicated. The heart of a hydroelectric system is a fan-shaped device called a **turbine** (Fig. 11-26). Water falling through a pipe in the dam flows past the blades of the turbine. The blades are forced to rotate, thereby driving an electric generator. No fossil fuels are used, and the power supply is renewable continuously as water evaporates from the ocean and falls to the Earth to collect in the mountains.

Today, about 5 percent of the total world consumption of energy is supplied by hydroelectric generators. However, there are large regional differences in the use of this resource. In the United States, 4 percent of the total energy is supplied by falling water, whereas this total is 50 percent in Norway and around 75 percent in parts of China. Figure 11-27 shows that only a small percentage of the total worldwide potential for hydropower has been exploited. On the one hand, engineers are starting to plan huge projects such as damming the Amazon in Brazil, the Yangtze in China, and many of the rivers flowing into the Arctic such as the Yukon, the Mackenzie, the Ob, and the Lena. On the other hand, environmentalists have raised many serious questions about large-scale hydroelectric projects. Some of the pros and cons of the use of this energy source are listed opposite. (See also Chapter 19.)

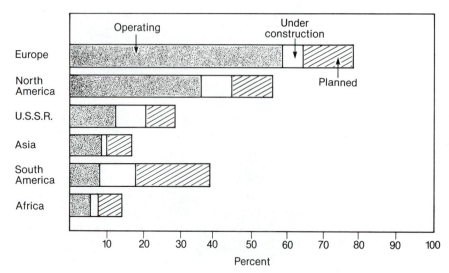

Figure 11-27 Hydropower use in various regions. The numbers represent the percentage of the total potential hydropower in the specific region.

Benefits and Problems of Large-Scale Hydroelectric Development

BENEFITS	PROBLEMS
Cheap, pollution-free, renewable energy is made available.	Silt from soil erosion upriver can fill in the lakes behind large dams, reducing their utility within a few decades. Then the expensive dam projects become useless.
Water stored behind dams can be used for irrigation, thus increasing productivity of local agriculture.	Dams flood valuable farmland, reducing food production in many areas.
Cheap electricity provided by dams can be used to manufacture fertilizer.	In a free-flowing river, soil nutrients released from regions upstream are carried to the valleys below. During flood time, some of these nutrients are spread out on lowland farms, renewing the fertility of the soil. When the river cycle is disrupted by the dam, the flow of nutrients is cut off, reducing soil fertility near the river mouth.
Dams establish lakes that provide a habitat for fish, which in turn can be used for a food supply.	The traditional flow of nutrients to the sea also fertilizes ocean estuaries, providing food for salt water fish. When this flow is reduced, populations of ocean fish have been disturbed. For example, the sardine catch on the mouth of the Nile River declined by 18,000 tonnes annually when the Aswan dam was built. In some cases, dams disrupt migration of spawning fish such as salmon. Also, water taken from considerable depths below the surface of the dammed lake is often much colder than surface water and therefore changes the biota of the stream below the dam.
Lakes produced by dams are prime recreational sites.	Great aesthetic loss occurs when beautiful natural canyons are obliterated.
In many areas a dependable water supply created by dams has increased public hygiene.	In many areas disease organisms breed in dam-produced lakes or irrigation canals, leading to sickness and death.
Hydroelectric energy reduces the need for fossil fuel, which produces carbon dioxide, which may disrupt world climate. In this way, world climate is protected.	There is a fear that major dams planned for the future might disrupt the flow of warm water into the Arctic, thereby disrupting world climate.

In view of the environmental problems outlined here, many people believe that a better answer might be construction of many small-scale hydroelectric facilities rather than a few large dams (Fig. 13–17). Small-scale hydroelectric power was popular in the United States in the late 1800s. Today these facilities are used extensively in parts of Asia, and they are slowly gaining popularity in the developed world because they are cheaper than the construction of large dams (see Chapter 13).

Tidal Power

Ocean waves, tides, and currents are global in extent and extremely powerful. To date, the most significant attempts at harnessing the energy of the ocean have been related to tidal power.

In many coastal regions, the flow of the tide naturally funnels through narrow entrances into bays and estuaries, and strong currents are established. Twice a day the water flows inland, and twice a day it rushes outward with the ebb tide. Some of the energy from this movement of currents can be harnessed if a tidal dam is built and a turbine is installed.

During the 1800s tidal power was popular in the United States. At that time it was an attractive alternative to bulky steam engines that consumed large quantities of wood. As a result, many grainmills and sawmills were located along tidal races. One of these old mill sites can be seen today on the island of Vinalhaven, off the coast of Maine (Fig. 11–28). During the last century, two turbines were placed in a tidal dam and used to power a large sawmill in Vinalhaven. Today, a motel sits on part of the foundation of the old mill, and most of the machinery has long since been abandoned. In the early 1970s the town electric generator, powered by coal, began to show signs of wear, and the city needed a new source of electricity. Three alternatives were suggested: (1) Rebuild the local generator; (2) repair the tidal dam and install new turbines to be used to produce electricity; or (3) lay an undersea transmission cable to the

mainland and connect it to the public utility grid. The last alternative was the cheapest at the time and the one that was eventually employed.

Certainly it would have been technically feasible to use tidal power. In fact, many local residents now regret their decision, because recently the price of electricity has risen with the rising price of fuel. Now that the town is drawing energy from the mainland, however, it is not at all easy to convert back to a tidal system. There are mortgage payments on the undersea cable, and it will be many years before the mortgage is paid back. Change in our energy structure occurs slowly. Once a decision is made, it often becomes entrenched in physical and economic structures, and it may be years or decades before change is feasible.

Power from Ocean Waves and Currents

Since 1876, some 150 patents have been issued in the United States alone for devices to harness the energy in waves (Fig. 11–29). Two of these devices are shown in Figure 11–30.

Figure 11–28 An old tidal dam and mill along the coast of Maine. In the early 1900s this mill was used to saw logs, but at the present time the dam has deteriorated and the mill is no longer in use. (Courtesy of the Bath Maritime Museum.)

However, as of 1984, there are no large scale wave energy generators in use.

The steady movement of winds, the spin of the Earth, and other factors combine to create large mass movements of ocean water called **currents.** These currents, which can be likened to large rivers within the sea, could be harnessed to create usable energy. The problem is one of concentration. When engineers want to obtain

Figure 11–29 *Wave power. The Great Wave at Kanajawa, by Katsushika Hokusai. (Courtesy of the Metropolitan Museum of Art.)*

A

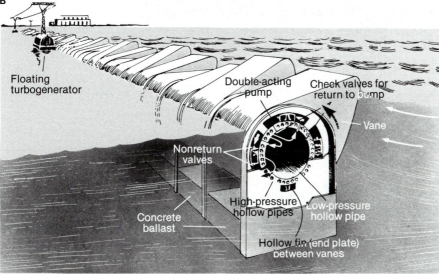

B

Floating turbogenerator

Double-acting pump

Cheek valves for return to pump

Vane

Nonreturn valves

High-pressure hollow pipes

Low-pressure hollow pipe

Concrete ballast

Hollow fin (end plate) between vanes

Figure 11–30 *Two devices to harness energy from waves. A, Flexible rafts: As raft flexes with waves, hinged outer floats operate double-acting hydraulic cylinders in central section. Cylinders pump fluid to hydraulic motor driving an alternator. B, Salter's Duck consists of a series of segmented cam lobes that are rocked on a large spine by waves, which are shown coming from the right. Pumps connected to the lobes send high-pressure water through small pipes to drive a generator. The back sides of the cams are cylindrical to minimize wave regeneration by the device. (A and B from* Popular Science, *reprinted with permission.)*

electricity from the movement of a river, they first build a dam, which in effect concentrates the energy of the moving water. There is a large amount of potential energy in a current such as the Gulf Stream, but it is so broad and deep that little energy is available per square meter; and it is impossible to dam the ocean. Inventors have suggested the possibility of using turbines or some sort of paddlewheel system to trap some of this potential energy. The problem is that any device would have to be so large that it would be unreasonably expensive.

Despite these drawbacks, some planners believe that the oceans could provide us with considerable energy. Figure 11–31 shows one author's estimate of the potential harnessable power from the oceans. Although it looks encouraging, don't expect this potential to be realized within the next few years. Some projects, such as local tidal dams, are certainly practical today, but the engineering problems and capital investments needed for others are enormous.

Ocean Thermal Power

It is also possible to use the heat of the ocean to produce electricity. The Second Law of Thermodynamics tells us that a heat engine can be built to derive energy whenever there is a temperature difference between two bodies. For example, a coal-fired generator runs because the temperature of the steam on one side of the turbine is hotter than the temperature of the exhaust. In certain tropical regions the surface of the ocean is approximately 20°C warmer than the subsurface layers. The warm water is hot enough to vaporize a pressurized liquid such as ammonia. The gaseous ammonia can drive a turbine just as hot steam drives a turbine in a coal-fired plant. In an ocean thermal generator, the gaseous ammonia is cooled by the subsurface water and reused, as shown in Figure 11–32. The theoretical thermodynamic efficiency of an engine that operates between 5° and 25°C is only 7 percent. In practice, engines always operate at less than theoretical efficiency, so the practical efficiency is somewhere around 3 percent. Proponents of the plan to use ocean thermal power plants argue that even if the efficiency is low, there is so much water in the oceans that a large amount of energy is available. Opponents argue that the capital costs of the power plants are so high that it will be decades at least before ocean thermal power is economically feasible.

Energy from the Wind

The power of wind has been used since antiquity to drive ships, pump water, and grind grain. What is its potential in modern society? There is more than enough wind energy available to supply the world's energy needs, and it

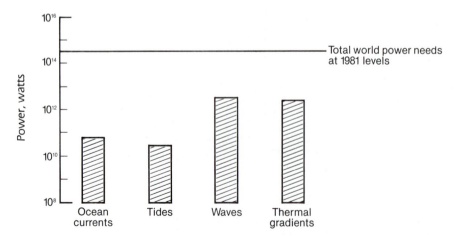

Figure 11–31 Global potential for practical harnessable power from the ocean. Note that this figure represents one author's opinion of what can be achieved. To date, very little useful energy has been extracted from the oceans. Also note that this graph is drawn on a logarithmic scale (multiples of 10). (From Isaacs, J. D., and Schmitt, W. R.: *Science, 207:* 265, January 18, 1980.)

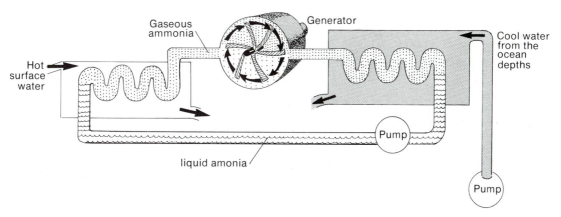

Figure 11–32 *Schematic of an ocean thermal power plant. Pressurized ammonia is boiled by the warm surface waters of tropical oceans. The ammonia gas expands against the blades of a turbine, and the spinning turbine drives a generator to produce electricity. The gases are cooled and condensed by colder subsurface waters that are pumped into the power plant.*

is technically feasible to build windmills capable of producing electricity. These can be built either as small, home-sized units or as large central generators (Fig. 11–33). In the United States at the beginning of the 1980s, energy from wind was, in some instances, more expensive than that from fossil fuels but cheaper in others. An accurate accounting is difficult because of var-

A   B

Figure 11–33 *A,* On July 11, 1979, the world's largest wind turbine generator, with blades as long as the wingspan of a Boeing 707, was dedicated near Boone, North Carolina. At winds of 18 km per hour the giant experimental wind turbine begins to generate electricity, and at winds of 40 km per hour and above it can produce 2000 kilowatts (2 megawatts) to serve part of the needs of Boone while supplying important test data to the U.S. Department of Energy. This is approximately the amount of electricity used by 500 average homes. (Courtesy of Department of Energy. Photo by Dick Peabody.)
   B, A small windmill for generation of electricity for a single household. These units are economical in such places as small communities in Alaska where electricity is expensive.

Figure 11–34 This oil tanker, built by Nippon KoKan Company of Japan, uses sails to assist the conventional diesel engines. The sails are set and furled by remote control from the bridge, and reduce fuel consumption. (Courtesy NKK, American, Inc.)

ious governmental policies that subsidize or favor one system or another. But if windmills were mass produced, then perhaps energy from the wind could be cheaper than energy from nuclear power plants. The "catch-22" of the problem is this: No industrialists want to mass produce windmills because they are not economically attractive, but they are not economically attractive because they are not mass

Figure 11–35 Section of the geothermal steam field where Pacific Gas and Electric Company generates 396,000 kilowatts of electricity from underground steam.

produced. To be economical, a wind turbine for home use must generate electricity at a cost of about 6.5¢ per kWh (kilowatt hour). In 1974, windmill electricity cost 20¢ per kWh; in 1982 the cost was reduced to 7.5¢. If the cost can be reduced still further, or if the price of fuel continues to rise, wind turbines will become competitive.

In recent years, commercial interests have again been examining the economics of sailing ships. At present, some traditional sailing ships are used in the fishing industry or in local island areas such as the South Pacific or the Caribbean. In addition, designers are interested in the construction of large, technically advanced sailing vessels for commercial trade between major industrial ports. One test vessel is the Japanese-built oil tanker shown in Figure 11–34. The sails on this tanker are operated hydraulically and can be controlled by a single pilot on the bridge. Using sails combined with commercial engines and employing advanced hull designs, sailing-freighters operating between major industrial cities could save 50 percent of the fuel currently used by commercial ships.

Geothermal Energy

Energy derived from the heat of the Earth's crust is called **geothermal energy.** In various places on the globe, such as in the hot springs and geysers of Yellowstone National Park in Wyoming, hot water is produced near the surface. Although no one suggests harnessing Old Faithful for generating electricity, there are several places where hot underground steam is available. The power company must simply dig a well and pipe the free steam into a turbine. Geothermal energy has been used in many different regions. The Pacific Gas and Electric Company produces some electricity from a generator connected to wells in central California (Fig. 11–35), an Italian facility at Larderello has been in operation since 1913, and parts of New Zealand have been supplying all their electricity from geothermal plants for many years.

At the present time only 0.1 percent of the energy used in the United States is met by geothermal sources. Even optimistic supporters of the program do not expect a significantly larger portion of our power to be supplied in this manner in the future. There are not enough hot springs at or near the surface of the Earth. Con-

tinuous exploitation for more than a century or two is expected to exhaust the water or heat content of these wet wells. Geothermal energy is not always free from pollution. Underground steam or hot water is often contaminated with sulfur compounds, which must be removed before they are discharged to the air or to a lake or river.

Summary

The United States uses considerably more energy per capita than other developed countries do.

Oil and gas shortages are expected around the year 2000. Coal reserves are expected to be plentiful for 200 years or more. Additional supplies of fuels are available from heavy oils.

Solar energy in the form of passive solar design or solar collectors can be used to heat air or water for domestic use or to produce high-temperature steam for operating turbines. Solar cells can be used for the direct generation of electricity.

Plant energy can be used in the form of wood, energy plantations (sometimes for production of alcohol), burning of wastes, and production of biogas.

Additional energy is available from hydroelectric sources, from the tides, waves, and currents, from temperature differences in the ocean, from the wind, and from geothermal sources.

Questions

Energy Use and Choices

1. Discuss the qualitative and quantitative differences in energy consumption among people in the United States, Sweden, and Bangladesh. Can you explain how these differences may ultimately affect global politics?

2. Imagine that you were a Peace Corps advisor to a small village in the mountains in India. What energy sources would you encourage the people to develop? Remember, capital for modern machinery would be extremely limited.

3. Analyze the potential for non–fossil fuel energy resources in your home town. What energy sources could you use that would eventually be cheaper than fossil fuels?

Fossil Fuel Energy

4. Discuss the problems inherent in predicting the future availability of fossil fuel reserves. What is the value of the predictions?

Energy from Water

5. Discuss the potential advantages and disadvantages of hydroelectric energy. Would similar problems evolve if large-scale tidal energy sites were developed?

Energy from Plant Matter

6. Speak to a local municipal politician about the feasibility of collecting methane from the city sewage treatment plant. Report the results of your discussion to your classmates.

7. Go to your local supermarket and determine the number of cardboard boxes discarded daily. What is done with this cardboard? Can you suggest other uses for it?

Solar Energy

8. Discuss the utility and limits of using solar collectors and passive solar design to reduce the energy consumption in the United States.

9. Suggest some practical cost-effective improvements that could be added to your school, dormitory, or home that would use solar energy effectively.

10. Discuss the advantages and disadvantages related to the use of solar cell–operated generating stations to produce electricity. What could be done to supply energy at night?

11. Explain how wind, waves, ocean currents, plant, and hydroelectric energy resources are all forms of solar energy.

Suggested Readings

The following general references provide an overview of the energy issues:

Philip H. Abelson, ed.: *Energy: Use, Conservation, and Supply.* Washington, D.C., American Association for the Advancement of Science, 1974. 154 pp.

Barry Commoner: *The Poverty of Power.* New York, Alfred A. Knopf, 1976. 314 pp.

Allen L. Hammond, William D. Metz, and Thomas H. Maugh, II: *Energy and the Future.* Washington, D.C., American Association for the Advancement of Science, 1972. 184 pp.

S. S. Penner, and L. Icerman: *Energy.* Reading, Mass., Addison-Wesley, 1974. 373 pp.

Marion L. Shephard, Jack B. Chaddock, Franklin H. Cocks, and Charles M. Harmon: *Introduction to Energy Technology.* Ann Arbor, Mich., Ann Arbor Science Publishers Inc., 1976. 300 pp.

H. Stephen Stoker, Spencer L. Seager, and Robert L. Capener: *Energy from Source to Use.* Glenview, Ill., Scott, Foresman and Company, 1975. 337 pp.

Some specific references that apply to individual sections are given below.

Introduction

John U. Nef: An early energy crisis and its consequences. *Scientific American,* November 1977, p. 140.

11.1 Modern Energy Use

Joy Dunkerley, W. Ramsay, L. Gordon, and E. Cecelski: *Energy Strategies for Developing Nations.* Baltimore, Johns Hopkins University Press, 1981. 265 pp.

Baclaz Siml: Energy flows in the developing world. *American Scientist,* 67:522, September-October, 1979.

11.2 Fossil Fuel Sources and Availability

Richard A. Dick, and Sheldon P. Wimpfen: Oil mining. *Scientific American,* October 1980, p. 182.

Richard C. Dorf: *The Energy Fact Book.* New York, McGraw-Hill Book Company, 1981. 226 pp.

Andrew R. Flower: World oil production. *Scientific American,* March, 1978, p. 42.

Ford Foundation: *A Time to Choose, America's Energy Future.* Cambridge, Mass., Ballinger Publishing Company, 1974. 512 pp.

Gerard J. Manjone: *Energy Policies of the World,* Vols. 1 and 2. New York, Elsevier Publishing Company, 1977.

Robert S. Pindyck: *The Structure of World Energy Demand.* Cambridge, Mass., MIT Press, 1979. 299 pp.

An excellent source of energy statistics is available in the following two periodicals:

U.S. Department of Energy, Energy Information Administration: *Monthly Energy Review.* Washington, D.C., U.S. Government Printing Office.

U.S. Department of Energy, Energy Information Administration: *1980 International Energy Annual.* Washington, D.C., U.S. Government Printing Office.

11.3 Energy from the Sun

Three how-to-do-it books on conserving energy in your own environment are:

Bruce Anderson: *Solar Home Book.* Harriville, New Hampshire, Cheshire Books, 1976. 298 pp.

George S. Springer, and Gene E. Smith: *The Energy-Saving Guidebook.* Westport, Conn., Technomic Publishing Co., 1974. 103 pp.

Carol H. Stoner: *Producing Your Own Power.* Emmaus, Pennsylvania, Rodale Press, 1974. 322 pp.

Daniel K. Reif: *Solar Retrofit: Adding Solar to Your Home.* Andover, MA, Brick House Publishing Co., 1982.

An excellent book on the history of solar energy is:

Ken Butti, and John Perlin: *A Golden Thread—2500 Years of Solar Architecture and Technology.* Palo Alto, Calif., Cheshire Books, 1980. 283 pp.

An interesting article on passive solar cooling is:

Mehdi N. Bahadori: Passive cooling systems in Iranian architecture. *Scientific American,* February, 1978. p. 144.

A text on solar technology is:

J. Richard Williams: *Solar Energy Technology and Application.* Ann Arbor, Mich., Ann Arbor Science Publishers, 1974. 119 pp.

11.4 Energy from Plants

David A. Tillman: *Wood as an Energy Resource.* New York, Academic Press, 1978. 252 pp.

Good articles on gasohol production and energy plantations are:

Lester R. Brown: Food or fuel: New competition for the world's cropland. *Worldwatch Paper 35,* March 1980.

Lester R. Brown: Food versus fuel. *Environment, 22* (4):32, 1980.

R. S. Chambers et al: Gasohol: Does it or doesn't it produce positive net energy? *Science, 206:789,* 1979.

Gasohol: Does it save energy? *Environmental Science and Technology, 14(2):1402, 1980.*

An excellent article on biogas is:

Edgar J. Dasilda: Biogas: Fuel of the future? *Ambio, 9(1), 1980–1982.*

11.5 Energy from Wind, Water, and the Earth

Daniel Deudney: An old technology for a new era. *Environment, 23(7):17, September, 1981.*

John D. Isaacs, and Walter R. Schmitt: Ocean energy: Forms and prospects. *Science, 207(4428):265, January 18, 1980.*

Gerald W. Koeppl: *Putnam's Power From the Wind.* New York, Van Nostrand Reinhold Co., 1982. 470 pp.

12

Nuclear Energy and the Environment

Nuclear energy burst onto the public consciousness in 1945, when two fission bombs were used against Japan. The reaction everywhere was that a new kind of energy had been unleashed. The public understood that the "new energy" could be used for human betterment as well as for war. There was much hope for the "peaceful use of atomic energy," meaning that nuclear energy could be utilized in power plants rather than in bombs.

The nuclear energy program was launched with widespread public approval and support. However, the facts that radioactive materials could be deadly, that fission fuels could also become explosives, and that the development of nuclear weapons had not stopped with the end of the war led to a gradually widening sense of unease. Furthermore, accidents started to happen—one at Chalk River in Canada in 1952 and another at Windscale in England in 1957. In both cases, various amounts of radioactivity were released to the environment.

By the 1960s, as the biological dangers of radiation came to be better understood, the nuclear debate became a recognized public issue. Concern about nuclear safety became more urgent, but the development of nuclear energy and the plans for its further rapid expansion continued.

Despite these developments, neither the public nor the "experts" were prepared for the events of the 1970s. That was the decade of frightening near-catastrophes at the Browns Ferry Nuclear Power Plant and at Three Mile Island. For the first time, the debates about the advisability of continuing the nuclear program, or even the possibility of scrapping it, reached significant political levels.

In the early 1980s, the nuclear industry in the United States actually stopped growing—for an unexpected reason. It seemed that it was no longer profitable to generate electricity by nuclear energy. In September, 1981, for example, the Boston Edison Company abandoned plans for its Pilgrim II nuclear plant in Plymouth, Massachusetts. When this plant had first been proposed in 1972, it was supposed to cost $420 million and to begin producing electricity in 1979. By October, 1981, the cost estimate had reached $4 billion, and start-up was projected for 1990. Such costs meant that nuclear energy had little if any economic advantage over coal. Furthermore, the demand for energy did not grow as rapidly as expected. The rate of growth of electrical consumption in the United States, which was about 7 percent per year in the early 1970s, had fallen to about 3 percent by 1982. Because of such factors, investment firms have suggested that the power industry simply scrap some of the reactors for which construction permits have been issued.

Serious discussions about the future of the nuclear industry continue. Some people continue to argue in favor of a nuclear future; others wish to abandon the industry entirely. How, then, can you make your own decisions about such questions? Surely you should at least start with a background of reliable information. This chapter will not teach you how to operate a nuclear plant or how to make a bomb. Rather, it will describe, in principle, where nuclear energy comes from and how it can be used to generate electricity. The potential dangers of the nuclear energy program will also be considered.

12.1 Elements and Atoms

Chemical elements are considered to be the stuff of which all other substances are composed. There are about 105 known elements. Some common ones are hydrogen, carbon, nitrogen, iron, and uranium.

Atoms are the fundamental units of elements. An atom consists of a small, dense, positively charged center called a **nucleus** surrounded by a diffuse cloud of negatively charged **electrons** (Fig. 12–1). The unit of positive charge in the atom is the **proton,** and the **atomic number** is the number of protons, or unit

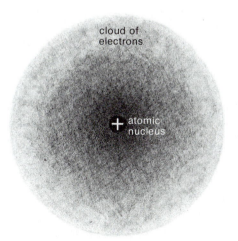

cloud of electrons

+ atomic nucleus

Figure 12–1 An atom consists of a positive nucleus surrounded by an electron cloud.

TABLE 12–1 Particles in the Atom

Particle	Electrical Charge	Mass Number
Electron	-1	0
Proton	$+1$	1
Neutron	0	1

positive charges, in the nucleus. It is also the number of electrons in the neutral atom. A hydrogen atom, atomic number 1, consists of a proton with a charge of $+1$ as its nucleus, surrounded by one electron with a charge of -1. An atom of carbon, atomic number 6, contains six protons and six electrons.

However, the protons account for only about half the mass of the carbon atom. The remainder is attributed to neutral particles, called **neutrons,** in the nucleus. The neutron has about the same mass as the proton. The **mass number** of an atom is the total number of protons and neutrons in its nucleus.

The properties of the three fundamental atomic particles are shown in Table 12–1. The atomic number and the mass number of an element are often written with its symbol, as shown below for carbon.

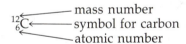

$_{6}^{12}$C ———— mass number
———— symbol for carbon
———— atomic number

Table 12–2 gives the nuclear compositions, atomic numbers, and mass numbers of atoms of carbon and uranium. Note that a given element may have more than one mass number. Atoms of the same element (that is, atoms with the same atomic number) that have different mass numbers are called **isotopes.** Thus, $_{6}^{12}$C and $_{6}^{14}$C are carbon isotopes. Both $_{6}^{12}$C and $_{6}^{14}$C represent carbon atoms (or carbon nuclei), because they both have six nuclear protons. They are isotopes because they have different mass numbers. This difference results from the fact that there are different numbers of neutrons in the two nuclei. Isotopes of an element are chemically equivalent (or very nearly so): they have the same ability to combine with other atoms. Thus, $_{6}^{12}$C and $_{6}^{14}$C are both chemically the same element, carbon. If the carbon is converted to some other compound, such as carbon dioxide, the isotopes themselves

TABLE 12–2   Nuclear Compositions of Some Isotopes of Carbon and Uranium

Name and Symbol		Nuclear Composition		Atomic Number (Protons)	Mass Number (Protons + Neutrons)
		Protons	Neutrons		
Carbon-12	$^{12}_{6}C$	6	6	6	12
Carbon-14	$^{14}_{6}C$	6	8	6	14
Uranium-235	$^{235}_{92}U$	92	143	92	235
Uranium-238	$^{238}_{92}U$	92	146	92	238
Uranium-239	$^{239}_{92}U$	92	147	92	239

are not altered but retain their separate identities in the carbon dioxide molecules.

The general formula that relates these terms is

$$\text{mass number} = \text{protons} + \text{neutrons}$$

or,

$$\text{mass number} = \text{atomic number} + \text{neutrons}$$

Because all isotopes of an element have the same atomic number but different mass numbers, the symbol and mass number are enough to identify the isotope. Thus $^{235}_{92}U$ may be written as ^{235}U or as uranium-235, or even just U-235.

12.2 Radioactivity

The French physicist Henri Becquerel discovered in 1896 that uranium minerals spontaneously emit energy in the form of radiation. Thus, if a piece of photographic film is held near a uranium mineral in the dark, the film becomes exposed, just as it would be if it were held near a light. This emission of radiation from an element was called **radioactivity**. A series of careful, tedious separations carried out by Marie and Pierre Curie (Fig. 12–2) resulted in the discovery of new radioactive elements, the most important of which was **radium.**

Some elements exist both as nonradioactive **(stable)** isotopes and as radioactive isotopes **(radioisotopes).** Radioactive elements undergo spontaneous **radioactive decay** in which they decompose, giving off particles and radiation and leaving another isotope behind. The products of radioactive decomposition may be radioactive in their turn or they may be stable. Both the particles and the radiation emitted during the decay process carry energy. Therefore, *all radioactivity is a source of energy.* This fact was recognized from the very beginning. (After all, it takes energy to darken photographic film.) However, naturally radioactive materials give off energy at very low rates. What's more, radioactive isotopes are very thinly distributed in the Earth's crust. Therefore, it was evident that natural radioactivity could never be a *practical* source of energy for human needs.

Figure 12–2 Madame Curie's laboratory. (From Weeks and Leicester: *Discovery of the Elements,* 7th ed. Easton, Pa., Journal of Chemistry Education, 1968.)

12.3 Half-Life

If you observed just one atom of say, radium-226, containing one nucleus, when would it decompose? This question cannot be answered because any particular radium nucleus may or may not decompose at any time. But observation has shown that a collection of radium nuclei decomposes at a particular rate, derived from its half-life. The **half-life** is the time it takes for half the nuclei in a sample to decompose. The half-life of radium-226 is 1600 years. Therefore, if 1 g of radium-226 were placed in a container in 1985, there would be only 1/2 g left after 1600 years (in the year 3585) and only 1/4 g after another 1600 years (in the year 5185), and so on. Each radioisotope has its own characteristic half-life.

The concept of half-life does not imply that after 1600 quiet years half of the radium will suddenly decompose. The half-life is an averaged value for all the radium nuclei. This means that there is a chance for some decompositions to occur in *any interval of time*. Since there are many atoms in a sample of radium, some will be decomposing every second, and any nearby Geiger counter (a device that responds to radiation) will be clicking all the time.

The rate at which the radiation is emitted from a sample of radium-226 depends on its quantity. Since each atom of radium-226 that decomposes is converted into another element (radon-222), the quantity of radium-226 in any sample constantly decreases. However, the radon it produces is also radioactive, and it, too, decomposes to produce another radioisotope. This series of radioactive disintegrations goes on through a number of "generations" until finally a stable isotope, lead-206, is produced. These radioisotopes have various half-lives, ranging from fractions of a second to about 20 years. Therefore, the total radioactivity produced by a sample of radium together with its radioactive waste products is more than that produced by the radium alone.

A final question is: If nuclei of radioactive elements are unstable, why are there any left on Earth? One conceivable answer is that these survivors are all descendents of radioisotopes with very long half-lives. The half-life of natural uranium-238, for example, is 4,500,000,000 years. The radiations from such materials plus the effect of radiation that comes to the Earth from outer space is called the **background radiation.**

We cannot invent anything to stop radioactivity. It slows down by radioactive decay at a rate determined by the half-lives of the radioisotopes involved.

12.4 Nuclear Reactions

Radioactivity and other nuclear changes can be represented by simple equations. An example is

$$^{226}_{88}\text{Ra} \rightarrow {}^{4}_{2}\text{He} + {}^{222}_{86}\text{Rn}$$
radium   helium   radon
nucleus

This equation tells us that a $^{226}_{88}$Ra nucleus decomposes to give one $^{4}_{2}$He and one $^{222}_{86}$Rn nucleus. (A $^{4}_{2}$He nucleus is called an **alpha particle.**) Note that both the mass numbers (226 = 4 + 222) and the atomic numbers (88 = 2 + 86) are balanced.

Nuclear reactions may also involve neutrons. We will use the symbol $^{1}_{0}$n to represent the neutron. (Its charge is 0 and its mass number is 1.)

The first artificial nuclear reaction was carried out by Ernest Rutherford and his co-workers in 1919, when they bombarded ordinary nitrogen with helium nuclei to produce oxygen-17, which is nonradioactive:

$$^{14}_{7}\text{N} + {}^{4}_{2}\text{He} \rightarrow {}^{17}_{8}\text{O} + {}^{1}_{1}\text{H}$$

Fifteen years later, in 1934, Irène and Frédéric Joliot-Curie, Mme. Curie's daughter and son-in-law, converted boron to nitrogen-13, which is radioactive. This was the first artificially produced radioisotope.

$$^{10}_{5}\text{B} + {}^{4}_{2}\text{He} \rightarrow {}^{13}_{7}\text{N} + {}^{1}_{0}\text{n}$$

The idea that neutrons might be used to bombard and alter atomic nuclei was exciting to all the scientists who were studying nuclear reactions. Of particular interest was the possibility that neutron bombardment of uranium (then the heaviest known element) might yield still heavier elements. In 1939, through the work of Otto Hahn, Fritz Strassman, and Lise Meitner, it was realized that neutrons caused the splitting, or **fission,** of uranium nuclei. Further studies showed that extra neutrons were also released in the fission reaction. If the fission reaction is *started* by neutrons and then also *releases* neutrons, a new possibility arises. It is the opportunity for a **chain reaction.** This discovery changed nuclear science from a study of purely theoretical interest to an issue of utmost importance to everyone.

A chain reaction is a series of steps in a process that occur one after the other, in sequence, each step being added to the preceding step like the links in a chain. Chemical chain reactions can also undergo branching. An example of a branching chemical chain reaction is a forest fire. The heat from one tree may initiate the reaction (burning) of two or three trees, each of which, in turn, may ignite several others. If the lengthening of one chain proceeds at a given rate, the production of 10 branches means that 10 reactions are going on at the same time, so that the rate has increased tenfold. A chemical chain reaction that continues to branch can produce an explosion. The condition under which a chain

reaction just continues at a steady rate, neither accelerating nor slowing down, is called the **critical condition.**

The production of the atomic (fission) bomb and of nuclear reactors depends on branching nuclear chain reactions. The process is initiated when a neutron strikes a uranium-235 nucleus and can proceed in any of several different ways. For example, as shown in Figure 12–3, uranium-235 plus a neutron forms barium-142 and krypton-91, while releasing three neutrons. In equation form:

$$^{235}_{92}U + ^{1}_{0}n \rightarrow ^{142}_{56}Ba + ^{91}_{36}Kr + 3\,^{1}_{0}n$$

Note the following important points about this equation:

1. The reaction is started by one neutron but produces three neutrons. These neutrons can initiate three new reactions, which in turn produce more neutrons, and so forth. This is, therefore, a branching chain reaction.

2. The uranium-235 nucleus is split in half (roughly) by these reactions. The splitting of a nucleus is called **atomic** or **nuclear fission.** Fission releases energy because the uranium nuclei have more energy than their breakdown products. The amounts of energy involved are very

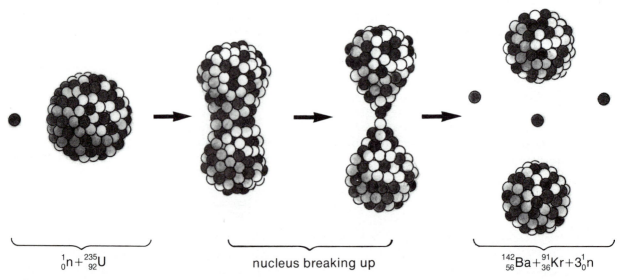

$^{1}_{0}n + ^{235}_{92}U$ nucleus breaking up $^{142}_{56}Ba + ^{91}_{36}Kr + 3^{1}_{0}n$

Figure 12–3 Nuclear fission.

large compared with those of chemical reactions. If the branching chain reaction continues very rapidly, it produces an atomic explosion. If the chain branching is carefully controlled, energy can be released slowly and used to drive turbines to generate electricity.

3. Fission reactions produce radioactive wastes. Barium-142 and krypton-91, the products shown in the preceding equation, are both radioactive. Furthermore, the reaction represented by this equation is only one out of many that occur in atomic fission. Many different radioisotopes are produced by atomic fission. Also, the fission products, as a group, are much more radioactive than the uranium from which they are produced. A compound with a short half-life decomposes quickly, but in its early stages it emits radiation at dangerously high levels.

12.5 Nuclear Fission Reactors

First, recall that the function of *any* power plant is to drive a turbine to generate electricity. The only differences among power plants are in the sources of the energy to drive the turbine. In a nuclear plant the source of energy is the nuclear fission reaction. A coal-fired plant releases chemical energy. A wind power or hydroelectric plant utilizes the mechanical energy of wind or falling water (Fig. 12–4).

Nuclear fission reactors require fuel, and the fuel must be a substance whose nuclei can undergo fission. There are two significant nuclear

fission fuels, uranium-235 and plutonium-239. These are not the only known fissionable isotopes, but they are the ones on which currently operating power plants and most designs for future plants are based.

Uranium-235 occurs in nature; it constitutes 0.7 percent of natural uranium. The remaining 99.3 percent is the heavier isotope, uranium-238, only very little of which undergoes fission in a reactor.

The second fuel, plutonium-239, does not occur in nature; it is produced by the reaction of uranium-238 with neutrons. Thus, the two important naturally occurring sources of fission energy are uranium-235 (fissionable but not abundant) and uranium-238 (abundant but not fissionable until it is converted to plutonium-239).

Nuclear reactors require another essential ingredient besides fuel, namely, neutrons. In fact, the chain reaction is initiated by a source of neutrons. There are four possible events that can happen to a neutron in a reactor. The design and operation of reactors, as well as their safety, depend on how the neutrons are managed and controlled. The four possibilities are as follows:

First, a neutron could undergo *fission capture* by uranium-235. This reaction produces fission products, provides more neutrons to branch the chain, and yields energy. But there are problems. Fission capture is favored by slow neutrons, but the reaction releases *fast* neutrons. Some medium is therefore needed to slow down the emitted neutrons. Such a medium is called a neutron **moderator.** The first moderator used was graphite (a form of pure carbon). Water is

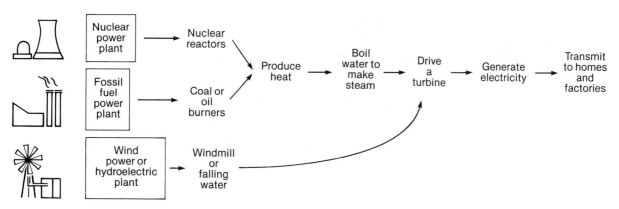

Figure 12–4 *Systems for energy conversion.*

also suitable. Another problem arises from the meager abundance of uranium-235 (0.7 percent) in natural uranium. The rate of fissions, and hence the production of energy, can be speeded up by increasing the proportion of uranium-235 in the fuel. Such enrichment must be limited, however, because the chain reaction must not be allowed to branch out of control.

Second, a neutron can undergo *nonfission capture* by uranium-238, thereby producing plutonium-239:

$$^{238}_{92}U + ^{1}_{0}n \rightarrow ^{239}_{94}Pu + 2 \text{ electrons}$$

Plutonium, in turn, can undergo fission. This capture requires fast neutrons, not slow ones, and will therefore be impeded by moderators used to slow down neutrons for uranium-235 fission capture. On the other hand, the production of plutonium-239 is, in effect, a "breeding" of new fuel and therefore is very attractive as a source of energy for human needs. The choice of whether or not to favor the breeding of plutonium determines, in large part, the design of the reactor.

Third, a neutron can undergo *nonfission capture by impurities*. This causes loss of neutrons and dampening of the chain, but it also offers a convenient means of controlling the reactions.

Fission products accumulate as impurities and, because they soak up neutrons, the fuel elements that they contaminate must eventually be removed and purified. But it is necessary to have a controlled means of absorbing neutrons to regulate the reaction. The most direct method is to insert a stick of neutron-absorbing impurity. The more impurity that is pushed in, the more the reaction is slowed down. Devices that do this are called **control rods;** they usually contain cobalt or boron and other metals, and they can be inserted into or withdrawn from the reactor core to regulate the neutron flow with great precision.

Finally, a neutron, traveling as it does in a straight line, may simply miss the other nuclei in the reactor and *escape*. Obviously, the larger the reactor, the more atoms will stand in the way, and the greater will be the chance of neutron capture and the less the chance of escape. This circumstance imposes lower limits on reactor size—there will never be pocket-sized fission generators, nor even fission engines for motorcycles. This escaping tendency also demands adequate shielding to prevent neutron leakage.

How, then, can a reactor be constructed to satisfy all these requirements? The first decision to be made is whether the reactor is to be designed only to generate energy from uranium-

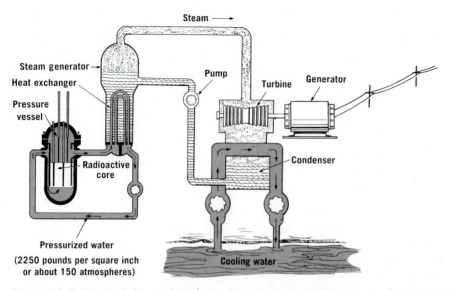

Figure 12–5 Schematic illustration of a nuclear plant powered by a pressurized water reactor.

235 (a "nonbreeder") or whether it is also to "breed" plutonium-239.

Nonbreeders

Refer to Figure 12–5 as you read the following description. The radioactive core contains fuel, moderator, and control rods. The fuel is typically a ceramic form of uranium dioxide. This compound is much better able than the pure metal to retain most fission products, even when overheated. The uranium fuel itself is prepared by conventional (non-nuclear) chemical treatment and consists of the natural nonfissionable isotope, uranium-238, enriched with fissionable uranium-235 by a factor only three or four times above its naturally occurring level. This low level of enrichment sets an upper limit on the maximum rate of the fission reaction. The fuel is inserted into the reactor core in the form of long, thin cartridges (Fig. 12–6). These fuel cartridges are covered with stainless steel or other alloys.

The fuel elements are surrounded by the moderator, whose function, remember, is to slow down the neutrons so that they will undergo fission-capture. Water is advantageous because it is both a moderator and a coolant. The energy released by fission reactions heats the water, which then releases its energy to another body of water in a heat exchanger. The mode of generation of electricity is the same as in fossil fuel plants. The steam drives a turbine that powers an electric generator; the waste steam is then cooled and returned to the heat exchanger.

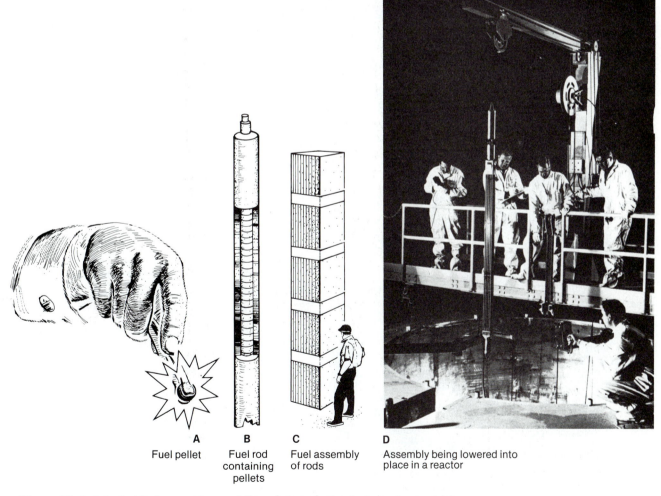

A	B	C	D
Fuel pellet	Fuel rod containing pellets	Fuel assembly of rods	Assembly being lowered into place in a reactor

Figure 12–6 A typical fuel assembly, consisting of many fuel rods, being lowered into place in a reactor.

Interspersed into the matrix of fuel cartridges, moderator, and coolant are the control rods. They serve not only to regulate neutron flow but also as an emergency shut-off system (Fig. 12–7). To speed up the chain reaction, the rods are partially withdrawn; to slow it down, they are inserted more deeply. In the event of malfunction, the rods are pushed rapidly all the way into the core to capture as many neutrons as possible and quench the chain reaction.

Breeders

Recall that neutron capture by abundant uranium-238 can produce fissionable plutonium-239. This reaction makes it possible for one re-

Figure 12–7 Control-rod driving mechanism of a pressured water reactor, partially disassembled. (From D. R. Inglis: *Nuclear Energy: Its Physics and Its Social Challenge.* Reading, Mass., Addison-Wesley, 1973.)

actor to provide fuel for another. (Hence the reactors "breed.") The uranium-238 is called a "fertile" material, to maintain the biological metaphor.

The breeding reactions require fast neutrons. This means that a moderator, which slows down neutrons, must be excluded. The reactor core, then, consists of uranium-238 that is highly enriched with uranium-235 or plutonium-239 (the sources of the fast neutrons), and no moderator. The space vacated by the moderator can accommodate additional enriched fuel; thus, the overall concentrations of *both* the fertile and fissionable materials in the breeder core are much greater than they are in a nonbreeder. This situation is inherently more dangerous because energy is released in a more concentrated form. Thus, in case of a malfunction, there is greater danger of overheating and melting of the core, which would concentrate the fuel still further and release radioactive products still more rapidly.

The compactness of the breeder core demands a very rapid removal of heat. Water is disadvantageous because it is a neutron moderator, which must be avoided. Furthermore, water boils at relatively low temperatures even under high pressures, and steam is a poor heat conductor. The coolant of choice is liquid sodium. Sodium is a silvery, soft, chemically active metal. It reacts with water to produce hydrogen

BOX 12.1 CURTAILMENTS OF BREEDER REACTOR PROGRAMS

In 1960, a 375-megawatt prototype breeder reactor, which is about one third the size of a full scale reactor, was proposed for construction at Clinch River, Tennessee. The project was approved by President Nixon in 1971 at a proposed cost of $500 million and scheduled for completion by 1982. The project was repeatedly delayed because of problems with technical design, uncertainties about the political risks associated with the production of plutonium, and runaway costs, which the General Accounting Office predicted could reach $8 billion. As of 1983, the project is "on hold."

The French breeder reactor program, too, has been curtailed; the original plan for eight breeders of the Superphenix type has been reduced to two.

gas; if air is present, the heat of the reaction can spark the explosion of the hydrogen. The sodium becomes highly radioactive when exposed to the reactor core. But its saving virtue is its ability to carry heat away from the reactor core rapidly, since it is an excellent heat conductor and it remains in the liquid state over a very wide temperature range, from 98°C to 890°C at normal atmospheric pressure.

The heat exchanger in which steam is produced to drive the turbine must be shielded from the radioactive sodium. This is accomplished by an intermediate loop of nonradioactive sodium. The entire arrangement is shown schematically in Figure 12–8A. A more detailed illustration of a breeder reactor is depicted in Figure 12–8B.

12.6 The Nuclear Fuel Cycles

From mine to factory to disposal to ultimate death by radioactive decay, nuclear fuels will become more and more intimately involved with human activities if the nuclear industry continues to grow. It is therefore important to understand the steps involved in the processing, utilization, and disposal of nuclear fuels.

There are two different sequences to consider. The first applies to nonbreeder reactors and is really better characterized as a "once-through" process rather than a cycle (except that the wastes *do* eventually return to earth). The second applies to breeder reactors, which may represent the nuclear economy of the future, sometimes called the "plutonium economy." Both cycles are described below, and are illustrated in Figure 12–9.

The "Once-Through" Uranium Cycle

Uranium ore is mined in various areas of the Earth as a black deposit containing perhaps 0.3 percent uranium. It is concentrated and then converted to a brilliant orange oxide, UO_3. (Many American homes contain old orange-colored kitchen pottery prepared from this uranium oxide pigment. Get them out of your kitchen and donate them to the nearest university.)

The next step is **enrichment,** which is carried out by gaseous diffusion. The uranium must first be converted to the gaseous uranium hexafluoride, UF_6, then enriched in the fissionable uranium-235 isotope. The enrichment process is very complex and costly to operate, and few countries can afford to do it. The enriched material is reconverted to an oxide and fabricated into the fuel pellets used by the power plant.

After a year or more in a reactor, an appreciable portion of the uranium-235 is consumed, and fission products have accumulated. This fuel is no longer useful in the reactor, and the assembly, containing unused uranium, waste products, and newly formed plutonium, is removed. At this time the waste products are in their most intensely radioactive state, and they are too dangerous to ship. They are therefore stored underwater at a site on the plant premises for a few months to allow the most highly radioactive components to decay. The partially decayed fuel is then shipped to a fuel processing plant. Here the pellets are cut up, dissolved, and chemically processed to recover uranium and plutonium (which was produced during the time the fuel was in the reactor).

The uranium can be reconverted to UF_6 and recycled for enrichment, but this is not advantageous as long as rich uranium ores can be mined. Meanwhile, most of it is stored. Plutonium is also stored for possible future use in reactors or as an explosive. Some of the radioisotopes that have special applications in science, medicine, or industry are also separated and set aside for such uses. The remainder is a solution of **radioactive waste** (Section 12.8).

The Plutonium Breeder Fuel Cycle

The breeder *produces* plutonium from uranium-238. This circumstance affects the fuel cycle in two important ways. First, the uranium enrichment step can be skipped because the reactor breeds its own fuel. Second, the fuel it breeds, which is plutonium, must be recovered and incorporated again into pellets. Since these operations do not all take place at the same location, the breeder fuel cycle has the effect of introducing plutonium in a highly enriched form into channels of commerce and transportation.

246

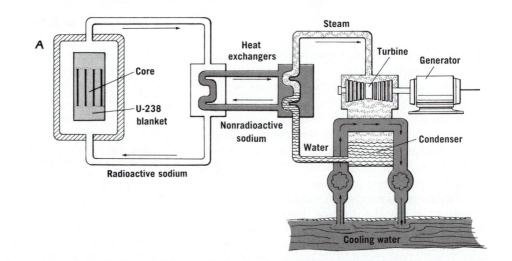

A

Core

U-238 blanket

Radioactive sodium

Heat exchangers

Nonradioactive sodium

Steam

Turbine

Generator

Water

Condenser

Cooling water

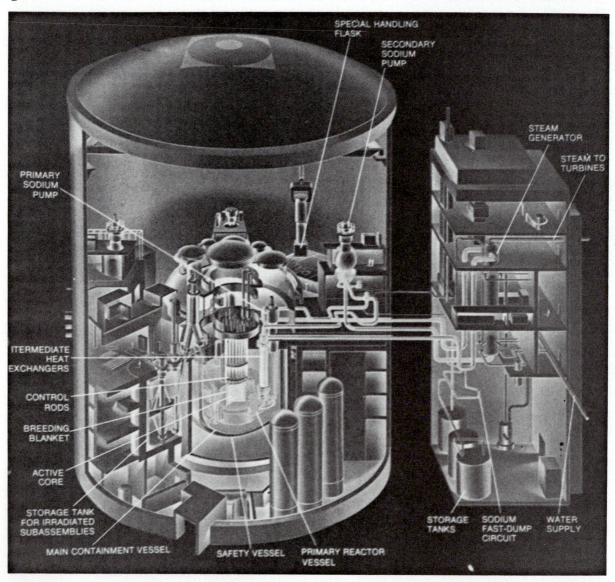

B

SPECIAL HANDLING FLASK

SECONDARY SODIUM PUMP

STEAM GENERATOR

STEAM TO TURBINES

PRIMARY SODIUM PUMP

ITERMEDIATE HEAT EXCHANGERS

CONTROL RODS

BREEDING BLANKET

ACTIVE CORE

STORAGE TANK FOR IRRADIATED SUBASSEMBLIES

MAIN CONTAINMENT VESSEL

SAFETY VESSEL

PRIMARY REACTOR VESSEL

STORAGE TANKS

SODIUM FAST-DUMP CIRCUIT

WATER SUPPLY

◀ Figure 12–8 *A,* Schematic diagram of a fast breeder reactor. *B,* Vertical section of the French Superphenix reactor building and one of the four identical steam-generating buildings shows the main operating components of the plant. Superphenix is classified as a pool-type breeder reactor, which means that the active core, the primary sodium pumps, and the intermediate heat exchangers are all located within a single large vessel; in this particular design the main steel containment vessel, which is hung from a steel-and-concrete upper slab, is 21 meters across, and is filled with 3300 tons of molten sodium. A cylindrical structure welded to the main vessel supports the control-rod mechanism and the fuel subassemblies, which constitute the active core of the reactor. The four primary pumps convey the sodium upward through the core. The primary reactor vessel separates the "cold" sodium, which enters at the bottom of the subassemblies at a temperature of 395°C, from the "hot" sodium, which leaves at the top at 545°C. The hot sodium then flows downward through the eight intermediate heat exchangers, which form part of a secondary circuit of nonradioactive sodium, inserted for reasons of safety between the primary sodium circuit and the water-steam circuit. Each of four secondary loops consists of two intermediate heat exchangers, a secondary pump installed inside a spherical expansion tank, and a steam generator in the adjacent building. (From George A. Vendryes: "Superphenix: A full-scale breeder reactor." *Scientific American*, March, 1977, p. 28. Copyright © 1977 by Scientific American, Inc. All rights reserved. Drawing by George V. Kelvin.)

12.7 The Practice of Safety in Nuclear Plants

Questions about safety and hazards are at the heart of the nuclear controversy. It will be helpful, in approaching this very complex issue, to itemize the five basic principles of safe practice that apply to any industry. The applications of these principles in nuclear plants will then be described in more detail.

1. *Safe operation must be part of the original design.*
2. *There should be "back-up" or duplicating systems that will take over in cases of failure.* This approach is sometimes called **redundancy.**
3. *There should be a system for warning* of possible accident if something starts to go wrong.
4. *A schedule of inspection and maintenance* should be provided.
5. Finally, if all these systems fail and an accident does occur, the design should *provide features that prevent or minimize injury to people.*

(1) Safe Design. Recall that in a nonbreeder reactor the uranium-238 is only modestly enriched with the fissionable uranium-235, so that the fuel is nothing like an atomic bomb. The control rods are inserted by pushing them down into the core, so that if power fails, they could simply fall. Ordinary water is both a coolant and a mod-

erator. If excess heat should boil the water out, the loss of moderator would stop the chain reaction. Design specifications require that the materials of construction be of the highest engineering quality and be fully tested before use. The breeder reactor is inherently more dangerous, but this means only that safe design is, if anything, even more critical.

(2) Redundancy. The system for which a backup is most important is the one that cools the reactor core. If that should fail, there are generally at least *two* other independent cooling systems. If the power system on which the emergency measures depend should fail, an off-site source of power can be used. If *that* fails, on-site diesel generators or gas turbines can take over. Secondary systems of this type are quite complex and are interrelated in such a way that their responses are specifically appropriate to the nature of the emergency. Furthermore, these responses are fully automatic; they do not have to be initiated by a human operator.

(3) Warning. The control room of a nuclear plant displays a panorama of gauges, dials, lights, buzzers, and bells (Fig. 12–10). Individual workers are supplied with badges that are sensitive to radiation and that will monitor the degree to which the wearer has been exposed. Detection devices are distributed throughout the plant and are also set outdoors at various distances from the plant.

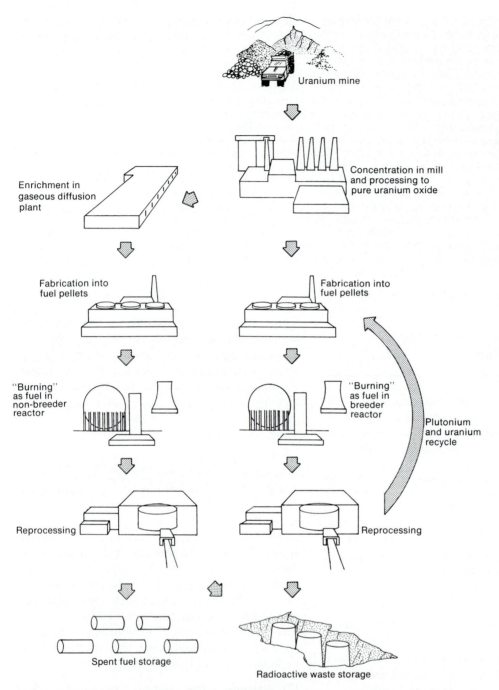

Uranium mine

Concentration in mill
and processing to
pure uranium oxide

Enrichment in
gaseous diffusion
plant

Fabrication into
fuel pellets

Fabrication into
fuel pellets

"Burning"
as fuel in
non-breeder
reactor

"Burning"
as fuel in
breeder
reactor

Plutonium
and uranium
recycle

Reprocessing

Reprocessing

Spent fuel storage

Radioactive waste storage

Figure 12–9 Nuclear fuel cycles. *Left,* once-through uranium cycle. *Right,* plutonium cycle.

(4) Inspection and Maintenance. Reactor operators must go through strict licensing procedures, with periodic renewal. The plants themselves are inspected several times each year, penalties are applied to violators of regulations, and listings are kept of any defects or failures.

(5) Protection in the Event of Accident. The reactor vessel, made of thick steel, is itself surrounded by an antiradiation shielding several feet thick. As a final barrier, the entire system is surrounded by a vapor proof, steel-lined, reinforced concrete **containment structure** (Fig. 12–

Figure 12–10 *Control room of a nuclear plant.*

11). This barrier is designed to withstand earthquakes and hurricanes and to contain all matter that might be released inside, even if the biggest primary piping system in the reactor were to shatter instantaneously. The Soviet Union and some other countries do not require this last barrier, a fact that nuclear proponents have cited to emphasize the high priority given to safety in the United States.

If all of the above sounds very comforting, remember that things do go wrong in nuclear plants.

12.8 Environmental Hazards from Fission Reactors

Five sets of question must be discussed in assessing the environmental hazards associated with nuclear fission plants: (1) What are the chances of a serious accident? (2) What are the extent and the environmental effect of the routine emissions of radioactive materials under normal conditions of operations? (3) What are the problems associated with the disposal of radioactive wastes? (4) What are the dangers of terrorism and sabotage? (5) What are the environmental effects of the waste heat released from the nuclear plants? The last of these effects,

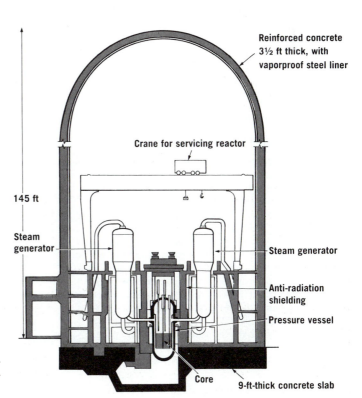

Figure 12–11 Containment structure for a nuclear power plant. (The New York Times Company. Reprinted by permission.)

which is thermal pollution, is dealt with in Chapter 13. Here we consider the first four.

The Chances of a Serious Accident

The nightmare of the reactor safety engineer is the possible overheating of the radioactive core. If this should happen, the reactor vessel could melt, and the molten mixture of fuel and fission products might spread out and possibly escape to the environment. Such a disaster would be the result of a series of accidents, each of which has its own probability of occurring. To analyze the overall danger, therefore, it is necessary to consider each event in the series separately and also to judge how these events might influence each other.

Imagine, first, that a pipe carrying cooling water to the reactor bursts. Since the water is also the neutron moderator, the chain reaction in the core would be terminated. However, there is another source of heat in the reactor that cannot be turned off—the energy released by the radioactivity of the accumulated fission products. It has been predicted that parts of the reactor would heat up to 1480°C in about 45 seconds, at which temperature water would react with the metal surrounding the fuel to release hydrogen gas, which can explode in air.

To prevent such an occurrence, an **emergency core cooling system** automatically introduces another stream of water. Even if this works as expected, large quantities of water and steam carrying radioactivity will still spray out of the reactor at high pressure. The containment structure is designed to prevent any of this material from escaping to the outside. However, if the emergency cooling system should fail to operate—or if it should fail to hold the reactor below the 1480°C danger point—the fuel rods may buckle or even break and thus block the passage of water completely. In such an event, the temperature would continue to rise until, in about half an hour, the fuel would melt. Within a few hours the molten fuel would break through the reactor vessel, and tons of white-hot radioactive material would melt its way into the ground. (This scenario is sometimes jokingly called the **China Syndrome,** meaning that the molten mass is moving toward the other side of the globe.) Large quantities of radioactive matter could be released from the ground to the air, to subsurface water or, if the containment structure itself failed, directly from the reactor core to the atmosphere.

What, then, are the chances that such a series of accidents, leading to a disaster, could occur? Consider the first step, the bursting of a water pipe. It is possible to survey places where industrial pipes have been in service for many years to find out how many *have* burst. Furthermore, radiation can make metals more brittle and hence more liable to failure. Such information may then be used as a basis for calculating the chance of a loss of cooling accident in a nuclear plant. Remember, however, that such an accident triggers the emergency core cooling system, and that, too, would have to fail before a catastrophe could occur. Again, one can obtain a history of failures of pipes, valves, motors, and monitoring instruments (they are all susceptible) and, in like manner, apply the results to the nuclear reactor. Finally, if a radioactive cloud *does* escape, where will it go?

The numbers game now involves the chances that the wind will blow the cloud toward populated areas, or that it will be washed down by rain, and so on and on. Taken together, such procedures require considerable time and effort (not to mention guesswork). It is important to recognize, however, that the United States government estimates of nuclear safety are based on just such an approach. In

---

BOX 12.2 ATOMIC BOMBS

A nuclear reactor is nothing like an atomic bomb. Opponents of nuclear energy have complained that no one ever made such an accusation, and therefore the statement diverts attention from more credible hazards. Nonetheless, the reader should recognize that a bomb contains highly concentrated fissionable materials, which leaves only two significant fates for neutrons— fission capture or escape. The factor that determines which of these two fates will predominate is size, or mass; the minimum mass required to support a self-sustaining chain reaction is called the **critical mass.** To set off an atomic bomb, therefore, subcritical masses of uranium-235 or plutonium-239 are slammed together by chemical high explosives to make a supercritical mass. The chain reaction instantly branches, and the mass explodes. The fuel in a nuclear reactor contains no such concentrations.

fact, the estimates were prepared under the direction of Professor Norman C. Rasmussen of the Massachusetts Institute of Technology. The study, which required two years and cost three million dollars, was released by the Nuclear Regulatory Commission (NRC) in 1975. The key conclusions are shown in the following table.

EVENT	CHANCE OF OCCURRENCE
Complete fuel meltdown of any one reactor in any one year.	1 in 17,000
An accident killing as many as 1000 people by acute radiation sickness in any one year.	1 in a million
The "worst" accident occurring to any one reactor in any one year, which would cause 3300 deaths, 45,000 "early illnesses," over 1500 latent fatal cancers, and an approximately equal number of genetic defects.	1 in a billion

Rasmussen claimed that many of the risks we normally accept as part of living—risks associated with riding in automobiles or airplanes, with working in factories, or with burning coal, to name but a few—are far greater than those associated with the nuclear program. However, opponents of nuclear power rejected the Rasmussen report on various grounds but mainly because (1) they did not agree with its assumptions, and (2) they maintained that the report did not cover risks involved in the *entire* fuel cycle. In 1978 the NRC issued a report that severely criticized the Rasmussen study. The essence of the criticism was that the statistical analysis used by Rasmussen was so deeply flawed that we still do not really know what the chances are of a severe reactor accident.

"Routine" Radioactive Emissions

Even with the best design and with accident-free operation, some radioactivity is routinely released to the air and water outside the plant. In the boiling-water reactor, for example, the water passes directly through the reactor core and thus circulates around the fuel elements. Some of the fuel claddings, which are very thin (about 0.05 cm) inevitably develop small leaks, which permit direct transfer of radioactive fission products to the water. Even in the absence of leaks, however, some neutrons do get through to the water

and make some of its impurities radioactive, and this effect, too, is a source of "routine" emissions to the watercourses that serve as the ultimate coolants for the power plant.

Radioactive material can also be gaseous. Krypton-85, for example, is a radioactive fission product (half-life, nine years) that is insoluble in water and escapes to the atmosphere through a tall stack. These emissions, taken together, are so small that their effect may be compared with the background radiation from cosmic rays and naturally radioactive materials in the Earth's crust to which all life is subject. Higher levels of radioactivity can be found in the emissions from the stacks of coalfired plants. This fact is often cited to support the contention that routine radioactive emissions are trivial.

The Disposal of Radioactive Wastes

Refer back to Figure 12–9 to note the items labeled "radioactive waste storage." Recall that different radioisotopes have different half-lives. Among the radioactive wastes, these half-lives range from a fraction of a second to thousands of years. Furthermore, these wastes produce various generations of new radioisotopes, one following the other, before a final stable isotope is produced. Thus, as any one isotope in the series is decomposing, it is also being continuously replenished by the decay of its "parent." The shape of the decay curve of a hypothetical isotope with a half-life of one month is shown in Figure 12–12.

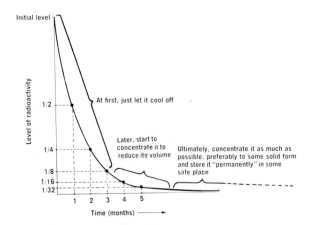

Figure 12–12 Disposal of a hypothetical radioactive waste product with a one-month half-life.

The handling of radioactive wastes involves a sequence of three steps. First, they must be stored somewhere temporarily to allow the initial, very intense radioactivity to die down and also to allow time to decide what to do with them next. Second, the wastes must be "reprocessed" to convert them into a form suitable for the final step. Third and (it is hoped) last, the reprocessed wastes must be "permanently" stored somewhere. Various problems are associated with each of these steps, as described in the following paragraphs.

The "temporary" storage of radioactive wastes from the U.S. nuclear weapons program has been going on since 1947; some of the 335 million liters accumulated by 1982 had been in the same holding tanks for 35 years. At the same time, spent-fuel rods from nuclear power plants had reached a total of over 7000 tonnes. The old tanks are nearing the end of their useful lives. Radioactive waste liquids are made by dissolving the solid wastes in strong acid. The resulting solution is hot, radioactive (the radioactivity keeps the temperature up), and corrosive. Eventually, some of the tanks leak. It has been estimated that some 1.7 million liters (450,000 gallons) have leaked out of the tanks at the government nuclear facility in Hanford, Washington. New tanks (Fig. 12–13) are being built to accept transfers from old, leaky ones and to hold the new

material that is constantly being generated. Even so, capacity is being strained to keep up with the demand.

The next step, the reprocessing of the spent fuel from nuclear plants, involves political decisions as well as technical problems. Consider a 1000-megawatt nuclear plant using uranium fuel enriched to 3.3 percent uranium-235. Such a plant will use about 28 tonnes of fuel per year and will produce 28 tonnes of waste fuel, as shown in Figure 12–14. The political aspect of the problem involves a decision about whether to recover the uranium and plutonium or to dispose of them. Plutonium represents a particular danger because it is intensely carcinogenic and because, if stolen, it can be used to make illicit fission bombs. The technical aspects of reprocessing are not theoretically difficult, but the U.S. experience has not been good.

The first such reprocessing plant, near Buffalo, New York, opened in 1963 but was closed in 1972 when reprocessing was determined to be economically unfeasible. To clean up this facility, about 2.3 million liters (600,000 gallons) of highly radioactive waste liquids stored in underground tanks must be solidified and removed. In 1982 this operation, estimated to cost $400 million, was undertaken by the federal government. Other plants have also experienced difficulties. In view of such circumstances, critics of

Figure 12–13 Three one-million-gallon capacity, double-shell tanks (shown during construction) were completed in the spring of 1977 and tied into Hanford waste management operations. After completion the tanks were covered with a minimum of six to seven feet of soil. The tanks employ the latest monitoring and leak detection equipment and are linked by computer to the integrated tank surveillance system.

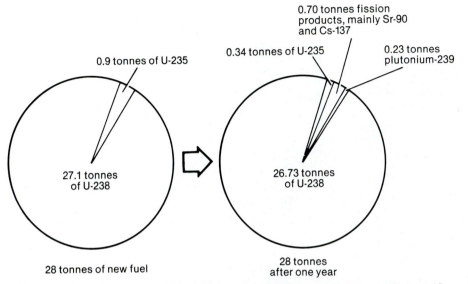

0.70 tonnes fission
products, mainly Sr-90
and Cs-137

0.9 tonnes of U-235

0.34 tonnes of U-235

0.23 tonnes
plutonium-239

27.1 tonnes
of U-238

26.73 tonnes
of U-238

28 tonnes of new fuel

28 tonnes
after one year

Figure 12–14 Conversion of uranium fuel in a 1000-megawatt reactor operating at 62 percent capacity, in tonnes per year. (Data from B. J. Skinner, and C. A. Walker: *American Scientist 70*:180, 1982.)

the nuclear program do not consider the technology of fuel reprocessing to be completely safe and reliable.

The third and final step concerns safe, "permanent" storage. How long is permanent? It has been suggested, as a reasonable target, that the wastes should be reduced to the same radioactivity level as that of a natural uranium mine. (Not everyone would consider that sufficient; uranium miners do get lung cancer.) If that target is accepted, then the time needed for permanent storage depends on whether reprocessing is done. If the uranium and plutonium are removed, then only the fission products need be permanently stored. These wastes decay more rapidly and would reach the radioactivity level of a uranium mine in about 500 years. If unreprocessed spent fuel is to be stored, it is estimated that about 7000 years would be necessary. There have been various suggestions for storage places—deep in the sediments under the ocean floor, under the Antarctic ice cap, or even blasted out to space. All major current effort, however, is devoted to developing mined, geologically stable, repositories. The medium of choice is rock salt (Fig. 12–15), although other types of rock, such as granite and basalt, have been considered. Various types of impervious containers have been designed. After the con-

tainer is sealed in the rock, it continues to give off heat from its radioactivity. If the container holds unprocessed spent fuel, its temperature, and that of the surrounding rock, remain elevated for about a thousand years. There are two ways in which the waste could get out: It could be physically exposed (by geological or by human activities), or it could be carried out by groundwater. The risks of damage to the host rock by the evolved heat, of erosion by streams or glaciers, of transport by groundwater, and of human intervention have all been studied, and scientists who work on these problems have concluded that the risks are small and acceptable. Critics of the nuclear program have disagreed; they point out that political systems, on which we depend for reliable continuity of any public policy, do not last as long as radioactive isotopes. Furthermore, the assumptions about the permanent segregation of the radioactive canisters in salt caverns or rock formations are not *proved*; they are predictions.

A nuclear power plant itself has only a finite lifetime. After a few decades unavoidable deterioration progresses to a point where the facility can be no longer used. This problem is only now becoming significant, as the useful lifetimes of the first generation of power plants, built in the early years of the nuclear energy program, are

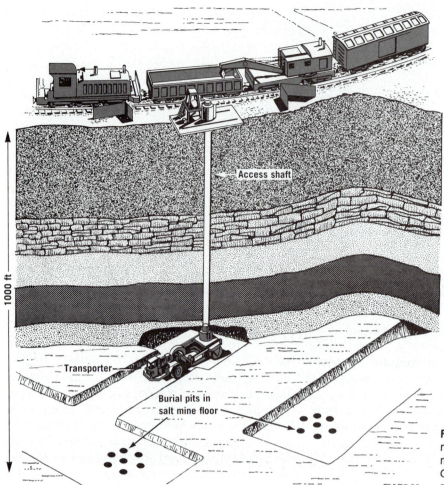

Figure 12–15 Final disposal of radioactive wastes in a salt mine. (The New York Times Company. Reprinted by permission.)

coming to an end. Just how this problem will be handled has not yet been resolved.

There is no question that the risk of any single accident during any single day in the life of any particular batch of radioactive waste is very, very small. But it is also true that the consequences of an accident (for example, the contamination of a large supply of groundwater) could conceivably be very great. Furthermore, many batches of wastes are involved and there will be very, very many days. To judge the overall risks, very small numbers (the risk of any one accident) must be multiplied by very large ones (serious consequences; long periods of time for storage of wastes). Mathematicians point out that such exercises approach indeterminate, or unknown, answers.

Terrorism and Sabotage

Uranium and plutonium inventories are not always fully accounted for. There are documented cases of substantial quantities of missing materials. Such discrepancies may not be real but may only be the results of errors in bookkeeping. Nevertheless, the question remains, "Can they be stolen, and, if so, can amateurs convert them into bombs?" A related question is, "Can saboteurs damage nuclear plants so that radioactive matter is released to the environment?" Some writers have speculated on this matter; some physics majors have written term papers on how to make a bomb (but they didn't make one); and some terrorist acts have actually been threatened and attempted. All these circumstances have led to increased security measures around nuclear installations. Since our past experience is (fortunately) so limited, we can only guess at the possibilities.

12.9 Case History: Three Mile Island and the China Syndrome

The Three Mile Island nuclear plant in Middletown, Pennsylvania, is a subsidiary of the General Utilities Corporation. Starting at 4 A.M. on March 28, 1979, a series of mishaps occurred. The sequence is enormously complicated—some forty different events have been identified during the subsequent investigations. This case history highlights the key factors and addresses the two major questions:

1. What was the underlying cause of the accident? Was it the fault of the operators, or did it come about because the plant was incorrectly designed?
2. Was there ever a serious risk that the accident might have gone out of control, causing a meltdown (China Syndrome) that would endanger the people over a wide area (the state of Pennsylvania)?

The operation that seems to have started the trouble was a routine one—the change of a batch of water purifier in a piping system. For a nuclear plant, this job is considered to be as routine as, say, changing the oil filter in an automobile. But a problem developed; some air got into the pipe, causing an interruption in the flow of water. Of course, the back-up systems in a nuclear plant are designed to respond automatically to such an event, but, in this instance, several other things went wrong.

• Two spare feedwater pumps were supposed to be ready to operate at all times. However, the valves that control the water from these pumps were out of service for routine maintenance; therefore, the spare pumps could not deliver water. The controls for these valves were provided with tags to indicate that they were being repaired. The tags hung down over red indicator lights that go on when the spare pumps are not feeding water. Since the lights were obscured by the tags, the operators did not see that they were lit and did not realize that no water was flowing.

• As a result, pressure built up in the reactor core. A relief valve in the primary coolant loop then opened automatically (as it should have) to let out superheated steam. *But the relief valve failed to close,* causing a dangerous drop in pressure. This malfunction is considered to be the crucial failure of equipment in the entire sequence.

• When the emergency core cooling system came on automatically, the pressure gauges in the control room gave a false reading, leading operators to think the water level was still above the fuel rods. It wasn't. Instead, bubbles of gas from below were pushing the water up, leaving part of the core exposed.

• The primary and the emergency cooling pumps, which should have been left on, were turned off twice by operators misled by the faulty pressure gauges.

The net result of this confusing series of mishaps or errors was that the nuclear core overheated. The temperatures inside the reactor vessel climbed off the recorder charts. For 13½ hours the situation was very unclear. Subsequent investigations showed that the entire core *was* exposed for some time. This means that only steam, not liquid water, was circulating through the core to remove excess heat.

The "void" or "bubble" that caused the problem was something entirely unexpected—it was a 28,000-liter volume of steam and hydrogen gas. The plant management as well as the federal regulators were utterly unprepared for this possibility. Where did the hydrogen come from? There were two possibilities, both of which probably played a part. One was **radiolysis,** which is a chemical change produced by radiation. In other words, the radioactivity in the core chemically decomposed the water and produced hydrogen (as well as oxygen). The other possibility was the chemical reaction of water with metals, which also produces hydrogen (but not oxygen). Mixtures of hydrogen with air or oxygen are explosive, and therefore the production of hydrogen introduces a new danger. Later investigation revealed that the NRC was profoundly ignorant of the radiolytic chemistry in the reactor core during the accident.

The net result of this confusion, in the words of Roger J. Mattson, director of the Division of Systems Safety, was that "We saw failure modes the likes of which have never been analyzed."

The superheated steam released into the atmosphere early in the emergency was radioactive, and for this reason Pennsylvania's Governor Thornburgh ordered the evacuation of children and pregnant women from the area near the plant. Others left of their own accord. As the problem subsided, the evacuees returned, with misgivings one can only imagine.

Now consider the two major questions previously referred to:

1. *What was the underlying cause of the accident?* One of the first public reports on the incident was that of the *President's Commission on the Accident at Three Mile Island,* issued on October 30, 1979. It is known informally as the "Kemeny Report" because the Commission was chaired by John G. Kemeny, then president of Dartmouth College. The Commission found enough fault to go around. Much of the blame was directed at the complacency of the nuclear industry and the NRC with regard to safety. The attitude of the industry had been that nuclear power is essentially safe. The Commission said that such an attitude is wrong and must be reversed. That is to say, nuclear power must be regarded as potentially dangerous by its very nature, and "one must continually question whether the safeguards already in place are sufficient to prevent major accidents." The Commission believes that the complacent attitude resulted in such lapses that "an accident like Three Mile Island was eventually inevitable."

Of course, the *immediate cause* of the accident was a combination of malfunction of equipment and errors by the operators. The Commission was especially critical of the confusing design and arrangement of the indicators in the control room. Operators should not shoulder the blame for faulty equipment or gauges. But what about the fact that the operators turned off the cooling pumps when the core was partially exposed by bubbles below? One of the scientists who was on the original investigating team a few days af-

ter the accident commented, "Well, if you were filling a bucket with water, and it started to overflow, wouldn't you turn the water off, even if there were a big bubble below that you couldn't see?" Of course, hindsight will always show how the operators could have done better. Certainly, they should have been more highly trained. In fact, the Kemeny Report recommended that accredited centers be set up for training operators and that graduates be given a degree. This would become a prerequisite for employment at a nuclear plant.

2. *Was there ever a serious risk of a complete meltdown?* During the event, some people certainly thought so. But later analysis indicated that, when the feedwater flow stopped, the fission reaction was automatically shut down (by the control rods) within 12 seconds. Of course, the radioactivity continued—that cannot be shut down. Furthermore, as mentioned previously, the core actually *was* completely exposed for some time. This evidence is both frightening and, to some people with another point of view, comforting. It is frightening because an accident of a kind that the nuclear industry implied was too unlikely to worry about did really happen. Many people consider that their confidence and trust have been betrayed. On the other hand, the fact that an exposed core did not melt down may mean that a nuclear plant is more resistant to catastrophe than its severest critics believe. Furthermore, the impact of the accident on public health was negligible, according to the findings of the Kemeny Commission. (This does not take into account the mental distress suffered by people in the nearby communities.) The Commission also predicted that the effect in terms of increased cancer will be undetectable.

12.10 Nuclear Fusion

Another source of nuclear energy is that which can be released by the combination, or **fusion,** of nuclei of certain light elements, particularly certain of the isotopes of hydrogen. The great attraction of nuclear fusion is that it promises abundant energy with much less environmental danger than is faced from fission reactors. However, no useful fusion reactor has yet been developed.

Any fusion reactor would utilize hydrogen nuclei. There are three isotopes of hydrogen: **protium, deuterium,** and **tritium.** Some information about them is given in Table 12–3. Fusion reactions can occur between any two hydrogen isotopes. The heavier isotopes fuse more readily, but the lighter ones are more abundant.

Unlike the fission reaction, however, fusion cannot be triggered by neutrons. What is required instead is that the fusing nuclei be fired at each other at very high speeds, which means at very high temperatures. The resulting fusion is therefore called a **thermonuclear reaction,** the type of reaction by which the Sun produces energy. If a large mass of hydrogen isotopes fuses in a very short time, the reaction cannot be contained and it goes out of control; this is the explosion of the "hydrogen bomb." On the other hand, useful energy could be extracted from fusion if it were possible to devise what is called a controlled thermonuclear reaction.

Two deuterium nuclei can fuse to form a helium-3 nucleus plus a neutron:

$$^2_1H + ^2_1H \rightarrow ^3_2He + ^1_0n$$

To start this reaction, the temperature would have to be raised to about 400 million degrees Celsius. The problems imposed by this requirement are so severe that it is not even being attempted. Instead, all efforts to control fusion are being directed to a cooler reaction ("only" 40 million degrees Celsius)—the fusion of deuterium with tritium to give helium-4 plus a neutron:

$$^2_1H + ^3_1H \rightarrow ^4_2He + ^1_0n$$

This process requires a source of tritium, which is not naturally available on Earth.

Tritium is produced by the reaction between neutrons and the light isotope of the metal lithium, ^6_3Li. This isotope constitutes 7.6 percent of natural lithium and is about 10 times as abundant in the Earth's crust as uranium-235.

$$^6_3Li + ^1_0n \rightarrow ^3_1T + ^4_2He$$

Thermonuclear reactions are extremely difficult to control because no materials can withstand the required temperatures. Instead, all substances decompose into their free atoms, and the atoms themselves decompose into a mixture of positive nuclei and free electrons that is called a **plasma.** No rigid container exists that can survive long enough to confine a plasma for the useful production of thermonuclear energy. Instead, what is envisaged is a sort of "magnetic bottle," which does not consist of a physical substance at all, but rather is a magnetic field so designed that it will confine the charged particles of the plasma in which the thermonuclear reaction is going on. The useful energy will have to be extracted in the form of the kinetic energy of the evolved neutrons. Since the neutrons carry no charge, they will pass through the magnetic field and escape from the plasma. The energy of the speeding neutrons can then be ex-

TABLE 12–3   Isotopes of Hydrogen

Symbols	Names	Radioactive?	Natural Abundance (%)
^1_1H	"Ordinary" hydrogen "Light" hydrogen Hydrogen Protium	No	99.985
^2_1H, or ^2_1D	"Heavy" hydrogen Deuterium	No	0.015
^3_1H, or ^3_1T	Tritium	Yes (12-year half-life)	Almost none

tracted by a moderator, just as in a fission reactor. If the moderator is water, the energy will create steam that can drive a turbine. The entire fusion reactor would be encased in a sheath or blanket in which molten lithium is continuously circulated. The lithium would ab-sorb the neutrons, supply the tritium and then release its heat to water in a heat exchanger (Fig. 12–16).

Recent efforts have been directed to an alternative approach—**laser fusion.** The object of this method is to create miniature thermonuclear ex-

A

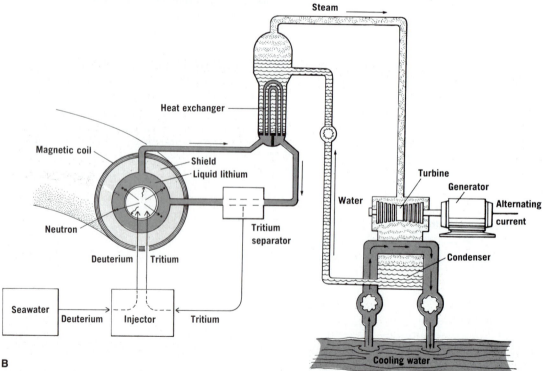

B

Figure 12–16 *A,* Nuclear fusion reactor (*cut-away view*). *B,* Schematic drawing of a thermonuclear power plant.

plosions by hitting frozen pellets of fuel with converging laser pulses of enormous power. The combined laser pulses compress the hydrogen pellet and heat it to the point of thermonuclear fusion.

Estimates of the time it will take to develop a practical fusion reactor of either type—magnetic bottle or laser—range from a few decades to a century or more.

Could a fusion reactor get out of control and go off like a hydrogen bomb (Fig. 12–17)? Nuclear scientists are entirely confident that the answer is no, an explosion could not occur. The reason is that the hydrogen isotopes are continuously fed into the reactor and are continuously consumed; they do not accumulate. The total quantity of fuel in the plasma at any one time would be very small—about 2 g or so—very far below the critical mass required for a runaway reaction. If the temperature were to drop, or the plasma somehow dispersed itself, the reaction would stop; in effect, the fusion would turn itself off. The situation is rather analogous to that of a burning candle; if something goes wrong, the flame goes out, the candle does not explode.

Would there be a problem of environmental radioactivity? The answer here is yes, because both tritium and neutrons could be released. Tritium is radioactive (half-life, 12 years) and can combine with oxygen to form radioactive water. However, the energy released by tritium has so little penetrating power that it is virtually harmless to living organisms as long as its source is outside the body.

Neutron release is another potential hazard. Neutrons are absorbed by atomic nuclei, and the new atoms that are thereby produced may be radioactive. As a result, there could well be substantial quantities of radioactive matter to be disposed of, but in general, the problem should be much less difficult than that of wastes from fission reactors.

Figure 12–17 A hydrogen bomb explosion. Note that the giant battleships are dwarfed by the cloud. (Photography by H. Armstrong Roberts.)

12.11 Nuclear Fuel Resources

Oil shales, tar sands, and chemical conversions will undoubtedly extend the life of the Fos-

Figure 12–18 Mining uranium ore underground near Grants, New Mexico. (Courtesy of Ranchers Exploration and Development Corporation.)

sil Fuel Age, but such stopgaps cannot be the basis for a long-range continuation of our technological existence. Instead, we must turn either to renewable resources or to nuclear energy.

In 1980, only 4 percent of the energy used in the United States was provided by nuclear fuels. Any significant increase in this percentage will make large demands on the availability of uranium (Fig. 12–18). Both the abundance of ores and the concentration of uranium that they contain must be taken into account. Table 12–4 estimates the amounts of ore available at various prices. These figures imply that the cost of nuclear energy would rise very steeply around the turn of the century. But there are many unknown factors in such predictions. The major uncertainty involves the question of whether the technical, economic, and political problems that now beset the nuclear energy program will be successfully resolved, or whether nuclear energy will be remembered only as a noble effort that failed.

TABLE 12–4   Estimated Uranium Reserves (As Uranium Oxide, U_3O_8)

Cost of Uranium Oxide, $/kg*	Thousands of Tonnes Available	Year of Exhaustion
18	600	1986
22	700	1988
33	900	1990
65	1000	1992
110	4000	?
220	7000	?
450–1100	more than 5,000,000	?

*Dollar value in 1974.

Summary

Elements with the same atomic number but different mass numbers are **isotopes.** Some isotopes are unstable, giving off particles and radiation. Such decomposition, which is called **radioactive decay,** or **radioactivity,** may occur in a series of steps, ending when a stable isotope is produced. All radioactivity releases energy.

The **half-life** of a radioactive isotope is the time required for half of the nuclei in a sample to decompose.

In **nuclear fission,** an isotope of a heavy element is split into lighter elements, releasing a large amount of energy. The important naturally occurring fissionable isotope is uranium-235. The fission is triggered by a neutron, and each atom releases two or three neutrons. The result can be a chain reaction, in which a series of steps occurs in sequence.

Nuclear fission can be used in an atomic bomb or in a nuclear power plant. The difference between the two applications depends on the neutrons. There are four things that can happen to a neutron:

1. *Fission capture by uranium-235 to yield energy and fast neutrons.* But the reaction is favored by slow neutrons. Therefore, a **moderator,** which is a substance that slows down neutrons, can be used.

2. *Nonfission capture by uranium-238, thereby producing plutonium-239, which can undergo fission.*

3. *Nonfission capture by impurities, which causes loss of neutrons.* This action can be used to control the fission process.

4. *Escape.* A neutron might miss everything and get lost.

In a **nonbreeder reactor,** cartridges of uranium-238 containing three or four times the natural abundance of uranium-235 are surrounded by a moderator such as water. The heat of the reaction produces steam, which drives a turbine to generate electricity.

In a **breeder reactor** the fuel contains a higher percentage of uranium-235 and no moderator. The fast neutrons are captured by the uranium-238 to produce fissionable plutonium-239. The heat is removed by a flow of liquid sodium, which eventually transfers its heat to water to make steam.

In a **bomb** a large enough quantity of pure fissionable material supports a branching chain reaction.

The nonbreeder reactor requires a **once-through uranium fuel cycle,** in which the mined uranium ore is processed, enriched in the uranium-235 isotope, used in the reactor, and then stored as spent fuel.

The breeder reactor utilizes the **plutonium fuel cycle,** in which the uranium, instead of being enriched in uranium-235, is mixed with plutonium-239, which is produced from uranium-238 in the reactor. The uranium is thus in effect recycled.

Safety in a nuclear plant is exercised by (1) safe design of the entire system; (2) back-up or **redundant** safety systems; (3) warning devices, (4) inspection and maintenance procedures, and (5) final protection in case an accident does happen.

The environmental hazards from fission reactors include (1) the possibility of a serious accident that releases a large amount of radioactivity, (2) the hazard of even the usual small radioactive emissions with normal operations, (3) the problem of disposal of radioactive wastes, (4) the potential for terrorism or sabotage, and (5) the thermal pollution resulting from the waste heat of the nuclear plant.

In nuclear fusion, the nuclei of the light isotopes deuterium and tritium combine to produce helium and release energy. Because of the extremely high temperature required for the reaction, it has not yet been possible to design a useful fusion reactor.

Uranium reserves can last for hundreds of years, but as the most concentrated ores are used, the cost of uranium will continue to rise.

Questions

Radioactivity

1. (a) A Geiger counter registers 256 counts per second (cps) near a sample of polonium-210; 276 days later the counter registers 64 cps. What is the half-life of polonium-210? What will the counter register after another 276 days? (b) Polonium-210 decays in one step to lead-206, which is not radioactive. If you were asked to give a rough estimate of the length of storage time needed to reduce the radioactivity of polonium-210 to a safe level, would you say it is a matter of months, years, decades, or centuries? Would you be concerned about any radioactive progeny that might be produced?

2. Iodine-131 is a radioactive nuclear waste product with a half-life of eight days. How long would it take for 2000 mg of iodine-131 to decay to 125 mg? Would it be correct to say that iodine-131 is no environmental hazard because its half-life is so short? Defend your answer.

Nuclear Reactors

3. What are the essential features of a nuclear fission reactor? Explain the function of each feature.

4. List the possible fates of neutrons in a fission reactor. Which of these events should be favored and which should be inhibited in order to (a) shut down a reactor, (b) breed new fissionable fuel, and (c) produce more energy?

5. Explain how breeder reactors differ from nonbreeders in fuel, moderator, coolant, and any other features.

Nuclear Safety

6. (a) What are the five principles of industrial safe practice? (b) How are they applied to the construction and use of automobiles? of the place where you live? of a nuclear plant?

7. Outline the general concept and approach to safety used in nuclear power plants. Can you think of any specific series of events that would cause all of the safety features to fail and a radioactive cloud to be released to the atmosphere? If so, describe them.

8. Do you think it would be reasonable to set safety limits in nuclear power plants that would prohibit *any* release of radioactive matter? Defend your answer. If your answer is no, what criteria would you use to set the limits?

9. Explain why a critical mass of pure fissionable material must be attained if an explosion is to occur. Why is it thought that a nuclear reactor could not explode in the manner of a bomb?

Nuclear Fuel Cycles

10. (a) Outline the steps in the "once-through" uranium fuel cycle. Do you think it is appro-

priate to use the word "cycle" in this context? Defend your answer. (b) Outline the steps in the plutonium fuel cycle. What are its major differences from the uranium fuel cycle? What new environmental hazards would the plutonium cycle produce?

Radioactive Wastes

11. Describe the three steps involved in the handling of radioactive wastes.

12. Is the storage time for radioactive wastes that contain uranium and plutonium measured in years, hundreds of years, or thousands of years? If the wastes are reprocessed by removing the uranium and plutonium, is the required storage time increased, decreased, or unchanged? Explain.

13. Sandlike radioactive leftovers from uranium ore processing mills, called "mill tailings," have been used to make cement for the construction of houses in Colorado, Arizona, New Mexico, Utah, Wyoming, Texas, South Dakota, and Washington. These tailings contain radium (half-life 1620 years) and its daughter radon (a gas, half-life 3.8 days), as well as radioactive forms of polonium, bismuth, and lead. Radon gas seeps through concrete but is chemically inert. Are the following statements true or false? Defend your answer in each case:
(a) Since radon has such a short half-life, the hazard will disappear quickly; old tailings, therefore, do not pose any health problems.
(b) Even if the radon gas is present, it cannot be a health problem because it is inert and does not enter into any chemical reactions in the body.
(c) Continuous ventilation that would blow the radon gas outdoors would decrease the health hazard inside such a house.

14. List the various proposals for "permanent" storage of radioactive wastes. Suggest some advantages and some problems or uncertainties associated with each.

Fusion

15. Suppose that someone claims to have found a material that can serve as a rigid container for a thermonuclear reactor. Would such a

claim merit examination, or should it be ignored as a "crackpot" idea not worth the time to investigate? Defend your answer.

16. Outline the reasons why fusion reactors are expected to be far less serious sources of radioactive pollutants than fission reactors.

Suggested Readings

For a general overview of nuclear reactors, written at a level that is not too technical, the best choice is:
Anthony V. Nero, Jr.: A Guidebook to Nuclear Reactors. Berkeley, University of California Press, 1979. 289 pp. Paperback.

Another general overview is:
Walter C. Patterson: Nuclear Power. London, Penguin Books, 1976. 304 pp. Paperback.

The pronuclear viewpoint is very clearly presented in:
Bernard L. Cohen: Nuclear Science and Society. Garden City, N.Y., Anchor Press/Doubleday, 1974. 268 pp. Paperback.

Antinuclear viewpoints are given in:
John J. Berger: Nuclear Power—The Unviable Option. Palo Alto, Calif., Ramparts Press, 1976. 384 pp. Paperback.
Union of Concerned Scientists: The Nuclear Fuel Cycle. Revised ed. Cambridge, Mass., 1974. 291 pp. Paperback.
Robert D. Pollard, ed.: The Nugget File. Cambridge, Mass., Union of Concerned Scientists, 1979. 95 pp. Paperback.

Books that integrate social and technical aspects of nuclear energy include:
Nuclear Energy Policy Study Group: Nuclear Power Issues and Choices. Cambridge, Mass., Ballinger, 1977.
David Rittenhouse Inglis: Nuclear Energy: Its Physics and Its Social Challenge. Reading, Mass., Addison-Wesley Publishing Co., 1973. 395 pp.
Henry Foreman, ed.: Nuclear Power and the Public. Minneapolis: University of Minnesota Press, 1970. 272 pp.

Specific discussion of nuclear hazards may be found in:
Geoffrey G. Eichholz: Environmental Aspects of Nuclear Power. Ann Arbor, Mich., Ann Arbor Science Publishers, 1976. 681 pp.

Problems of sabotage and terrorism are considered in:
Mason Willrich and Theodore B. Taylor: *Nuclear Theft: Risks and Safeguards.* Cambridge, Mass., Ballinger, 1974. 252 pp.

The Rasmussen Study is published under the following title:
Reactor Safety Study. U.S. Nuclear Regulatory Commission, October 1975. 198 pp. Copies can be purchased from the National Technical Information Service. Springfield, Virginia. A separate *Executive Summary* (12 pp.) is also available.

The Three Mile Island Accident is taken up in:
Thomas H. Moss, and David L Sills, eds.: *The Three Mile Island Nuclear Accident: Lessons and Implications.* New York, Academy of Sciences, 1981. 343 pp. Paperback.

Mike Gray and Ira Rosen: *The Warning: Accident at Three Mile Island.* New York, W. W. Norton & Co., 1982. 287 pp.

Discussions of the options for handling of radioactive wastes appear in:
American Scientist, Vol. 70, March-April, 1982.
Fred C. Shapiro: *Radwaste: A Reporter's Investigation of a Growing Nuclear Menace.* New York, Random House, 1981. 288 pp.
Charles A. Walker, Leroy Gould, and Edward Woodhouse, eds.: *Too Hot to Handle?* New Haven, CT, Yale University Press, 1983. 220 pp.
Finally, the excellent periodical *Bulletin of the Atomic Scientists* carries many articles on the general problems of nuclear energy and nuclear weapons.

Use of Energy and Its Consequences

In many ecosystems, some of the animal species face alternating cycles of feast and famine—times of plenty and times of shortage. Historically, human beings have been no exception. Tribes of hunter-gatherers ate well when the game was plentiful and starved when it was scarce. With the introduction of agriculture, cities, and manufacturing, the times of plenty have become closely linked with the availability of energy as well as of food. Energy crises have been recorded in many civilizations, including ancient kingdoms of the Middle East, the Roman Empire, and medieval Europe. Perhaps the last great time of seemingly unlimited plenty occurred when Europeans started to colonize North and South America. The continents were so large, the forests so dense, and the prairies so bountiful that it seemed as if people could take whatever they wanted, whenever they wanted, and however they wanted, without destroying or diminishing the productivity of the land. This feeling gave rise to what the economist Kenneth Boulding has called "cowboy economics," which is in effect the economics of limitless, uncontrolled consumption.

The age of cowboy economics is over. True, North Americans live in greater wealth and comfort than any other people on Earth ever have. But today, most of the virgin forests are gone, the bison are gone, mines and fuel deposits are being depleted, and the threat of scarcity has become a reality. In addition, the environmental disruptions caused by human actions have begun to alter seriously the ecosystems of the Earth.

13.1 Environmental Problems Caused by Mining and Drilling

(a) Oil. Oil drilling on temperate farmlands or deserts does little environmental damage. An oil well requires the use of only a few hundred square meters of surface land (Fig. 13–1). In many instances, wells are located in farmers' fields or even in people's backyards with minimal environmental problems. In recent years, however, oil companies have begun to search for petroleum in more hostile environments, such as in the Arctic or under the ocean floors.

In Alaska, major problems arose in transporting the oil from a region far above the Arctic Circle to large cities farther south. After a lengthy court battle, oil companies were given

Figure 13–1 An oil well in a temperate ecosystem such as this one near Bakersfield, California, does little environmental damage.

permission to build a pipeline across Alaska (Fig. 13–2). The completed pipe crosses previously untracked wilderness, delicate icy tundra, mountains, rivers, and active earthquake zones. In winter, outside temperatures drop to −50°C (−58°F). In the summer they soar to +35°C (+95°F). Environmentalists have argued that remote wilderness areas have been altered. Roads were built where none existed before, allowing improved access for hunters and tourists. Tundra and caribou migration routes have been disturbed.

During the first five years of operation, from 1977 to 1982, the pipeline ruptured twice, spilling oil over the landscape. The most serious rupture released over 5,000 barrels of petroleum onto the tundra and into local rivers. Overall, the oil spill was less ecologically damaging than some people had feared. One of the reasons for this is that the contaminated rivers were not economically important salmon spawning grounds. If a break had occurred along the Yukon or the Copper rivers, the disruption could have been much more severe.

In many places around the world, valuable oil fields are located under the sea. To obtain

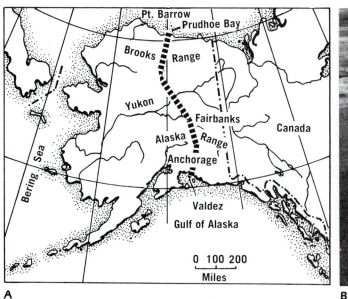

A

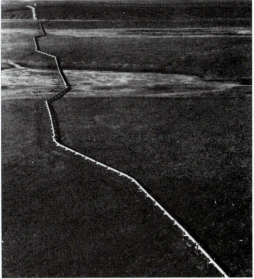

B

Figure 13–2 The Alaska pipeline. A, Route of the pipeline across Alaska. B, Pipeline zigzags over the tundra so that the pipe will resist fracture if the earth moves. Despite these precautions, the pipe fractured twice between 1977 and 1980. (World Wide Photos)

Figure 13–3 Offshore oil drilling operations are planned and built with great care, but nevertheless oil spills have occurred in all major drilling areas.

this oil, engineers must first build fixed platforms in the ocean (Fig. 13–3). Drilling rigs are then mounted on these steel islands. Despite great care, accidents occur in all types of drilling operations. Broken pipes, excess pressure, or difficulty in capping a new well have repeatedly led to blowouts, spills, and oil fires. When these accidents occur on shore, small areas of farmland are destroyed, but entire ecosystems are not affected. When accidents occur at sea, millions of barrels of oil can be dispersed throughout the waters. Fish and other marine life are poisoned, and ocean ecosystems are disrupted. Significant oil spills have occurred in virtually all underwater drilling areas—in the waters of southern California, Louisiana, and Mexico and in the North Sea in Europe.

Another problem arises when oil is transported over the ocean. Large supertankers carry oil around the globe. Many accidents have occurred in recent years, and ecological damage

has been significant. This issue will be discussed in more detail in Chapter 20.

(b) Coal. Coal can be mined either in underground tunnels or in exposed open pits. **Tunnel mines** do not directly disturb the surface of the land, but the earth and rock dug out to make the tunnel must go somewhere. This material, called **mine spoil,** is often seen as ugly surface pollution of land near mining operations. Sometimes the land above the tunnel collapses, leaving gaping holes on the surface. Perhaps most significantly, underground mining is dangerous and unhealthy for the workers. Fires, explosions, and cave-ins claim many lives. In addition, fine coal dust suspended in the air enters the miner's lungs. This dust gives rise to a series of serious and often fatal ailments known as **black lung disease.**

Tunnel mines alter the flow of ground water and pollute underground streams. They are also often more expensive and less efficient than open pit mines. For these reasons, mining companies have been switching many of their operations to open pits, also called **strip mines** (Fig. 13–4). To dig a strip mine, the surface layers of topsoil and rock are first scooped off (stripped) by huge power shovels. This exposes the underlying coal seam. The coal is then removed, and a new cut is started. If the land is not reclaimed, strip mines leave behind vast holes and huge piles of rubble. Neither arguments nor figures are needed to convince anyone that open, rootless dirt piles are uglier and less useful than a natural forest, a prairie, or a wheat field. The ugliness is a loss to all of us. The dirt piles erode, clogging streams and killing fish. As shown in Figure 13–5, a mine may also disrupt the flow of groundwater. This in turn upsets river flows and disturbs drinking water supplies in nearby areas. Furthermore, sulfur deposits are often associated with coal seams. This sulfur reacts with water in the presence of air to produce sulfuric acid, which pollutes the streams and kills fish.

Additional problems arise because millions of hectares of coal lie under farmland in many parts of the world. For example, large deposits exist under the fertile wheat fields of the Great Plains in North America. Is the land more valuable as a wheat field or as a coal mine? A wheat field can provide needed food for many years to

A   B

Figure 13–4 Strip mining operations. *A,* Aerial view of a strip mine in southeastern Montana. To the left, piles of rubble are being reclaimed and blended into the terrain. *(Courtesy of Bureau of Reclamation, photo by Lyle Axthelm.) B,* Decker Coal Mine in Montana. *(Courtesy of Bureau of Reclamation, photo by Lyle Axthelm.)*

come. Coal is used only once and then it is gone.

Perhaps people can have both wheat and coal if old mines are reclaimed. After the coal is removed the holes can be refilled with subsoil and covered with topsoil. If the land is then fertilized, it is possible to plant wheat and corn over the mines after about five years. However,

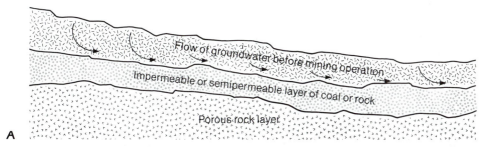

A

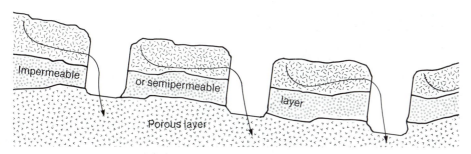

B

Figure 13–5 One possible sequence whereby strip mining can disrupt the traditional flow of groundwater. Groundwater flow before *(A)* and after *(B)* the mining operation.

the soil generally is not as rich as it was originally.

In the absence of environmental regulations, a mining company could buy a farm that lies over a coal seam, mine the coal, destroy the land, leave it, and still realize a large profit. But the government has realized that farmland is a national resource, a type of commons. The economic externalities of destroying this resource include pollution of waterways, rising food prices, and perhaps eventual food shortages. Therefore, laws have been enacted that require coal companies to reclaim the land (Fig. 13–6).

In the United States the Surface Mining Control and Reclamation Act was passed in 1977. It required that:

1.   Mining companies must prove that they can reclaim the land before a mining permit will be issued.

2.   After a mining operation is finished, the land must be restored so that it is useful for the same purposes for which it was used before the mining operation was started.

3.   Strip mining is not allowed in certain regions containing particularly prime agricultural land.

4.   Mining companies must use the best available technology to minimize water pollution and disruption of streams, lakes, and the flow of groundwater.

5.   A tax of 35¢ per tonne of strip-mined coal and 15¢ per tonne of coal mined in underground tunnels is levied. This money will be used to reclaim land that was mined and destroyed before the law was put into effect.

In 1981, Secretary of the Interior James Watt initiated proceedings to reduce regulations on strip-mine reclamation. He claimed that the $3,000 to $10,000 per acre cost for reclaiming land is hurting the mining companies and upsetting the economy. Before you make your own personal value judgment on the relative merits of mine reclamation, look at the map shown in Figure 13–7. Think of the environmental externalities that would result if all the untouched coal fields in the United States were reduced to bare, muddy, piles of dirt alternating with long rows of deep trenches.

13.2 Environmental Problems Caused by Production of Synfuels

Environmental problems do not end after a fuel is extracted from the ground; mining and drilling are only the first steps. Then the fuels must be processed, transported, and eventually burned. Each step represents a potential threat to the environment. For example, crude petroleum is a gooey, tarry fluid that must be refined to produce conventional fuels. The refining pro-

Figure 13–6 Reclaimed farmland in previously strip-mined coal area of central Illinois. The former agricultural productivity of such farms is not always completely recovered.

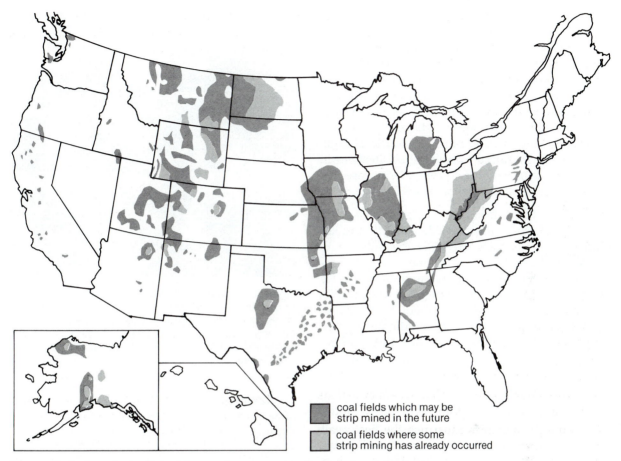

Figure 13–7 Areas in the United States where economically attractive coal deposits are found.

coal fields which may be
strip mined in the future

coal fields where some
strip mining has already occurred

cess consumes energy and produces significant air pollution. Any fuel, such as heating oil, gasoline, kerosene, or coal also produces air pollution when it is burned.

In recent years the search for quality fuels in the era of growing scarcity has led to many additional problems. In Chapter 11, the production of liquid fuels from oil shale, tar sands, and coal was discussed. Fuels produced in this manner are called synthetic fuels or **synfuels.**

(a) Liquid Fuels from Oil Shale. Oil shale can be mined either in open pits or in underground tunnels or caverns. In either case, environmental disruptions such as those related to coal mining are routinely encountered. But there are additional problems specific to shale development. Shale deposits are made of dense, heavily compacted rock. If the shale is dug up, broken apart,

and treated to extract the oil, the volume of rock actually increases even though oil is removed from it. If you have a hard time visualizing this, take a potato, slice it into thin strips, and make potato chips out of them. Now place the potato chips in a pile. Which occupies more volume, the pile of potato chips or the original potato? Some people suggest that this mine spoil could be simply used to fill in some of the scenic canyons in western Colorado and central Utah. What a price to pay for gasoline!

Another problem inherent in shale development is water consumption. Approximately two barrels of water are needed to produce each barrel of oil from shale. The western states where shale is found are semiarid. Today water is used for agriculture and industry in the area. Proponents of energy development claim that new dams can be built to conserve water and that industries can move elsewhere. Moreover, some of the farmers in the area use water inefficiently because much of the land is marginally productive anyway. Therefore, according to this argument, it would be in the national interest to use that water for other purposes. Even if all these sources of water were exploited, shale development would still be limited. Beyond that, one wonders, should rural ranchers and local business people be forced to forfeit their lifestyles in the interests of national efficiency? The case history in Section 13.3 personalizes this issue.

(b) Liquid Fuels from Coal. As discussed in Chapter 11, coal can be readily converted to liquid or gaseous fuels. In 1981, the only large-scale commercial coal-to-gasoline conversion facility in the world was operating in South Africa (Fig. 13–8). At that time, the plant was producing fuel at a price that was about $3 per barrel less than the cost of oil from Middle Eastern suppliers. The United States has vast coal reserves, enough to extend our fossil fuel age for several hundred years, and many people have advocated a crash program to develop synfuels. However, a great many problems arise:

1. Labor is much more expensive in North America than it is in South Africa. Therefore, the cost of synfuels would be several dollars per barrel higher in this country.
2. The capital investment needed to develop a significant synfuel program is enormous. Best estimates indicate that in the United States it would cost about $85 to $90 billion (in 1979 dollars) to construct enough conversion plants to cut the U.S. oil import level by 25 percent.

Figure 13–8 The SASOL (South African Coal, Oil, and Gas Corporation, Ltd.) refinery in South Africa. The plant produces commercial quantities of gasoline, diesel, and heating fuels and many different types of chemicals, using coal as the raw ingredient. (Courtesy of South African Consulate General.)

3. The engineering difficulties involved are also enormous. As one expert recently wrote,[1]

> Synfuel plants are very large enterprises—a 50,000 barrel per day coal (liquefaction or gasification) plant would cost, at a minimum, over 2 billion dollars and use several times as much coal as the largest electric generating plant. And no one has ever built even one at this scale. And we are talking about building 20 to 30 of them in the next 10 years. . . . The managers, engineers, and laborers will have to be trained or diverted from other activities, (and) new plants will have to be built to produce the equipment going into the synthetic fuels plants. . . . Some of these steps . . . can be compressed, but, taken together, the prospects for anything like 2.5 million barrels per day by 1990 seem to me slim unless as a nation we decide nothing much else is important.

4. The South African plant is a source of considerable air and water pollution. According to one reporter, "Indeed, smoke often hangs like a gray curtain for days over Sasolburg (where the plant is located)—people are now prepared to accept the air pollution. . ." Many of the by-

---

BOX 13.1 AN END TO SYNFUELS?

In April and May of 1982 some synfuel developments were abruptly ended. First, subsidiaries of Royal Dutch-Shell Ltd. and Gulf Oil Corporation dropped out of the world's largest synfuel project—the extraction of oil from tar sands in the Canadian province of Alberta. Next, the Exxon Corporation announced that it was closing its shale oil project in Colorado, the largest synfuel development in the United States. Several other projects were slowed down. These actions were taken because the estimated costs of the projects rose sharply while oil prices fell.

---

[1] Science, 205:978, Sept. 7, 1979.

products of coal gasification are carcinogenic, and radioactive trace elements are released as well. It is certainly possible to build a modern plant that emits less pollution than the South African model, but significant problems would always remain.

5. Coal is composed mainly of carbon, whereas natural gas or gasoline contains compounds of carbon and hydrogen. In order to gasify coal, hydrogen must be added, and the cheapest source of hydrogen is water, H_2O. However, the same problem that affects shale development also affects synfuels derived from coal—water is scarce in many regions where coal is being mined. Representatives of industry claim that in the southwestern part of the United States there is enough water for production of 500,000 barrels per day of synfuels. However, some environmental experts disagree and claim that beyond 250,000 barrels per day severe problems would be encountered. In the late 1970s the United States imported about 8.7 million barrels per day of petroleum. If more than 5 to 10 percent of this import level were to be replaced with synthetic fuels, water priorities would have to be adjusted drastically.

6. Hundreds of thousands of hectares of valuable farm land and scenic regions would be disrupted by a coal mining industry of the scale required for a significant synfuel program. Much of the insult can be reduced by extensive reclamation programs, but even if utmost care is taken, many years would be required for the Earth to recover fully.

None of the problems just outlined is totally insurmountable, but it is important to understand that a synfuel program would be associated with heavy economic and social costs.

13.3 Case History: Water and Coal in the Southwest United States

During the late summer of 1977, I (Jon) was driving through northeastern Arizona. This region is semiarid, and small herds of cattle and sheep graze amid sparse grasses and sage. Just outside of Kayenta my car broke down, and I

was forced to hitchhike to town to get help. An old, 1950 vintage flatbed truck slowed down but did not stop. The driver shouted, "Jump in, I don't have any first gear and can't stop!" I ran along and hopped onto the moving vehicle. The engine groaned, the bearings rattled, but we slowly accelerated onto the highway. The driver, a Navajo rancher, was hauling a load of hay south from Green River, Utah. In other areas with wetter climates, late summer is the time to gather in the hay. But there is too little rainfall in this region to grow much hay. Therefore, the Navajo and Hopi farmers must drive north to buy feed for their animals. The man who was driving the truck explained to me that it is very difficult to make a living raising cattle if you cannot grow your own hay. Every fall he sells many of his animals. Most of the money is used to buy hay to feed the rest over the winter. Calves and lambs are born in the springtime, and the animals are raised on grass until fall. None of the farmers of the region make much money. He was hoping that his animals would bring a high price at the auction this year so he could afford a rebuilt transmission for his truck.

Later I spoke with Keith Smith, the Navajo Chapter President in Kayenta, about the water problems in the region (Fig. 13–9). He told me, "The last few years the water table has been going down. During June and July, people's wells have gone dry. Ranchers have to drive to town to haul water in pick-up trucks so the cattle won't die." After a silence he continued, "A rancher can hardly break even if he has to haul water in a pick-up."

There is a large coal mine a few kilometers south of Kayenta. Some of the coal from the mine is shipped by rail to a power plant nearby. Most of the rest is sold to the Mohave Power Project nearly 480 kilometers (300 miles) away. There the fuel is burned to produce electricity for Las Vegas and southern California. The coal is shipped from Kayenta to Mohave via a coal-slurry pipeline (Fig. 13–10). Pipeline shipment is cheaper than shipment by rail. The problem is that it uses water. The coal is first ground into a fine powder. Then it is mixed with water. The whole mixture, called a **slurry,** is pumped through a large pipe. Water for the project is taken from several large wells near Kayenta. The pipeline uses approximately 7500 liters

Figure 13–9 "A man can ride under the power lines and feel the energy traveling through them. But the electricity is going to Phoenix, Tucson, Las Vegas, and Los Angeles. We can't tap into that power and many farms have no electricity. I have to drive eight miles in the pickup truck to get water because I don't have a pump and the water table is going down." Ernest Yellowhorse, Cove, Arizona.

(1900 gallons) of water *per minute* and operates 24 hours per day, 7 days a week.

As the water is removed, the water table for the entire region has dropped, and many of the shallower wells in the area have gone dry. If the coal for the Mohave Power project were shipped by rail instead of by pipeline, the price of fuel would rise slightly. However, the ranchers in the region would have enough water to maintain their stock.

As the energy crisis worsens, the coal companies will need even more water than they use

Figure 13–10 Dotted line shows the route of the coal slurry pipeline.

now. Recall that one solution to the oil shortage is to convert coal to liquid fuel. This conversion requires large quantities of water. Water is an absolutely essential chemical ingredient in the conversion. Where will the power companies find the water? Even now, the farmers near Kayenta cannot get what they need. Experts feel that even if other consumers were rationed more severely than they already are, there still might not be enough water to manufacture the needed fuel. In some regions, water, not energy, is becoming the limiting resource.

13.4 Electricity and Thermal Pollution

One of the major issues in environmental planning today concerns the problems of generating electric power. People who believe in a future based on coal and nuclear energy feel that a great many new electric generating facilities must be built in the next twenty years. Others disagree. They argue that the production of electricity leads to depletion of resources, water pollution, air pollution, solid wastes, radioactive wastes, land uglification, and thermal pollution. If people would learn to conserve electricity, they could benefit from the use of this convenient energy supply without building more power plants and adding an extra burden on the environment.

Most of our electricity is now produced in **steam turbines.** The operating principle here is uncomplicated. Some power source, such as coal, gas, oil, or nuclear fuel, heats water in a boiler to produce hot, high-pressure steam. This steam expands against the blades of a turbine. A turbine is a device that spins when air or water is forced against it. You can think of it as a kind of enclosed windmill. The hot, expanding steam forces the turbine to spin. The spinning turbine then operates a generator that produces electricity. After the steam has passed through the turbine it is cooled, liquefied, and returned to the boiler to be reused. Normally the steam is cooled with river, lake, or ocean water (Fig. 13–11).

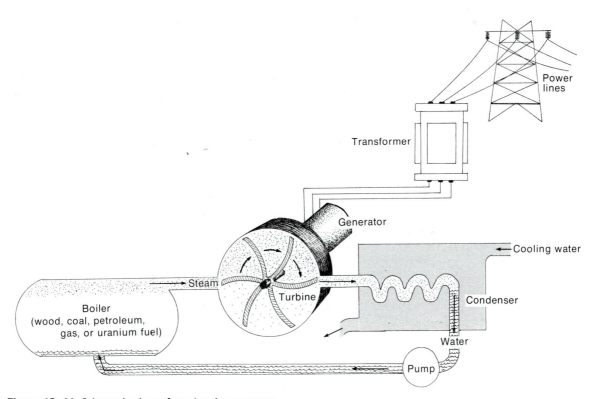

Figure 13–11 Schematic view of an electric generator.

The **cooling cycle** is essential to the entire process. Recall that the Second Law of Thermodynamics states, "It is impossible to convert all the energy of a fuel into work." But the efficiency can be improved if the exhaust gases are cooled. Even the best electric generators operate at only about 40 percent efficiency, but if old power plants are considered, the average efficiency in North America is closer to 38 percent. This means that for every 100 units of potential energy in the form of fuel, only 38 units of electrical energy can be produced. The other 62 units of heat energy are lost to the environment. Because additional energy is lost when electricity is transmitted through long-distance power lines, energy delivered to the home represents only about 34 percent of the original potential in a fuel. Although modern electrical production is only 34 percent efficient, it is more efficient than other common heat engines. An automobile is 25 percent efficient, and a steam locomotive is only 9 percent efficient. Therefore, electricity is an efficient way to perform work. Less fuel is needed to operate electric lawn mowers or cars than gas-powered machines.

Electric Heat.   In modern society, electricity is often used to provide heat. Advertisements advise people to "live better electrically," and buy stoves, toasters, space heaters, and water heaters. But electric heaters are thermodynamically inefficient. Sixty percent of the heat is discarded at the power plant. On the other hand, a gas-fired stove can be nearly 100 percent efficient, and home furnaces are 60 to 80 percent efficient. Electricity is essential for many functions but it is wasteful when used to produce heat. Much energy could be conserved if electricity were used only where it is needed and if fuel were used directly when heat is needed (see Fig. 13–12).

Thermal Pollution.   The amount of heat that must be removed from an electrical generating facility is quite large. A one-million-kilowatt power plant running at 40 percent efficiency heats 10 million liters of water by 35°C (63°F) every hour. It is not surprising that such large quantities of heat, added to aquatic systems, cause ecological disruptions. The term **thermal pollution** has been used to describe these heat effects.

What happens when the outflow from a large generating station raises the water temper-

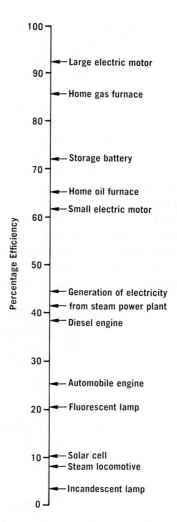

Figure 13–12 Maximum efficiencies of some common machines, devices, and processes. The values for electrical devices do not include the energy losses at the generating station.

ature of a river or lake? Fish are cold-blooded animals. This means that their body temperature increases or decreases with the temperature of the water. In natural systems, water temperatures are relatively constant. When the temperature is raised, all the body processes of a fish (its metabolism) speed up. As a result, the animal needs more oxygen, just as you need to breathe harder when you speed up your metabolism by running. But hot water holds less dissolved oxygen than cold water. Therefore, fish accustomed to cold water may suffocate in warm water. In addition, warm water can cause outright death through failure of the nervous system. In general, not only fish but also the entire aquatic ecosystems are rather sensitively affected by temperature changes. For example, many animals lay their eggs and plants disperse seeds in the springtime when the water naturally becomes warm. If a power plant heats the water in midwinter, some organisms may start reproducing, but if the eggs are hatched at this time, the young may not find the food needed to survive.

Not all power plants discharge waste heat directly into natural environments. Many use special lakes called **cooling ponds.** Hot water is pumped into the ponds, where evaporation as well as direct contact with the air cools it, and the cool water is drawn into the condenser from some point distant from the discharge pipe. Water from outside sources must be added periodically to replenish evaporative losses. Cooling ponds are practical where land is cheap, but a 1-million-kilowatt plant needs 400 to 800 hectares of surface, and the land costs can be prohibitive.

An alternative solution is a **cooling tower,** which is a large structure, typically about 180 meters in diameter at the base and 150 meters high (Fig. 13–13). Hot water is pumped into the tower near the top and sprayed onto a wooden mesh. Air is pulled into the tower either by large fans or convection currents and flows through the water mist. Evaporative cooling occurs, and the cool water is collected at the bottom. No hot

water is introduced into aquatic ecosystems, but a large cooling tower loses over 3.8 million liters of water per day to evaporation. Thus, fogs and mists are common in the vicinity of these units, reducing the sunshine in nearby areas. Reaction of the water vapor with sulfur dioxide emissions from coal-fired power plants can cause the resultant air to carry sulfuric acid aerosols, as described in Chapter 21.

13.5 Politics and Economics of Energy Supply

So far, fairly predictable limits and problems related to energy supply and use have been discussed. If a reasonable estimate can be made of how much fuel is available in the ground and how rapidly it is being used, it is not hard to predict how long the supply will last. All these predictions assume that the resources will be exploited in an orderly, uninterrupted fashion. Such order is not guaranteed. In 1973 and again in 1979 there were worldwide energy crises. The price of fuels rose rapidly in a matter of a few weeks, and even then severe shortages were felt in most countries. There was plenty of oil in the

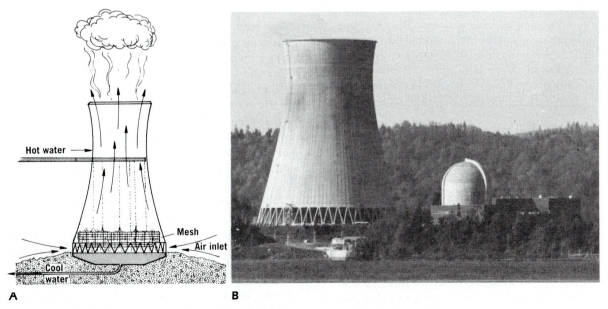

Hot water

Mesh
Air inlet
Cool water

A

B

Figure 13–13 Cooling towers. A, Schematic view of the operating principle of a cooling tower. B, This large cooling tower dwarfs the nuclear reactor at a reactor site in southern Washington.

ground, but global political and economic maneuvering led to temporary shortages. Early in 1983 energy was relatively available worldwide. The price of oil was actually declining, and reserves were so high that people were talking about an oil glut. At the same time, the global political climate was underlaid with tension. Some Middle Eastern oil-producing nations were already involved in warfare, and all the others were armed and ready for war. It appears that there is a very real potential for disaster. In the book, *Energy Future—Report of the Energy Project at the Harvard Business School* (see Suggested Readings), Robert Stobaugh and Daniel Yergin write:

> Political instability in the Middle East, supply interruptions, the extension of Soviet influence—such factors only make a very bad situation much worse. *This point must be underlined.* For the industrial nations to continue to depend on Middle Eastern oil in the way current trends indicate means heavy reliance on a region of high political tension and risk. In the last three decades, the Middle East has been subjected to a half-dozen wars, a dozen revolutions, and innumerable assassinations and territorial disputes. Dependence reinforces the twin vulnerabilities—interruption of supplies and major price increases.
>
> Yet today, as we enter the 1980s, even after the second oil shock that accompanied the fall of the Shah of Iran, the cause and consequences of the new era of oil have yet to be taken seriously in the United States. The key contradiction is this: While the declared aim of American policy is to reduce the use of imported oil, the United States is in fact becoming more and more dependent upon it. Between 1973 and early 1979, U.S. oil imports almost doubled, and had begun to provide half of the nation's oil. By current trends the United States will be even more dependent on imported oil in the 1980s.

In a similar tone, the author of *Energy and Security* (see Suggested Readings) wrote:

> Today, nearly two-fifths of the oil consumed by the free world's economy is vulnerable to terrorism, accident, warfare, and extortion. . . . The sudden loss of Persian Gulf oil for a year could stagger the world's economy, disrupt it, devastate it, like no event since the Great Depression of the 1930's.

The probability of Soviet tanks rolling across the North German plain is much lower than the likelihood of an interruption of oil supplies stemming from various conflicts in the Middle East. Yet our energy plans and our diplomatic strategy do not reflect those probabilities. We are far less prepared for an energy emergency than for a military attack.

Even in the absence of war or political instability, the cost of energy has a serious effect on the economy of most nations. In many developing countries, the price of fossil energy is so high that people cannot afford these fuels; instead they use manure and wood. The use of manure as a fuel robs the soil of a valuable fertilizer, and the destruction of forests leads to soil erosion. Since the price of commercial fertilizer is linked to the price of fossil fuels, many farmers are caught in a downward spiral from which there is no obvious escape (Fig. 13–14).

The economic situation is unhealthy in developed countries as well. Traditionally, high productivity and economic growth have been associated with high energy consumption. In 1977, then Secretary of Energy James S. Schlesinger said, "Restraining energy growth means restraining the growth of jobs. It means unemployment." Many people do not agree with this statement. Consider the alternative.

A country has an equal **balance of trade** when the total value of all imports equals the total value of all exports. A positive balance of trade means that there is a net flow of wealth into a nation and is naturally desirable for that country. On the other hand, a negative balance means that more resources are leaving the country than are entering it. One immediate effect of a negative balance of trade is that the country's money loses value in world markets, leading to inflation. Another effect is that there is less money available in the nation for loans, for capital investment, and for modernizing industry. In 1980, the United States spent approximately $40 billion on imported oil. This expenditure was one factor leading to a negative balance of trade. What would have happened if that $40 billion spent on foreign fuel had been kept within the country? On the average, it takes about $50,000 per year to create and support one job; therefore the oil money represented some 800,000 jobs. Furthermore, if more money be-

duction, but the automobile workers would not necessarily lose their jobs. They could be employed to manufacture buses and trains (Fig. 13–15). Workers who now specialize in road construction could build and repair rail systems. If cars and appliances were maintained longer and thrown away less frequently, more repair specialists would be needed. Plumbers and carpenters could be hired to install solar collectors and to add extra insulation to homes.

Conservationists argue that if energy consumption continues to grow until suddenly serious oil or coal shortages develop, the economy will collapse disastrously. If people start now to build a low-energy society, changes can be made gradually. The economy need not suffer; in fact it could prosper.

13.6 Planning an Energy Future in North America

A transition in the energy structure of society may occur in an orderly and well-planned manner or it may be disruptive and chaotic, but in any case it is inevitable. How can we plan for the future, and how should society adapt to the

Figure 13–14 A woman collecting cow dung for use as cooking fuel in India. The dung is needed for fuel because kerosene is too expensive, but the soil is impoverished when a valuable fertilizer is removed from the fields. (Photo by M. P. Kahl, Bruce Coleman, Inc.)

came available, interest rates would decline, and manufacturers would be in a better position to modernize their plants, which in turn would improve their competitiveness in world markets.

Certainly a sudden fuel shortage would cause widespread unemployment. Factories would be forced to shut down, gas station attendants and automobile mechanics would be out of work, shippers and truckers would become unemployed. On the other hand, a gradual conservation program would *increase* jobs. Suppose, for example, that people in the United States gradually shifted toward increased use of mass transit. Automobile factories would slow pro-

Figure 13–15 Construction of energy-efficient means of transportation, such as this subway system in San Francisco, creates jobs.

TABLE 13–1   Comparison of Various Energy Sources in the United States

Energy Source	Current Use	Future Availability of Energy Supply	Current Cost	Pollution Problems
Oil	44%	Shortages by year 2000	Relatively inexpensive	Pollution of the ocean from oil spills; air pollution when fuel is burned; global climate change possible from release of carbon dioxide
Natural gas	24%	Shortages by year 1990	About the same as oil	Cleanest fossil fuel
Coal	22%	200- to 300-year supply	Less expensive than oil	Land disruption from mining; air pollution from burning; global climate change possible from release of carbon dioxide
Hydroelectric	4%	Possible doubling of current supply (renewable)	Less expensive than oil	Disruption of farmland and scenic and recreational areas
Heavy oils and synfuels	<1%	As long as coal	More expensive than conventional fossil fuels	Large disruption of land surfaces; depletion of water resources; other pollution problems associated with burning of coal and oil
Nuclear	4%	Uncertain	More expensive than fossil fuels	Hazards from nuclear wastes and fears of potential radioactive contamination
Wood and plant matter	2%	Uncertain	The least expensive fuel in certain regions	Disruption of forests; preempts land that could be used for food production
Garbage	<1%	2.5% (renewable)	Variable	Air pollution problems but conserves resources and eliminates a solid waste problem
Solar	<1%	Excellent and renewable	Cheaper than fossil fuels in some applications, more expensive in others	Minimal
Wind	<1%	Excellent and renewable	More expensive than fossil fuel	Noise from windmills; use of land
Waves and tides	<1%	Excellent and renewable	Some tidal generators are competitive today—others unknown	Disruption of coastal habitats for fish and recreation
Geothermal	<1%	Small expansion	Inexpensive where available	Some pollution from released steam
Hydrogen fusion	0	Excellent and virtually limitless if reactors can be built	Technology not yet available	Serious thermal pollution

inevitable changes that will occur? These questions form the basis for one of the most highly contested debates of our times. Perhaps the best way to start looking at the problem is to review possible energy sources and their availability. Table 13–1 shows that today fossil fuels provide about 90 percent of the energy supply in North America, but that these fuels will be exhausted or depleted in the near future. Oil and gas, the two most versatile and widely used fuels, will be the first to go. What will replace them? There are two divergent views. These have been named the **hard** and the **soft** pathways by the physicist Amory Lovins. **Hard energy** systems are based on large centralized electric generating plants. A hard energy scenerio might involve central coal or nuclear-powered generators supplying enough electricity to heat homes, provide power for industry, and charge batteries to drive electric vehicles. The **soft pathway** is based on the use of smaller, home-sized, decentralized energy systems such as small windmills and rooftop solar collectors. Note that although in popular usage coal and nuclear energy are often considered to be synonymous with hard energy, and solar and wind with soft energy, the real differences lie not in the source of energy but in the way it is used. For example, if there is a coal seam on the edge of a town and the local residents dig out a few chunks to use in home furnaces, coal heat becomes soft technology. Alternatively, in some applications solar energy can be hard technology. For example, people have suggested that giant arrays of solar cells be orbited in space (Fig. 13–16). The electric energy produced would then be transmitted to Earth via microwaves and distributed to conventional transmission lines. The total cost of the system is estimated to be some $2 trillion. Such a centralization of the energy structure coupled with an astronomical concentration of capital investment is an example of the hard path.

Today, some 97 to 98 percent of the energy used in North America comes from hard sources. Many people point to the tremendous success of this approach; after all, it changed civilization in a matter of 100 years or so. According to this train of thought, the energy future should be a continued modernization of systems that have proved successful in the past. Others claim that hard energy technology will no longer be viable in the future. This argument contends that centralized systems are becoming too complicated and too expensive to build, operate, and maintain. Fifty years ago, even 25 years ago, there were ample supplies of cheap oil, the population was lower than it is today, and the promise of nuclear energy seemed to be just around the corner, so hard energy systems were cheap and convenient. Today, however, oil is expensive and is often found in distant places, and supplies are politically vulnerable. The population is greater than it has ever been, and nuclear energy has become expensive and possibly dangerous.

Perhaps the central argument in favor of soft energy is simply cost. There is a growing realization that new fossil fuel and nuclear energy systems are becoming increasingly more expensive. Recall that the true price of gasoline, if all economic externalities are considered, is far greater than the actual price now charged at the gasoline pump. A few other examples are listed below.

● In recent years the United States government has spent many billions of dollars to support the nuclear power industry. This support must be paid for in taxes even if it is not reflected in electric bills. Yet, even with government subsidies, nuclear-powered electric heat is more expensive in the long run than good solar construction. If the true cost of nuclear energy were tallied directly, alternative renewable energy sources would certainly be the cheapest way to heat a home.

● A new pipeline is being planned to transport natural gas from the deposits in northern Alaska to the central United States. The total cost of this pipeline is expected to exceed $43 billion. If completed, this will be the single most expensive private construction project in history, far surpassing the Alaskan oil pipeline, which cost a mere $9 billion. In order to meet expenses, the oil companies are authorized to raise energy bills before the pipeline is even started. Individuals are to be charged up to $190 per year and industrial users up to $25,000 per year, and these charges can be levied even if the pipeline is never built. This is yet another indication that the era of inexpensive fossil fuels is rapidly ending.

● Think of the proposed solar power satellite. As mentioned earlier, this project would cost a whopping $2 trillion. If this money were divided among all the people in the United States, each man, woman, and child would receive $8000, and a family of four would receive $32,000. If

Figure 13–16 *Artist's conception of the Solar-Power Satellite. (Courtesy of NASA.)*

Figure 13–17 An early American hydropower facility now reconstructed in Bantam, Connecticut. (Photo by George Betancourt.)

every household used this money to install the best solar and wind technological devices, there would be no need for the electricity from the solar power satellites.

According to its proponents, soft energy systems could benefit from the economics of mass production. Very simply, if a few large generators are built, they must be machined on a special order basis, whereas millions of small generators can be built with assembly-line efficiency. For example:

• Large dams for the production of hydroelectric power cost about $1000 per kilowatt of power available. This price includes the construction both the dams and the generators. However, small-scale dams and hydroelectric generators can be built for about $500 per kilowatt. Many small dams along a river would produce more electricity at a lower price with less environmental impact than a few large centralized facilities (Fig. 13–17).

• Fiat Motor Company in Italy has recently designed a home-sized power system. This consists of a small diesel engine to generate electricity. Waste heat from the engine is used directly in a home-heating system. The overall system produces energy at a cost that is about 15 percent less than electricity generated from a centralized coal fuel plant, and 20 percent less than a system powered with a nuclear reactor.

Another argument in favor of the soft path is that it is much less vulnerable in times of war or natural catastrophe. Large power plants and dams are particularly vulnerable to enemy attack or could be destroyed by an earthquake, whereas millions of home-sized power sources are decidedly less vulnerable.

13.7 Energy Planning for the Immediate Future (Present to 1995)

Significant changes in energy supply and consumption do not occur overnight. For exam-

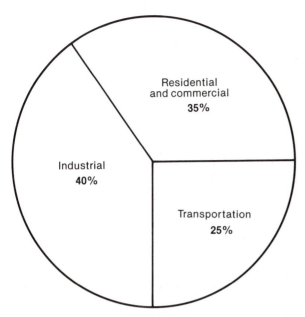

Figure 13–18 Uses of energy in the United States in 1981.

ple, time spans of 10 to 15 years are required to plan and build a new power plant or pipeline. Similarly, even if every new home were built to the best passive solar design, many old, inefficient homes would remain serviceable for many years. Even if windmills were economically competitive tomorrow, it would be a matter of decades before units could be installed in everyone's backyard. Planning must begin now for the future, but many of these plans will not bear fruit for years to come. What, then, can be done immediately to improve our energy balance?

Only conservation can have an immediate and significant impact on patterns of energy use. No new technology is required to turn off unused lights, drive to work in carpools, or turn the thermostat down and put on a sweater. These kinds of savings can extend the life of our fossil fuels and buy time for a smooth transition to the future.

Industry

Between 1972 and 1980, the energy required to produce industrial goods in the United States declined by 15.4 percent per unit of output. As a result, industry saved $1 billion in fuel bills in 1980 and pumped a total of $8.7 billion into the economy to upgrade the industrial processes.

These changes were not initiated for altruistic reasons; they came about because money will be saved in the long run.

Transportation

The two least efficient modes of transportation, the automobile and the airplane, are the two major transportation industries in the United States (Fig. 13–19).

The automobile in particular has modified our lives. Houses are far from places of work and from shopping centers. Many people live in the suburbs and must commute long distances daily. What can be done in the immediate future? No one can reorganize the cities overnight. The system is set in concrete.

People can't easily move existing buildings, but they can and are changing transportation patterns. As mentioned previously, carpooling is an ideal and effective short-term solution. It isn't necessary to buy new automobiles, move houses, stores, or factories, or build new mass transit systems. If people shared rides, they could reduce fuel consumption by 50 percent or more overnight. Consider another factor. In the United States, most automobile trips are for distances of less than 8 km. People drive to the corner market to pick up a newspaper and a liter of milk or to mail a letter. These trips are costly. Large amounts of gasoline are needed to start a cold automobile. As a rough approximation, a car that is capable of operating at 10 kilometers per liter on long trips operates at 2 kilometers per liter on short trips.* If people reduced the number of short car trips, walked, or rode bicycles, large quantities of fuel would be saved (Fig. 13–20).

Other changes cannot be implemented so quickly, but in future decades they will become increasingly important. For example, a current trend toward smaller, more efficient automobiles is encouraging, but there is a lag time of several years before most of the older, heavier "gashogs" wear out and are discarded.

There have also been some trends toward the redesign of cities and suburbs. Many large offices and manufacturing operations are moving into suburban regions so employees who live in the country don't have to commute to the

---

* To change km/L to miles/gal, multiply by 2.35.

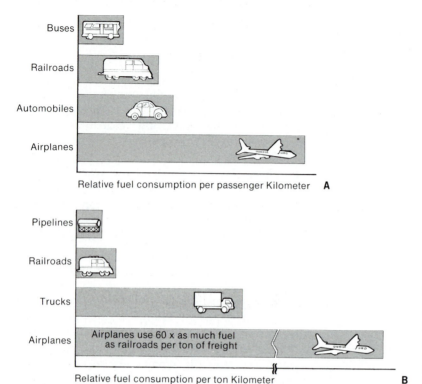

Relative fuel consumption per passenger Kilometer   A

Relative fuel consumption per ton Kilometer   B

Airplanes use 60 x as much fuel as railroads per ton of freight

Figure 13–19 Relative fuel consumption of various means of transportation. In each case, estimated average load capacities were used to compute efficiencies. (From Transportation's place in the energy picture, *Association of American Railroads,* August, 1979.) *A,* Intercity passenger carriers. *B,* Freight carriers.

cities to work. At the same time, there has been a renewed interest in centralized downtown shopping centers so people who live in the cities needn't drive to the suburbs to buy groceries and other essentials.

Figure 13–20 The bicycle is an energy-efficient means of transportation. (From D. Plowder: *Farewell to Steam.* Brattleboro, Vermont, The Stephen Greene Press, 1966.)

Surface Mass Transit. Surface mass transit—buses, trains, and trolleys—can provide clean, efficient transportation. At the present time, only 3 percent of urban transport and 4 percent of intercity traffic is carried by public ground transportation systems; the remainder is carried by airplanes and automobiles. The most recent trend is an overall reduction of urban mass transit rather than an increase. Although bus and subway systems are relatively popular in the largest cities in North America, the automobile still reigns supreme in most small or intermediate-sized urban centers. As a result, a dangerous situation has evolved. If fuel shortages were to develop quickly, the existing mass transit system could not possibly handle transportation needs. Millions of workers would be stranded and unable to get to work. Cities would be paralyzed.

Most mass transit systems in the United States and Canada are far behind their European counterparts. Intercity rail services have been reduced in the United States while the federal expenditure for new highways has escalated. In many cities a person may have to wait up to 20 minutes for a local bus, whereas in most European cities bus service is available every 5 minutes during rush hours. If mass transit systems were properly developed, they could provide an effective alternative to the automobile. At the present time, traffic is often stalled for long periods during rush hours and travel is slow; accidents are common. Moreover, parking spaces are often hard to find; rarely can people drive directly to their destination and park nearby. If quiet, efficient, comfortable mass transit systems were built in most cities, and if road building projects were curtailed, mass transit might become more efficient, more economical, and therefore more popular.

Residential and Commercial Use

Perhaps in no other sector of our society can more energy be saved with less need for social change than in the area of heating homes and commercial buildings. The easiest way to implement change is simply to turn the thermostat down in winter and to turn the air conditioner off and open the windows on hot summer days.

The decision to change the thermostat settings involves a sacrifice; people must adjust to living under what are considered to be less than optimally comfortable conditions. The sacrifice really is not that great, however, and can be largely overcome by changes in dress habits. For example, according to research carried out by The American Society of Ventilating Engineers in 1932, the preferred room temperature during the winter for a majority of subjects was 18.9°C. Similar research at later dates showed that the comfort range had risen to 19.3°C in 1941 and up to 20°C by 1945. In 1980, many homes were heated to 21° to 22°C. The increase in preferred indoor temperature is due in part to changes in fashion. Fifty years ago people naturally wore sweaters and long underwear indoors. (If one lives through several winters at 18°C, then 21°C seems uncomfortably warm.) Today men wear

BOX 13.3 SAVING ENERGY

In some areas of northern Japan, temperatures often fall well below freezing on winter evenings. Yet local residents use very little fuel for heating. After dark, family social life is centered around the dinner table. Tables are low to the floor and people sit on cushions rather than on chairs. A large, heavy tablecloth extends beyond the table and rests on people's laps to serve as a common blanket. The primary source of heat in the dining room comes from a small charcoal brazier located under the table. The heat from the fire is retained by the tablecloth, so the air around the people's lower bodies is warm. The remainder of the house is cool, and the people simply wear heavy sweaters to keep their upper bodies comfortable. After dinner and tea, they go to bed and snuggle down under warm quilts.

jackets, ties, and long pants in business offices in summer and then feel the need to turn the air conditioners on to "high." In the winter many women wear dresses and skirts and turn the heaters up. If fashion changed so that men wore short pants and light shirts during hot weather, and women wore wool pants in the winter, less fuel would be needed for heating and air conditioning (Fig. 13–21).

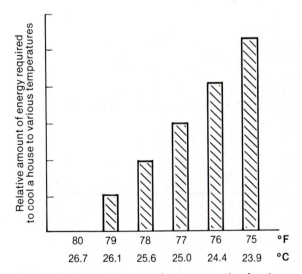

Figure 13–21 Relative extra fuel consumption for air conditioning to cool an average home to various temperatures *below* 80°F (26.7°C). For example, it takes three times as much extra fuel to cool the average home from 80°F (26.7°C) to 77°F (25°C) as it does to go from 80°F to 79°F (26.1°C).

BOX 13.2 ENERGY WASTE

Colorado's Governor, Richard D. Lamm, was asked what was the worst case of waste that he observed in the last year. "That is easy," he said. "It was one day last July when the outside temperature was at 96°F. The state building in Denver had the air conditioner on so cold that one of the secretaries had a heater plugged in alongside her desk."

The recent rise in the use of energy for air conditioning in particular is a measure of social demand. In 1952, only 1.3 percent of the houses in the United States had air conditioning; by 1979, over 50 percent of all homes were air conditioned. The United States, with a mere 5 percent of the world's population, uses over 50 percent of the energy consumed in the world for air conditioning. Houses, cars, shopping malls, and even entire sports arenas are cooled.

Changing the thermostat setting is really only a first step. Additional savings can be realized by more efficient construction. As discussed in Chapter 11, a well-built, well-insulated building employing passive or active solar heating systems or both, uses one quarter to one tenth of the energy needed to heat or cool a "conventional" house. Sound design and construction save money, and with such construction there would be no need to change the thermostat. However, even today, with the threat of rising prices and global energy shortages, a great many homes are being built with little consideration for energy conservation, and a great deal of fuel is wasted. Remember, an inefficient house built today may last fifty years or so (Fig. 13–22).

What about the conventional houses that have already been built? Significant savings can be realized immediately through simple and inexpensive conservation practices (Fig. 13–23). If an old, uninsulated house were insulated, storm doors and windows added, and leaky seams caulked, fuel consumption could be cut in half. Note, however, that such measures increase the concentration of indoor air pollutants, unless proper preventive methods are used (see Chap-

Figure 13–22 Energy loss from a poorly insulated two-story home with an open fireplace.

ter 21). Further savings can be realized by fitting solar design features onto existing structures. Two suggestions are shown in Figure 13–24.

13.8 Energy Planning for the Intermediate Future (1995–2020)

Looking ahead toward the end of the twentieth century and the start of the twenty-first, it seems inevitable that the cost of energy from fossil fuel and nuclear sources will continue to

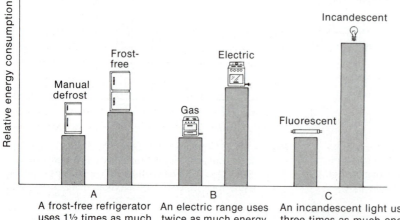

A
A frost-free refrigerator uses 1½ times as much energy as a conventional one

B
An electric range uses twice as much energy as a gas range

C
An incandescent light uses three times as much energy as a fluorescent light

Figure 13–23 Fuel consumption in the home could be reduced without any loss of comfort if more efficient appliances were used.

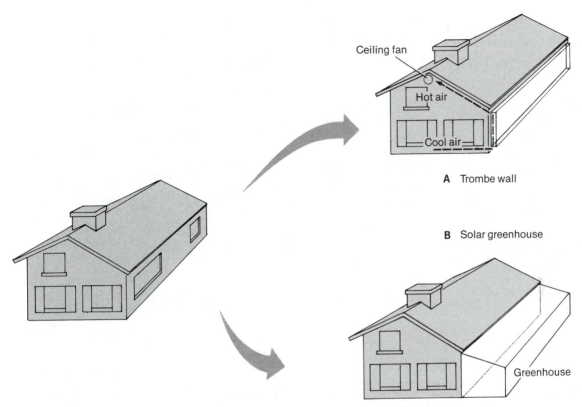

Figure 13–24 Two ways to fit passive solar design onto an existing structure. *A,* Trombe wall. A layer of glass is set 3 to 5 cm in front of the existing south wall, and holes are cut in the inside wall for the passage of air. The newly formed air space becomes a solar collection system. *B,* Solar greenhouse. A solar greenhouse is a passive solar collection system. The air in this space heats up, and hot air can be conducted into the main house. Plants can also be grown here, and thus the greenhouse can provide a household food source.

increase. Therefore, even larger changes in the physical structure of the energy system will become necessary. These changes may follow the hard or the soft strategies or, more probably, some of each. Renewable energy systems will undoubtedly become more popular. In addition, if planning starts today, advanced conservation techniques can be realized a decade or two from now. For example, in the field of transportation, an effort can be made to redesign our cities to create pleasant places to live and work in the same neighborhood so that people won't have to commute long distances. When commuting is necessary, advanced transportation systems could do the job. There are several attractive alternatives. A few examples are given below.

● Several prototypes of 42 km/L (100 miles/gal) commuter cars have been built and could be manufactured in the near future. This represents a sixfold increase of fuel efficiency over most cars built in the early 1970s.

● Significant improvements can be realized in mass transit as well. Commuter trains must start and stop frequently. In a conventional vehicle, large amounts of energy are needed to accelerate after picking up passengers, and this energy is simply dissipated as useless heat when the brakes are applied. In experimental vehicles designed in the early 1970s, a massive round disk, called a **flywheel,** is connected to an auxiliary electric generator (Fig. 13–25). Instead of applying conventional brakes, the engineer activates a lever that connects the wheels of the train to the flywheel. The energy inherent in the motion of the train is converted to energy of rotation of the spinning flywheel. In other words, the train slows down, causing the flywheel to speed up. While the subway train rests at the station to

BOX 13.4 HOW YOU CAN SAVE ENERGY

How You Can Save Fuel In Transportation

METHOD	FUEL SAVINGS COMPARED TO WASTEFUL PRACTICES
Walk or use a bicycle, stay at home	Saves 100%
Use mass transit	Saves 50% (or more)
Carpool	Saves 50% (or more)
Keep car well tuned	Saves 20%
Drive smoothly (no jerks, fast starts and stops)	Saves 15%
Purchase car without an air conditioner	Saves 10%
Keep tires inflated to proper pressure	Saves 5%

Energy-Saving Methods That Don't Cost Money

1. After taking a bath, open the bathroom door and let the water stand in the tub until it cools to room temperature. The heat from the water is enough to heat a small house for an hour.
2. Similarly, after doing the dishes, don't let the kitchen sink water drain out until it cools.
3. Use kitchen and bathroom exhaust fans sparingly, as they blow away a houseful of warm air in one hour.
4. Wear sweaters and long underwear in winter, and light clothes in summer. Then adjust the thermostat accordingly.
5. Close off unused rooms, walk-in closets and stairways, and save the fuel that would otherwise be needed to heat them.

How You Can Save Fuel In Home Heating

METHOD	HEAT SAVINGS COMPARED TO ENERGY-INEFFICIENT HOUSE
Have your furnace maintained, cleaned, and tuned properly	Saves 10%–20%
Add extra insulation in ceilings and walls	Saves 30%–50%
Add storm windows	Saves 10%
Caulk leaky windows and doors	Saves 5%–15%
Cover windows with drapes and shades at night	Saves 5%–15%
Close off and do not heat unused rooms	Variable
Turn thermostat down	Saves 3% per °F, or 1⅔% per °C

How You Can Save Fuel For Home Appliances

METHOD	SAVINGS (DOLLARS/YEAR)*
Place extra insulation around hot water heater	Saves $19/year
Place insulation around hot water pipes	Saves $20/year
Hang clothes on line in summer rather than use dryer	Saves $25/year
Disconnect drying cycle from dishwasher	Saves $25/year
Put covers on pans when cooking	Saves $5/year
Use wool or down blankets rather than electric blankets	Saves $8/year
Use proper size light bulbs, turn lights off when not in use	Variable
Turn off pilot lights on gas stove	Saves $10/year

*Energy savings here have been calculated using a value of $.05 per kilowatt hour.

pick up and discharge passengers, the flywheels would spin rapidly. Then, when power is needed to accelerate the train, the shafts of the flywheel could be connected to their electric generators. The motion of the flywheels would be converted to motion of the generators, electricity could be produced, this electricity would be used to power the train, and the train would speed up while the flywheels would slow down

BOX 13.5 MPG

In 1979, a contest was held in England to determine the maximum mileage possible with a motorized vehicle. The winner was a tiny torpedo-shaped "car" using a 50 mL engine and three bicycle wheels. It averaged 597 km/L (1403 miles/gal) over an oval-shaped race track.

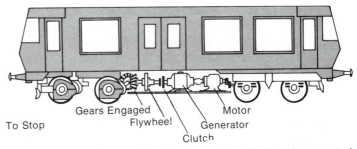

Gears Engaged Motor
Flywheel Generator
To Stop Clutch

When the train is stopping, the kinetic energy of the train is converted to rotational energy of the spinning flywheel. Notice that the clutch is disengaged and there is a gap between the flywheel and the generator. Thus the generator does not spin.

Figure 13–25 Flywheel-operated subway train. *To stop:* When the train is stopping, the kinetic energy of the train is converted to rotational energy of the spinning flywheel. Notice that the clutch is disengaged and there is a gap between the flywheel and the generator. Thus the generator does not spin. *To start:* When the engineer wishes to start the train, the clutch is engaged, connecting the flywheel to the generator. The spinning flywheel forces the generator to turn, thereby producing electricity. This electrical energy drives the motor and helps the train to accelerate.

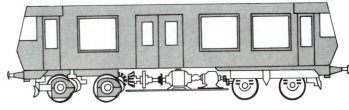

To Start

When the engineer wishes to start the train, the clutch is engaged, connecting the flywheel to the generator. The spinning flywheel forces the generator to turn, thereby producing electricity. This electrical energy drives the motor and helps the train to accelerate.

again. The train could produce some of its own electric power and therefore save energy.

• Other modern transportation networks could replace the airplane for intermediate distance travel. In France, a train called the *à grande vitesse* (of great speed) has been built and now operates at 260 km/hr (162 miles/hr); it is capable of speeds of up to 378 km/hr (236 miles/hr). This train (Fig. 13–26), operating from the center of one city to the center of another, is faster than an airplane for many journeys if commuter time to the airport is considered, yet far less fuel is consumed.

Long-term projections for energy savings are possible for home heating as well. Although complete conversion to soft technology is not probable for the intermediate period, other approaches that could operate during this time pe-

Figure 13–26 French train, *à grande vitesse.*

riod involve changes in conventional heating systems to use fuel more efficiently than present systems do. In order to understand some possible technologies, one must consider the entire pathway of fuel use from its beginning in the ground to its eventual consumption. Figure 13–27 shows different ways that petroleum can be used to heat buildings. Note that, however the fuel is eventually used, the petroleum must first be extracted, refined, and transported. These processes, especially refining, use tremendous quantities of fuel. Approximately 20 percent of every liter of crude oil extracted is used up before any of it is available at the retail level, leaving only 80 percent for consumption. Of course, drilling and transportation are absolutely essential. Although it would be possible to burn unrefined crude in a home furnace or an electric generating plant, the engineering problems involved in handling the thick goo and the pollution that would be released have discouraged direct consumption.

Once the fuel is available, the most direct form of usage is to burn it in a household furnace. Typically, a small furnace operates at about 60 percent efficiency. Thus, 40 percent of the remaining fuel is wasted as heat that escapes out the chimney. The result, as shown in Figure 13–27, is that 48 percent of the original energy in a barrel of petroleum is converted to useful heat; the remaining 52 percent is lost and dispersed into the environment.

This may seem shockingly inefficient, but other systems are worse. As explained in Section 13.4, electrical generation and transmission are only about 34 percent efficient. If losses in fuel extraction, transportation, and refining are added as well, electricity as delivered to a home represents only 27 percent of the original energy in a liter of petroleum.

A conventional electric heater, called a **resistance heater,** is simply a wire or filament that resists the movement of electrons. As an electric current is forced through the wire, the electrons collide with the atoms in the wire and produce heat. Electric ranges, toasters, hot water heaters, baseboard heaters, and a wide variety of other electrical appliances operate on this principle. There is no loss of energy in a resistance heater; the process is 100 percent efficient. Therefore, the overall efficiency of resistance heaters, as shown in Figure 13–27, is 27 percent.

Despite the fact that resistance heating is 100 percent efficient, it does not represent the best use of electricity for supplying heat. To understand how it is possible to obtain greater than 100 percent efficiency, consider the workings of an ordinary refrigerator. A refrigerator is actually a type of **heat pump;** heat is pumped from inside the appliance into the room. Thus the refrigerator heats the room while it cools the food. An air conditioner is another form of heat pump, except that it is designed to cool a room, not heat it. Of course, the energy is not lost, so the heat removed from the room is exhausted outdoors. Thus, the air conditioner heats the outdoors while it cools the inside of a building. If you took a conventional air conditioner unit and simply turned it around, it would heat a room and cool the outside environment. This is exactly the principle of the heat pump. As it turns out, an air conditioner turned backward is not always an efficient heating unit. Think about a house in Montana in January. Outdoor temperatures might typically be $-10°C$, and indoor temperatures are, say $+18°C$. A lot of energy is needed to pump heat across a difference of 28°C, and considerable amounts of electricity are required. As an alternative system, suppose a hole is dug under the house about 5 meters into the Earth. The subterranean soil and rocks, insulated from the air above, are surprisingly warm, approximately $+10°C$. If a heat pump is used to pump heat from this warm earth into the house, then the machine is only required to operate across a temperature difference of 8°C, not 28°C. This is, in a sense, a form of geothermal heating, because the heat from the Earth is used to warm a house. Assuming that the Earth's heat supply is free, the overall fuel efficiency of a heat pump can be 200 percent. Thus, as shown in Figure 13–27, a system of this type uses petroleum with an overall efficiency of 54 percent, twice that of resistance heating and slightly better than a common oil furnace.

Heat pump technology is well developed, and in most cases these units are the cheapest form of heat over the life of a house. However, heat pumps are not widely used. At the present time, the problem is twofold. First of all, most people are not familiar with this technology and its advantages. Second, most homes and apartments today are built by a construction firm and then sold or rented to someone else. The people

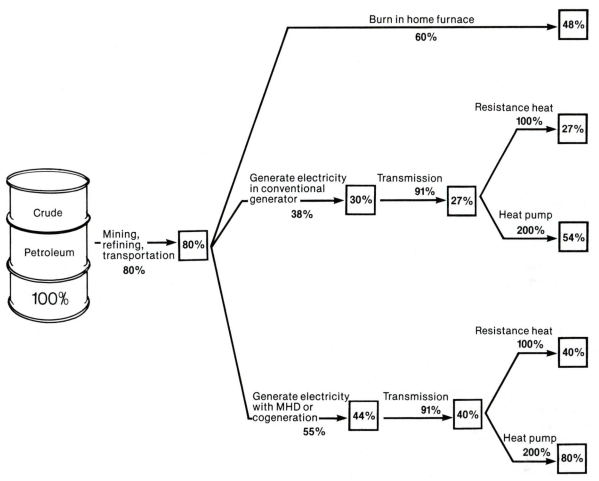

Figure 13–27 Energy efficiency of various ways of heating a building using petroleum as a fuel. In each case, the number under the arrow indicates the efficiency of the individual process, and the number in the box indicates the percentage of the original fuel remaining after the sequence of steps. (The number 80 percent for mining, refining, and transportation represents a rough average of widely divergent figures. Published values for the efficiency of refining vary from a low of 2 percent to a high of very close to 100 percent. The difference is not an index of efficiency necessarily, but reflects the different chemical composition of different fuels and the different end-products desired. The values, a 10 percent loss in mining and transportation and another 10 percent loss in refining, were chosen as a world average following a private communication with a representative from the Gulf Oil Corporation.)

who build the units try to offer them at the lowest possible price; they are not the people who must pay the heating bills for decades to come. The cheapest heating systems to build are baseboard electric resistance heaters, so they are the most common even though they are the least efficient and, in many cases, the most expensive in the long run. The consumer saves a few thousand dollars initially or has a lower rent, but then must pay twice as much for fuel. This is a classic example of short-sighted economics.

Cogeneration

The Second Law of Thermodynamics assures us that it is impossible to convert heat to work without the production of waste heat as a byproduct. In the United States nearly all of this waste heat is released into the environment. At the same time, homes and other industries need large quantities of hot water for processes such as heating, canning food, or refining petroleum. Thus, a system exists in which hot water is thrown away at one place, and fuel is burned to heat water a few kilometers away. What an incredible waste! Hot water from an electric generator can be sold to those who need it. One example of such a system is the relationship between the Bayway, New Jersey, refinery of the Humble Oil and Refining Company and the Linden, New Jersey, generating station. The Linden power plant is capable of producing electricity at 39 percent efficiency. This efficiency is lowered by a less than optimum cooling of the condenser, and the waste heat is sold as steam to Humble. Such a tandem operation is called **cogeneration.** If the two-plant operation is considered as a single energy unit, the overall efficiency of power production has been raised to a level of 54 percent. The process is beneficial to many: The companies save money, fuel reserves are conserved, and thermal pollution of waterways is reduced 15 percent.

Although cogeneration is not yet widely used in the United States, it is common practice in other countries. Figure 13–28 compares electrical generation in the United States with that in Sweden. As shown in the figure, only 47 percent of fuel used in Sweden for the generation of electricity is wasted as nonproductive heat, whereas in the United States 68.5 percent of such fuel is wasted. Even today, most new generating plants in North America are not designed for cogeneration. One may argue for or against the use of energy for luxuries in an energy-poor world, but it is difficult to justify energy waste through antiquated engineering.

MHD

Mechanical losses account for a 15 to 20 percent loss of efficiency in conventional power plants. Thus, thermal pollution could be reduced if power plants operated at higher efficiency. A promising new design that increases efficiency is based on the principle of **magnetohydrodynamics (MHD).** In this system, air is heated directly and is seeded with metals such as potassium or sodium, which lose electrons at high temperatures:

$$\text{K} + \text{heat} \longrightarrow \text{K}^+ + \text{e}^-$$
potassium atom / potassium ion / negative electron

This hot, electrified air stream is allowed to travel through a large pipe ringed with magnets (Fig. 13–29). The movements of these charged particles through a magnetic field generate electricity. Furthermore, the exhausted hot air can operate a conventional turbine, thus producing additional electricity. The advantages of this system are twofold. First, the overall efficiency of the system is expected to reach 55 percent, and second, most of the waste heat is dissipated directly into the air rather than into aquatic ecosystems. The Russians have shown that this technique is feasible by operating a 250,000-watt MHD generator near Moscow. In the United States, most of the $100 million funding for MHD research was cut in 1980, and as a result development has stagnated.

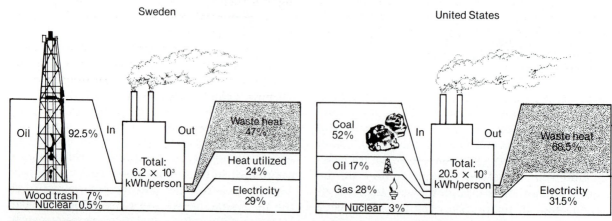

Figure 13–28 Comparison of the use of fuel to produce electricity in Sweden and the United States. Note that large quantities of waste heat are used in Sweden, leading to significantly higher fuel efficiencies. (After *Science,* 194:1001, Dec. 1976.)

The MHD program was cut because it was considered to be too expensive, but the government continues to support the ailing nuclear power industry. On the other hand, if all the fossil fuel–powered electric plants in the United States could improve their efficiency by 14 percent either through cogeneration or by switching to MHD, the energy savings alone would equal the total output of all the nuclear power plants in the country. Of course, it would be impossible to convert all conventional plants in a short period of time, but these comparisons are helpful in planning future priorities.

Return to Figure 13–27 for a final comparison. The bottom line illustrates an overall heating system that uses electricity derived from a cogenerator or an MHD generator to drive a heat pump. This system results in 80 percent efficiency, 2.8 times as much as ordinary resistance heating. In other words, by using the best available technology, the petroleum used for heating can be stretched by a factor of nearly three without mandating any sacrifice or inconvenience.

13.9 Energy for a Sustainable Society (2020 into the Future)

Looking ahead to the next generation and beyond, the final decline of oil and the slower decline of coal reserves are virtually certain. Then renewable resources, nuclear fission, and possibly fusion will be the only energy alternatives. The choices available have already been discussed, but it is impossible to predict what will actually happen.

Summary

Environmental problems associated with oil drilling are particularly severe for wells in the Arctic or under the seas. Coal can be mined either in tunnels or in open pits, also called **strip mines.** Strip mines deface the land.

Production of **synfuels** produces a wide variety of serious environmental problems includ-

Figure 13–29 An experimental MHD generator. (Courtesy of General Electric.)

ing destruction of land surfaces, depletion of water resources, and air pollution. The processes are also expensive.

Production of electricity is about 40 percent efficient at best, and therefore electric heating systems are inefficient. Waste heat from an electric power plant causes **thermal pollution** of natural ecosystems. Alternatives include **cooling ponds** and **cooling towers.**

In most countries, fuel costs strain the national economy, and supplies are vulnerable to war and political instability.

"Hard" energy systems are based on centralized generating plants, and "soft" systems are based on the use of smaller, decentralized power production. Traditionally, hard systems have been cheaper, but many people feel that soft energy pathways will be less expensive in the future.

In the immediate future, conservation is the most effective energy strategy. In the intermediate period, changes in the physical structure of our energy supply and consumption systems will become important. In the distant future, our society must base its needs either on nuclear fission and fusion, or on renewable systems.

Questions

Uses of Energy

1. List five ways to improve the heating efficiency in your home or apartment. If you live in a private home, estimate the cost of implementing these changes. If you live in an apartment house, do you think that such changes would be a worthwhile investment for the owner?

2. Give some reasons why it is often difficult to alter patterns of surface transportation.

3. Explain why thermal pollution is an unavoidable result of the Second Law of Thermodynamics.

Energy Politics and Economics

4. Recent reports indicate that many carcinogenic compounds are produced when liquid fuels are synthesized from coal. In addition, over one third of the fuel content of the coal is consumed during the conversion. Using this information and the material in the text, list the categories of direct costs and external environmental costs required to produce liquid fuel from coal.

5. Many people suggest that the government should provide low cost loans for solar collectors. Others disagree. They argue that solar energy would benefit only those living in suburbs or in the country. City dwellers would not benefit. They argue further that if solar energy were such a good idea, it would be profitable enough to work without government support. Discuss these arguments.

6. Gasoline taxes have traditionally been used for the construction of new roads. This practice has been considered fair because the roads are paid for by those who use them most. Increasingly, economists and social philosophers feel that many of our traditional concepts of fairness must be re-evaluated in the light of environmental problems. Do you feel that there should be a re-evaluation of road tax use? If so, how would you allocate funds? If not, explain.

Energy Futures

7. Explain how environmental problems related to extraction of fuels have changed both qualitatively and quantitatively in the past 20 years. How are they expected to change further in the next 20 years?

8. Prepare a class debate. On one side argue in favor of a soft energy future; on the other side argue in favor of the hard path.

9. In 1975, 97 percent of urban traffic was carried by the automobile, and only 3 percent was carried by mass transit systems. Thus, even if mass transit systems were to double the number of passengers they carry, there would be relatively little impact on overall fuel consumption rates. Should these figures be used as an argument against building new mass transit systems? Defend your answer.

10. Discuss the differences in energy planning for the immediate future, the intermediate period, and a sustainable society.

Suggested Readings

The general references listed for Chapter 11 also provide background material for this chapter. Additional materials that refer to specific sections are listed below.

13.1 Environmental Problems Caused by Mining and Drilling

Carroll L. Wilson: Coal—Bridge to the Future. Cambridge, MA, Ballinger Publishing Co., 1980. 247 pp.

13.2 Environmental Problems Caused by Production of Synfuels

J. L. Anastasi: SASOL: South Africa's oil from coal story—background for environment assessment. Environmental Protection Agency Report 600/8–80–002. Washington, D.C., U.S. Government Printing Office, January 1980. 36 pp.

K. W. Crawford, et al.: A preliminary assessment of the environmental impact from oil shale development. Environmental Protection Agency Report 600/7–77–069. Washington, D.C., U.S. Government Printing Office, July 1977. 173 pp.

E. J. Hoffman: Synfuels: The Problems and the Promise. Laramie, WY, The Energon Co., 1982. 547 pp.

Ronald F. Probstein and R. Edwin Hicks: Synthetic Fuels. New York, McGraw-Hill, 1982. 490 pp.

13.5 Politics and Economics of Energy Supply

Irvine H. Anderson: Aramco, the United States, and Saudi Arabia. Princeton, Princeton University Press, 1980. 259 pp.

Wilson Clark and Jake Page: Energy, Vulnerability, and War. New York, W. W. Norton & Co., 1980. 251 pp.

David A. Deese and Joseph S. Nye (Eds.): Energy and Security. Cambridge, MA, Ballinger Publishing Co., 1980. 489 pp.

Robert Stobaugh and Daniel Yergin: Energy Future—Report of the Energy Project at the Harvard Business School. New York, Random House, 1979. 353 pp.

Amory Lovins and Hunter Lovins: Brittle Power: Energy Strategy for National Security. Andover, MA, Brick House Publishing, 1982.

13.6 to 13.9 Planning for Energy Futures

Four books that present arguments for the hard and the soft energy pathways respectively are:

Jim Harding and the International Project of Soft Energy Paths: Tools for the Soft Path. Friends of the Earth Publishing, 1982.

Fred Hoyle: Energy or Extinction? The Case for Nuclear Energy. Salem, N.H., Heinemann Educational Books, 1977. 81 pp.

Amory B. Lovins: Soft Energy Paths. Cambridge, MA., Ballinger Publishing Company, 1977. 231 pp.

Simon Rosenblum: The Non-Nuclear Way: Creative Energy Alternatives for Canada. Regina, Saskatchewan, Regina Group for a Non-Nuclear Society, 1981. 112 pp.

Other excellent books on the energy future are:

Elihu Bergman, Hans A. Bethe, and Robert E. Marshak: American Energy Choices Before the Year 2000. Lexington, MA., D. C. Heath and Company, 1978. 152 pp.

Lester R. Brown, et al.: Running on Empty: The Future of the Automobile in an Oil-Short World. New York, Viking Press, 1980.

John H. Gibbons, and William U. Chandler: Energy, The Conservation Revolution. New York, Plenum Press, 1980. 258 pp.

Amory B. Lovins, and John H. Price: Non-Nuclear Future: The Case for an Ethical Energy Strategy. Cambridge, MA., Ballinger Publishing Company, 1975. 222 pp.

Marc H. Ross, and Robert H. Williams: Our Energy: Regaining Control. New York, McGraw-Hill Book Company, 1980. 354 pp.

SERI-Solar/Conservation Study: A New Prosperity: Building a Sustainable Energy Future. Andover, MA., Brick House Publishing Company, 1980. 454 pp.

Jon Van Til: Living with Energy Shortfall. Boulder, CO., Westview Press, 1982. 209 pp. Paperback.

Manas Chatterji (ed.): Energy and Environment in the Developing Countries. Somerset, N.J., John Wiley & Sons, 1982. 352 pp.

D. Allan Bromley: A Desirable Energy Future. Philadelphia, Franklin Institute Press, 1982. 255 pp.

Daniel Yergin and Martin Hillenbrand (eds.): Global Insecurity. A Strategy for Energy and Economic Renewal. Boston, Houghton Mifflin, 1982. 427 pp.

Three books on practical energy conservation are:

Farallones Institute: The Integral Urban House: Self-Reliant Living in the City. San Francisco, Sierra Club Books, 1979.

James W. Morris: The Complete Energy Saving Handbook for Home Owners. New York, Harper & Row, 1980.

U.S. Department of Agriculture: Cutting Energy Costs. Washington, D.C., U.S. Government Printing Office, 1980.

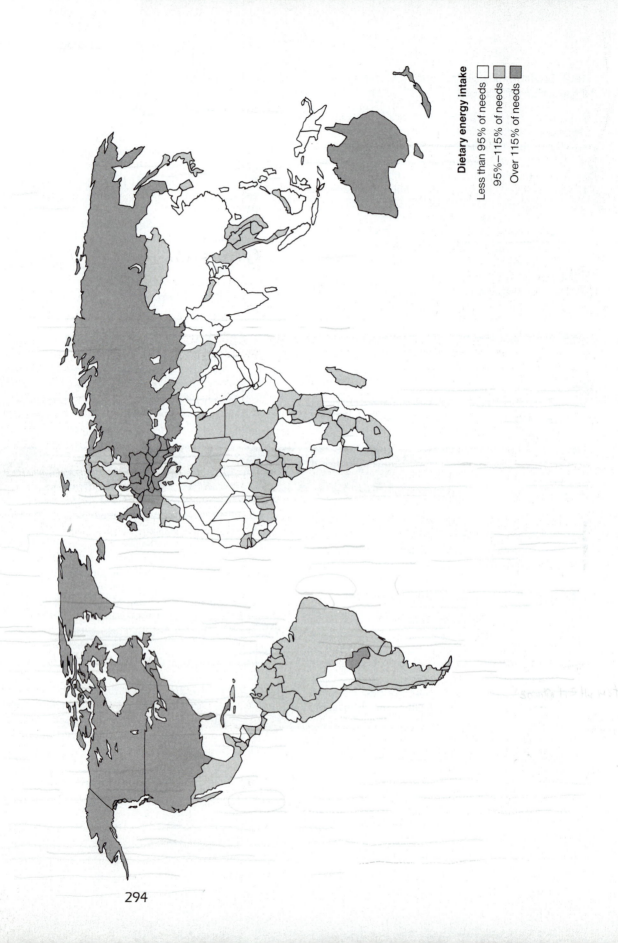

Dietary energy intake

☐ Less than 95% of needs
▨ 95%–115% of needs
▧ Over 115% of needs

Soil, Land, and Minerals

World dietary energy intakes. Countries within the
95–115% range are considered to have adequate
diets. Note that most of the developed countries
show excessive consumption (more than 115% of
needs). Overeating, like undereating, is a form of
malnutrition. Norway, Sweden, and Finland are the
exceptions in this category; their populations do not
overeat. Unfortunately, the countries with the most
people do not always produce the most food.
(From *Our Magnificent Earth.* © 1979 by Mitchell
Beazley as *Atlas of Earth Resources.* Published in
the U.S.A. by Rand McNally & Company.)

14

Soil and Agriculture

As human populations grow, the task of increasing agricultural productivity to provide enough food has become more and more difficult. In some areas the climate and soil are ideal for agriculture. Unfortunately, these areas tend to be the wide valleys of temperate zones, which are also the areas most in demand for housing and industrial development. More than a million hectares of prime agricultural land is converted from farmland to roads, houses, and factories every year in the United States alone. Thus a global condition has developed in which the availability of prime agricultural land is decreasing and the total human population is increasing. There has been a threefold response to this dilemma—(1) to increase productivity on existing farmland, (2) to farm less desirable "marginal" farmland, and (3) to provide political and economic incentives to protect prime farmland. The successes and failures of these responses form the subject matter of Chapters 14 through 17.

14.1 Plant Nutrients

Plants are autotrophs, which produce their own food from inorganic substances. The major

elements required by all living organisms, plant and animal, are carbon, oxygen, and hydrogen. Plants obtain carbon and oxygen from carbon dioxide and oxygen in the air, hydrogen and oxygen from water. In addition, plants need a great many different mineral nutrients, most of which they obtain from the soil. Some minerals, collectively called the **macronutrients,** are used in relatively large quantities. The important macronutrients are nitrogen, phosphorus, potassium, calcium, magnesium, and sulfur. Those needed in the largest amounts are nitrogen, phosphorus, and potassium. Commercial fertilizers are rated by the percentages of these elements that they contain; 5-10-5 fertilizer contains 5 percent nitrogen, 10 percent phosphorus, and 5 percent potassium by weight. ? so what is the rest?

Besides macronutrients, plants require tiny quantities of several different **micronutrients.** For instance, whereas crops use several hundred kilograms of nitrogen per hectare per year, the treatment for molybdenum-deficient soils in Australia is 140 g of molybdenum oxide per hectare, applied once every 10 years. An inadequate supply of any of these minerals results in general symptoms of mineral deficiency, a condition known as **chlorosis** (paleness) and poor growth (Fig. 14–1). On the other hand, some minerals are used in only part of a plant's development,

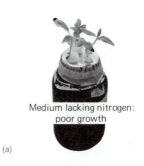

Medium lacking nitrogen:
poor growth

(a)

Medium lacking iron:
extreme chlorosis, poor growth

(b)

Medium lacking micronutrients:
short internodes at tops

(c)

Nutritionally complete
medium

(d)

Figure 14–1 *An experiment showing the effect of inadequate quantities of various mineral nutrients on the growth of sunflower seedlings. (Courtesy of Carolyn Eberhard.)*

[handwritten: Q, what is soil, what composed of crushed]

and their deficiency symptoms may be quite specific. Zinc, for instance, is needed to produce plant hormones called auxins, which cause growing stem cells to elongate. If a plant lacks zinc, its stem cells remain so stunted that there is little space between adjacent leaves, and the plant assumes a rosette shape. Some plants have special adaptations that demand particular nutrients.

Two examples illustrate these needs. Legumes (plants in the pea family) are unusual in needing cobalt, a component of vitamin B_{12} used by the bacteria that live in legume roots. (These bacteria are very important to agriculture because they fix nitrogen from the air into a form that plants can use.) Some grasses and cereals need silicon, which they incorporate into an abrasive material that strengthens their stems and protects them from some herbivores.

14.2 Soil Structure

When the Earth first formed, there was no soil. Soil is a mixture of small grains of mineral matter formed from decomposing rocks combined with organic matter resulting from the decaying bodies of dead organisms. Most of the minerals that plants use come ultimately from rock particles in the soil. Soils are classified by the average size of the particles they contain. The finest particles give rise to **clay** soils, larger particles to **silt**, and still larger particles to **sand** soils.

[handwritten margin: clay 1, silt 2, sand 3]

The size of its particles influences the soil's capacity to hold soil water. Some of the water that falls on soil drains through it, carrying dissolved nutrients away with it. The water that remains in the soil is held in three main ways. First, some of it forms a film around the soil particles. Second, wedges of water are held between soil particles by capillary attraction. These films and wedges together make up the capillary water of the soil. Third, water may be held by **imbibition,** that is, by the chemical attractions in which water becomes tightly bound up in the structure of clay particles and of organic matter (Fig. 14–2). When soil dries out, capillary water wedges are removed most rapidly, followed by the films around soil particles, and finally by the imbibed water, which in some cases is so tightly bound that plants may suffer water deficiency

[handwritten margin: How does water stay?]

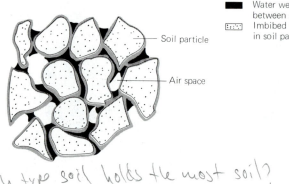

Water film around soil particles
Water wedges in smallest spaces between soil particles
Imbibed water, tightly bound in soil particles

— Soil particle

— Air space

Figure 14–2 Soil water is held in three ways—as capillary films around individual soil particles, as capillary wedges in small spaces between soil particles, and as imbibed water incorporated into the colloidal structure of particles of clay and organic matter.

Q³ which type soil holds the most soil?

even when considerable reserves remain in the soil.

Clay soils hold more water than sandy soils for several reasons. They have finer particles with more surface area to hold surface films. The spaces between the particles are smaller and thus have greater capacity for holding water by capillarity (Fig. 14–3). Finally, clay particles can imbibe large quantities of water. This water-holding capacity is not always advantageous, because clay soils easily become so saturated with water that gaseous oxygen in the soils is displaced. Oxygen is an important component of soil because plant roots need oxygen to survive, as do the microorganisms that release minerals in fertile soil. If excess water displaces the soil oxygen, many plants become waterlogged and die. In contrast, sandy soils tend to be well drained. This prevents waterlogging, but the rapid movement of water through the layers of sand washes minerals out of the root zone, so that, other factors being equal, sand contains fewer nutrients than clay. The best soil for plant growth is one that has a mixture of particle sizes. ∴ need mixture

The most fertile soils contain large quantities of organic matter, including both living organ-

isms and the decomposed bodies of dead ones. The living organisms are essential to the process of decay, and the accumulated decay products are vital to the chemistry of the soil.

The number of living organisms in fertile soil is staggering. One kilogram of rich farm earth contains up to 2 trillion bacteria, 400 million fungi, 50 million algae, and 30 million protozoa, as well as thousands of different worms, insects, and mites. Soil organisms decompose organic matter and convert many nutrients into forms that plants can absorb. The burrowing of larger organisms breaks up the soil and allows air to penetrate.

When a given piece of tissue such as a leaf, a stalk of grass, or a dead bird has decomposed sufficiently in the soil system so that its origin has become obscure, then the organic matter has been converted into a complex mixture called humus. Humus is an essential component of most fertile soils.

1. Humus retains moisture both by capillarity and by imbibition. It also insulates the soil from excessive heat and cold and reduces evaporation. Organic matter retains so much moisture that it swells after rain and then gradually

Clay   Sand

Figure 14–3 Comparison of water held by clay soils (small particles) and water held by sandy soils (larger particles).

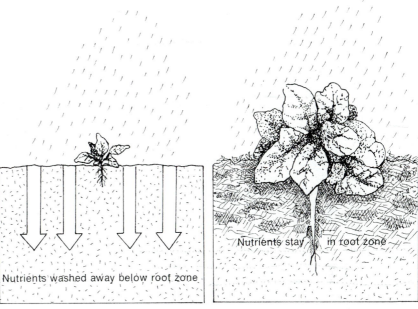

Figure 14–4 Humus retains nutrients and moisture in the soil.

A Soil with low humus content   *B* Soil with high humus content

Nutrients washed away below root zone

Nutrients stay in root zone

shrinks during dry spells. This alternate shrinking and swelling keeps the soil loose, allowing roots to grow through it easily.

2. Humus in soil is a reservoir of nutrients that are released slowly as the organic matter decomposes (Fig. 14–4). The storage and release of nutrients is regulated by a complex series of chemical reactions in the humus. For example, calcium ions, Ca^{2+}, can exist in water solutions from which they are usable by plants. However, atmospheric carbon dioxide reacts with water to form carbonate ion, CO_3^{2-}, which reacts with calcium in water to form the sparingly soluble compound calcium carbonate, $CaCO_3$. A striking consequence of its meager solubility has been the deposition of large masses of $CaCO_3$ on earth in the form of limestone and marble. But it is not *completely* insoluble; moreover, its solubility depends largely on the soil acidity, and under certain conditions, the calcium may be liberated. Under other conditions, it is very insoluble, and the calcium is not readily available to plants, even though it is present in the soil. Moreover, if dissolved calcium is not used immediately by plants, it may travel with water droplets down below the root zones, where it becomes unavailable. This movement of free ions into the subsoil and underground reservoirs is called **leaching.**

Richly humic soils provide metal ions with certain unique chemicals that are not available in simple inorganic solutions. These chemicals in the humus, which are known as **chelating agents,** react with inorganic ions such as Ca^{2+} to form a special class of compounds known as chelation complexes. Ions bonded in chelation complexes are held tightly under some conditions but are easily released under others. A calcium ion chelated by humus will not react readily with carbonate to form $CaCO_3$, nor will it leach easily. Rather, it will tend to remain bonded to the chelating agent and thus be retained in the humic matter.

Plants and soil microorganisms have evolved mechanisms whereby they can release chelated ions readily and incorporate them into living tissue, as shown schematically in Figure 14–5. The specific chelation chemistry is different for each soil nutrient, but in general, humus main-

BOX 14.1 VOLCANIC SOILS

Volcanic soils are an exception to the generalization that fertile soils must contain humus. Some types of volcanic ash contain such large quantities of minerals and are naturally so porous that they support lush plant growth even when the humus content is low.

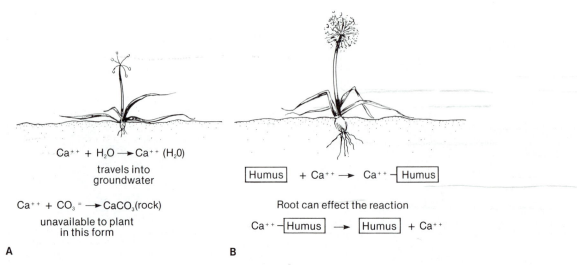

$$Ca^{++} + H_2O \longrightarrow Ca^{++} (H_2O)$$

travels into
groundwater

$$Ca^{++} + CO_3^= \longrightarrow CaCO_3 (rock)$$

unavailable to plant
in this form

Humus + Ca^{++} → Ca^{++} Humus

Root can effect the reaction

Ca^{++} Humus → Humus + Ca^{++}

A   B

Figure 14–5 Schematic illustration of pathways for transfer of inorganic nutrients, A, Without humus. B, With humus.

tains a nutrient reservoir and enhances the ease with which nutrients are used by the soil organisms.

When soil humus is not maintained, the efficiency of converting fertilizer to plant tissue is low and the ability of soils to store reserves of nutrients is poor. As a result, large quantities of nitrogen or mineral fertilizers are leached into surface and ground waters or tied into the soils in chemically unusable forms. This loss of nutrients represents a waste of energy and ore.

14.3 Soil Treatments

Tillage

People treat soil in various ways to improve the conditions for plant growth. One of the oldest agricultural practices is **tilling** the soil, turning it over before planting seeds and turning it over between plants as they grow (Fig. 14–6). Tilling has two main functions. First, it mixes up the nutrients and loosens soil particles, making it easier for roots to penetrate the soil. Second, it gives crop plants, which are left alone, a competitive advantage over weeds, which are deliberately disturbed to damage their root systems. Tilling also introduces oxygen into the soil.

Figure 14–6 Tilling the soil before spring planting damages the weeds that would compete with crop plants for light, water, and nutrients. Tilling also loosens the soil particles, allowing oxygen to enter the soil and making it easier for seedlings to become established. (Courtesy of Paul Feeny.)

On the negative side, tillage exposes deep layers of the soil to the air and therefore speeds up the oxidation and eventual decomposition of humus. Although exposure of soil also leads to more rapid evaporation of water, the destruction of the plant cover reduces water loss from leaf surfaces. Therefore, sometimes tillage leads to

Figure 14–7 Navajo farmer George Blue-eyes works far harder than his neighbors because he tills his fields entirely by hand. But traditional farming methods limit moisture loss and erosion and bring him consistently higher yields per hectare. (Photo by Janet Bingham.)

> **BOX 14.2 HYDROPONICS**
>
> Plants do not need soil to grow. All they need is water, mineral nutrients, and air. The cultivation of plants without soil is becoming increasingly common. It is called **hydroponics.** The plants are usually grown in plastic sacs of water and mineral nutrients. Because their roots are in water and not in soil, plants grown hydroponically require support, either from their plastic containers or from frameworks over the containers.
>
> The advantage of hydroponic culture is that the nutrients supplied to the plants can be regulated precisely. The disadvantage is cost. It is much more expensive to buy land and hydroponic tanks for plants than merely to plant the crops straight in the ground. Large-scale hydroponic agriculture may always remain economically impossible, but hydroponics has an increasing role to play nevertheless. For instance, there are islands in the Pacific that are almost all rock and have little soil. Importing food to these islands is expensive, and it is cheaper to grow food on the islands using hydroponic techniques. Similarly, small quantities of particular plants may be produced efficiently by this method. The San Diego zoo grows plants hydroponically indoors under lights to provide food for some of its animals. The zoo can produce the precise quantities it needs of plants that are not commercially available.

net loss of water and sometimes it increases soil moisture. In times of heavy rainfall or dry winds, bare soil is vulnerable to erosion because the root structure that normally anchors the soil particles has been disrupted.

Many primitive societies have traditionally practiced no-till agriculture (Fig. 14–7), primarily because plowing is difficult or expensive. A benefit of this approach is that soil quality is maintained. In recent years, no-till agriculture has become popular in developed countries as well. The farmers who practice it often find that they need less irrigation water and fewer fertilizers. The machinery required is expensive, however, and conversion to this technique is slow.

Fertilization

Farmers fertilized their crops with manure, straw, or dead fish long before they understood the chemistry of fertilization. Today, mined and manufactured fertilizers are used so extensively that they have become a major component of most agricultural systems.

Of the mineral elements needed by plants, nitrogen is most often deficient. Plants need nitrogen in large quantities because it is a major constituent of proteins and nucleic acids. Nitrogen makes up about 78 percent of the atmosphere by volume, and the supply is therefore virtually unlimited, but gaseous nitrogen cannot be used by most plants. Nitrogen can be "fixed," that is, converted to usable forms, by bacteria that live on the roots of plants of the legume family (which includes alfalfa, beans, and peas). Thus soil fertility can be maintained by rotating crops. One typical rotation scheme would be to grow corn, a plant that depletes soil nitrogen, one year, and then cultivate a nitrogen-enriching crop such as alfalfa the following season. Unfortunately, there is a greater market for corn than for alfalfa, and in many regions the farmer would realize more profit if corn were planted every year. To offset this problem, chemists have learned to convert atmospheric nitrogen to synthetic plant fertilizers by producing ammonia:

$$N_2 + 3H_2 \rightarrow 2NH_3 \text{ (ammonia)}$$

Ammonia can be used in soils directly or converted to other usable compounds such as nitrates (NO_3^-). Taking all manufacturing steps into account, the manufacture of nitrogen fertilizers requires large expenditures of energy. Therefore, in modern times the price and availability of nitrogen fertilizers are tied to the cost and supply of coal and oil.

Large quantities of phosphorus and potassium are also added to commercial farms. Both are mined commercially from mineral deposits. Of these, the world's supply of phosphorus is more limited (see Chapter 18), and the price of this fertilizer is linked to the availability of the deposits.

In addition, soil is often treated to adjust its pH. (See Appendix C for an explanation of pH.) Lime is applied to acid soils to raise the pH, whereas sulfur or ammonium sulfate is applied to alkaline soils to lower the pH. These treatments have other effects besides changing the pH. For example, lime provides calcium, and ammonium salts contribute nitrogen. Changing the pH many also change the solubility of different minerals in the soil water, making them more or less available to plants. For example, if the pH of soil is raised too much, it may cause iron to precipitate and become unavailable to plants. One way to overcome this effect is to apply chelated iron (iron bound to organic molecules so that it cannot be precipitated) along with the lime.

Unquestionably, commercial fertilizers have increased agricultural yields all over the world and thus have helped feed the expanding human population. But several problems have arisen as well. For one, fertilizers are expensive and are therefore unavailable to farmers in many less developed countries. For another, soil chemistry is complex. If too much fertilizer is applied, or if the fertilizer is not chemically compatible with the soil, large quantities of it will leach out into streams and groundwater supplies. Leaching represents not only a financial loss but also a significant source of water pollution. It is difficult to assess the amount of pollution resulting from agricultural fertilizers. According to the United States Public Health Service, water containing more than 10 mg of nitrate per liter is unfit to drink. High nitrate concentration is especially harmful to infants, who may die from drinking formula milk prepared with water polluted in this manner. In a few ag-

ricultural areas in the United States, domestic wells and municipal reservoirs have been contaminated by nitrates. However, it is difficult to blame agricultural runoff unequivocally because the same materials may leach from sewage disposal systems or open dumps.

Perhaps the most serious concern about the use of chemical fertilizers is related to the quality of the soil itself. In a system where land is tilled every year, where crop residues such as straw are removed, and where only chemical fertilizers are used, the soil humus is continuously depleted and is not replaced. Many people fear that even though agricultural yields are high in chemical farming systems today, the loss of humus could lead to a decline in productivity in the future. If humus is destroyed, water losses will increase, leaching problems will become more severe, trace minerals will be removed, and the agricultural system will become totally dependent on applications of large quantities of fertilizer.

One solution to this negative spiral is to formulate increasingly sophisticated fertilizers that include synthetic chelating agents and soil conditioners. Another solution involves using agricultural wastes as fertilizers.

14.4 Organic Farming

In the twentieth century, most farmers in developed countries have used ever-increasing amounts of inorganic fertilizers, coupled with herbicides and pesticides. Others have felt that these practices not only are harmful to the environment but don't even make sound economic sense. This argument states that the use of herbicides and pesticides is an expensive and inefficient way to control pests (see Chapter 15) and that the loss of humus associated with a program of chemical fertilizers ultimately reduces soil productivity. An **organic farmer** is one who uses only organic fertilizers such as manure, bone meal, and waste plant products and applies no synthetic pesticides or herbicides.

In recent years, some large-scale commercial farmers in the United States have switched to organic techniques even though organic fertilizers are heavy and therefore expensive to spread, and the threat of pest invasions is always a concern. A study of these farms showed that the switch to organic practices led to a decline in energy use by 15 to 50 percent, depending on the crop. In some cases, yields improved with the change, but in many other regions total yields decreased somewhat. However, in many cases in which yields were diminished, the farmer's profit went up because operating costs decreased. According to a report by the U.S. Department of Agriculture, if all the farms in the United States reverted to organic techniques, great quantities of energy would be saved and the American people would still be adequately fed, but overall yields would decrease. This decrease would lead to a reduction in the quantity of food available for export. This conclusion supports the assumption that, if only short-term factors are considered, agriculture based on fossil fuel energy is the most productive system available. On the other hand, the problems raised by this system are potentially so severe that many believe that the gain of extraordinarily high yields at present is more than offset by the threat of a serious decline in productivity in the future.

14.5 Soils in the Tropics, the Temperate Region, and the Arctic

Since the structure of the soil varies dramatically from region to region, it is obvious that some areas will be better suited to intensive agriculture than others. Cereal crops—mainly wheat, rice, and corn—are the most important human foods. Cereals are cheap to produce, remain fresh for a long time when they are stored, and have high nutritional value. Most of the world's cereals are grown in a wide belt of land that stretches through the temperate zone. These zones cover the middles of North America, Europe, and Asia in the Northern Hemisphere and include Argentina, southern Africa, Australia, and New Zealand in the Southern Hemisphere. The temperate belt is much larger in the Northern than in the Southern Hemisphere (see endpaper).

The natural ecosystems in this huge farming belt are mainly temperate forest and, even more important, the grassland that makes up the prairies of North America, the steppes of Asia, the pampas of South America, and the veldt of southern Africa. The soil in temperate areas may be very deep. Parts of the North American prairie had soil more than 2 meters deep before it was farmed, and forests may have soil nearly a meter deep. This soil lies on top of a layer of subsoil, which contains little organic matter. In turn, the subsoil lies over solid rock. This rock slowly weathers and decomposes, adding minerals first to the subsoil and then to the soil above. The reason for the great depth of the soil in the temperate regions is that in most cases organic matter accumulates faster than it decomposes. During the winter the temperature is so low that bacterial and fungal action is slow. A deep layer of partially decomposed organic matter accumulates, forming fertile soil that can support agriculture for hundreds of years even if very little fertilizer is added.

In contrast to the fertility of the temperate zone, soil in tropical and polar regions is usually poor. If decomposition is slow in the temperate zone, it is even slower at higher latitudes, and one might expect that the soil would be correspondingly deeper. Such is not the case, however, because the summer in these areas is so short that plant growth is restricted, and little organic matter is produced. Even further north, where permafrost occurs in the tundra, only a few centimeters at the surface of the soil ever thaw out at all during the summer. A layer of ice, frozen subsoil, and rock lie just beneath the surface. The only topsoil present exists in the thin layer of decaying organic matter on the surface of the ice (Fig. 14–8). This soil has very little mineral content because it lies on ice instead of on rock. There is no rock to decompose and add minerals to the soil as it does in other places.

In the tropics, precisely the opposite conditions prevail. Conditions are warm and moist, ideal for decomposition throughout the year. Organic matter decomposes rapidly after it falls to the ground, releasing minerals that are quickly absorbed by plant roots. As a consequence, only a narrow layer of partially decomposed organic matter remains. In many regions, when a tropical forest is cut down, the thin layer

Figure 14–8 Permafrost accounts for the thin soil of the tundra. This section through the ground in Alaska shows soil that has formed on top of a permanent ice cap, which extends an unknown distance (probably about 150 m) to the rocks beneath. The soil is thin and nutrient-poor because it cannot be replaced from below by the erosion of rocks. Tundra plants are shallow-rooted and are often flooded in summer when some of the ice beneath them melts. The jackknife shown against the ice layer gives the scale of this picture (Courtesy of J. C. F. Tedrow.)

of soil may support crops for only a year or two. In primitive agricultural systems, people native to these areas have traditionally farmed a small area for a few years, abandoned it, and moved on to farm a new area. If the abandoned area is small enough, forest will rapidly re-establish itself in the clearing. On the other hand, if a large area of forest is cut down, the heavy rainfall rapidly leaches any remaining minerals out of the thin soil. In some cases, the ground may be left covered with bare rock. A common soil in other tropical areas is **laterite,** which contains a large proportion of aluminum and iron oxides, and only a thin surface layer of organic matter. When this surface layer is washed away, the remaining clayey soil can bake to a bricklike texture during a subsequent dry period. Such soil is useless for agriculture. In recent years, many billions of dollars were spent in Brazil to develop agricultural land where tropical rain forests had been cleared. However, only about one third of

this type of land is capable of supporting agriculture. As a result, large developments have been abandoned, huge amounts of money have been lost, and in some instances, barren wasteland has replaced lush forests.

One of the reasons why many less developed countries are so poor is that their soils are poorly suited to agriculture. As a result, it is expensive or in some cases impossible to feed the population adequately. Unfortunately, there is no simple remedy for this problem.

14.6 Arid Lands and Irrigation

Most of the ancient civilizations of the Western world originated in a zone between 25 degrees and 35 degrees north latitude. The hot summers and mild winters must have made the struggle for bare survival less intense and provided the free time needed to advance science and culture. On the other hand, the climate in this zone is not idyllic, because these are the horse latitudes, which contain drying winds and

---

SPECIAL TOPIC A
An Agricultural Success Story

One of the most remarkable success stories among the developing nations comes from the Ivory Coast in Africa, an area with sufficient rainfall for agriculture and with soil that is superior to that of either the dense forests or the sandy deserts of much of Africa. About 1960, the Ivory Coast decided that agricultural development was a vital precursor to modernization. Laws were passed to encourage soil conservation and to provide education and the necessary financing to peasant farmers. Money was invested in research and in transport and distribution systems that carry crops to market. Today the Ivory Coast exports food. Almost alone among the developing nations it has a trade surplus and money to spend on population control and education. The per capita income is about five times the African average, and the life expectancy of the people is 50 percent greater than that of people in most other African nations.

This success story is unusual, however. It depended on an enlightened government and the possession of land that is better suited to agriculture than that in most of the tropics.

desert or semiarid conditions in a great many regions. Today, the centers of world military and economic power have shifted largely to the temperate zones, but some 640 million people, or about one seventh of the population of the Earth still live in arid or semiarid lands.

The impact of large human populations on some tropical and subtropical ecosystems has been severe. The loss of croplands in the ancient civilizations of Greece, Rome, and parts of the Fertile Crescent was documented in Chapter 8. Even today, with modern agricultural technology, the process is continuing at an alarming rate. Approximately 5.7 million hectares (14 million acres) of arid but productive land turn into desert every year (Figs. 14–9 and 14–10). Deserts are expanding in Africa, the Middle East, Iran, Afghanistan, northwestern India, and the United States. New deserts, totally created by short-sighted agricultural practices, have recently appeared in Brazil and Argentina. The situation in the Sahara in North Africa is particularly well studied. The Sahara Desert has recently been growing outward in all directions, towards the north, south, west, and southeast. For example, between 1958 and 1975 it expanded

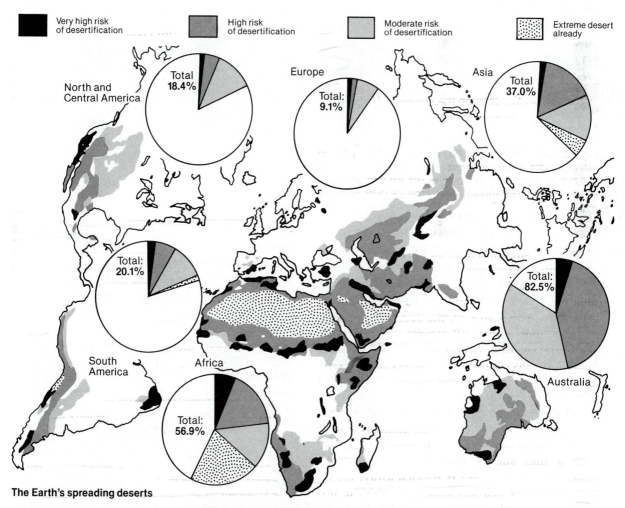

The Earth's spreading deserts

Figure 14–9 The Earth's spreading deserts. Map shows existing deserts and areas into which deserts may spread. The "desertification" hazard is based on a calculation of rainfall and evaporation in a given place. Where rainfall is low and evaporation is high, the risk of desertification is great. Circles show percentage of total land area on each continent that is or that may become desert.

Figure 14–10 The effects of overgrazing and drought in the Sahel region of north Africa. (Courtesy of Alain Nogues-Sygma.)

about 100 km eastward, and today it continues to grow at a rate of up to several km per year. In countries where **desertification** (the creation of deserts) is particularly severe, notably Algeria, Iraq, Jordan, Lebanon, Mali, and Niger, per capita food production declined by an alarming 40 percent between 1950 and 1980. The source of the problem can be separated into two categories—overgrazing and improper use of irrigation.

(a) Overgrazing. One hundred years ago, most of the semiarid range lands bordering the Sahara Desert were populated by nomadic herders. For centuries these people survived by adapting to their harsh environment. They moved across the continent with little regard for national boundaries, travelling with the seasons and abandoning an area after it had been grazed. This constant movement allowed rejuvenation of affected areas and prevented overgrazing. In addition, population levels of these

nomadic tribes were stable and relatively low. In recent years, however, their lifestyle has changed. The demographic transition has been accompanied by decreased infant mortality and a rapidly rising population. In addition, enforcement of national borders and frequent hostilities in some areas have led to the demise of nomadism. As people have become settled and populations have grown, land has been overgrazed, plant systems that normally conserve the sparse rainfall have been destroyed, and the desert has grown. This process has been repeated in many other regions of the world as well (see Fig. 14–9). Even in the United States, where food is relatively abundant, overgrazing and desertification are a problem in the Southwest. In a wealthy nation such as the United States, it is feasible to initiate a program to halt the spread of deserts on public lands. In 1976, the Federal Land Policies and Management Act was passed, mandating the Bureau of Land Management to care for the land and ensure that grazing rights are not overextended. Some argue that the Bureau of Land Management is still favoring short-term profits for ranchers rather than the long-term stability of the range, but the region is subject to at least some control and regulation. In many less developed countries, the situation is far more desperate. Millions of people are literally starving, and it is very difficult for a government to order people with hungry babies at home to reduce their herds or to save a range for future generations. Yet the destruction of the land today may lead to a downward spiral of increasing starvation followed by even more overgrazing and acute desertification the following season.

(b) Improper Irrigation. Since antiquity, people have known that semiarid land can be made to "bloom" if water is imported for irrigation (Fig. 14–11). The food supplies of ancient civilizations in the Tigris–Euphrates valley were dependent on irrigated farmland. Today, the practice is common throughout the world. In the United States, most of the vegetables grown in California depend on irrigation systems, and in recent years, widespread use of irrigation has been extended to field crops as well (Fig. 14–12). For example, in Nebraska, yields of corn can be tripled and profits increased if large mechanized sprinkling systems are installed.

Once again, stories of phenomenal successes are mixed with reports of ecological problems.

Figure 14–11 *Laborers digging an irrigation culvert in Nigeria. (U.S. Agency for International Development.)*

Irrigation water contains more minerals than rainwater since it invariably comes from underground reservoirs or from lakes and rivers in contact with rock. When irrigation water is applied, particularly in dry areas, much of it evaporates, leaving the minerals behind. As these minerals accumulate, the salinity (salt content) of the soil increases. Since most plants cannot grow in salty soils, the productivity of the land decreases. Salinity is often accompanied, and even augmented, by waterlogging. In most farming systems, irrigation water is not carefully controlled, and in many situations more water is applied than is needed. If the practice continues, the water table in the region will rise. A rising water table drowns roots. It also carries added

Figure 14–12 *Irrigation system in the United States.*

salts to the surface and therefore impairs plant growth even more.

In 1977 the United Nations reported that there were 210 million hectares of land under irrigation throughout the world. Of this total, half had been developed after 1950, and about one fifth was already suffering serious loss of productivity due to increased salinity and waterlogging. Some of this damaged land has been irrigated for centuries, but much of it has been heavily watered for only a few decades. If soils can be depleted in such a short period of time, serious concerns are raised about the future of global agricultural productivity.

Salty and waterlogged soils can be reclaimed, but the price is high. Pakistan is a less developed country that is almost entirely dependent on irrigation for food production. In one region, an ambitious project that was started in 1901 ended in disaster by 1960 because close to 90 percent of the soils in the region had been waterlogged and rendered nonproductive. Massive amounts of foreign aid were applied for and received. Fields were left temporarily idle, salty groundwater was drained, and eventually productivity was restored.

In the United States, many farmers have installed perforated drainage pipes under their fields to remove excess salty water. These projects are also expensive; the United States Department of the Interior has started work on a $300 million drainage project in the San Joaquin Valley. However, in many areas of the world, capital investments of this sort are impossible.

Another problem threatening irrigation systems is the depletion of water reserves. This topic will be discussed in Chapter 19.

14.7 Soil Erosion

In a natural temperate ecosystem, new soil is formed by the decomposition of rock at a rate varying between 2 and 11 tonnes per hectare per year, depending on the region. At the same time, some of the soil is carried away by wind and water. In most temperate ecosystems, the rate of production is equal to or greater than the rate of removal, so soil depth and fertility increase slowly with time. When natural forests and prairies are cut and plowed, soil is exposed for at least part of the year and is susceptible to erosion.

In 1977 the United States Soil and Water Resources Conservation Act called for a detailed survey of soils in the United States. According to the report, approximately one third of the natural topsoil in North America has already been lost since the start of agriculture on the continent. At the time of the study, erosion was continuing in the United States at an average rate of about 10.5 tonnes per hectare per year. However, in some regions the figure was considerably higher (Fig. 14–13). Average yearly losses of 31 tonnes per hectare were reported in parts of Tennessee, 25 tonnes per hectare in New Jersey, 33 tonnes per hectare in Texas, and 22 tonnes per hectare in Iowa. A little arithmetic puts these numbers in perspective.

- A 1-cm layer of soil weighs approximately 200 tonnes per hectare.
- In many agricultural areas, topsoil is about 30 cm deep.
- At an erosion rate of 30 tonnes per hectare per year the net annual loss is only about 25 tonnes per hectare because on the average 5 tonnes are replaced by natural processes. This loss amounts to approximately one-eighth cm of soil every year, which would lead to complete eradication of the topsoil in about 240 years.

Figure 14–13 Soil erosion. Water washes 20 tonnes of soil per hectare per year off this farm in Wisconsin. Soil particles, pesticides, and fertilizers washed into waterways cause water pollution. (U.S.D.A. Soil Conservation Service.)

Figure 14–14 The better practice—terraced fields on this hillside in Nepal prevent soil erosion. This land has been passed on for generations and has remained productive. However, in other parts of Nepal, soil erosion is a serious problem resulting from extensive deforestation. (U.S. Agency for International Development.)

(a)

(b)

Figure 14–15 Wind erosion—the Dust Bowl and part of the cure. A, When the wind stopped, this road in Idaho was covered with soil, which in places was 2 feet deep. B, Wind-breaks of willow prevent soil erosion on this farm in Michigan. (U.S.D.A. Soil Conservation Service.)

IF this "B" is condensed in area function =7

On a national average, soil erosion in the United States destroys an equivalent of 1.2 million hectares (3 million acres) of prime farmland every year. (This number was derived by calculating the partial soil loss over all the cropland in the country and mathematically concentrating it to a total loss over a smaller area.) This loss amounts to the destruction of the agricultural capacity of an area the size of the state of Rhode Island every four years.

Elsewhere in the world, erosion rates are quite variable. In some regions, agriculture has been practiced for centuries and the soil is still fertile (Fig. 14–14). In others, erosion has become an even more serious problem than it is in the United States. People living in urban centers do not see the immediate effect of soil erosion, and the problem is much less obvious than other environmental problems such as air or water pollution. But the soil is the basic resource needed to produce food; without it we could not survive.

Soil erosion contributes to problems other than the loss of agricultural land. Its effect on water pollution has been discussed in Section 14.3. Soil particles may also block out much of the light needed by producers in aquatic ecosystems.

Soil erosion can be prevented by simple measures. First, land should not be left unplanted for long, since plant roots hold soil in place. Terracing and contour plowing prevent soil from washing down slopes, and planting windbreaks checks erosion by high winds (Fig. 14–15). Also, since organic debris can bind a great deal of water, soil with a high content of organic matter is less likely to be washed away; thus the use of manure instead of chemical fertilizers is a considerable deterrent to erosion.

The soil erosion problem is largely economic: an individual farmer can profit more by planting crops instead of rows of windbreak trees, by cultivating land that is flooded periodically, or even by leaving land without plant cover for a period. Although governments in many countries are waking up to the disastrous effects of continued soil erosion, they face the difficult task of designing economic incentives that will encourage farmers to conserve soil.

14.8 Case History: Wheat Farming in the Northern Great Plains

One hundred and fifty years ago the northern Great Plains of North America were covered by a varied mix of prairie grasses. In this region, rainfall is too sparse to support forests, so the grasses represent the stable, self-sustaining plant communities. Sometime in the late 1700s, a few white people began to travel in the region, hunting and trapping for a living. In 1804, the U.S. government sent an expedition led by Meriwether Lewis and William Clark to travel to the Pacific Ocean and explore the Northwest Territories. In one of their journals, Lewis and Clark reported that in a few places there were small salty patches in the plains where no grasses grew.

Today, soil scientists can explain the formation of these dead salted regions. Several meters underneath the northern plains there is a vast layer of impermeable clays and shales. The subsoil immediately above this impermeable layer consists of **glacial till**. Glacial till is a mixture of rock, soil, and sand that was left behind when the last great continental ice cap retreated some 10,000 years ago. Above the glacial till lies a thinner layer of fertile topsoil. The prairie ecosystem was traditionally subject to seasonal cycles. In the springtime, melting snows and heavy rains saturated the soil. In some years, excess water leached downward into the glacial till. As water leached through the till, it dissolved many of the natural salts found there. This salty water then collected on the impermeable layer of clay and shale. Eventually it flowed along this layer in underground streams and rivers until it collected in natural depressions on the plains, as shown in Figure 14–16. This process is called **saline seep**. During the hot dry summers, the water in the salty pools eventually evaporated, leaving behind concentrated salt deposits that killed all the grasses in the area and left the salty patches recorded by Lewis and Clark (see Fig. 14–17).

Western agricultural methods were introduced in the northern Great Plains in the late 1800s and early 1900s. The bison were killed, the land fenced, and cattle and wheat became the dominant crops. The natural prairie had been a diverse system containing many different species of annual and perennial plants. During times of drought, the deep-rooted perennials survived and prevented wind erosion, and when the rains fell, quick-growing annuals exploited the abundant resources, protecting the soil with their extensive system of surface roots.

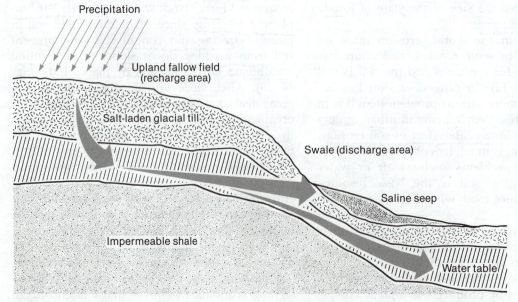

Figure 14–16 Saline seep formation.

Figure 14-17 The lower part of this wheat field in Montana has been affected by saline seep and no longer supports agricultural crops. (Photo courtesy of James Krall, Montana State University.)

When this natural system is plowed under and replaced with a single plant species such as wheat, problems arise. Wheat is an annual grass. If there is not enough moisture in the soil, the wheat will die and the surface of the land will become barren. One solution to this problem is an agricultural practice called summer fallowing. In this technique, a field is plowed over early in the spring, thereby killing all the grasses, and then left barren all summer. Since there are no plants to absorb and transpire the moisture, the soil acts like a sponge, collecting and storing the water that enters the system. When enough moisture has accumulated, the farmer can plant a new crop, either in the fall or in the following spring. This system leaves the land idle for long periods of time but gives greater assurance that a crop, once planted, will succeed.

If summer fallowing operates exactly as planned, the new crop is planted just when the soil moisture has reached optimal levels. In practice, the amount of rainfall hardly ever exactly matches the farmer's needs. Usually there is more or less than the optimum. If there is too little moisture, the farmer may elect to plant the crop and take a chance or wait another season until the soil stores more water. On the other hand, if rainfall is too heavy, the soil becomes saturated because there are no plants to absorb the excess moisture, and the excess water filters down into the glacial till. Here it picks up salts, collects on the impermeable layer of shale, and then runs off into natural depressions on the prairie. The important point to emphasize is that in a summer fallow system, this salty runoff is much greater than it is in a natural prairie.

The problem is further augmented by the use of chemical fertilizers. Chemical fertilizers per se are not harmful, but the practice of using chemicals instead of organic matter results in a gradual loss of humus. As a result, the water-holding capacity of the soil decreases, and saline seep increases.

In the natural prairie, a few thousand hectares of grassland had been destroyed by saline seep. Today 80 million hectares of farmland have been ruined, and 8 million additional hectares are being lost every year. In some communities the salt water has crept into wells, making the domestic water supply unfit to drink.

Once soil has become saline, it is expensive and time-consuming to reclaim it. Therefore, technical remedies are aimed at preventing the problem from becoming worse. Some experts recommend that waterlogged fields should be

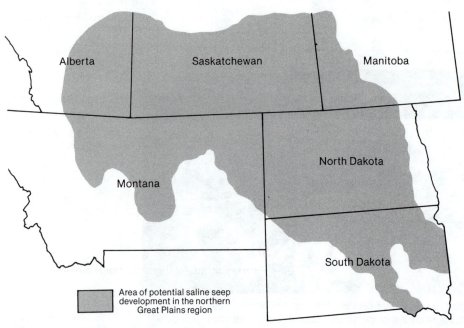

Figure 14–18 Northern Great Plains region, showing area of potential saline seep development.

2.

planted with deep-rooted perennials such as alfalfa. Others suggest that farmers practice no-till agriculture, but the machinery for such a conversion is expensive. Meanwhile, serious and long-lasting destruction of farmland continues in the United States and Canada, the two largest grain-exporting nations in the world (Fig. 14–18).

Summary

The minerals required by plants are divided into two groups, macronutrients and micronutrients, depending on the quantities needed. Some nutrients are needed only by plants with special adaptations.

Nutrient deficiencies of plants may result from lack of nutrients in the soil, a soil pH that makes nutrients unavailable to plants, unfavorable proportions of one nutrient to another, or genetic deficiencies that make the plant unable to handle nutrients properly.

Plants take up most of the water and minerals they need from the soil. The nature of the soil is determined mostly by the type of rock from which it is derived. Rainfall, organic matter, soil organisms, and oxygen are also important in determining the quality of the soil. All of these components of soil interact and affect the availability of water and minerals in the soil solution. Agricultural practices such as tilling, fertilizing, or liming can improve the soil to meet the needs of crop plants.

Organic farming produces a slightly lower yield per acre than farming with conventional fertilizers and pesticides but can be economically just as profitable.

The world's best farmland is found in the temperate zone, most of which lies in North America, Europe, and Asia, with lesser areas in the Southern Hemisphere in South America, Australia, and Africa. The tundra and tropical forest have very shallow soil unsuited to sustained agriculture. A main reason for the poverty of the less developed nations today is almost certainly the very poor soil found in these countries.

Irrigation has brought vast areas of arid land into cultivation. Sustained agriculture is often impossible, however, because irrigation increases the salt content of the soil. The world's deserts are expanding as a result of poor agricultural practices. Overgrazing semidesert, in particular, turns these areas into complete desert in which nothing will grow.

A major concern today, particularly in fertile temperate agricultural areas, is the loss of soil as a result of erosion by wind and water. Such erosion can eventually destroy the soil completely or reduce its depth to the point where agriculture is no longer profitable, as has happened in parts of the United States.

preserving soil quality, and then sell the fields after the soil is depleted. What type of economic externalities are ignored in this type of accounting? Who pays for these external costs?

7. Discuss the benefits and dangers of irrigation in modern agricultural systems.

8. American farming, as practiced today, is essentially mining a nonrenewable resource rather than exploiting a renewable one. Defend or criticize this statement.

Questions

Soil Structure

1. Why is vegetation often sparse in soils with large numbers of pebbles and boulders?

2. The mineral composition of rocks collected from the surface of the Moon is similar to that of many rocks found on Earth, but there is no life on the Moon. How would lunar soils differ from those found on Earth?

Soil Treatment

3. What are the advantages of organic rather than inorganic fertilizers? What fallacies are involved in statements that organic fertilizers are more nutritious for plants than inorganic fertilizers?

4. Explain why areas with very rapid or very slow rates of decomposition contain poor soils. Give examples.

5. Discuss the tradeoff between short-term profit and long-range stability in agriculture. In many regions, farming is marginally profitable at best, so local farmers must maximize their return in a particular year. Do you feel that it is in a nation's interest to provide economic support for soil conservation by subsidizing farmers? Discuss.

6. In some regions where local land prices for residential or commercial development are high, it is profitable to buy agricultural land, farm it for a few years with little regard to

Suggested Readings

Information on soils and soil fertility is available in several standard texts:

R. Akeny: *Conservation Tillage: Problems and Potentials.* Iowa Soil Conservation of America, 1977. 75 pp.

Martin Alexander: *Introduction to Soil Microbiology.* New York, John Wiley & Sons, 1977. 407 pp.

Heinrich Bohn, Brian McNeal, and George O'Connor: *Soil Chemistry.* New York, John Wiley & Sons, 1979. 329 pp.

A. D. Bradshaw, and M. J. Chadwick: *The Reforestation of Land: The Ecology and Reclamation of Derelict and Degraded Land.* Boston, Blackwell Scientific Publications, 1980. 317 pp.

Hans Jenny: *The Soil Resource.* New York, Springer Verlag, 1979. 377 pp.

Philip Royston, and Charles Morgan: *Soil Erosion.* New York, Longmans, Green and Co., 1979.

Samuel Tisdale, and Werner L. Nelson: *Soil Fertility and Fertilizers.* 3rd ed. New York, Macmillan, 1975. 694 pp.

Information on rates of soil erosion in the United States was obtained from:

Ann Crittenden: Soil erosion threatens U.S. farms' output. *New York Times,* October 26, 1980. p. 1.

Information on irrigation and desertification can be found in:

James Aucoin: The irrigation revolution and its environmental consequences. *Environment,* 21(8):17, October, 1979.

Lester R. Brown: *Building a Sustainable Society.* New York, W. W. Norton & Co., 1981. 433 pp.

15

Control of Pests and Weeds

Throughout all ages, farmers have had difficulty protecting their crops from small animals and disease organisms. If a cow wanders into a field to eat the corn, the farmer can chase her away. An insect, a field mouse, the spore of a fungus, or a tiny root-eating worm is more difficult to deal with. Since these small organisms reproduce rapidly, their total eating capacity is very great. Small pests may also be carriers of disease. Malaria and yellow fever, spread by mosquitos, have killed more people than have all wars.

Not all insects, rodents, fungi, and soil microorganisms are pests. Most do not interfere with people, and many are directly helpful. Millions of small animals live within a single cubic meter of healthy soil. Most are necessary to the

Body louse

process of decay and hence to the recycling of nutrients. Fungi, too, are essential to the process of decay in all the world's ecosystems. Rodents are part of natural ecosystems. Thus squirrels help to spread pine seeds, and lemmings provide a staple food for almost all carnivores in parts of the Arctic. Many carnivorous insects eat various pest species and therefore are directly helpful to humans. In addition, bees are essential to the life cycle of most flowering plants. In their search for food, bees transfer pollen from flower to flower and thus fertilize the plants.

Pests have lived side by side with people for thousands of years. At times pest species have bloomed and brought disease and famine. But most of the time, natural balance has been maintained, and humans have lived together with insects in reasonable harmony.

Rat flea

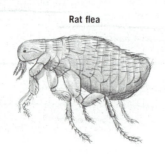

The periodic invasion of some African villages by driver ants is a fascinating illustration of insect ecology. Many disease-carrying rodents and insects live in the village houses and pose a constant threat to the human population. Occasionally, millions of large driver ants invade the villages. They chase away the inhabitants, and eat everything that remains. When the people return, they find that their stored food supply is gone. But so are all the cockroaches, rats, and other pests—everything has been eaten.

In modern times, people are no longer willing to accept these natural cycles. Perhaps even more critical is the fact that the human population is now so large that tremendous quantities of food are needed. One way to increase crop yields is to reduce competition from insects.

15.1 Insecticides—A Great Debate

A farm is a carefully controlled environment. Fields are plowed, planted, and harvested. Usually only one species of plant is cultivated in a large field. Yet a great number of different species of animals exist within these artificial systems. Scientists studying a cabbage field in upstate New York found 177 different species of insects. Of these only 5 species were significant pests. Many were predators that ate the pest species. Some ate other types of food besides cabbages.

Agricultural systems are subject to the normal checks and balances of a natural ecosystem. If left alone, pest species are usually—but not always—kept under control by their enemies. According to one reliable estimate, insects ate 10 percent of the food crops in the United States in 1891. At that time very few pesticides were being used. The pest populations were controlled by insect predators, parasites, and disease. In 1970, after years of intensive use of pesticides, crop losses to insects rose to 13 percent. Does this mean that chemical sprays have increased the pest problem? Would farmers be better off if no pesticides were used at all?

No one knows the answer to this question. First, national averages often mask local problems. Since agriculture began, local pest epidemics have occasionally completely destroyed crops in certain areas (Fig. 15–1). Farmers naturally wish to protect themselves against such disas-

Figure 15–1 A locust invasion in Morocco. Local infestations of this type can completely destroy crops. (Courtesy of Food and Agricultural Organization of the United Nations, photo by Studios du Souissi.)

ters. But perhaps even more important is the fact that times and agricultural practices have changed. In the 1800s farms were relatively small, and many different crops were raised in a given area. Today fields are much larger, and hundreds or even thousands of hectares of a single species of plant may be grown in one region. This type of monoculture provides a favorable environment for pest insects because their food supply is much more concentrated than it would be in a series of small farms or in a natural environment. Some plant scientists and most growers claim that modern agricultural practices must be accepted if the people of the world are to be fed, and therefore pesticides are a necessity. Without such controls, insect populations would flourish and consume a large percentage of the world's food supply. Other scientists disagree and feel that chemical pesticides are harmful to the environment and actually increase pest problems.

The modern pesticide era started about 1940 when a chemical called **DDT** was discovered to be a potent insecticide. DDT was far cheaper and more effective against almost all insects than the previously known chemicals. The use of DDT led to dramatic early successes. It squelched a threatened typhus epidemic among the Allied army in Italy. Antimosquito programs saved millions from death from malaria and yellow fever. Pest control, leading to increased crop yields all over the world, saved millions more from death by starvation.

Enthusiastic supporters of DDT predicted the complete destruction of all pest insects within the foreseeable future. Paul Müller, the chemist who first discovered its insecticidal properties, received a Nobel Prize. But within 30 years the promise of insect-free abundance had been broken. The "miracle" chemical that was to have achieved it had fallen from grace. On January 1, 1973, all interstate sale and transport of DDT in the United States was banned except for use in emergency situations. Following the ban on DDT, several other organochloride pesticides were also banned—aldrin and dieldrin between 1972 and 1974, and chlordane and heptachlor in 1975.

A few decades ago, many people believed that chemical pesticides would liberate people from the whims of a changing environment.

These chemicals were viewed as a road to plenty. But pest insects have not been removed from the face of the Earth. In fact they have survived and thrived despite an intense and continued program to annihilate them.

Many other pesticides besides DDT have been in use during the past few years. Some of the most common of these are listed in Table 15–1. Note the three classes of compounds—**organochlorides**, **organophosphates**, and **carbamates**. The chemical formulas of some important pesticides are given in Appendix D.

15.2 Two Problems with Chemical Pesticides— Resistance and Broad-Spectrum Poisoning

Chemical insecticides were initially received with great enthusiasm because they are inexpensive, easy to use, fast-acting, and effective against a wide range of pests. Their promise engendered an uncritical optimism. It was imagined, for example, that tomato farmers faced with a midseason invasion of some insects would not have to identify the pest. They would simply call in an aerial spray company and would expect 90 percent destruction of the pests within the next day or two. A simple problem, a simple solution—or so it seemed.

Yet if the problem is examined more closely, complexities emerge. A tomato field under attack by some insect is not merely a two-species system. The tomatoes and pests are but two members of a large agricultural ecosystem of thousands of species that include predator insects, bacteria, parasites, and many types of soil dwellers, as well as carnivorous, herbivorous, and omnivorous birds and other migratory animals. Therefore, despite the undeniable fact that innumerable successes of spray programs have been recorded during the past 30 years, it is important to look more closely into the intricacies of the problem.

The use of nonselective sprays has often led to the destruction of the natural controls on relative population sizes. As an example, when DDT and two other chlorinated hydrocarbons

TABLE 15–1   Some Common Chemical Insecticides

Compound Class	Examples	Persistence in the Environment
Organochlorides (also called chlorinated hydrocarbons)	Aldrin Chlordane DDD DDT Dieldrin Endosulfan Endrin Heptachlor Kepone Lindane Mirex Toxaphene	High—5 to 15 years
Organophosphates	Azodrin Diazinon Malathion Parathion Phosdrin	Intermediate—1 week to several months
Carbamates	Carbaryl (Sevin) Matacil Temik Zectran Zineb	Low—2 weeks or less

*Chemical formulas of various pesticides are shown in Appendix D.

were used extensively to control pests in a valley in Peru, the initial success gave way to a delayed disaster. In only four years, cotton production rose from 490 to 730 kilograms per hectare. However, one year later the yield dropped precipitously to 390 kilograms per hectare, 100 kilograms per hectare less than before the insecticides were introduced. Studies indicated that the insecticide had destroyed predator insects and birds as well as insect pests. Then, with natural controls eliminated, the pest population thrived better than it had ever done before.

A second example illustrates the same point. In the San Joaquin valley in California, chemical manufacturers sold the organophosphate Azodrin for use against the cotton bollworm, a pest of cotton plants. After several heavy spray dosages, it was discovered that the predator populations were killed more effectively than the pest species. Losses to the bollworm grew more severe. Sales representatives recommend that more pesticide should be used. (Of course this would lead to greater profits for the manufacturer.) An independent research team at the University of California determined that if no pesticides were used *at all*, bollworms would consume approximately 5 percent of the crops.

After three spray applications, so many predators had been killed that 20 percent of the crop was destroyed.

How could the pests stage a comeback? Why didn't the predators stage an equal comeback? Why couldn't the farmers combat the pest resurgence with more spraying?

One of the major problems with chemical insecticides is that many insects become resistant to the poisons. In other words, a given insecticide at a given concentration often becomes less effective after some years of use. It appears as though the chemical has diminished in potency, although the composition has, of course, remained unchanged. To understand the reason for this phenomenon, recall that the chemistry of plants and animals changes from time to time as a result of random mutations of the hereditary material in their reproductive cells. A mutant has a good chance of survival if its particular mutation protects it from a hostile environment. This mechanism of random mutation has allowed insects to adapt to their environment for millions of years, and it is this process that protects insects from pesticides. In areas where spraying is heavy, strains of insects have evolved that are genetically resistant to a particular chemical. Genetic resistance to insecticides is an extremely serious problem. By 1945 at least a dozen species had developed some resistance to DDT; by 1980 the number had increased to over 200 species. About 35 of these resistant species carry disease and about 80 others are serious agricultural pests. Since resistant parents can pass this characteristic on to succeeding generations, the old pesticides are rendered ineffective. This effect was directly demonstrated in an experiment in which DDT-resistant bedbugs were placed on cloth impregnated with DDT. They thrived, mated, and the females layed eggs normally. The young, born on a coating of DDT, grew up and were healthy. Attempts to change pesticides have in many cases simply produced strains of insects that are resistant to more than one chemical.

The problem of resistance of insects to poisons can be compounded if the pest becomes resistant and various predators (Fig. 15–2) do not. If this happens, the pests have a new biological advantage and can thrive in greatly increased numbers. This is favored by three factors:

1. *Insect pests are often smaller and reproduce at a greater rate than their predators.* More frequent reproduction promotes more frequent mutation and thus improves the chances that resistant

strains will arise. Furthermore, once a resistant population of insect pests appears, it can repopulate its ecological niche much faster than can a larger, more slowly reproducing species. In effect, the pest species develops resistance faster than the predator.

2. *There are always fewer predators than herbivores (including pests) in an ecosystem.* A small population runs the risk of local extinction because it is more likely that a small group containing only a few individuals will be wiped out by some disaster than it is for a large group to be eliminated. Therefore, the species of herbivores (the pests), which are more numerous, have a greater chance of survival than the species of predators. In addition, in a large pest population there is a statistically greater chance that some individuals will develop resistant mutations compared with the same probability in a smaller predator population.

3. *Predators generally eat a diet richer in insecticides than that of the original pests.* Because chemical poisons are not immediately excreted by herbivorous insects (or any other organism for that matter), the concentration of the chemical in their bodies becomes greater as more contact is made with the poison either directly or through the ingestion of sprayed leaf tissue. Since death

Figure 15–2 Many insects are predators. Here a praying mantis is eating a Satyrid butterfly. (Courtesy of Emeritus Professor Alexander B. Klots, Biology Department, the City College of the City University of New York.)

may be delayed for some time after poisoning, many poisoned but living insects will be eaten by their natural enemies. In this way predators eat a diet that is more concentrated in poison than the diet of the herbivores, the original pests.

The series of events that has occurred in many agricultural systems can now be reconstructed. Although the pesticide causes a rapid decrease in the pest population at first, resistant mutants soon displace the susceptible individuals. The situation then becomes worse than it was originally because natural predators do not achieve immunity as well as the pests. As a result, the controls on the pest population actually diminish, and the pests grow and multiply faster than ever.

Another effect of broad-range pesticides is that destruction of one species sometimes leads to the bloom of another pest that hadn't been a problem in the first place. Consider the story of the spider mite in the forests of the western United States. The spider mite feeds on the chlorophyll of leaves and evergreen needles. Because in a normal forest ecosystem predators and competition have kept the number of mites low, mites have never been a serious problem. However, when the United States Forest Service sprayed with DDT in a campaign to kill another pest, the spruce budworm (Fig. 15–3), complications arose. The budworms were effectively killed, but the insecticide also poisoned such natural enemies of spider mites as ladybugs, gall midges, and various predator mites. The next year the forests were plagued with a spider mite invasion. Although spraying had temporarily controlled the spruce budworm, the new infestation of spider mites proved to be more disastrous.

Wild animals and even livestock have also been affected. For example, when several communities in eastern Illinois were sprayed aerially in an effort to stop the westward movement of the Japanese beetle, many species of birds were almost completely annihilated in the sprayed area, ground squirrels were almost eradicated, 90 percent of all farm cats died, some sheep were killed, and muskrats, rabbits, and pheasants were poisoned. These unwanted side effects might have been considered a necessary

Figure 15–3 Spruce budworm larva and adult.

price to pay for pesticidal success, but the cost did not yield the desired benefit; the Japanese beetle population continued its westward advance.

In a natural ecosystem, bees pollinate many plant species as they travel from flower to flower. But bees are also killed by pesticides. When the bee population is decimated, crop losses arise from lack of pollination. In some cases these losses have been more damaging than the insect attack itself.

15.3 Why Was DDT Banned?

All the insecticides listed in Table 15–1 are broad-spectrum poisons. They all kill nonpest species and have led to disruption of ecosystems. Why, then, have DDT and other chlorinated pesticides been banned, while other types of pesticides are still used freely?

(a) Persistence. If an organophosphate is sprayed on a field, it will decompose rapidly in the environment. For example, when parathion, an organophosphate, is sprayed onto loamy soil, over 98 percent will decompose after four months. These materials are said to be **biodegradable**. The organochlorides are not biodegradable. They persist for long periods of time in the environment. For example, in some soils DDT and aldrin have been detected 15 years after a single spray application. Such persistence leads to a series of severe ecological problems.

For example, if a field is sprayed once, predators are subject to continuous exposure to poisons for many years. If a field is sprayed many times, the insecticide accumulates in the environment. Therefore, all the problems outlined in the previous section may persist for many years.

(b) Physical Dispersal of Insecticides. A chemical pollutant that is nonbiodegradable not only persists in the immediate environment but also lasts while it is being mechanically and biologically transported throughout the biosphere. The physical dispersal of organochlorides has been particularly well studied. To understand the mobility of insecticides, some of their physical characteristics must first be reviewed. Organochlorides are relatively insoluble in water, evaporate slowly, have a strong tendency to adhere to tiny particles of dust, dirt, or salts and are soluble in the fatty tissues of living organisms.

Now consider the fate of an application of DDT or one of its sister compounds sprayed from an airplane onto a crop, as shown in Fig. 15–4. Not all of the spray will hit the exact target. The accuracy of dispersal depends on such factors as the skill of the pilot, the number and

Figure 15–4 Crop duster applying a fungicide to an orange grove in Florida. Some of the sprays are carried by the wind, and they eventually land in nontarget areas. (Courtesy of U.S.D.A.)

position of electrical wires and trees that must be dodged, and the wind direction, velocity, and turbulence. Some of the insecticide that misses the desired target lands on nearby houses, roads, streams, lakes, and woodlots, while some is carried into the air either as a gas or as a "hitchhiker" on particles of dust or on water droplets. Therefore, a great deal of the insecticide travels long distances.

Although the use of DDT in the United States was banned in 1973, it was still being manufactured in this country in the early 1980s and exported for use in other nations. In fact, many of the fruits and vegetables imported into the United States have been sprayed with some type of organochloride pesticide. Thus, long-term global contamination continues. Today, DDT can be found in all major rivers of the world. There are traces in the ocean, in the fat of penguins in Antarctica, in people's bodies—just about everywhere. Pesticides disrupt the growth of phytoplankton in the ocean. They kill microorganisms in the soil. They have been known to disrupt forest ecosystems. In short, if a persistent chemical poison is sprayed on one field it not only may affect the ecological balance of that field but may also travel into other ecosystems. It may continue to upset the ecological balance in many regions for years to come.

(c) Effects on Nontarget Species—Concentration in Biological Systems. Plants and animals are able to wash many poisons out of their systems. Most wastes are removed from the body in water solutions of urine, sweat, or pus. But organochlorides do not dissolve in water. Instead, they dissolve in fat. Therefore, they tend to remain in the body for very long periods of time. If an organochloride pesticide is sprayed onto a hay field and a cow eats the hay, the cow consumes small concentrations of poison every day. The pesticide then remains in the cow's body; each daily intake is stored. Therefore, the pesticide residues slowly accumulate in the animal (Fig. 15–5). A person who then eats the cow's meat will consume the large concentrations of this accumulated chemical. In most cases, pesticide concentrations in meat do not cause acute illness, and the consumer might not be aware of the problem at all. Yet there is concern that the long-term effects of these com-

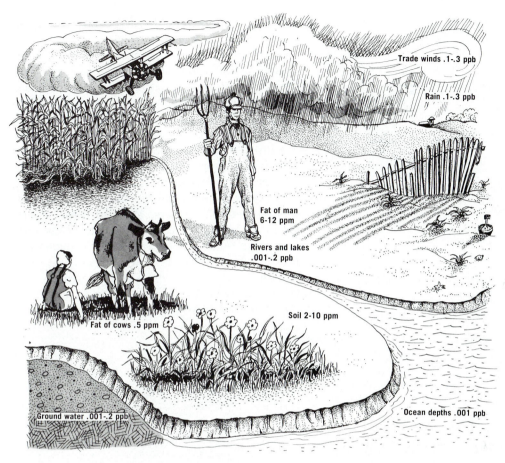

Figure 15–5 *Physical dispersal of insecticides with some average values for DDT levels, as recorded in 1970, before the ban on DDT.*

pounds may be detrimental to human health (see Section 15.4).

Clear Lake in north-central California is a popular recreational area. But the shallow, calm waters have been a breeding place for annoying colonies of mosquitos and gnats. In the 1950s government workers sprayed Clear Lake with the organochloride DDD to control the insect pests. After the project was completed, the water contained 0.02 ppm* of DDD. Plankton that live in Clear Lake concentrated and stored some of the pesticide until there were 5ppm in their tissues. Plant-eating fish ate the plankton and

further concentrated the poison. As a result, their bodies contained 40 to 300 ppm of DDD. Some predatory fish and birds had as much as 2000 ppm of DDD in their tissues (Fig. 15–6). Thus, the original concentration of the pesticide in the water seemed innocently low, but higher in the food chain, fish and birds were poisoned. The problem persisted for a long time. A year after the spray application, no pesticide could be detected in the water at all. Yet the poison still existed in the plants and animals.

The effects of high levels of insecticides on carnivorous fish, mammals, and birds have been particularly severe. In many cases the animals die soon after being poisoned. In other situations, small doses may not kill the animals directly, but they may upset the normal activities of the body and cause delayed death. In some

---

* Recall that ppm means parts per million. One ppm is one ten-thousandth of one percent. One ppb (part per billion) is one thousandth of a ppm. (See Appendix A.)

Figure 15–6 *Biological magnification of DDD (Clear Lake, California).*

cases, the adults may survive but the infants will die. For example, when mink were fed fish that were contaminated with DDT the adults remained healthy, but 80 percent of the newborn infants died within a few days.

The evidence seems undeniable that DDT poisoning was responsible for the sharp decline in populations of many predatory birds (Fig. 15–7). Birds that have been poisoned by DDT cannot use calcium properly. This has led to the production of thin-shelled eggs. Often these weakened eggs crack and break in the nest, resulting in death of the unborn chicks. Before several of the organochloride pesticides were banned in the United States, the populations of several species of birds—among them, the peregrine falcon, the pelican, and some eagles—were declining so rapidly that many conservationists feared that they were on the verge of extinction.

Pesticides can be fatal to animals that store food energy in fat for use during winter months. Trout build up a layer of fat during the summer months, when food is plentiful. In areas where spraying has been done, this fat contains high concentrations of DDT. During winter, the fat is used as a source of energy. The DDT released into the bloodstream upon fat breakdown has been known to kill the fish. The eggs of fish also contain a considerable amount of fat, which is used as food by the unborn fish. In one case,

700,000 hatching salmon were poisoned by the DDT in their own eggs.

The deaths of majestic animals like trout and falcons attract considerable attention, whereas the effects of pesticides on other organisms are often overlooked. Even one kilogram of fertile soil may contain a few trillion different decomposer organisms. These organisms are vital for continued fertility of the soil. They fix nitrogen, they break down rock and thus make minerals available to the plants, they retain moisture, they aerate the soil, and they bring about the essential process of decay. Without these organisms, the plants above ground usually die out. The effect on these organisms of an increasing concentration of poison in the soil is largely unknown. In many heavily sprayed areas of the world, farmers are harvesting more food per acre than ever before. Yet some facts are coming

Figure 15–7 *Biological magnification of DDT after spray application for Dutch elm disease.*

to light that may presage future disaster. Studies in Florida have shown that some chlorinated pesticides seriously inhibit nitrification by soil bacteria. Termites have not been able to survive in soils that were sprayed with toxaphene 10 years previously. Similarly, endrin present in as low a concentration as 1 ppm caused significant changes in the population of soil organisms and consequently in the relative concentrations and availability of important soil minerals. As a result, beans and corn grown in soils treated with endrin contained different nutrient values than the same crops grown in untreated soils—some minerals were increased, some decreased. The net result, however, was a decrease in the total plant growth. In another instance, aldrin routinely sprayed on a golf course depleted the number of earthworms.

As is the case for many types of ecological disruptions, the long-term effects of insecticides in soils are not known. Perhaps many soil organisms are or will become resistant to the pesticide accumulations. But the stakes in the gamble are high, for if life in the soil dies, plants will soon succumb.

Not all pesticides in the soil remain fixed there. Some fraction of the total seeps into groundwater reservoirs and hence into drinking water supplies. A government study of private wells in Illinois in 1971 showed that all those surveyed contained water contaminated with insecticides. A large fraction of soil-based insecticides is carried into the world's surface water supply along with the tonnes of sediment eroded from agricultural systems. As a consequence, even though several organochloride pesticides have already been banned, all major rivers in the United States still contain measurable insecticide concentrations in the parts per billion (ppb) range. In river water at 7.2°C (45°F), which is normal temperature for trout, 1.4 ppb of endrin will kill half of a population of rainbow trout in three days. Similarly, many other species of fish cannot survive insecticides in concentrations greater than about 1 to 10 ppb. Trout in the United States today live in the more mountainous regions, which are not so polluted, but those that were once native to the Great Lakes water systems or the lower Missouri River can no longer be found. Under severe conditions, such as those that occurred during 1950 in parts

of the South when heavy rains followed heavy spray applications, agricultural runoff was so high that the residual pesticide concentrations were raised a thousandfold. Nearly all of the fish in many watersheds were killed.

Inevitably, if insecticides are present in major rivers, they must also be present in the ocean, and the concentrations must be highest in estuary systems, that is, at river mouths and in coastal bays. One of the most serious problems in estuary systems is that insecticides reduce photosynthesis carried out by plankton. DDT at a concentration of 1 ppb can reduce phytoplankton activity by 10 percent and at 100 ppb by 40 percent compared with plankton grown in unpolluted waters.

15.4 Insecticides and Human Health

Most chemical pesticides are poisonous to humans as well as to insects. The organophosphates, which have been used extensively in North America since 1973, are much more poisonous than the banned DDT that they replaced.

Since the mid-1940s, many thousands of people have been sickened or died from severe pesticide poisoning every year. At present, more than half of these are children who are exposed to the toxic chemical through carelessness in packing or storage. Most of the others are workers who handle these materials in the factory or on farms. (Fig. 15–8).

The case of the workers in a kepone factory deserves mention. Kepone is an organochloride pesticide developed in the early 1950s. It was manufactured by a small corporation closely associated with Allied Chemical Company. Safety regulations at the factory were amazingly lax. Pesticide dust was scattered over floors, equipment, clothes, and even the employees' lunch area. In 1975, several workers began to complain of severe illness. Their muscles would tremble uncontrollably at times. People suffered blurred vision, loss of memory, and pains in their joints and chest. In all, 70 of the 150 employees were poisoned by the pesticide. In addition, some of the workers brought the pesticide dust home on their clothes and poisoned their families as well.

Figure 15–8 Migrant farm workers near Salinas, California. These people are not told of past spray applications and are sometimes exposed to high concentrations of pesticides.

In July of 1975, the plant was shut down, and Allied Chemical paid millions in damage suits. Unfortunately, cash settlements do not make people healthy again.

It is relatively easy for most people to avoid exposure to large doses of insecticides, but it is impossible to avoid exposure to trace contaminants in food, in the air, and in drinking water. We all carry measurable quantities of insecticides in our bodies. What are the chronic (long-term) effects of these chemicals? Unfortunately, it is impossible to answer this question with certainty. If there were two large groups of people and one group ate contaminated food while the second group ate pesticide-free food, and if these groups were studied for twenty or thirty years, then perhaps the long-term effects of pesticides could be measured. However, it is morally unjustifiable and impossible in practice to control the diets of large groups of people over many years. Instead, scientists must rely on studies of small groups of people and of laboratory animals. As explained in Chapter 10, such studies are always uncertain. In addition, most experimental subjects are examined for only a few years, although it is known that cancer may appear ten to twenty years after exposure to a carcinogenic chemical.

Scientists have fed various pesticides directly to people. They have studied men and women who work in pesticide factories or with these chemicals in the field. None of these studies has *proved* that pesticides cause cancer. Extensive experiments have been conducted with rats, mice, and other laboratory animals. Some of the results have shown that various pesticides induce cancer in certain species of animals. Although some of these experiments suggest that the same pesticides may induce cancer in humans, other experts have questioned the validity of the experiments. In some instances extremely high doses of pesticide were used. In others, the strains of rats and mice that were studied showed an abnormally high susceptibility to cancer without exposure to pesticides. All in all, it can be safely said that after a decade and a half of study, no one is certain whether pesticides do or do not cause cancer in humans.

15.5 Other Methods of Pest Control

Broad-spectrum pesticides have been effective in many situations, but they have serious flaws as well. They destroy populations of predators, upset ecosystems, and poison people and animals. In addition pesticides may possibly cause cancer. What are the alternatives?

Use of Natural Enemies

Pesticides have caused problems where they have poisoned natural predators. Why not use the opposite treatment and, instead of poisoning predators, import them? This type of approach has been quite successful in many cases.

The Japanese beetle was imported to North America from the Orient and has become a serious agricultural pest. In Japan, the beetle population is controlled by natural enemies. These predators do not live in North America. One of the natural predators is a type of wasp. A female wasp paralyzes a Japanese beetle grub and attaches her eggs to it. When the young wasp hatches, it eats the grub as its first food. These wasps lay their eggs only on Japanese beetle grubs. They do not attack other species. In the United States, spray applications against the Japanese beetles have caused serious ecological disruptions. In recent years, people have realized higher levels of success simply by importing natural enemies.

Insects can also be controlled through the use of certain strains of bacteria and viruses. A virus effective against the cotton bollworm and the corn earworm has been utilized as a pesticide in the United States. It has been approved by federal regulatory agencies and is now ready for mass production. Viral strains that combat several other species of pests should be commercially available in the near future.

There are many advantages to importing enemies of pests. Because these agents are living organisms, they reproduce naturally, and one application can last for many years. Most insect parasites and disease organisms are very specific and do not interfere with the health of large animals. No harmful or questionable chemicals are introduced into the environment.

Sterilization Techniques

Pests can be controlled without killing them directly if the adults of one generation are sterilized. The sterilized adults cannot produce healthy babies, so the pest population will soon die off.

In theory, the technique is simple and should produce rapid results. For example, as shown in Table 15–2, if there is a population of 2 million insects, the release of an additional 2 million sterile males for five successive generations should completely eliminate the pest. Although this arithmetic seems encouraging, in practice, no pest has been eliminated in this manner.

The screwworm fly (Fig. 15–9) which is metallic green, is a bit larger than a housefly. Unlike the housefly, however, its larvae feed on living tissue. They can kill a mature steer in ten days. In 1976 they killed an invalid elderly woman in San Antonio who was unable to help herself. Some years ago, the United States Department of Agriculture initiated a program in the southeastern states to raise male screwworm flies, sterilize them by irradiation, and release them in their natural breeding grounds (Fig. 15–10). If a sterile male mates with a normal female, she will lay eggs but they will not hatch.

For several years, millions of sterilized males were released annually to mate with healthy females living in the area. Initially, the program was a spectacular success, and screwworm infestations were completely eliminated from 1962 to 1971. But in 1972 the screwworm population in-

TABLE 15–2 Mathematical Population Decline When a Constant Number of Sterile Males are Released Among an Indigenous Population of One Million Females and One Million Males

Generation	Number of Virgin Females	Number of Sterile Males Released	Ratio of Sterile to Fertile Males	Number of Fertile Females in the Next Generation
1	1,000,000	2,000,000	2:1	333,333
2	333,333	2,000,000	6:1	47,619
3	47,619	2,000,000	42:1	1107
4	1107	2,000,000	1807:1	less than 1

A

Figure 15–9 *A*, A female screwworm fly—the one that lays the eggs and does all the damage. (Courtesy of U.S.D.A.) *B*, Screwworm larvae infestation in ear of a steer. An untreated, grown animal may be killed in 10 days by thousands of maggots feeding in a single wound. (Courtesy of U.S.D.A.)

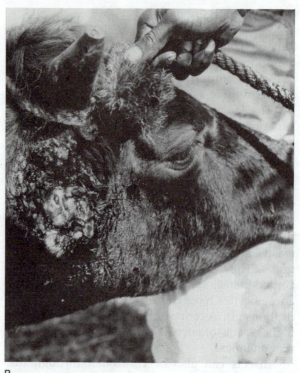

B

creased dramatically and infested nearly 100,000 cattle. During the following three years, screwworms continued to cause serious losses of cattle. What had happened?

Researchers reasoned that perhaps the screwworms raised in captivity evolved into a domestic strain that was no longer well adapted to natural conditions. For some reason these domestic irradiated flies no longer mated successfully with wild females. Perhaps the males could not fly so far or fast, perhaps they lost resistance to natural predators or disease organisms. In 1977 the program was renewed using a new strain of male flies. In May of that year, pilots dropped 400 million sterilized males a week in Mexico and parts of Texas. The infestations were stopped in the control regions, but the following year, the screwworm fly opened new fronts in New Mexico, Arizona and California. Air drops of sterile flies were raised to 500 million a week and extended to New Mexico. At that time, Dr. Carl R. Watson, a staff scientist with the eradication program said, "This is a war. You may lose a battle here and there, but we won't lose the war."

Dr. H. Q. Sibley, the program director, predicted, "I'd say that we'll eventually eradicate the screwworm fly in the continental U.S.A. and Mexico, and hope we can do it sometime in '81, '82—along in there."

By the fall of 1981, however, the flies continued to survive in many regions. Biological selection had favored those females that were able to recognize sterile males and to reject them in favor of the wild fertile ones. Thus, an extremely expensive program proved unsuccessful.

Other attempts at control through sterilization have been even less successful. When sterilized males were released against the codling moth, a serious pest of apple and pear orchards, the initial control was successful, but migration of fertile males from nearby areas spoiled the early gains.

In 1981, an imported pest of fruits and vegetables, the Mediterranean fruit fly, nicknamed the **medfly,** appeared in California. Local agricultural experts feared that because there were few natural enemies of the medfly, populations might flourish and destroy billions of dollars worth of produce. As part of the program to control the pests, "sterilized" males were re-

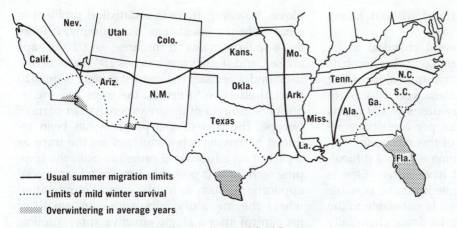

Figure 15–10 Screwworm distribution in the United States.

Key:
— Usual summer migration limits
······· Limits of mild winter survival
▨ Overwintering in average years

leased in several target areas. A few days later scientists learned that the released males were fertile after all. Therefore the "control" procedure proved to compound the problem rather than to solve it. The story has a happy ending, however, for medflies did not adapt well to the environment in Calfornia, and populations never expanded to crisis levels.

These examples should not be used as an argument against new techniques, but they do serve as a reminder that insect control is a formidable problem and that quick and permanent solutions are not yet available.

Control by Hormones

Many insects begin their lives in some larval stage and later metamorphose into a mature adult. As an example, a caterpillar is a larva that later matures to become a moth or butterfly. When an insect is in the larval state, it continuously produces a chemical called the **juvenile hormone.** As long as sufficient quantities of juvenile hormone are in the insect's system, it remains a larva. It is only when the flow of that biochemical agent stops that the insect metamorphoses (Fig. 15–11). If an insect larva is artificially sprayed with the juvenile hormone specific to its species, it will never mature into its adult form. Because insects can neither mate nor survive long as larvae, such a spray application is eventually lethal. Therefore, juvenile hormones can be used as insecticides. It is likely that widespread use of these chemicals or their analogues would produce minimum environmental insult,

because they are biodegradable and active only against specific insects. With the aid of careful

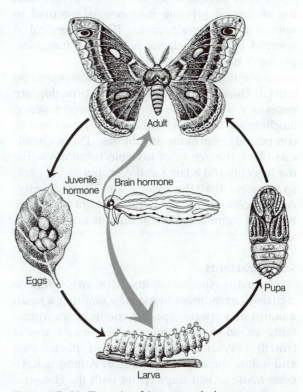

Figure 15–11 The role of hormones in insect metamorphosis. The juvenile hormone, secreted mainly during the larva stage, keeps the caterpillar in this immature state until it is ready to metamorphose into a pupa and adult. The hormones must be secreted in the right amounts at the right time. (Redrawn from Fortune, July, 1968.)

timing, pests can be destroyed without killing their predators.

The first hormone control chemical to be marketed commercially is an agent that combats three species of floodwater mosquitos, including *Anopheles albimanus*, the malaria vector in South and Central America. Spray doses of 150 grams per hectare (about 2 ounces per acre) have resulted in complete control of this mosquito.

Despite this success, some technical difficulties with hormone control have arisen. One is that although natural juvenile hormone is stable in the body of a caterpillar, it is not stable in the environment and often breaks down chemically before it can act. This problem has been circumvented by the discovery of organic chemicals that are structurally similar to natural juvenile hormone and are biologically active but more stable. Another difficulty is in choosing the best time for the spray application. Juvenile hormone is an effective insecticide only during the relatively brief period of an insect's life cycle when the *absence* of juvenile hormone is essential to normal insect development. If the chemical is sprayed before the larva is ready to metamorphose, it has no effect.

Since the timing of spray applications is so crucial, considerable expertise and precision are necessary for effective control. In recent years, biochemists have been studying a new class of compounds, **hormone inhibitors.** These chemicals block the action of juvenile hormone while the insect is still a larva and disrupts its life cycle so severely that the insects are killed directly. Active research on hormone inhibitors is under way, but commercial distribution is not yet available.

Sex Attractants

In many species of insect, a virgin female signals her readiness to mate by emitting a small amount of a species-specific chemical sex attractant, called a **pheromone.** The males detect (smell) very minute quantities of pheromone and follow the odor to its source. Attempts have been made to bait traps either with the chemical or, more simply, with live virgin females. In this program, only minute quantities of natural chemicals are released into the air by evaporation, and environmental perturbations are nonexistent. As in the case of the sterile male technique, however, there is a statistical problem in areas of heavy infestation. Very simply, if there are a million males in the area, and 1000 traps are set, and if each trap catches 100 males before they find a female, the control project is only 10 percent effective.* Further, the traps and the labor of setting them up are expensive. In parts of Europe, the cost of this program has been reduced by having schoolchildren set the traps as a part of an educational program. Still, the trapping technique is probably best suited for special applications, such as for the control of migration where the migratory populations are small, or for control after a single effective spray application.

Some successes have been registered by the confusion technique. In one application millions of cardboard squares were impregnated with pheromone and dropped from an airplane. With a high ratio of cardboard squares to females, males become confused and were observed to try to copulate with the cardboards. Another approach has been to spray the area with pheromone so that the entire atmosphere smells of virgin females (to the male insect, that is, not to the human nose). Then the males are unable to track down a mate.

Use of Resistant Strains of Crops

Some plants are naturally resistant to pests because they synthesize their own insecticides. For some time, plant breeders have been actively developing more resistant strains and have succeeded in a number of instances, such as with a variety of alfalfa that is resistant to the alfalfa weevil and various strains of cereal crops that are resistant to rust infections.

The successes achieved have been encouraging, and further research deserves support, but it should be remembered that the technical problems are difficult. The new plant variety must produce high yields as well as maintain the natural resistance to other diseases. Moreover, genetic adaptation is not stagnant, for throughout biological time genetic defense in one species traditionally has been met by genetic changes in the attacking organism to neutralize the defense. Thus, many resistant crops lose

---

* $\dfrac{1000 \text{ traps} \times 100 \text{ males/trap}}{1{,}000{,}000 \text{ males}} \times 100\% = 10\%$

their immunity after several years, and new resistant varieties must be developed.

Control by Cultivation

Uniformity is not typical of virgin land masses. The systems created by modern agriculture differ from natural systems, for farms tend to specialize in a very few species of plants, whereas areas untouched by people do not. For instance, in Kansas and Nebraska thousands of acres of land are planted almost exclusively in wheat. The result of such specialization is that a mold, fungus, or insect that consumes wheat has a vast food supply and an extremely hospitable environment. No barriers to spreading exist, so the pest can grow quickly in uncontrolled proportions. A fungus that attacks wheat will spread more slowly if half the plants in a field are something other than wheat, because the spores have a reduced chance of landing on a wheat plant. Moreover, if the disease spreads slowly, there is more time for the development of natural enemies of the fungus or of naturally resistant strains of wheat. Therefore, one solution to the problem of pests is to grow plants in small fields, with different species grown in adjacent fields. Unfortunately, the mechanization that is essential to modern agriculture is best suited to large fields of single crops.

Simply planting small fields of different crops is effective in itself, but by judiciously choosing the companion plants, additional success can be realized. For example, the grape leafhopper, a pest of vineyards, can be controlled by a species of egg parasite that winters in blackberry bushes. Knowledgeable grape-farmers therefore maintain blackberry thickets. Conversely, one variety of stem rust that attacks North American cereal crops must live part of its life cycle on barberry bushes, and selective destruction of these plants will reduce rust infestations.

Other methods of cultivation that have been successful in controlling pests include (1) crop rotation so that a given pest species cannot establish a permanent home in one field, (2) specific plowing and planting schedules to favor predators over pests, and (3) planting certain weeds that some omnivorous insects, which prey on pests during part of their life cycle, need for food during other stages.

Changes in Consumer Attitudes

In the United States, consumers have come to expect blemish-free produce. Yet the hidden cost of "flawless" fruits and vegetables is an increased use of biologically potent chemical poisons. As an example, San Jose scale attacks apples as well as other fruits such as pears, peaches, plums, and apricots. This insect pest produces many pin-sized red dots on the skin of the fruit. An apple blemished in this way is still as edible as any other apple but is unappealing to the consumer. The grower cannot sell such apples, and San Jose scale must therefore be controlled by spraying with a broad-spectrum insecticide at petal fall and then on a 10- to 14-day schedule until harvest. If the consumer could be educated to accept produce that is blemished but edible, the use of pesticides could be reduced.

Integrated Control

If any lesson has been learned during the DDT era, it is that insects are not passive recipients of control measures, and that any approach to pest management is subject to resistant reactions from the insect world. Therefore, many scientists now believe that unilateral attacks are not the answer at all; instead, that farmers should use as many control measures as possible in an integrated and well-planned manner. Integrated control will never be so conceptually or technically simple as straightforward chemical control can be. It will be slower acting, more labor-intensive, and more expensive initially, although experts hope that it will be more effective and cheaper in the long run.

Recall from Section 15.2 how the yields of cotton in a valley in Peru were reduced after several years of insecticide use. Following the crop failure, a new program was initiated. First, predacious and parasitic insects were imported from nearby valleys. Second, all cultivation of marginal soils was banned in an effort to cull out weak plants, which are often breeding centers for disease. Third, the cycles of planting and irrigation were adjusted to interrupt the life cycles of the pests most effectively. Last, all spraying with synthetic organic insecticides was banned except under the permission and supervision of a panel of scientists. The result was that, despite the fact that some land was purposely not

planted and very little insecticide was used, cotton yields were higher than had ever been recorded in the area.

15.6 Politics and Economics of Pesticides

There are four groups of people involved in
the pesticide controversy (Fig. 15–12): (1) Farmers want to grow as much food as possible so as
to earn a living. (2) Pesticide manufacturers
want to sell pesticides and realize a profit. (3)
Government agencies have the responsibility to
regulate and control dangerous chemicals. (4)
Consumers want a plentiful supply of food at
reasonable prices, but they want to avoid environmental pollution that endangers their health.

Any pesticide policy results from a balance of
these four viewpoints.

Think for a minute of an "ideal" pesticide.
Perhaps it would be a virus or bacterium that
would attack a specific insect pest. These microorganisms would infect only one pest species.
They would not attack nontarget species. No
foreign chemicals would be introduced into the
environment. Moreover, bacteria and viruses reproduce themselves. One spray application
would protect a field for years to come. With
these encouraging advantages, one may wonder
why insect enemies are not used more frequently.

One problem is that this type of control measure acts too slowly. A chemical pesticide kills
pests within a few hours. A disease organism requires weeks or months to act. Although effective pest control may eventually be realized, results cannot be expected immediately.

Another problem is that in most cases biological pest management initially requires more
labor and expertise than chemical control techniques. At the very least, some knowledgeable
person must walk through the fields, study the
insect life cycles, and decide when to spread the
biological agents. Bad timing may lead to failure.
For example, predators that are released when
there are few pests in the area may migrate to
other areas in search of food. Alternatively, a
parasite that thrives only on the larvae of a species may die out if released when the pest is in
the adult stage.

Even if a biological pest program is carefully
monitored, difficulties may arise. A farmer who
does try biological control may find that the expense and effort are lost when chemical sprays
from a nearby field blow across the unsprayed
area and kill the predator populations.

Many farmers operate with a tight yearly
budget. Most are unwilling or unable to gamble
a high expense in one year on the chance that it
will more than pay for itself in the future. This
reluctance is actually encouraged by the pesticide industry. An ideal pesticide program would
not be likely to be highly profitable to the manufacturer. Many millions of dollars are needed to
develop, produce, test, and market a new product. If the product can be used against many
pest species, the company can expect large sales.
On the other hand, if the insecticide is effective

Figure 15–12 There are four groups of people involved in the pesticide controversy. (1) Farmers want to grow as much food as possible to earn a living. (2) Pesticide manufacturers want to sell pesticides and realize a profit. (3) Government agencies have the responsibility to regulate and control dangerous chemicals. (Courtesy of U.S.D.A. Soil Conservation Service.) (4) Consumers want a plentiful supply of food at reasonable prices, but they want to avoid pollution that endangers their health.

against only one species of pests, sales will be much less, and profits will be lower. In addition, recall that parasites and disease organisms reproduce naturally. Therefore, a farmer can buy one application of the pesticide and guarantee protection for many years. Obviously, sales of a truly effective pest program will be much lower than sales of chemicals that must be used every year. Therefore, it is not surprising that industry has traditionally been more interested in broad-spectrum poisons than in biological control. Government support of truly effective pesticide measures would be a valuable environmental goal.

The bans on various pesticides have been both praised as farsighted and damned as shortsighted. One critic of the bans has been Dr. Norman E. Borlaug, the winner of the 1970 Nobel Peace Prize for his work in developing high-yield wheat strains. Borlaug has emphatically denied that such chemical pesticides significantly contribute to the deterioration of the environment. Furthermore, he fears that other pesticides will soon be banned, leading to failures in agriculture and possible mass starvation.

On the other hand, many scientists support the ban on DDT and other pesticides. They argue that any substances as biologically active, persistent, and foreign to natural food webs as the organochloride insecticides can be *presumed* to be harmful to human health. They point out that cigarette smoke and asbestos dust cause cancer several decades after exposure. Perhaps it would be slightly better to be cautious now than to risk a large-scale epidemic of cancer in the next decade. Supporters of the ban also note that so many insect species had developed genetic resistance to many chlorinated pesticides by the early 1970s that they were losing their usefulness anyway. In addition, government restrictions may serve as a drive to develop other, more environmentally sound methods of pest control.

15.7 Case History: The Boll Weevil

Many of the most serious agricultural pests are immigrants that have moved to a new and hospitable environment relatively free of the predators and pathogens native to their original home. The boll weevil (Fig. 15–13) is a small insect—about 0.65 cm (¼ in) long—that first migrated to the United States from Mexico about 1880. Weevils feed on the buds and bolls of young cotton plants, thereby destroying the valuable fiber. Since there are few natural enemies of the weevil in the United States, population blooms of these insects are common. If left uncontrolled, a light infestation of 10 weevils per hectare can increase to 10,000 weevils per hectare within a single growing season and completely consume an entire crop. The weevils' reproductive and eating capacity is so great that they are responsible for approximately $300 mil-

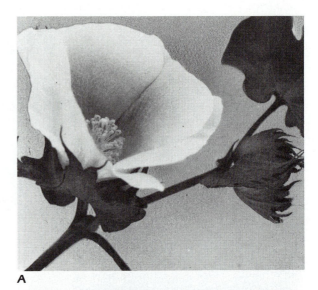

A

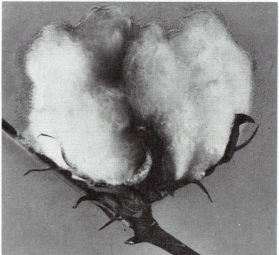

B

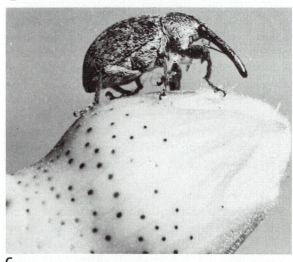

C

Figure 15–13 *A,* Cotton flower. *B,* Open cotton boll. *C, Boll weevil on a cotton plant. (A, B,* and *C* courtesy of U.S.D.A.)

BOX 15.2

One time I seen a Boll Weevil, he was settin' on a
    square.
Next time I seen the Boll Weevil, he had his whole darn
    family there;
He was lookin' for a home
He was lookin' for a home.

The farmer take the Boll Weevil, put him on the ice;
Boll Weevil said to the farmer, You's treatin' me mighty
    nice;
And I'll have a home.
And I'll have a home.

                                        Folk ballad

lion worth of crop losses in the United States annually.

In the early 1900s farmers used cultural controls to combat the weevils. Cotton was planted and harvested as early in the year as possible, and the plant stalks and leaves were burned immediately after harvest. If a large portion of the weevil population could be destroyed by fire, they reasoned, perhaps an early crop could be harvested the following year before the weevils could repopulate.

This nip and tuck race with growing weevil populations was often unsuccessful, and many crops were completely destroyed. In the 1930s a new pesticide, calcium arsenate, was introduced, but it was so highly toxic to nontarget species (including soil microorganisms) that it was largely replaced by DDT in the 1940s. At first DDT was extremely effective—so effective, in fact, that farmers largely abandoned the cultural controls that had been used earlier. However, by 1960, many DDT-resistant strains of boll weevil had evolved. Disaster threatened. In response, farmers did not return to cultural control techniques but instead switched to other chemical pesticides, the organophosphates.

Broad-spectrum pesticides were effective in partially controlling the population of weevils, but many secondary problems were raised. Cotton is not the only crop grown in the South. Tobacco, corn, peanuts, soybeans, and vegetables are also commercially important. The insect pests of many of these other major crops have traditionally been controlled by natural enemies. When cotton fields were sprayed heavily with broad-spectrum chemicals, adjacent regions were contaminated also, and many natural predators were killed. As a result, epidemics of tobacco and corn pests became a problem for the first time. During the 1970s, a new concern was raised. What would happen if the boll weevil became resistant to organophosphates just as it had become resistant to DDT?

Obviously, a new strategy was necessary. A Special Committee on Boll Weevil Eradication was formed with government support, and a five-pronged attack on this insect pest was outlined. Farmers were asked to (1) spray lightly with chemical pesticides during the growing season; (2) spray again in the fall just before the weevils hibernate for the winter; (3) destroy all plants, leaves, and stalks just after harvest to kill any remaining insects; (4) set traps in the early spring baited with a sex pheromone to catch and destroy males emerging from hibernation; and (5) release sterile male weevils to prevent fertile matings of any remaining females.

A pilot project was initiated in 1971 to determine the overall effectiveness of such a program. Some optimistic scientists even predicted that perhaps the boll weevil could be eradicated permanently. The project was effective in controlling weevil populations in the target area, but total annihilation was not realized. Research teams found that 95 percent control was feasible, but it is extremely difficult to destroy every weevil in a region. Despite an intense campaign, small pest populations survived in several safe havens. One farmer who was cheating on his cotton allotment and income taxes did not tell authorities about a hidden field. Weevils survived there and later migrated outward. Another farmer claimed that pesticides were killing his chickens and also refused to cooperate. Many owners of roadside stands and restaurants kept small plots of cotton to attract Yankee tourists, and for the most part these smalltime growers did not participate in the program, so more weevil breeding populations survived. Finally, wild cotton plants grow throughout the South, and many weevils breed in small cotton patches in the woods and swamps, far from agricultural areas.

There are two different interpretations of the pilot project. Many scientists argue that the eradication of a species of insect is a totally impracticable objective. Instead, the issue should be viewed as a cost/benefit question (see Chapter 2). If no controls are used, crop losses will be unacceptably high. By the same token, if an attempt is made to kill every weevil in a specific area, then the control measures become exorbitantly expensive. This logic leads to the conclusion that the real problem is to develop a balanced program that minimizes the total cost (crop losses plus control measures) to the farmer.

Other scientists have argued that with more money and effort the weevils could, in fact, be completely eradicated. Therefore, the extra expense of the program would more than pay for itself in the future, when no control costs or crop losses would burden the farmer. Government funding supported this group of experts and a second pilot project was started in 1978, but again, no unequivocal results were obtained. After an intensive control program was completed, a few weevils were still found in the region. Supporters of complete eradication claimed that the weevils found in the target area must have been immigrants from outlying areas and that the concept of annihilation is sound. Most biologists, however, believe that total destruction of the species is impossible and that the cost of such an attempt, estimated at between $240 million and $4.9 billion, would not be worth the money.

As the eradication study was being conducted by one team of scientists, a separate research group obtained data that not only might lead to effective boll weevil control, but also provide an interesting commentary on insect ecology. Fire ants are an insect pest that inflicts painful bites on people and live in many of the same regions that are infested by boll weevils. For years, the government has spent millions of dollars to rid the area of these annoying creatures. During the late 1970s, it was learned that fire ants are an effective predator of boll weevils. If both insects are allowed to live in the same region and no sprays are used at all, economic damage to the cotton crop remains low.

15.8 Herbicides

Have you ever grown a garden? If you have, you know how much work is involved in hoeing and picking weeds (Fig. 15–14). Yet if the weeds are not killed, they will choke out your flowers and vegetables and reduce yields. Farmers have battled weeds for as long as they have battled insect pests. People have known for centuries that certain chemicals may kill plants. The challenge in modern times is to develop chemicals that will kill weeds but not valuable crops. Such formulations are called **herbicides**.

Figure 15–14 Many workers with hoes are needed to clear a field of weeds, whereas it is much cheaper to spread herbicides. But chemicals pollute the environment.

Herbicide development and production has been quite successful in recent years. A hired laborer spends an average of 10 to 25 hours hoeing the weeds between the rows of 1 hectare of cotton. Today an airplane can spray hundreds of hectares in a few hours for $5 to $15 per hectare. Agriculture in North America and Europe depends on small inputs of labor and a heavy use of machinery and chemicals. Certainly herbicides are an important part of such a program. As a result, in 1978 approximately 400 million kilograms of herbicides were manufactured in the United States.

Whenever large quantities of synthetic chemicals are spread throughout the environment, serious ecological problems are likely to arise. Although herbicides are not nearly as persistent in the soil as organochloride insecticides are, they pose difficulties nevertheless. Some herbicides kill earthworms and other organisms that contribute to a healthy soil. Others have been known to kill certain insect predators. Of course, if the predators are killed, pest populations will rise again. Then, farmers are likely to use more pesticides along with the herbicides, resulting in a spiraling pollution problem.

The herbicide 2,4,5-T has been used for a variety of agricultural applications. It was also an ingredient in Agent Orange, a defoliant used in Southeast Asia during the Vietnam War. After the war, the herbicide was used by the United States Forest Service to kill brush that interferes with the growth of valuable timber. 2,4,5-T is poisonous to humans. Perhaps even more frightening is the fact that commercial preparations of this herbicide are contaminated with minute quantities of a chemical called **dioxin.** Dioxin is one of the most potent poisons known, and concentrations in the parts per billion range have been known to kill laboratory animals. In addition, this chemical is suspected as a cause of birth defects and cancer. Vietnam War veterans have claimed that exposure to Agent Orange has led to a variety of adverse effects, including muscular weakness, loss of sex drive, numbness, liver damage, and loss of sleep. When the government sprayed forests in Oregon, reports indicated that an abnormally high number of women in the region had miscarriages within a few months after the spray application. No one denies that 2,4,5-T in large doses is an acute poi-

son, but all other allegations about its chronic low-dose effects have been disputed by the chemical manufacturing industry. The battle between individuals who believe that they were harmed by commercial or military spray applications and the government and herbicide manufacturers has been long and bitter. Studies have been made and refuted, lawsuits have been filed and fought. Some scientific data presented on both sides have been discredited as false or misleading. In 1979 the Evironmental Protection Agency issued an emergency order banning domestic use of 2,4,5-T. This order has been challenged, and the court battles continue.

Agricultural scientists are faced with a dilemma. If herbicides are banned, the price of food will rise sharply. Farmers might not be able to find the laborers needed to work long hard hours in the fields picking weeds. Then food production would decrease. Therefore, the use of herbicides continues. However, chemical pollution of the environment is also increasing, and no one knows how this pollution is affecting human bodies and world ecosystems.

Summary

There are three major classifications of chemical pesticides—organochlorides, organophosphates, and carbamates. The use of DDT, an organochloride, led at first to some dramatic successes, but now the chemical is banned in many areas.

Broad-spectrum pesticides kill predators as well as pest species and thus disturb natural controls. In some cases, pest species have evolved to be resistant to the poisons and various predators have not. Then the pest populations have bloomed.

DDT and other organochloride insecticides were banned because (1) They are not biodegradable. (2) They are easily transported throughout the environment. (3) They are concentrated in biological systems and poison many nontarget species.

After a decade and a half of study, proof for or against the carcinogenicity of pesticides in humans is still lacking, but the very fact that these materials are potent poisons makes caution advisable.

Other methods of pest control include (1) natural enemies such as predators, parasites, and disease organisms; (2) sterile animals (usually males); (3) hormones; (4) sex attractants; and (5) resistant strains of crops. The most effective controls utilize an integrated approach.

Species-specific biological control methods are environmentally sound but are not highly profitable to the manufacturer. Farmers are wary of them because their action is slower than that of chemical pesticides. Pesticide bans have been both praised as farsighted and damned as nearsighted.

During the past ten years, the use of herbicides has become increasingly popular. Some commonly available herbicides are known to cause birth defects. Others are suspected carcinogens. Weed control is necessary in agriculture, but indiscriminate use of herbicides may lead to ecological imbalance.

Questions

Effects of Chemical Pesticides

1. What are the harmful effects and the benefits that insects bring to people? How did people cope with insect problems before the introduction of modern insecticides?

2. An amateur ecologist studying wildlife populations before and after a heavy spray application determined that since no animals were directly killed by the spray, no harm had resulted. Would you agree? Defend your answer.

3. Do you think that some bird species might become resistant to DDT? Would you think that resistance might save the birds from extinction? Defend your answers.

4. Imagine that there is an impending malaria epidemic in a region and that an aerial spray program using DDT would be likely to kill the mosquitos and avert the epidemic. Proponents of DDT may argue that we know that malaria debilitates and kills many people. Furthermore, since the chronic effects of DDT on human populations are not known, it would be wise to protect people against malaria by making use of the pesticide. Opponents of the program, on the other hand, might argue that DDT is a potent environmental poison and should not be used under any circumstances. Discuss your feelings on the subject.

5. Explain how one indiscriminate spraying with DDT could destroy the effectiveness of an integrated control program.

Transport and Dispersal of Pesticides

6. Describe three ways in which DDT is transported through the environment by natural means. Why is an organophosphate less likely to be transported?

7. Explain why carnivorous fish are generally more susceptible to low levels of pesticides in the water than are herbivorous fish.

8. If no measurable quantities of DDT were found in the water of a pond, would that necessarily mean that the aquatic ecosystem was unpolluted by DDT? Explain.

Other Methods of Pest Control

9. Explain why it is important to spray pests with hormones at a precise time. If a spray application were successful on May 1 of one year, would it be safe to assume that farmers could spray successfully on that date every year?

10. Birds often become major pests in vineyards because they eat the grapes.(1) Which of the following control programs would you recommend for bird control: (a) spreading poison; (b) broadcasting noise from a loudspeaker system to scare them; (c) shooting; (d) covering the vineyard with some fencing material? Defend your choices. (2) Do you think that it might be wise to initiate research directed toward (a) developing a sterilization program against the birds, or (b) developing new strains of grape that would be unpalatable or poisonous to birds? Explain.

Health

11. Prepare a class debate. Have one side argue that a pesticide should not be banned until

it is proved to be carcinogenic. The other side should argue that even a suspected carcinogen should be banned.

Herbicides

12. Ragweed is considered to be a major plant pest because its pollen causes misery to hay-fever victims. In natural systems ragweed is characterized as an early successional plant. (An early successional plant is one that grows in areas that have been stripped of vegetation.) A few years ago, the state of New Jersey initiated a program to eliminate ragweed. Thousands of acres of roadways and old fields were sprayed with herbicides. Why do you think this program failed?

Economics and Politics of Pesticides

13. Explain how some peoples' jobs (for example, that of a crop duster pilot, a pesticide salesperson, a government inspector, or a farmer) may affect their opinions on pesticide use.
14. Discuss the statement: "Pest control techniques that are truly effective would be a disaster for the pesticide industry."
15. Sometimes an integrated control program may not become effective until the second season of its application. Moreover, crop losses during the first season may actually be greater than average. Do you feel that it would be good policy for the government to subsidize such losses in an effort to improve environmental quality? Defend your answer.
16. A worm that thrives in the core of an apple does not usually eat a large proportion of the apple, but people do not like wormy apples anyway. Discuss the social and economic factors influencing the choice between wormy apples and apples that might contain pesticide residues.

Suggested Readings

The book that started much of our current concern about pesticides, and a more recent sequel to it, are the following:
Rachel Carson: *Silent Spring*. Boston, Houghton Mifflin Co., 1962. 368 pp.
Frank Graham, Jr.: *Since Silent Spring*. Boston, Houghton Mifflin Co., 1970. 333 pp.

Several general references on pesticides include:
A. W. Brown: *Ecology of Pesticides*. New York, John Wiley & Sons, 1978. 523 pp.
W. W. Fletcher: *The Pest War*. New York, John Wiley & Sons, 1974. 218 pp.
Rizwanul Haque, and V. H. Freed (Eds.): *Environmental Dynamics of Pesticides*. New York, Plenum Press, 1975. 387 pp.
C. B. Huffaker, (Ed.): *Biological Control*. New York, Plenum Press, 1971. 511 pp.
David Irvine, and Brian Knights, (Eds.): *Pollution and the Use of Chemicals in Agriculture*. Ann Arbor, Mich., Ann Arbor Science Publishers, 1974. 136 pp.
W. W. Kilgore and R. L. Doutt (Eds.): *Pest Control*. New York, Academic Press, 1967. 477 pp.
Fumio Matsumura: *Toxicology of Insecticides*. New York, Plenum Press, 1975. 503 pp.
Robert L. Metcalf and William H. Luckman, (Eds.): *Introduction to Insect Pest Management*. New York, John Wiley & Sons, 1975. 587 pp.
David Pimentel: *Ecological Effects of Pesticides on Nontarget Species*. Washington, D.C., U.S. Government Printing Office, 1971. 219 pp.
U.S. Department of Health, Education, and Welfare: *Report of the Secretary's Commission on Pesticides and Their Relationship to Environmental Health*. Washington, D.C., 1969. 677 pp.
John H. Perkins: *Insects, Experts, and the Insecticide Crisis*. New York, Plenum Press, 1982. 304 pp.

16

Food Production and World Hunger

Well-watered, fertile farmland can be quite productive. An intensively cultivated garden that is only 0.07 hectares (one sixth of an acre) can supply a family of four with plentiful amounts of green vegetables for a year. Yet for many people in the world, green vegetables are a rare luxury; in fact, even the basic grains necessary for the barest existence are often not available.

It is difficult for people in the developed countries to appreciate the despair of not having enough to eat. Yet reports show that in the world today, 400 million people—nearly twice the population of the United States—are starving. A reporter in a refugee camp in East Africa recently wrote,

> Once seen, it can never be forgotten: more than 70,000 people, 90% of them women and children, clustered together on a barren hillside, their only shelter small huts made of thorn bush branches, animal skins, and pieces of cloth. No one in the camp had received food rations in two days, and it was uncertain when the next food supply truck would arrive . . .

That reporter was observing only one refugee camp of many in Africa, and people are hungry in many other parts of the world as well. In this chapter the problem of world hunger will be examined and possible solutions investigated.

16.1 Human Nutritional Needs

Humans, as well as other animals, need food to serve three functions:

1. Food provides energy for maintaining body temperature and for doing work. The amount of energy needed depends on the outside temperature and on the rate at which the individual works. On the average, an adult needs about 2500 to 3000 kilocalories of food energy every day. (Remember, 1 Calorie, when spelled with a capital C, equals 1 kilocalorie.)

2. Food provides essential materials for growth, for replacement of body tissues that are lost as wastes or in injury, for development of the fetus in a pregnant woman, and for production of milk for breast feeding.

3. Food provides materials that are needed for the regulation of life processes such as transport

338

of oxygen in the blood and maintenance of the nervous and digestive systems.

The first requirement, energy, is met mainly by intake of carbohydrates and fats. These nutrients contain only carbon, hydrogen, and oxygen. Carbohydrates are sugars and starches and occur mainly in fruits, cereals, vegetables, nuts, and milk products. Fats (which include oils) are widely distributed in both plant and animal foods. In addition to supplying energy, fats are stored in the body as a reserve supply of energy and also are essential constituents of nerves and of cell membranes.

The second requirement, the need for material for growth and repair of body tissues, is supplied mainly by proteins. Proteins are very large molecules made up of combinations of smaller units called **amino acids**. There are about 20 natural amino acids, 8 of which cannot be synthesized by the human body and must therefore be obtained from the diet. Two more are produced only meagerly by the body, especially in children, and should also be obtained from food. The 8 (or 10) amino acids that are thus essential to the human diet are called **essential amino acids**. All amino acids contain carbon, oxygen, hydrogen, and nitrogen, and two amino acids also contain sulfur. Proteins occur in both plant and animal foods but in widely varying concentrations. Proteins also differ in their *quality*, which means that they differ in their content of essential amino acids. An **incomplete protein** is one that does not contain all the essential amino acids; a **complete protein** is one that does. Gelatin is an example of an incomplete protein; if gelatin were the only protein in a human diet, it could not support life. An ideal protein contains all the essential amino acids in the proportions needed by the body. A mother's milk provides ideal protein for her infant. Eggs provide protein of very high quality. However, different incomplete proteins can be combined to make a complete protein meal. Examples are found in various traditional ethnic food combinations, usually containing beans combined with either nuts, rice, or bread.

The third requirement, the need for materials for regulation of life processes, is supplied by a large number of nutrients that are needed in small, sometimes even trace quantities. These nutrients include vitamins and minerals, which are distributed very unevenly in a wide variety of foods. A good diet should therefore be a varied one, and should include foods of different types (proteins, grains, dairy foods, vegetables, and fruits) as well as foods from different geographical areas in which the soils have different mineral contents.

The opposite extreme of a plentiful, nutritious diet is **starvation**, which means death from lack of food. However, most people who are inadequately fed do not actually die because they take in too few calories to sustain life. Instead, because the food is deficient in essential amino acids and vitamins, they are malnourished, and death occurs because they have little resistance to diseases that would not be fatal to the properly fed.

Many diseases may result if a person is not fed an adequate diet. Protein deficiency diseases such as **kwashiorkor** are particularly damaging; they lead to mental retardation, particularly when they occur in young children. Vitamin and mineral deficiencies can lead to weak bones, loss of teeth, blindness, or failure of any of a number of vital organs. Children who do not receive either sufficient protein or calories develop characteristic bloated bellies, thin arms and legs, wide eyes, and shriveled skin (Fig. 16–1). Perhaps even more sinister is the fact that severe malnutrition in young people leads to early and irreversible brain damage. This results in a negative feedback cycle, for if undernourished and retarded children do survive to become adults, they will have decreased learning ability. Therefore, when they grow up, they will be likely to have a hard time finding work, and if work is found it is often of the kind that pays the least money. When these impoverished adults in turn have children, their young are likely to be undernourished as well, thereby perpetuating the tragic cycle.

Other diseases caused by nutritional deficiency are common throughout the world. By one estimate, a quarter of a million children become permanently blind every year because their diets are deficient in Vitamin A. Another 200,000 people per year become deaf owing to a lack of iodine. An additional uncounted

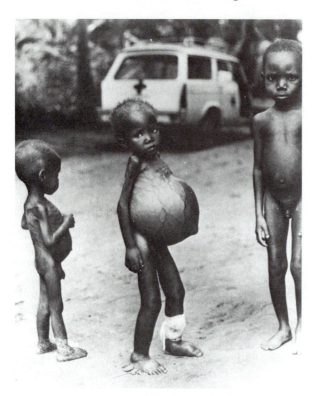

Figure 16–1 These starving children in Nigeria during the Biafran war show the advanced stages of emaciated limbs and bloated bellies characteristic of severe malnutrition. (World Wide Photos.)

number of individuals die of infectious diseases because their bodies and immune systems have been weakened by hunger and lack of proper nutrients. All told, some 15 million people starve to death or die indirectly from malnutrition every year.

16.2 World Food Supply—An Overview

Figure 16–2 shows that plant products make up about 90 percent of the world's agricultural output. Plants are so important in modern agriculture because people feed themselves most efficiently when they eat plants or plant parts. From 30 to 40 percent of a plant's net primary productivity can usually be harvested, and 70 to 80 percent of such a harvest can be digested by human beings.

In many ways, grains—often called cereals—are the ideal food staple. They contain about 10 percent protein, which is sufficient for an adult human, and they produce the highest yield of carbohydrates per hectare of farmland. As a result, of all the plants commercially cultivated, three grains—wheat, corn, and rice—account for about 57 percent of the total production. Despite their value, a diet of grains alone is deficient in several essential vitamins and contains insufficient proteins for infants and young children.

Eating animals is much less efficient than eating plants. The animals must eat plants first, and animals convert only about 15 percent of their food calories into calories that are available to their predators (see Chapter 3). In no case is the efficiency of conversion higher than the 25 percent attained in milk and egg products. As societies become wealthier, their consumption of animal products increases. This makes it even more difficult for food production to surpass population growth as the number (if not the proportion) of people with a reasonable income increases. On the other hand, since some grazing land is unsuitable for growing plant crops, growing meat in these areas is the most effective way to use this land to increase the human food supply.

The total global grain production as well as the grain production per person from 1950 to 1980 is shown in Figure 16–3. Note that from 1950 to 1971 both curves showed a steady increase. This continued upward trend was halted in 1972. In that year a serious drought struck India, Pakistan, and the grazing lands just south of the Sahara Desert. At the same time, there was a poor harvest in the wheat fields of the U.S.S.R., and for the first time in 25 years famine on a global scale became a reality. Since then, worldwide food production and population growth have been in a nip-and-tuck race. In some years, good harvests in key grain producing areas resulted in increased availability of food, whereas years of drought, flood, or crop failure brought local or even global scarcity.

Looking over at Figure 16–3 you can see that from 1970 to 1980 the per capita production of grain fluctuated widely—up some years and down others. (Note that the curve is smooth up to 1970 only because the figures are given only

Annual production (millions of tonnes)

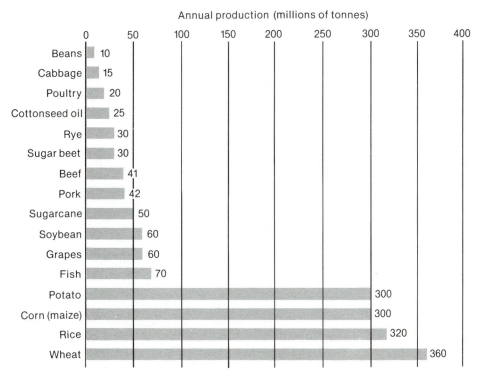

Figure 16–2 *Annual world production of major food sources. Note the overwhelming importance of the major cereals.*

for every tenth year—1950, 1960, and 1970; the year-to-year fluctuations in this interval are not shown.) All of these figures, however, are *global* averages. Even in years when global food pro-

duction rose, crops may have failed in some areas or even in some entire continents. After the poor harvests of 1972, there was a sudden demand for surplus grain. Within a matter of a

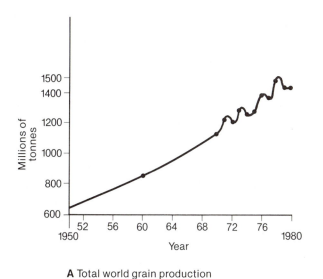

A Total world grain production

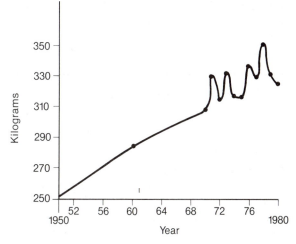

B Per capita grain production

Figure 16–3 *World grain production, 1950 to 1980.*

few months, the price of wheat doubled. As a result, the poorest people could no longer afford to eat, and many millions suffered severe malnutrition or starved to death (Figure 16–4).

In recent years, there has been an increasingly intimate relationship between energy and agriculture. In nations where energy is readily available, it is possible to manufacture fertilizers and pesticides at an affordable price, and farmers can produce high yields. On the other hand, farmers in less wealthy nations cannot afford these inputs, and raise less food per hectare. The difference is substantial. In the 1980s agricultural yields in the developed nations were an astounding 70 percent higher per hectare than yields in the less developed countries. As the gap between the rich and the poor continues to widen, the only regions *in the world* that are self-sufficient with respect to food production are those with ample money and a low population density. As shown in Table 16–1, the only significant food-exporting regions in 1980 were in North America, Australia, and New Zealand. The people in every other region survived on net imports. What a vulnerable system! Think of what would happen to people all over the world if a drought crippled the wheat crop in North America for even one year.

At the present time, the human population is increasing faster than its food supply. Thus, if current trends continue, the outlook for the human condition is indeed bleak. Perhaps the central question is whether the future will be a simple extrapolation of the past decade or whether radical changes are likely to occur.

Figure 16–4 A hungry girl searches through a pile of garbage for something to eat in a slum in New Delhi.

16.3 Food Production, Land, Weather, and Climate

As described in Chapter 14, productive agricultural land is being destroyed at an alarming rate throughout the world, even as populations continue to expand. Additional problems arise because farmers have always been at the mercy

TABLE 16–1   The Changing Pattern of World Grain Trade

Region	1934 to 1938	1948 to 1952	1960	1970	1980
North America	+5	+23	+39	+56	+131
Latin America	+9	+1	0	+4	−10
Western Europe	−24	−22	−25	−30	−16
Eastern Europe and Soviet Union	+5	0	0	0	−46
Africa	+1	0	−2	−5	−15
Asia	+2	−6	−17	−37	−63
Australia and New Zealand	+3	+3	+6	+12	+19

The results are expressed as millions of tonnes of grain; plus signs indicate net exports; minus signs, net imports.

of the weather. If temperature and rainfall are favorable, food will often be plentiful, but frost, flood, or drought may lead to crop failures. In one sense, people today are less vulnerable to changes in the weather than they were 100 years ago. Rapid communication and transportation systems make it possible to transport food large distances across the globe. Thus, for example, a poor harvest in the U.S.S.R. can be counterbalanced by a bumper crop in North America simply by shipping food from one continent to the other. In this way trade can minimize the effect of local disasters.

In another and perhaps larger sense, however, people are much more vulnerable than they ever were before. In the past, famine was often alleviated by migration. As long as the human population was low, farmers could often offset crop failures by moving to a new area and clearing previously virgin land. Today, this option is rapidly drawing to a close.

From the beginning of agriculture until about 1950, most increases in worldwide food production resulted from an expansion of the total area of farmland under cultivation. By 1950, nearly all of the naturally highly productive farming area in the world had been cleared, plowed, and planted. Therefore, in recent years,

the new farms that have been started have been located in the Arctic, in jungle regions, in deserts, and in alpine environments (Figure 16–5). Although these farms can be highly productive in good years, they are potentially vulnerable in bad ones. For example, if the temperature in Kansas falls by a degree or two, the wheat crop will probably not be seriously affected. But a small drop in temperature on a barley farm near Fairbanks, Alaska, may mean that the last spring frost arrives two weeks later than usual and the first killing frost in the fall is early enough to severely stunt the growth of the crop. Similarly, a 10 percent decrease in rainfall won't spell disaster for a pea farmer on the rainy northwest coast of the state of Washington, but it could lead to starvation in the Sahel region of North Africa, which is near-desert even in the best of times.

The problem that causes great concern is this: Historical records show that the 1950s and 1960s were years of unusually plentiful rainfall and mild climate. During that time period, food production and population were pushed to the limit. If the climate should change adversely, then farms in many regions of the world could fail, and famine would spread. Therefore, it is natural to ask whether the climate will change, and if so, whether it will change for the better or

A   B

Figure 16–5 Many new farms are being started in areas that were once considered too harsh to farm. These can be productive in good years but are extremely vulnerable to small changes in weather and climate. A, This gladioli plantation is being grown on land reclaimed from the Negev desert in Israel. (Courtesy of Israel Government Tourist Office.) B, Farming a steep mountainside in the Andes in Peru.

for the worse. Of course, no one knows the answer. It is difficult enough to predict what the weather will be next week, let alone to project temperature and rainfall patterns 5, 10, or 25 years from now. Scientists do know that climates have changed radically during the geological history of the Earth. For example, within the past 100 million years the area that now includes the fertile wheat fields of Colorado have been alternately covered by swamps, thick glaciers, and dry deserts. This historical information reminds us that global climates do change, but it doesn't provide information useful for forecasting future trends. In the late 1970s many climatologists believed that the Earth was cooling. In the early 1980s there was increased speculation that a warming trend would occur within a decade. This prediction is based on the fact that carbon dioxide released when fossil fuels are burned absorbs infrared radiation and warms the air (see Chapter 21).

16.4 Energy and Agriculture

A wild oat plant in an unfarmed prairie and a domestic oat plant in a farmer's field live in very different environments. To survive, the wild oat must compete successfully with its neighbors for sunlight, moisture, and soil nutrients (Fig. 16–6). A plant that is tall, that sprouts early, or that has an effective root system has a competitive advantage. The energy a wild plant needs to grow a tall stalk or a deep root comes only from the Sun. On the other hand, a cultivated oat is supplied with external aid to help it survive. The farmer waters it when necessary, removes competitive plants (weeds), and loosens the soil to stimulate the growth of root systems. Since all the seeds in the field are planted at the same time and are of the same variety, and unwanted weeds are removed, competition is minimal. The plant does not need as tall a stalk or as specialized and fast-growing a root system as a wild plant, and there is less competition for available nutrients. These advantages are not free. A farmer adds auxiliary sources of energy (in the form of fertilizers, pesticides, and so on) to encourage the growth of those plants that provide abundant food for human consumption.

In the poorest farming communities in the world, energy for working in the fields is derived directly from human labor. More fortunate farmers use various types of animals to plow, to plant, or to transport supplies. Modern industrial agricultural systems use large quantities of fossil fuels.

In a natural system, grazing animals wander around a prairie eating indigenous plants. For

Figure 16–6 Plants compete for space, light, water, and nutrients in a natural grassland.

every 10 calories of energy a plant receives from the Sun in such a natural system, however, only about 1 calorie is available for the production of animal tissue by a herbivore that consumes the plant. Some of the "lost" calories are required by the plant for its metabolism and some are needed by the animal for its metabolism, and to search for more food. By modern standards, this type of system is uneconomical.

Modern mechanized production of meat operates very differently. The growth of grasses and grains is accelerated by the addition of fertilizers and pesticides. These plants are then fed to a steer in a feedlot (Fig. 16–7). An animal in a modern feedlot need not move but may stand in front of a feedbin, eating the hay or grain that the farmer, with the help of a tractor, baled or threshed and brought in from the field. In addition, food additives and growth hormones, synthesized in factories powered by coal or oil, help to increase growth. In a feedlot, about 5 calories of fossil fuel energy are used for every 10 calories that the plant matter receives from the Sun. Plant yields are increased, more food is available to the cattle, and the energy available to the animal under these conditions corresponds to

about 4 calories (Fig. 16–8). However, the feedlot requires large quantities of fossil fuels.

Mechanized food production is based on a technological cycle. High crop yields are needed to feed urban workers, who in turn provide the technology required to maintain the high food production. The individual components of the cycle are interdependent. Large populations of human beings depend on high agricultural yields, while at the same time these high yields depend on high industrial outputs. If the cycle is broken, farmers will be forced to revert to more primitive forms of agriculture. This may lead to lower yields, a lower carrying capacity for humans, and ultimately, mass starvation.

What is our agricultural future? In 1981 there were about 4.5 billion people in the world. Some scientists have suggested that if modern agricultural techniques were widely available and food distribution were more equitable, our planet could easily support 10 billion inhabitants or more. But others disagree. They say that when fuel supplies become limited, agricultural productivity will decrease; the real carrying capacity of the Earth, according to them, is closer to 2 billion, only one half the number that are alive today.

Chemical Inputs

Chemical fertilizers and pesticides are widely used inputs in modern mechanized agriculture. There is very little unfarmed but potentially prime agricultural land left on the Earth today. The unfarmed land that exists is mostly marginally productive. On the other hand, much existing farmland is being either depleted by erosion or eliminated entirely by urbanization. Therefore, in recent years, the major solution to the problem of feeding the world's people has been to increase productivity on existing farmland. Between 1950 and 1971, 80 percent of the increase in food production was realized through increasing productivity per hectare. These gains have been realized largely through addition of fertilizers and water (Figs. 16–9 and 16–10). Thus, humans have, in effect, traded fossil fuels for food.

What will happen when the oil runs out? The worst case imaginable is one in which there

Figure 16–7 Cattle in a feedlot.

Figure 16–8 Comparison between meat production on the range and in a feedlot. The feedlot system produces more meat for humans but consumes fossil fuel energy.

is very little fuel available of any kind. Then potential fertilizers for next year's crop, such as manure or crop residues, will be used as fuel to cook this year's food. Today hundreds of millions of people are caught in this type of downward spiral, which results in increasingly severe soil depletion leading to deeper poverty and to continued exhaustion of the soil. For example, in Nepal the dung used for fuel every year repre-

sents a nutrient loss severe enough to reduce grain output by 25 percent by the year 2000. In a more hopeful scenario, alternative energy sources may be used to manufacture fertilizers or to cook food and to heat homes so that agricultural wastes can be used to improve the quality of the soil.

Pesticides and herbicides are another form of chemical input, and these too require energy to

Figure 16–9 A truck is loaded and ready to spread liquid fertilizer on a field in the Sacramento Valley, California.

manufacture and disperse. Efficient integrated pest programs would reduce this energy requirement.

Mechanization

High agricultural yields require specific varieties of seeds, an active fertilizer program, and pest control. In many regions, irrigation is also essential to high yields. In some places, irrigation systems are designed so that water flows downhill from the river into the fields, but often energy supplied by people, animals, or engines

Millions of tonnes

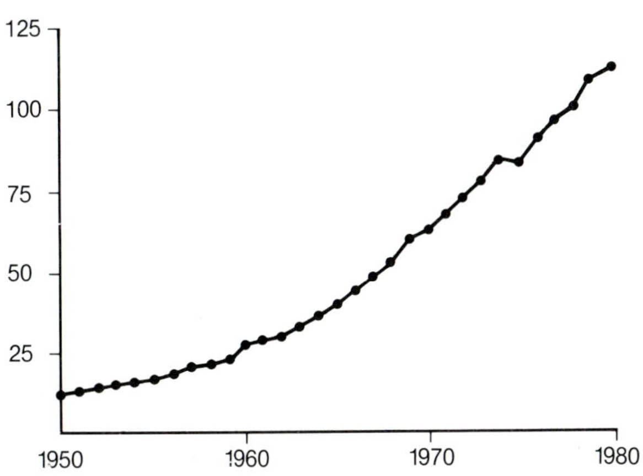

Figure 16–10 World fertilizer consumption, 1950 to 1980. The steady increase shows that our food production systems are highly dependent on auxiliary energy supplies.

is required to pump water from one place to another. On the other hand, the use of tractors does not improve the yield per hectare. On the contrary, yields are often *less* in mechanized agriculture than they are in high intensity nonmechanized systems. The reasons for this reduction are twofold. For one, in many cases crops must be planted farther apart if machinery is to be driven between the rows than if the field is worked by hand. Thus, crop density will be less and consequently the yield will be smaller. Secondly, machinery is not capable of handling food as precisely as a human laborer can. This difference is especially critical for fruits and vegetables that are easily damaged. For example, a mechanical tomato picker moves through a field once, picks the fruit before it is ripe and destroys the plants. (Ripe red tomatoes would be bruised by the machine; the green tomatoes are ripened with ethylene gas after picking.) If the crop is harvested manually, laborers can move through the field several times, picking only the fruit that is ready at the moment. The plants are thus allowed to keep producing for a longer time period, and more tomatoes can be harvested. Or, as another example, there are machines that shake apples from trees onto collecting frames, but more apples are bruised with this method than when picking is done by hand.

Despite these disadvantages, nearly all commercial farmers in the developed countries rely heavily on the use of machinery. The reason for this choice is that in regions in which fuel and

Figure 16–11 *Mechanization in agriculture. Left,* Combine. *Right,* Tomato picker. (Photos courtesy of Rorer-Amchem, Inc., Ambler, Pennsylvania.)

equipment are relatively inexpensive, the use of machinery saves money. If farmers were to revert to labor-intensive practices, the price of food would rise significantly. In addition, farmers using tractors can cultivate or harvest large tracts of land quickly and thereby take advantage of short periods of favorable weather.

The advantages and disadvantages of mechanization can be illustrated by comparing farming practices in North America with those in Japan. In North America, where tractors are used extensively, consumers spend a smaller proportion of their income on food, on the average, than any other people in the world. In addition, large quantities of grains are raised for export. Although this abundance is caused by many different factors acting in concert, the inexpensive availability of machinery certainly contributes. In contrast, in Japan, many farmers own small, two-wheeled rotary tillers, but very few own four-wheeled tractors, and heavy farm machinery is virtually unknown. Even though the use of heavy equipment is minimal, large quantities of fertilizer and much manual labor are used. Crop density is high, fields are cared for intensively, and, as a result, the average yield in Japanese agriculture is higher than that in North America.

Energy Use in Food Processing and Transportation

In its simplest form, energy for food processing involves cooking a pot of rice in a simple stone fireplace. In many less-developed countries even the small quantity of fuel needed for this task is not always available.

Table 16.2 shows the energy used in food processing from farm to table in the United States. Note some interesting facts. Over twice as much energy is used to package food products as is used to manufacture fertilizers. Such a large allocation of resources for packaging is unique in the history of human food production

Figure 16–12 *Using animal power to harvest wheat in India.* (Courtesy of the Agency for International Development.)

An example of extreme energy consumption for food production is processed cereal or snack foods made from rice grown in the San Joaquin Valley in California. The rice paddies are first prepared using heavy tractors and grading equipment. Maximum efficiency is realized if the field is absolutely flat. Therefore, the most mechanized farms use a laser-computer leveling system. A rotating laser transmitter is first established in the center of the field. This device beams an electronic reference elevation. A tractor then drags a leveling plow with a laser receiver mounted above it. A small computer on the tractor determines whether the elevation of the field at any point is above or below the required reference level. The computer then operates a motor which raises or lowers the leveling blades behind the tractor. Fertilizer is applied heavily. Water, piped in from hundreds of kilometers away, is used to flood the fields. Rice seeds, previously treated with chemicals to control seedling diseases, are spread by airplane. Further aerial spraying of pesticides and herbicides protects the crop, which is finally harvested by large combines.

Growing and harvesting is only a first step. The rice is then cleaned, milled to remove the nutritious outer hull, refortified with vitamins to replace some of the removed nutrients, and then puffed, blown, cooked, packaged, and transported to produce cold cereal products or snack foods available at a supermarket shelf. The efficiency of the entire process depends on how you look at it. In terms of total yield (kilogram of grain per hectare of land), U.S. rice culture is nearly twice as efficient as the world average. But in terms of energy input to energy output, the system is woefully inefficient. In the entire process, approximately five calories of fossil fuel energy are needed to produce one calorie of food energy.

TABLE 16–2 Use of Energy from Farm to Table in the United States

Use	Energy Consumed (trillions of kcal)	
On farm		
Fuel for tractors	232	
Electricity	64	
Energy to manufacture fertilizers	94	
Energy to manufacture farm machinery	101	
Irrigation	35	
	526	Subtotal
Processing industry		
Food processing industry	314	
Packaging industry	207	
Transportation	321	
	842	Subtotal
Commercial and home		
Refrigeration and cooking	804	
Grand total	2172	

the land was developed for residential or commercial use. In the early 1980s, farming was still more efficient in centralized agricultural systems, but in many cases the cost of energy needed to transport the goods had shifted the overall balance back in favor of the neighborhood farms. Unfortunately, many of these farms are now covered with concrete or asphalt.

This economy of local production has not been overlooked by individuals, and millions of families have converted portions of suburban lawns or rural backyards to vegetable gardens.

and presents an attractive target for energy conservation.

Another large allocation is the energy used to transport food from the farm to the market, which is more than the energy needed to fuel the tractors used directly on the farm. This circumstance is a result of various economic factors. During the 1960s and 1970s, there was an increasing trend in the United States toward centralization of farming practices. For example, economy of scale made it more profitable to raise milk on a large corporate farm in Wisconsin and ship it to Massachusetts than to operate a small dairy farm in a Massachusetts community. As a result, many local farmers sold their farms, and

BOX 16.2 HOW TO REDUCE ENERGY CONSUMPTION IN AGRICULTURE

1. Use less mechanization, more manual labor.
2. Use fewer commercial fertilizers, more recycled waste.
3. Use fewer pesticides and herbicides, more integrated management.
4. Use less packaging.
5. Locate farms closer to urban areas.
6. Improve the dietary efficiency of individuals, reducing overconsumption.

Note: These suggestions are easy to propose; to implement them would require a significant readjustment of many segments of society. They are not "answers" for tomorrow.

16.5 Social and Political Barriers to an Adequate Diet

In 1980, world grain production amounted to about 0.9 kg per person per day. If distributed evenly, this would provide every person in the world with the caloric intake of the average American adult. In addition, an uncounted number of cattle, sheep, and goats were grown on marginal range lands, and millions of tonnes of vegetables and fruits were produced as well. Faced with these statistics, one must ask why so much hunger and starvation still exist.

The question is easy to answer, but the problem is extremely difficult to solve. There is an unequal distribution of resources. This unequal distribution may take several forms.

(a) **Exploitation.** In many cases, the rich simply exploit the poor, and extreme poverty and extreme wealth are often very close neighbors. Some examples follow:

● In 1974 floods destroyed many rice fields in Bangladesh. Millions of poor farming families starved. Yet in other parts of the same country harvests were good. Approximately 4 million tonnes of rice were stockpiled in the central cities. The rice did not reach the hungry because they were "too poor to buy it."

● In 1978 a large American food corporation owned 23,000 hectares (57,000 acres) of prime farmland in Guatemala. Of this land, only 3650 hectares (9000 acres) were planted. The rest was left unused. Yet thousands of peasants went hungry because they could not grow the food they needed. Even people who worked on the plantation barely had enough to eat. Pay was low, and most of the food grown was exported.

● Great famines took a large toll of human lives in Africa in the early 1970s. Yet, throughout the famine years large quantities of cotton were grown in the drought area and exported to Europe.

● In 1973 and 1974 fuel shortages existed throughout the world. As a result, fertilizer shortages became acute. Most developed countries responded by restricting fertilizer exports. Wealthy people could buy fertilizers for their ornamental lawns and gardens, but many poor farmers could not purchase the fertilizer they needed to feed their families. By one estimate, the fertilizers used for ornamental purposes in the United States could have increased crop production in the less developed countries by enough to feed 65 million people.

● In many places around the world starving people live near large plantations of coffee, tea, and tobacco. In other situations food crops such as bananas are exported to foreign countries while local peasants again have little to eat. It is not easy to change present economic and political patterns. But such changes would reduce suffering considerably.

(b) **Overconsumption.** Part of the problem of uneven distribution of resources arises from overconsumption. Many people in wealthy societies are overweight and would be healthier if they ate less. Even more food could be conserved if people whose diet consists largely of meat would eat more grain instead.

In Chapter 3, it was explained that an omnivore utilizes its food resources most efficiently if it eats vegetable proteins rather than meat. In the example given in that chapter, a given quantity of soybeans and corn could feed 22 vegetarians or 1 person who eats nothing but meat. The average person in North America uses 900 kilograms of grain per year. Only 90 kilograms are consumed directly as rice, bread, or cereals. The other 810 kilograms are used as animal feed to produce meat, milk, and eggs. In India, by contrast, the average person uses only 180 kg of grains per year, almost all of which is eaten directly. These figures do not mean that everyone should become a vegetarian. Plants generally do not contain sufficient quantities of essential amino acids for human health, and some vitamins needed by humans can be obtained only by eating other animals or their products, such as milk and eggs. Young people especially need the protein found in animal products. In addition, many areas of land, such as steep hillsides or semiarid prairies, cannot support conventional grain agriculture but do serve as range for various grazing animals. Therefore, this land is used most efficiently for meat production. In many wealthy nations, however, especially in North America, cattle are housed in feedlots and fed prime grains. It is estimated that the feedlot

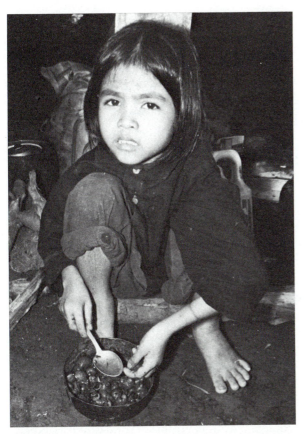

Figure 16–13 *Before intense warfare spread throughout the region, Southeast Asia was self-sufficient in food production. Today millions are starving, and rice must be imported. This photograph shows a girl eating a meal at a Thai military-run camp near the Cambodian border. The girl said she fled from an area under the control of soldiers of the fallen Phnom Penh regime (1979). (World Wide Photos.)*

grains used yearly in the United States could feed 400 million vegetarians. If North Americans reduced their feedlot meat consumption to the level of the people in Sweden, and the grain saved were distributed equitably, then 200 million starving people would have plenty to eat (in calories, if not in protein).

Related to the general problem of overconsumption is the issue of food for pets. Domestic dogs and cats provide comfort for many, but in North America they consume enough grain to feed 20 million people. One of us (A.T.), while traveling in a less developed part of the world, remarked to a companion that the area seemed quite poor, as evidenced by the presence of many half-starved dogs. His companion replied,

"No. If it were as poor as you think, there wouldn't be any dogs at all."

(c) War. In many parts of the world, warfare has been a way of life for years, decades, or even centuries. Families are uprooted from their homes, crops are destroyed or confiscated, and in many instances relief convoys are hijacked. Many of the most serious famines in recent times, for example those in Cambodia in Southeast Asia and in Somalia in East Africa, affected refugees displaced from their self-sufficient lifestyles by the horrors of war (Fig. 16–13).

Even in the absence of direct hostilities, the threat of war continues to take a terrible economic toll. According to the U.N. Conference on Disarmament, the nations of the world spent $1 billion *per day* on arms and warfare in 1977. In every nation in the world, fuel, machinery, technical expertise, and human effort are diverted from feeding people to supplying armaments. Given enough fertilizer, fuel, and machinery, the crop yields in the Ganges River delta in India could be increased sixfold. The impoverished Sudan in northeast Africa could be a bread and cereal basket for the world. The rice paddies of Southeast Asia could feed the hungry in that region. But great quantities of fuel and machinery are used for weapons and there is not enough left over for farming.

16.6 Increases in Crop Yields and the Green Revolution

Next time you are walking in the country, look closely at the wild grasses growing in meadows and fields. Notice that each plant has a slender stalk, a few leaves, and a small cluster of seeds at the top. These seeds are rich in starch, protein, and vitamins, but they are so small that they are not harvested and processed for human consumption. Ancient farmers living in the Stone Age probably started cultivating grains that were only slightly more productive than these modern "weeds." As various farmers replanted only the seeds of the largest, healthiest plants, modern strains of wheat, rice, and other grains were developed gradually over the centuries (Fig. 16–14). These grains were gener-

Figure 16–14 The wheat stalk on the right supports much larger quantities of grain than the native grass shown on the left.

ally well adapted to local growing conditions. They were genetically adjusted to peculiarities in soil conditions, water supply, length of growing season, and seasonal temperatures, and they were at least partially resistant to local diseases and pest infestations. However, as population increased in the nineteenth and twentieth centuries, it became apparent that traditional farming practices were not adequate to feed the world's people. Therefore, agriculturists searched for ways to increase crop yields. It was obvious that heavy doses of fertilizer could augment food production, but there appeared to be a limit to the quantity of fertilizer that could be utilized. When a native grain plant is fertilized heavily, the leaves grow broader and larger, thereby shading nearby plants, and the stalk grows to be long and thin. The heavy grain causes the elongated stalk to break and bend, and the grain falls to the ground and rots.

In the mid-1960s an interdisciplinary team of scientists working in Mexico and the Philippines developed new varieties of wheat and rice that were adaptable to tropical climates and were capable of producing higher yields than any native grains (Fig. 16–15). These varieties have short, upright leaves so plants can be grown close together without shading each other. In addition, their stalks are short and thick so that they will not bend and break when the plant is heavily

fertilized. The potential yields of these new varieties are so spectacular that many people have heralded their introduction as the **Green Revolution**. For example, in Mexico in the 1940s the yield from wheat fields averaged 750 kg/hectare, whereas yields from the new strains of seeds in the 1970s averaged 3200 kg/hectare. In India and Pakistan, massive shipments of Green Revolution wheat in the late 1960s raised the wheat harvest between 50 and 60 percent during a period of two growing seasons. In Colombia, rice production increased by a factor of 2½ despite the fact that the area of cultivation remained nearly constant (Fig. 16–16). Such increased food production across the globe has reduced famine and increased the well-being of millions.

On the other hand, figures of national grain production do not always reflect the fate of all people within that nation. To understand the total impact of the Green Revolution it is important to understand that a "wonder seed," by itself, does not produce large quantities of food. The seed only carries the genetic potential for high-grade production. If this production is to be realized, the plant must be fertilized heavily, watered, and protected from disease and insects. The new grain varieties planted in an impoverished soil and dependent on variable rainfall for growth produce equal or smaller yields

Figure 16–15 Workers planting rice at the International Rice Research Institute, Los Baños, Laguna, Philippines.

than the native grains, which have been cultivated in such poor areas for centuries. In addition, the new grain varieties are less resistant to insect pests and fungal diseases than the traditional plants. As a result, farmers who invest in the seed, the fertilizer, and the advanced irrigation systems must also invest in pesticides.

If a farmer has the money to buy the new vigorous seeds as well as the necessary fuel, fertilizer, pesticides, and irrigation facilities and knows how to manage a modern integrated farming program, then high yields and an eventual profit can be realized. Otherwise, the Green Revolution may provide no help.

Many individuals have actually been harmed by the introduction of the new high-yield vari-

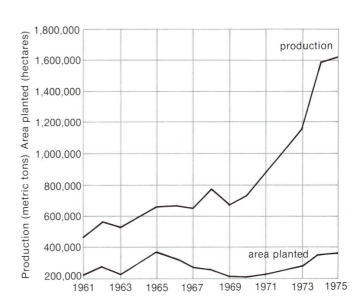

Figure 16–16 Rice production in Colombia following introduction of improved Green Revolution seeds in 1969.

SPECIAL TOPIC B
The Green Revolution: The Other Side

The use of Green Revolution seed has led to many auxiliary problems, such as

1. The taste and texture of hybrid rice IR-8, for example, differ from those of traditional rice, and it commands a lower price. In Pakistan, many wholesalers refuse to handle this variety.

2. In many rice cultures, paddy farmers have raised fish in irrigation canals. Now fertilizer and pesticide pollution of the canals have killed fish. Thus, increased total caloric content of many people's diets has been associated with decreased protein content.

3. The problem of protein deficiencies has been aggravated by the curtailment of production of soybeans and other protein-rich legumes, which are now less profitable for some farmers than grains. Attempts to develop new high-yield varieties of these crops have thus far been unsuccessful.

4. The new varieties of grain mature earlier than the old ones. Although this enables the prosperous farmer to grow two crops a year instead of one, many other farmers are finding that since harvest time now coincides with rainy weather, the grain can no longer be dried in the sun. Instead, mechanical driers are often needed, and mechanical dryers must be powered by fossil fuels.

5. National economic problems have arisen because of frequently insufficient foreign exchange to purchase imported pesticides and chemical fertilizers. Even if the developing nations were to build their own chemical factories, many would have to buy fuels, phosphate rock, and other raw materials.

SPECIAL TOPIC C
Monoculture and the Green Revolution
(contributed by Professor Susan Wilson)

Ecological diversity within a large community is of great importance. Plants differing in size and season of development are often complementary in their use of environmental resources. In **monoculture** (the practice of growing a single crop to the exclusion of all others), every plant is making the same demands upon environmental resources at the same season. In addition, with the elimination of the checks and balances of a complex ecosystem, insect pests and fungi become more prevalent, and the loss of natural controls makes large applications of insecticides and fungicides necessary. This is especially true in grain fields and the large one-crop tree farms designed for lumber and paper production. In the latter, parasitic infestations become a major problem; furthermore, wildlife is considerably reduced in the monotonous regularly planted forests. In addition, lack of rainfall can have a disastrous effect on a monoculture crop, as happened recently to the corn, soybean, and wheat crops of the American midwest.

In a diverse ecosystem, a change in climate or biotic conditions is likely to damage only certain species; the remaining species dampen the effects. Insect pests, parasites, and fungi are generally host-specific, attacking only a few species each.

Agricultural practices could be modified to accommodate cultivation of several noncompeting crops together; however, such practices eliminate the use of mechanical harvesting machinery. In underdeveloped areas, where manpower is the most abundant resource, multicrop agriculture is recommended by experts such as Barbara Ward and René Dubos. Labor-intensive farming lowers total energy consumption because use of machinery, agricultural chemicals, and so on is reduced. Unfortunately, implementation of the Green Revolution is based upon the technique of monoculture. Unless careful attention is given to the management of the land, such highly intensive single-crop agriculture can lead to irreversible loss of fertility, especially in the tropics. For these reasons many experts are fearful of applying the practices of the Green Revolution in underdeveloped areas.

eties of grain. If more grain is grown by wealthy farmers, the price of food will drop. The poor, who cannot afford to plant the new seeds, grow the same amount of food as they did in past years but receive less money for their harvest when it is time to sell.

Despite many individual hardships, the Green Revolution has increased food production throughout the world. Without these gains, hunger and starvation would be much more severe than they are. Yet even with these encouraging improvements the sad reality is that nearly every country "revolutionized" by new varieties

of grain is importing food today. Increases in yields have not been sufficient to surmount the twin hurdles of population growth and increasing scarcity of cheap energy.

Perhaps the key question in agricultural research today is: Is there some way to breed plants that will produce large quantities of high-grade food without requiring quantities of fossil

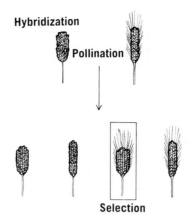

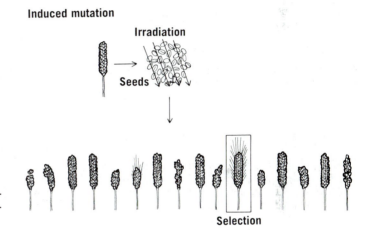

Figure 16–17 *Traditional cross-breeding techniques that were available before the development of genetic engineering.*

fuels? The search for the solution to this problem has already started. Although the quest has so far been unsuccessful, the hope that it engenders has led people to call the research the **Second Green Revolution**.

The Second Green Revolution is based on more advanced plant-breeding techniques. Individual wheat plants in a given field are not all exactly alike. They differ from each other in various characteristics just as individuals in a human population do. Traditionally, farmers have replanted seeds only from those plants that produce the most grain. In this manner, the yields of traditional grain crops have increased slowly throughout history. During the development of seeds for the first Green Revolution, scientists accelerated these processes by two techniques—

cross-breeding and **artificial mutation**. Suppose one wheat plant has a heavy cluster of grain but a thin stalk, whereas another has a thick stalk and less grain. A scientist will cross the two with the hope that a plant in the next generation will have a thick stalk and a heavy grain cluster (Fig. 16–17). Another possibility is to change the plant's characteristics by irradiating it with ultraviolet light or X-rays. Radiation randomly alters the genes of any organism, leading to mutation. A mutated plant may contain characteristics never before seen in the species, and occasionally these characteristics may be beneficial.

These experimental techniques are slow and unpredictable because they rely on chance. In recent years, a new field of biology called **genetic engineering** has been developed. Plant biologists are trying to learn how to remove a specific gene from the DNA of one individual or species and plant it artificially in another. In one spectacular experiment, scientists removed a gene that controls protein production from a bean plant and introduced it into a sunflower. The potential of this breakthrough is that it might enable people to raise sunflowers that produce protein as efficiently as beans do.

Think of the possibilities of this technique. Legumes such as alfalfa harbor bacteria on their roots that can "fix" atmospheric nitrogen thus providing nitrogen fertilizer for the plant. If this capability could be genetically engineered into a rice or wheat plant, a farmer would no longer need to apply expensive nitrogen fertilizer to these grains. As another example, many plants aren't attacked by certain fungi that destroy wheat. If someone could find the gene in a plant that can provide the necessary chemical defenses against the fungus and then implant this gene into the wheat plant, the need for pesticides would be reduced. The list of dream plants could go on and on. Why not engineer a self-fertilizing, drought- and insect-resistant plant with the tuber of a potato, leaves of spinach, stalks of sugar cane, and fruit of a tomato?

As of the end of 1983, no Second Green Revolution seeds were available on a large-scale commercial basis. However, many scientists are confident that such production will begin in the near future.

16.7 Food from the Sea

People have often regarded the sea as a boundless source of food, especially protein. After all, over 70 percent of the planet is ocean. Moreover, the ocean has traditionally been an unregulated commons, open to any individual or nation caring to exploit it. Shortly after World War II, with the seas again free of wars and warships and the first pinch of global food shortages being felt, global fishing industries boomed. From 1950 to 1970, the world fish catch more than tripled, and in 1970 fish supplied more protein for human consumption than all the herds of land-based cattle combined. Despite an ever-increasing effort, fish harvest barely rose from 1970 to 1979, and, as shown in Figure 16–18, per capita production of fish declined significantly in those years. What has happened?

Most of the central oceans are biological deserts. The highly productive parts of the oceans are concentrated along the coasts and continental shelves. Yet today the productivity of these areas has been reduced by overfishing and by pollution.

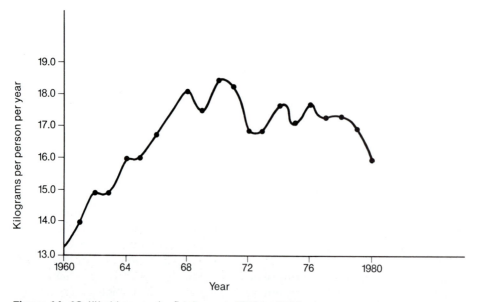

Figure 16–18 World per capita fish harvest, 1960 to 1980.

(a) Overfishing. Garret Hardin's essay on the tragedy of the commons (Chapter 1) explains that a single individual stands to gain in the short term by grazing as many cows as possible on a common pasture. The oceans are a type of commons, and unregulated fishing, like unregulated grazing, is profitable in any given year. International maritime laws have always been very complex and rather ineffectual, opening the way for wholesale exploitation of many ocean fisheries. As a result, fish stocks have declined, and the potential food production of many of these areas has diminished.

In response to this threat, individual nations have recently claimed jurisdiction over an increasingly large area of coastal water. Traditionally, a nation's border was considered to extend 3 miles (4.8 km) beyond its shore. This practical limit was defined by the range of cannon on the warships of the time. As artillery became more efficient, it seemed reasonable to demand that potentially hostile warships should remain further off shore, so a new coastal border of 12 miles (19 km) was defined. In recent times, maritime nations have realized the economic value of their fisheries and have decided that their territories extend 200 miles (320 km) out to sea. This extended limit allows individual governments to regulate fishing and prevent overexploitation. In some areas regulation has been poor and fish populations continue to decline, but in others fish harvests have been carefully controlled and are increasing.

(b) Pollution. Protected coastal areas are prime breeding grounds for fish, but they are also ideal sites for industrial development. The disputes have usually been resolved in favor of heavy industrialization of the coastlines and the seemingly inevitable disposal of sewage, oil, pesticides, and other chemicals into the sea. The result has often been the poisoning of marine organisms. For example, Chesapeake Bay on the east coast of the United States was once one of the world's richest estuaries, but it lies right in the heartland of a huge commercial zone with easy deep-water shipping access to Europe. In recent times the bay has become polluted, and although fishing has been regulated to control overexploitation, the traditional catch of oysters, shad, and bass has declined precipitously.

Ecologically, the destruction of an estuarine marsh is disastrous because estuaries are nursery grounds for many fish that spend their adult lives in deeper waters. It has been estimated that 1 hectare of estuary produces enough young per day to grow into 270 kg of marketable fish.

Figure 16–19 Shrimp fishing near Galveston, Texas. (Courtesy of National Oceanic and Atmospheric Administration.)

16.8 Other Methods of Increasing Food Production

The Cultivation of Algae

Recall from Chapter 3 that terrestrial plants, which operate at about 1 percent efficiency, are poor converters of solar energy into the nutrient energy of leaf tissue. Underwater algae do the job much more efficiently. Therefore, there is significant potential for harvesting algae for food. But because these plants also provide the food base for fish, algae harvests will necessarily diminish fish harvests.

What about artificial algae farms? If the efficient underwater cultivation could be maintained in concentrated cultures at full sunlight, an area of half a square meter could feed one person on a sustained basis. Even so, it does not follow that algae farming will solve food problems in the future, because algae do not grow as well in concentrated surface cultures as they do underwater. As the light intensity increases, the conversion rate (the percentage of solar energy that is converted to nutrient energy) decreases. In full sunlight, algae are only about four times

as efficient as wheat or rice. Furthermore, the advantage of this efficiency is offset by the fact that algae are not well adapted to concentrated growing conditions, and tremendous inputs of energy are required to maintain high yields (Fig. 16–20).

Processed and Manufactured Foods

Food chemists have pursued three different concepts in developing new food products—(1) alteration of good natural foods to make them more acceptable or more marketable, (2) conversion of agricultural wastes to edible food, and (3) chemical synthesis of food. These approaches are discussed in the following paragraphs.

Alteration of Natural Foods. Perhaps the most familiar example is the conversion of vegetable oils into the butter substitute, oleomargarine. Food conversion is becoming increasingly important in both the developed and the developing nations. In some areas such high-quality protein sources as soybeans, lentils, and other legumes, though readily available, are unpopular. One solution is to use concentrated vegetable protein as a base for manufactured foods that taste like something else. Thus, soy protein "hamburger," soy "milk," soy "bacon," and other simulated

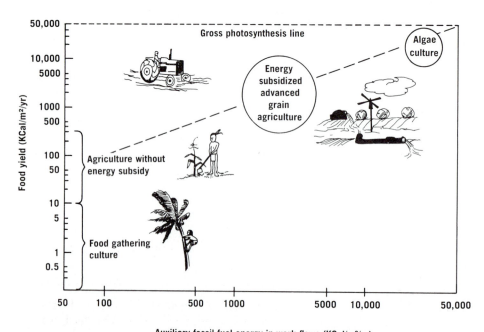

Figure 16–20 The relation between food yields and auxiliary energy.

products are readily available in retail stores in many parts of the world (Fig. 16–21). In some areas, undernourished people spend their money on soft drinks rather than on protein-rich foods. Recently, soy protein has been used as an ingredient in soft drinks, and various sweetened, carbonated, soy beverages are now sold in Asia, South America, and Africa.

Conversion of Agricultural Wastes. When cooking oils are extracted from peanuts, soybeans, coconuts, cottonseeds, corn, or any other vegetable seed, the residue is a waste mash high in vegetable protein. At present, most of these residues are fed to cattle or discarded. It is relatively easy to incorporate the proteins into synthetic foods.

Sawdust, straw, old newspapers, and other inedible plant parts also represent a major source of waste materials. The major chemical component of sawdust and straw is cellulose. Cellulose consists of very large molecules that, when heated with dilute acid and water, decompose into a mixture of various sugars. Yeasts can thrive on this sugar mixture if other nutrients such as urea (a source of nitrogen) and mineral

Figure 16–21 *Simulated ground meat made from soybean flour. (Courtesy of Archer-Daniels-Midland Company.)*

salts are added to the culture medium. In turn, yeasts are sources of high-quality protein and can be eaten directly. Alternatively, yeast protein extract can be used as an additive in manufactured or processed foods. For example, whole wheat is deficient in the amino acid lysine, which is necessary for protein synthesis in the human body. Lysine can be extracted from yeast cultures; in India, it is added to dough in government-owned bakeries for the production of enriched bread. Thus a process can be envisioned whereby people shred their daily newspapers, use the paper to culture yeast, and then eat the product. As novel as it sounds, the process is unlikely to provide a magic way out of the problem of world hunger, because yeast culture, like algae culture, requires large quantities of auxiliary energy to synthesize the fertilizers and to maintain the growth chambers. Once again, the limiting factor appears to be not technological knowhow or total agricultural potential, but the availability and cost of energy.

Chemical Synthesis. The high cost of maintaining yeast organisms or feedlot cattle is prompting chemists to bypass the heterotrophic organisms completely and synthesize amino acids directly from coal and petroleum. Synthetic lysine still costs about as much as lysine extracted from yeast cultures and requires about as much energy to produce. Other synthetic foods include vitamins used for food additives and artificial pie fillers derived from petrochemicals.

It is encouraging to know that bread in New Delhi is more nutritious than ever before, but such dietary advances are concomitant with the trend from the range cow to the feedlot cow to synthetic amino acids. Our existence is thus becoming increasingly dependent on technology and fossil fuel reserves.

16.9 World Food Outlook—An Overview

The analysis of the world food outlook may seem to be a pessimistic scenario of declining reserves, increasing scarcity, and mushrooming population. Are there any solutions? It is simple enough to write a list with a pencil and paper, but of course a list by itself doesn't feed hungry people. Therefore, the deceptively short, easy "solutions" that follow are not answers for to-

morrow. Rather, you may think of them as a set of goals that would mandate drastic, sweeping global changes in society if they were implemented.

REASONS WHY THERE IS NOT ENOUGH FOOD	SOLUTIONS
Overpopulation (Chapter 9)	Limit population growth
Soil depletion through erosion (Section 14.4)	Educate farmers and encourage them to use the best available agricultural technology
Agricultural land preempted for industrial development (Section 17.1)	Better land use planning to save farmland
Agricultural land preempted by mining (Section 13.1)	Reclamation of land after mine is abandoned
Uneven distribution of food	Reduce overconsumption and excess meat consumption in developed nations and improve international mechanisms for food distribution
Loss of food-growing potential through war	Initiate international disarmament
Loss of food to pests (Chapter 15)	Improve pest management; use integrated techniques
High cost of energy for fertilizers and pesticides	Use recycled organic fertilizer; use renewable energy sources such as solar energy for many tasks, thereby saving fossil fuels for situations where needed
High cost of energy for farm machinery, transportation, packaging, and so on	Use more labor-intensive agriculture; encourage decentralized agriculture; eliminate waste in packaging and processing

Summary

People need both quantity (calories) and quality (proteins, vitamins, and minerals) in their daily food intake.

During the decade between 1970 and 1980, food production kept pace with population growth in some years but not in others. At the start of 1980, approximately 400 million people were starving.

Agricultural systems have extended to marginal croplands in many areas of the globe, and a small change in climate would be disastrous in these regions.

Energy is used in agriculture for fertilizers, pesticides, machinery, processing, and transportation.

If food were divided equally, there would be enough for everyone to eat, but overconsumption, unequal distribution of resources, and war lead to pockets of extreme poverty.

In the mid-1960s new grain varieties were developed, leading to the **Green Revolution.** Green Revolution seeds lead to high yields only if sufficient water, fertilizer, and pesticides are applied. The **Second Green Revolution,** if it is actually realized, may promise high yields with low energy inputs.

Other approaches to increasing food production are all limited. They include increasing crop acreage, obtaining more food from the sea, cultivation of algae, and processed and manufactured foods.

Questions

Human Nutritional Needs

1. (a) What three functions are served by the human diet?
   (b) Which is the main function provided by (i) potatoes; (ii) a vitamin tablet; (iii) egg white; (iv) iodized salt?
   (c) What *combination* of the three functions is served by a meal of (i) beans, nuts, enriched or brown rice, and orange juice; (ii) a jelly doughnut, multivitamin pill, and black coffee?

World Food Outlook

2. Explain why a small shift in climate could seriously affect world food supplies.
3. Discuss some problems inherent in farming deserts; jungles; temperate forest hillsides; the Arctic.
4. Prepare a class debate. One side argues that it would be possible to feed 10 billion people by the year 2100, and the other side holds that only 2 billion people could be supported.
5. Many current events may affect the food supply of groups of people, even though food per se is not a topic of the news headlines. Select an article from the daily paper, and discuss the impact of the news on local or global food supplies.

Energy in Agriculture

6. One hundred years ago most farmers used horses or oxen to plow fields and harvest crops. Today tractors are commonly used. Imagine that the fossil fuel crisis became so severe that farmers were forced to rely on animal power again. Would the total food production increase, decrease, or remain constant? Defend your answer.

The Green Revolution

7. Briefly discuss the relative importance of each of the following characteristics to a plant species existing (a) in a natural prairie; (b) in a primitive agricultural system; (c) in an industrial agricultural system—(i) resistance to insects, (ii) a tall stalk, (iii) frost resistance, (iv) winged seeds, (v) thorns, (vi) biological clocks to regulate seed sprouting, (vii) succulent flowers, (viii) large, heavy clusters of fruit at maturity, (ix) ability to withstand droughts.
8. Explain why the Green Revolution has not helped the very poorest farmers.

Social and Economic Factors in Agriculture

9. Many authors have pointed out that if all the fertilizer used in the United States for lawns and ornamental gardens were shipped to farmers in the developing nations, many millions of people could grow more food to feed their starving families. Imagine that you would normally fertilize your lawn, but this year you chose not to so that there would be more fertilizer for the poor. Would the fertilizer that you chose not to buy actually get shipped to people in the developing nations? Would your action do any good? What policy, private or public, would you recommend? Defend your position.
10. What types of agriculture, if any, are currently practiced within a 50-km radius of your school or home? What types of agriculture were common in the region 25 years ago? What changes, if any, have occurred? If agricultural productivity has declined, where is the food in your area now coming from? If productivity has increased, where is the food being sold? How have these changes affected centralization and energy consumption in agriculture?
11. Interview a farmer in your region. Ask the following questions. Is farming profitable? Why does he or she choose to be a farmer? Would it be more profitable to sell the land

and retire? What can be expected to happen to the land in a generation or two? Report on your results.

Suggested Readings

Two general books on agriculture and food resources are:
John Harte and Robert H. Socolow: *The Patient Earth.* New York, Holt, Rinehart and Winston, 1971. 364 pp.
Kusum Nair: *The Lonely Furrow: Farming in the United States, Japan, and India.* Ann Arbor, University of Michigan Press, 1970. 336 pp.

A discussion of the role of biological diversity of crops in primitive agriculture and the significance of the loss in modern systems may be found in:
O. H. Grankel and Michael E. Soulé: *Conservation and Evolution.* New York, Cambridge University Press, 1981. 328 pp.

Specific references, by section, include:

16.2 The World Food Supply

Wes Jackson: *New Roots for Agriculture.* Friends of the Earth Basic Book, 1982.
Terry N. Barr: The world food situation and global grain prospects. *Science,* 214:1087, Dec. 4, 1981.
Lester R. Brown: *Building a Sustainable Society.* New York, W. W. Norton & Co., 1981. 433 pp.
Lester R. Brown: World population growth, soil erosion, and food scarcity. *Science,* 214(27):995, 1981.
Conservation Foundation Letter: Food resources threatened by mismanagement. Conservation Foundation, September 1980.
Conservation Foundation Letter: Long-range threats stalk U.S. farming. Conservation Foundation, August, 1980.

16.3 Food Production, Land, Weather, and Climate

Margaret Biswas and Asit Biswas (Eds.): *Food, Climate, and Man.* New York, John Wiley & Sons, 1979. 285 pp.
National Research Council: *Climate and Food.* Washington, D.C., National Academy of Science, 1976. 212 pp.
Stephen H. Schneider: *The Genesis Strategy: Climate and Global Survival.* New York, Plenum Press, 1976. 419 pp.

16.4 Energy and Agriculture

David Pimentel and Marcia Pimentel: *Food, Energy and Society.* New York, Halsted Press, 1979.
Marta Braiterman et al.: Energy-saving landscapes. *Environment,* 20(6):30, July/Aug., 1978.
William J. Jewell: *Energy, Agriculture and Waste Management.* Ann Arbor, MI., Ann Arbor Science Publications, 1975. 540 pp.
Howard T. Odum: *Environment, Power, and Society.* New York, Wiley-Interscience, 1971. 331 pp.
Robert C. Oelhaf: *Organic Agriculture.* New York, John Wiley & Sons, 1978. 269 pp.
J. Neil Rutger and D. Marlin Brandon: California rice culture. *Scientific American,* February, 1981. p. 42.
John Steinhart and Carol E. Steinhart: *Energy: Sources, Use, and Role in Human Affairs.* North Scituate, MA, Duxbury Press, 1974.

16.5 Social and Political Problems

Frances M. Lappe and Joseph Collins: *Food First.* Boston, Houghton Mifflin Co., 1977. 466 pp.

16.6 Green Revolution

Lester R. Brown: *By Bread Alone.* New York, Praeger Publishers, 1974. 272 pp.
Lester R. Brown: *Seeds of Change: The Green Revolution and Development in the 1970's.* New York, Encyclopaedia Britannica, Praeger Publishers, 1970. 205 pp.
Francine R. Frankel: *India's Green Revolution.* Princeton, NJ, Princeton University Press, 1971. 232 pp.
Peter Steinhart: The second Green Revolution. *New York Times Sunday Magazine,* Oct. 25, 1981. pg. 46 ff.

16.8 Other Approaches to Food Supply

See references to Section 16.2 and also:
Scientific American, the entire issue of September, 1976.
Philip H. Abelson (Ed.): *Food: Politics, Economics, Nutrition, and Research.* (A compendium of articles from *Science.*) Washington, D.C., American Association for the Advancement of Science, 1975. 202 pp.

16.9 World Food Outlook

Martin M. McLaughlin: *World Hunger or Food Self-Reliance? A U.S. Policy Approach for the 1980s.* Washington, D.C., Overseas Development Council, 1982. 56 pp.

Land Use

If there is a vacant lot or untended field near your house, visit it and look at it closely. Most probably there are various trails or paths cutting through the property. These trails were not planned by an architect or a city engineer; they were formed as people walked from one place to another. The very fact that people will form a trail rather than walk randomly across the land indicates that natural factors favor some routes over others. Some of these factors include the topography, the vegetation, the location of nearby buildings, and the most efficient routes between them.

At its best, land use planning incorporates the most efficient placement of buildings, farms, roadways, and other facilities to preserve the beauty of the land as much as possible while at the same time minimizing human effort. At its worst, development of the land can be terribly ugly and can create a system that encourages wasted energy, resources, effort, and time.

This chapter is concerned with the human settlement and use of the land surfaces of our planet. Both ideal and real patterns are considered.

17.1 The Ideal—Design with Purpose

Most land use planning must evolve around existing cities and towns, but in a few places, entirely new towns are built. Imagine that you were hired to construct a new community in a previously vacant area of land (Fig. 17–1). As a first step toward formulating such a plan, consider the immediate and direct needs of any community.

1. Food
2. Water
3. Waste disposal
4 Living spaces
5. Industrial and commercial zones
6. Electrical energy
7. Recreational zones
8. Transportation system
9. Raw materials (fuels, minerals, timber, and so on)

Figure 17–1 What factors would you consider if you were asked to build a new model community on a previously undeveloped piece of land? (Photo courtesy of Keith Gunnar.)

It is obvious from this list that all of the community's needs cannot possibly be met within a local area. There is no place on Earth that contains natural deposits of fuels, minerals, timber, water, and all the other necessities of modern life. Therefore, no city or town can be completely self-sufficient. But at the same time it is inefficient to import more goods and services than are necessary. Assume for the sake of argument that no minerals, fuels, or timber resources are available in the area, and that the plan must incorporate only the first eight items on the list.

If you were asked to design a new community, your job would be to divide the land in order to leave space for each of these eight requirements and at the same time to maximize the efficiency of the system and the beauty of the land. The first rule of land use planning is to consider the natural character of the area and take it into account in the design of any system. Most land is not a flat planar surface like a sheet of paper but includes hills, valleys, forests, rivers, marshes, or coastlines. Maximum efficiency is not achieved by overpowering nature but by designing structures to go with it. Design with nature involves an understanding of the type of land that is most suited for a specific need.

1. **Food.** The most productive farmland is generally found on level plains or in river valleys. In arid or semiarid zones, farmland should be located near a source of irrigation water.

2. **Water.** Rainwater collects in natural basins and then filters through the earth to concentrate in rivers, ponds, or lakes. To ensure a permanent supply of water, the surface topography and subsurface structure of soil and rock in a watershed either should not be altered at all or should be altered only with great care. Key watersheds might become recreational areas where no major construction or earth-moving projects are allowed.

3. **Waste disposal.** In a well-planned community wastes are recycled. Products such as scrap newspapers, metals, and glass can be recycled by industrial processes. Sewage and food scraps make excellent fertilizer. Therefore, waste disposal systems should be located near both industrial and agricultural zones.

4. **Living spaces.** Maximum efficiency can be realized if living spaces are (a) located near places of work and recreation to minimize the need for transportation, and (b) positioned favorably to receive maximum sunlight for solar heating.

5. **Industrial and commercial zones.** These areas should be located near enough to living spaces to minimize transportation needs but not so close that noise and pollution become bothersome. Industrial and commercial zones should also be located near water, intercity or international transportation systems, and sources of electricity.

6. **Electrical energy.** Power plants must be located near a source of cooling water. If plants are located near industrial zones, waste heat losses can be minimized through cogeneration (see Section 13.8).

7. **Recreational zones.** Parks, woodlands, waterways, and mountains should be as accessible as possible.

8. **Transportation systems.** If the rest of the community is well designed, transportation links connecting the individual components will be short and centralized. If population centers and industrial centers are both clustered, mass transit becomes practical. If the two zones are

sprawled out, however, people prefer to travel privately.

These general guidelines may sound simple enough, but many inherent conflicts arise. For example, the banks of a river provide the ideal location for industry, commerce, highways, and electrical energy production because the river offers intercity transportation systems and provides water for cooling, production of steam, and waste disposal. But the same area is generally the most fertile agricultural land. The riverside is also an ideal site for a park. Another problem arises over the choice of living space. People would like to have individual family homes in beautiful natural settings, but if the most beautiful settings are cut into subdivisions, then open spaces are destroyed and there is little room left over for recreation. Many planners suggest that people should live in central apartment complexes surrounded by green belts and parks. These areas would be a type of commons that everyone could enjoy and would in the long run be more pleasurable than thousands of tiny fenced-in backyards.

Even in an ideal situation, these problems are difficult to solve, but in the real world, land use planning is even more complicated. Planners seldom have the luxury of starting a new community; most often they must work with a city or town that already exists. Think of something simple like planning the width of a city street. Although there would be no problem in designing street width in a new community, most cities were first settled decades or even centuries ago. In many places, streets were first built for pedestrians or horses and today are woefully narrow for cars and trucks (Fig. 17–2). To widen them it is necessary to tear buildings down and redevelop the area, and this is horribly expensive. These and other land use patterns have grown haphazardly and are not easy to change.

17.2 Urbanization and Urban Land Use

The first cities arose along the Tigris and Euphrates rivers roughly 6000 years ago. Since

Figure 17–2 A street in Cuzco, Peru. This city was designed mainly for travel by pedestrians, donkeys, and llamas, and conversion to multiple-lane automotive traffic would involve a complete and expensive redevelopment plan.

then, great cities have grown and fallen in many areas of the world. Most people, however, have traditionally lived in rural areas. In Europe, historically the most urbanized continent, only 1.6 percent of the population lived in cities of over 100,000 in 1600; the figure rose to only 1.9 percent by 1700 and 2.2 percent by 1800. In fact, before 1800, no country was predominantly urban. Between the time of the fall of the Roman Empire and the beginning of the nineteenth century, no European city had 1 million inhabitants. Thus, on the eve of the Industrial Revolution, Europe was essentially an agrarian continent. Archeological evidence suggests that the first city population to exceed 1 million persons may have been the capital of the Khmer Republic (Cambodia) (Fig. 17–3). Tokyo and Shanghai may have had populations of over 1 million by 1800. Moreover, by 1800, many parts of East Asia boasted very large central cities surrounded by highly productive agricultural areas.

In the hundred years from 1800 to 1900, cities grew rapidly. In 1900, at least 12 cities had populations of over 1 million—London, Paris, Berlin, Vienna, Moscow, St. Petersburg (now Leningrad), New York, Chicago, Philadelphia, Tokyo, Shanghai, and Calcutta. By 1980, nearly 40 percent of the inhabitants of the world lived in urban areas. By the year 2000, over 50 percent of the world's population will probably live in

Figure 17–3 A view of Ankhor Wat, ancient capital of the Khmer Republic (Photo courtesy of Picon A. P.)

urban places, and there will probably be more than 250 cities of over 1 million inhabitants.

A pattern has developed, obviously, whereby villages have rapidly mushroomed into towns, towns into cities, and cities into complex urban centers called **megalopolises**.

Two types of cities should be distinguished. In the industrialized countries, people have traditionally flocked to the cities for available jobs. The 1960s and early 1970s were years of economic expansion. People demanded more space for housing, and eventually many moved from the central city to less densely populated suburban areas. The typical city in the developed world has a densely populated core containing both luxury and slum housing, as well as commercial and manufacturing areas. Surrounding the central city is a rapidly expanding peripheral suburban region, growing both in population and in area. The combined effect of these patterns has led to the growth of **urban sprawl**.

This development has led to many problems. The growth of urban sprawl gained momentum in an era when fuel was cheap and seemingly plentiful. For millions of people commuting became an attractive alternative to life in the city. Today, commuting is expensive, slow, and frustrating; traffic jams are common, parking is difficult, and in the event of fuel shortages, entire communities can become paralyzed. During the two "energy crises" of the 1970s, many commuters had to wait long hours in gas lines and others were simply unable to get to work.

Figure 17–4 shows schematic representations of the growth of urban areas in the developed world. In the first case, where nearly everyone lives in a central city, people can easily walk, ride bicycles, or use public transportation to move about the city. In the second case, populations are more dispersed. Walking and even cycling are impractical because the distances are too great. In addition, public transportation systems are harder to operate because people must be funneled together from a wide area, and in the suburbs the population density is often so much lower than it is in a central city that mass transit routes become expensive and complex. Cases III and IV are even more complicated, because mass transit systems must be even more diffuse. In an urban megalopolis, close neighbors or even two people in the same house may

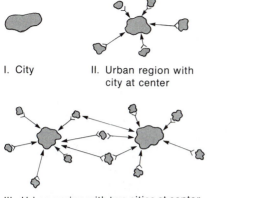

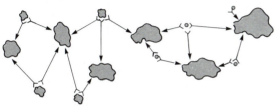

I. City

II. Urban region with
    city at center

III. Urban region with two cities at center

IV. Urban megalopolis

Figure 17–4 *Schematic diagram of types of urban places. Direction of arrows shows major commuting directions. Tail represents place of residence; head, place of work. I. City. II. Urban region with city at center. III. Urban region with two cities at center. IV. Urban megalopolis.*

work in different cities. Therefore, the transit network must be doubled. Even car pooling is difficult, and in many cases each person in the community drives alone in an automobile to work and back. Not only is fossil and human energy wasted, but large land areas are preempted for transportation systems.

Many other problems have arisen with urban sprawl. For example, in many cases electric power plants are located so far from potential customers who might use their waste steam that cogeneration is impractical. As local agricultural areas are paved and developed, farming has been pushed far away from urban areas, causing the costs and energy consumed for transporting food to rise also. In addition, because urbanization has spread to many local open spaces, recreation has suffered. People either spend large amounts of time indoors or must drive long distances to have a family picnic or take a walk in the woods.

The problem, for which there is no simple solution, is that the system is set in concrete; in

most cases it isn't practical to tear down buildings, dig up superhighways, or move factories.

Cities expand in the less developed world, too. Indeed, urbanization is proceeding at a faster rate in the less developed world than it is in the industrialized nations, although the less developed countries are still much less urbanized than Europe, North America, Australia, New Zealand, and East Asia. The land area of a city in the less developed world grows slowly in relation to the increase in population. The reasons are several. First, because jobs are often not available, people do not earn enough money to pay for adequate housing. Second, systems of transportation are not developed, so that the inhabitants must live close to the center of the city. As a result, these cities become overurbanized; that is, there are too many people for the economic base. As a result, the city suffers from poor sanitation and increasing social problems. The fringes of cities in the less developed countries are often characterized by dismal living conditions.

Even though the living conditions of many twentieth century cities are deplorable, people migrate there voluntarily (Fig. 17–5). The reason must be that conditions in rural areas are even worse. Migration to the city has been called a push-pull phenomenon—people are pushed out of rural areas by hopeless poverty and pulled by the lure of the city. India is characterized by what is termed a push-back phenomenon: Conditions in the city are even worse than in many rural areas, so many migrants return to their homes. This reversal is now beginning to make itself felt even in more developed countries, where the economies of scale once offered by life in the city are diminishing. Those who live in the country, on the other hand, are benefiting from advances in communication that make the countryside less isolated than it once was.

17.3 Urban Land Use and Agriculture

As mentioned in the last chapter, a person living in a rural community can raise enough

Figure 17–5 *Many people live in small tents on the side-walks in the slums of New Delhi, India.*

Figure 17–6 *The South Platte River basin is potentially one of the richest agricultural regions in Colorado, yet much of the farmland has been pre-empted by commercial interests.*

vegetables on 0.07 hectares of land to support a family of four. In contrast, the average urban American uses almost twice as much land per person (0.13 hectares) for nonagricultural purposes—for homes, lawns, roadways, parking lots, shopping centers, and factories. Much of this urban development has occurred in regions that once were prime farming areas. The loss of farmland to urban sprawl has become a serious concern in recent years.

In the United States, a total of 1.2 million hectares of farmland is paved with concrete and asphalt or replanted with ornamental lawns every year (Fig. 17–6). This is equivalent to an area nearly the size of the state of Connecticut. Unfortunately, the loss is concentrated in many of the most productive agricultural areas. If current trends continue, some 20 percent of the most fertile farmland in the United States will be urbanized by the year 2000. For example, unless

our priorities change, all of Florida's citrus groves, 16 percent of the vegetable-producing regions of southern California, and 24 percent of the prime agricultural land in Virginia will all be removed from production.

The same trend is occurring in the less developed countries, although at a less rapid rate. Although many people are urbanized in these countries, they use less space per person because many live in crowded slums, and elaborate roadways and shopping centers are practically nonexistent. Yet, estimates show that the cropland that would be urbanized in these areas between 1980 and the year 2000, if present trends continue, represents the growing capacity to feed 84 million people. Is there any way that this trend can be reversed? In a well-planned society, the rate of change could certainly be reduced. Two examples illustrate the point.

Suburban subdivisions are characterized by single story, one-family dwellings on small plots of land. Suburban shopping centers are often single story sprawling stores and malls. If these were replaced with multistory apartments and

Figure 17–7 Compared with mass transit systems, freeways represent an inefficient use of land surfaces. (Wide World Photos.)

department stores, a considerable area of land could be saved.

Transportation systems are similarly inefficient. A two-track local subway uses a roadbed 11 meters wide and can carry 80,000 passengers per hour. On the other hand, an eight-lane superhighway is 38 meters wide and carries only 20,000 people per hour under normal traffic conditions (Fig. 17–7). A superhighway capable of carrying 80,000 people per hour would have to be 152 meters wide (approximately 1½ times as wide as the *length* of a football field). Therefore, a shift to mass transit would conserve land surfaces.

In the modern world, what happens when a prime site for a new harbor is also the mouth of a river in which fish come to breed? Or suppose a coal mine lies under a wheat field? Should a vegetable farm be converted to a more profitable shopping center? When such conflicts arise, who decides how the land is to be used? In most nations today, economic factors control these decisions. In the last example cited above, if a developer offers the vegetable farmer enough money, the farmer may sell. The developer will then bulldoze the topsoil away and cover the land with concrete and asphalt. Bit by bit, many prime agricultural areas have been converted to

nonagricultural uses. What will happen in the future? Will continued conversion of agricultural land to other functions eventually lead to worldwide famine? If so, is there any way to reverse current trends and stabilize the system?

Legal and economic systems vary from nation to nation. In many regions of the United States, land is taxed according to its market value. If the land is valuable, the taxes are high. Imagine, then, what happens in some agricultural communities. One farmer sells fields to a manufacturing corporation, and a factory is built. The presence of the factory boosts land prices in the surrounding areas. Rising land prices lead to rising taxes. The high taxes then become a great burden to neighboring farmers. As profits from farming decline and the market value of farmland skyrockets, many more farmers sell their land. The spiral continues (Fig. 17–8). Eventually an agricultural region is converted into an industrial one.

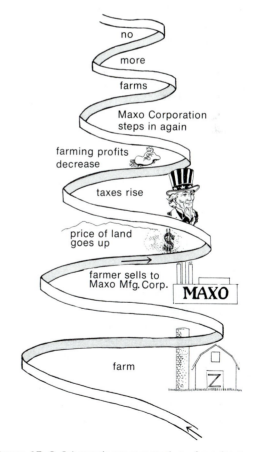

no
more
farms

Maxo Corporation
steps in again

farming profits
decrease

taxes rise

price of land
goes up

farmer sells to
Maxo Mfg. Corp.   **MAXO**

farm

Figure 17–8 Schematic representation of a price-tax spiral that leads to destruction of agricultural land.

In some regions, legislation has made farming economically more attractive. For example, farmland may be taxed at a lower rate than that applied to land used for commercial or urban development. Other laws protecting farmland are even more far reaching. In several states on the eastern seaboard of the United States, state governments have begun to subsidize the maintenance of agricultural land. If a farm is worth, say, $200,000 to a developer but only $100,000 if it remains farmland, the state will pay the farmer the difference if the land is set aside permanently for agricultural purposes.

17.4 Coastlines and Marine Estuaries

Although coastal zones make up only a small portion of the world's land surface, environmental pressures on these regions are particularly intense. For example, in the United States, over 50 percent of the population, 40 percent of the manufacturing plants, and 65 percent of the electrical power generators are located within 80 km of the ocean or the Great Lakes. These waterways provide many resources including international transportation systems, water for industrial cooling and waste disposal, food in the form of fisheries, and recreation. Unfortunately, a given section of coast cannot be used for all these purposes simultaneously, and conflicts arise. The most serious problems are encountered because industrial development is often incompatible with coastal food production and recreation. Estuaries provide nursery grounds for deep-water fish and therefore represent a significant food-producing region, but to a developer, an estuary is often seen as a protected bay or inlet that would provide an ideal harbor or industrial zone (Fig. 17–9).

Many attempts have been made to initiate coastal land use laws and to partition the coastal zones equitably among the various special interest groups. Few of these efforts have been particularly successful. For example, the Fisheries Conservation and Management Act passed in the United States in 1976 required that federal agencies must propose management plans to ensure optimum yields and guaranteed perpetua-

Figure 17–9 This estuary near the Mississippi Delta has been heavily polluted by industrial development. (Courtesy of National Archives.)

tion of all commercial species of coastal fish. Many state laws have been enacted for the same purpose. Yet the fishing industry does not have enough political power to confront the oil, electrical utility, shipping, and manufacturing industries simultaneously. As a result, nearly every coastal area is becoming increasingly industrialized while recreational opportunities and fishing productivity are declining. In the United States, the only exception to this trend exists along parts of the Alaskan coast. The Alaskan fisheries are among the most productive in the world simply because in many places there is no industrial competition for the estuaries and fish-breeding grounds. The healthy condition of these systems doesn't represent a victory for far-thinking land use management; rather, the natural productivity of the area has been maintained by default.

In many areas careful land use management that could be realized by designing structures with nature has been replaced by a mechanized confrontation against nature. The problem with

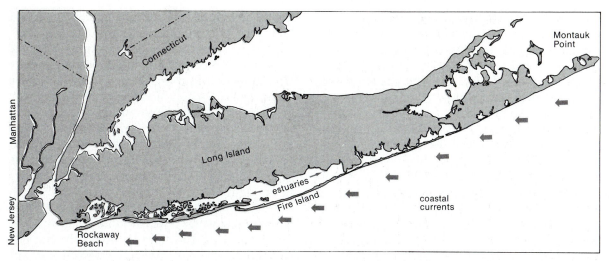

Figure 17–10 Map of Long Island and vicinity showing predominant movement of coastal currents.

this approach in coastal zones can be illustrated by examining the southern shore of Long Island, just east of New York City. When waves strike a beach, sand particles and small pieces of rock are dislodged from the shoreline and carried along the coast, where they are later redeposited. This continuous movement of material creates many distinct land forms and ecological habitats. Along the southern coast of Long Island, movement of sediment has formed a series of thin, low-lying **barrier islands** (Fig. 17–10). Between the barrier islands and the mainland there lies a series of quiet, protected salt marshes. These marshes represent an estuary system that provides home and shelter for many forms of aquatic life, including the young of many species of deep water fish. The predominant movement of water along this coastline is from east to west, as shown in Figure 17–10. On the eastern end of Long Island, near Montauk Point, there is a region of tall cliffs composed of sand and gravel originally deposited by the glaciers of the great ice ages. At the present time, the ocean waters are eroding the sediment from these cliffs and carrying the material westward toward Rockaway Beach and New Jersey. If we studied a beach midway along the island, say Fire Island (shown on the map), we would find that the existing beach sand is constantly being eroded and carried westward, only to be replaced by sand from the eastern end of the island. The beach, in dynamic equilibrium, remains stable.

Superimposed on this east-west movement are variations caused by seasonal changes and severe storms. In the wintertime, series of steep, sharply undercut waves remove large quantities of beach sand, carry the sediment out to sea, and deposit it in a formation known as a sand bar, which lies parallel to the shore. Thus, in winter the beach erodes and grows smaller. This erosion is only temporary, however, for the gentler waves of summer carry sand from the bar inland and rebuild the beach. In this manner, long-term stability is maintained despite seasonal fluctuations.

In any coastal system, severe storms periodically strike the shore. In a natural barrier system such as the one in southern Long Island, storms completely overrun the low-lying outer islands, but the system is not permanently damaged by these periodic inundations. The high waves that wash across the beach and roll inland over the dunes and salt marshes flatten some dunes, build others higher, and move sand and rock here and there. When the storm is over, there is little significant long-lasting erosion. Beaches remain intact, salt marshes rejuvenate quickly, and the dune grasses that had evolved in such a system grow back within a few months.

We do not mean to imply that geological forces have no effect on coastlines, for of course they do. But the change is generally slow. When the sand deposits near Montauk Point become depleted and no new inflow of sand can travel westward, the island will be reshaped and beaches will erode. But such changes are generally measured in centuries or millennia rather than in months or years.

The biological-geological balance that exists on Long Island has given rise to a fertile estuary system. However, the ebb and flow of beach sand and the rise and fall of storms are not always compatible with human activity. A shore-sand-dune system is dynamic—always changing. It is also prime waterfront real estate. So people build houses, resorts, and hotels on the shifting sands. Then, in winter, the beach starts to recede. Although this recession is entirely natural, many property owners have tried to save their personal stretch of beach from these cycles of the ocean. This can be done (temporarily) by building a **groin**, commonly called a **breakwater**. As mentioned previously, sand steadily moves from east to west along this par-

ticular stretch of coast. If a large stone barrier is built just west of a person's property (Fig. 17–11A), it will trap sand moving from the east and keep that particular beach from receding in winter. But now the overall flow of sand has been impeded (Fig. 17–11B). The property just west of the breakwater receives no sand, but the currents that move the sand originate from farther out to sea and are not impeded. So behind the breakwater the beach is eroded as usual, but not replenished. The beach then recedes, as shown in the figure.

This unnatural recession aided by the upstream breakwater is much more extensive than the natural ebb and flow of the beach, and erosion could remove the entire beach and any beach house behind it in a short period of time. So now, the person living behind the first breakwater may decide to build another one, west of his or her land, and pass the problem on downstream (Fig. 17–11C). The situation can perpetuate itself indefinitely, with the net result that millions of dollars must be spent to attempt to

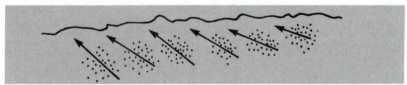

A Undeveloped beach; ocean currents (arrows) carry sand along the shore, simultaneously eroding and building the beach

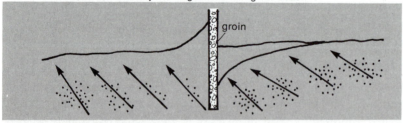

B A single groin or breakwater. Sand accumulates on upstream side and is eroded downstream

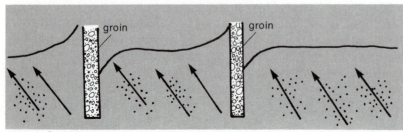

C Multiple groin system

Figure 17–11 Effect of breakwaters on a beach. *A*, Undeveloped beach; ocean currents (*arrows*) carry sand along the shore, simultaneously eroding and building the beach. *B*, A single groin or breakwater. Sand accumulates on the upstream side and is eroded downstream. *C*, Multiple groin system.

stabilize a system that was originally stable in its own dynamic manner.

Storms pose another dilemma. Periodic storms have flooded the dune lands every decade or so for millennia. A developer building a luxury home or an elite resort hotel cannot allow the buildings to be flooded or washed away every 10 or 15 years. Therefore large sea walls are constructed just inland of the beach itself. When a storm wave rolls over a low-lying dune, it dissipates its force gradually over the hills and pushes the beach sand inland. But if a sea wall interrupts this orderly flow, the waves will crash violently against the barrier. The steep breaking waves establish a circular turbulence that carries sand out to sea and erodes the beach.

An example of the futility of these human interferences is the case of Hurricane Ginger, which struck Long Island on September 30, 1971. Undeveloped islands were nearly completely flooded, and dunes were flattened, but within 10 months the beach had restabilized and was actually larger than it had been before the storm. By contrast, other beaches, which had been "protected" by sea walls, were severely eroded. In the spring of 1972 one community spent $500,000 to replenish the beach, but a storm later that year destroyed it again. In 1973, the National Park Service announced that "after spending $21,000,000 since the 1930's to erect and maintain artificial barriers against waves and storms, the agency has concluded that such work does more harm than good."

The social and political issues involved in beach erosion neatly illustrate some of the complications that develop when people try to occupy naturally delicate ecosystems. Perhaps a dynamic system such as a dune land should not be open to development at all. After all, the community as a whole benefits from unspoiled recreational areas, and the economic value of the estuaries as a breeding ground for commercial fish represents a valuable national asset. Furthermore, when an individual or corporation owns a section of beach, poor people are generally excluded. Since breakwaters and sea walls traditionally have been built with government support, this type of activity is a means whereby the poor pay taxes for the benefit of the rich. Therefore, many feel that shore reconstruction should be abandoned.

A

B

C

Figure 17–12 A breakwater system on Long Island. *A,* A beach house on Atlantic Beach, Long Island. The photograph was taken in February; note that natural rocks are exposed and there is no sandy beach. *B,* This photograph, taken from inside the house, shows a newly constructed breakwater. *C,* The same house the following winter has a sandy beach. However, the sand had to come from somewhere, perhaps from elsewhere along the coast. (Photos courtesy of Frank E. Karelsen.)

The counter argument is that significant development along the beach has, in fact, already occurred, and that the tourist business is an economic boon to local communities. In any event, the next hurricane must not be allowed to wash innumerable houses, businesses, and roadways to the sea. Therefore, new and better sea walls and breakwaters should be built (Fig. 17–12). Since breakwaters must be built along the entire stretch of coast to be effective, the government should support such projects.

17.5 Freshwater Wetlands

Many people consider swamps and marshes to be areas of wasted, useless land. After all, you can't grow wheat or tomatoes or build a house in a bog. Over the years 40 percent of the marshes in the lower 48 United States have been drained or filled in and replaced with farms, forests, or commercial developments. One key question in land use management concerns the value of an area if it is left untouched. The same question can be asked in another way—what are the economic externalities of destroying a natural habitat? An analysis of freshwater wetlands shows that their overall value is surprisingly great.

Flood Control and Maintenance of Groundwater Levels

The rate of flow of water across the land is vitally important to all terrestrial organisms. Consider a simple experiment in which two different materials, a sheet of metal and a sponge, were propped up at an angle to form two imaginary hillsides. Pour a measured amount of water over each one. The water would run off the metal quite rapidly, whereas it would be retained and stored in the sponge. In the real world, an asphalt parking lot can be compared to the sheet of metal; rainwater flows rapidly across the surface. At the other extreme, a natural wetland is more like the sponge; water is retained so that it can filter through the soil and move slowly downslope. Thus wetlands control rapid runoff and reduce flood potential during the wet season of the year and serve as a reservoir for water during the dry months (Fig. 17–13).

In many areas, wetlands have been drained, and engineers have then replaced them with artificial systems of reservoirs, dams, and flood control systems. Such projects are a type of technological fix. If well designed, the artificial system does work, and it has been the method of choice in many cases, but it is expensive. In 1972, the Army Corps of Engineers was asked to build a flood control system for the Charles River watershed in Massachusetts. Rather than build an expensive structure, the engineers decided to acquire about 3500 hectares of wetlands and simply leave them alone. The final report concluded:

> Nature has already provided the least-cost solution to future flooding in the form of extensive wetlands which moderate extreme highs and lows in stream flow. Rather than attempt to improve on this natural protection mechanism, it is both prudent and economical to leave the hydrologic regime established over the millenia undisturbed. In the opinion of the study team, construction of any of the most likely alternatives, a 55,000 acre-foot reservoir, or extensive walls and dikes, can add nothing.

The land acquisition provided recreational areas and wildlife habitats and saved the government $1.2 million every year.

Pollution Control. Many water pollutants, such as human sewage, can be readily consumed by a variety of different large and small organisms, both plant and animal. Once consumed, the pollutants are eventually converted by biological processes to nontoxic products such as carbon dioxide, water, or nitrogen (see Chapter 20). This process occurs naturally in all ecosystems, but some ecosystems cleanse themselves more rapidly than others. Bogs and marshes are areas of naturally rapid growth. When pollutants enter these areas, they fertilize many different organisms, populations expand quickly, and the pollutants are consumed and thereby removed. For example, in the mid-1970s a study of the upstream portions of the Alcovy

Figure 17–13 *A marsh is an important asset to natural systems, providing water storage and regulation, pollution control, recreation, and a habitat for many species of plants and animals. (Photo courtesy of Daniel Turk.)*

River in Georgia showed that the waterway was heavily polluted with human sewage and chicken offal. The river then passed through approximately 4.4 km of swamp-forest. This region purified the water so that the river downstream from the swamp was clean enough to support species of fish that could not survive in the polluted zone upstream.

Recreation and Wildlife Habitat. The tradeoffs between development and recreation—the difference between a parking lot and a marsh—are obvious. The importance of wetlands for wildlife, especially for migratory birds, is also important. Many aquatic birds such as ducks, geese, and cranes migrate between summer nesting grounds in the far north and winter homes in the south. In North America, many of the northern zones in Canada are relatively free of industrialization, and large areas of swampland still exist in the south. But in between lies the industrial heartland of North America. In many areas, especially along the east coast, wetlands are disappearing at an alarming rate. Loss of just a few hectares of strategically located swamps or lakes may mean removal of an ecosystem that is an essential resting zone for birds during their spring and fall migration. The waterfowl may occupy the stopover points for only a few weeks

out of the year, but without them the birds could not survive.

17.6 Case History—The Everglades

The **Everglades** is a broad, swampy region of southern Florida (see Figs. 17–14 and 17–15). The Seminole Indians called it *Pay-hay-okee*, the river of grass. Indeed it is a large, shallow, slow-moving river that starts from the shores of Lake Okeechobee and flows approximately 160 km (100 miles) into the ocean. The waters do not travel between well-defined river banks. Rather, they flow through a series of swamps that is nearly 60 km wide in places and slowly wind their way to the sea. One region of this swamp is covered mostly by low-lying plant matter called sawgrass.* Here and there in the sawgrass plain there are tree islands where dense stands of trees and bushes grow. These tree islands serve as nesting grounds and shelter for many species of birds, mammals, and reptiles.

South Florida has both a rainy and a dry season. During the rainy season the "river" level is high and the entire sawgrass plain is covered with about a meter of water. During the season of drought, some water flows through sloughs (miry channels), but the sawgrass plain is dry.

Animals and plants have adapted uniquely to the Everglade system. During the wet season plant growth is quite rapid. The sawgrass grows in abundance. If this growing season were to continue year round, the Everglade marsh would probably fill in with plant matter and evolve into a forest system. Instead, the water level starts to fall in the winter, and the grasses dry up. Almost every year, fires started by lightning or by people race across the plain. The timing of these fires appears to be crucial to the Everglade cycle. The sawgrass becomes dry and yellow during the start of the drought, even when there still may be a few centimeters of water left on the ground. If the grasses burn at this time the fire does little permanent damage to the

---

* Actually, the sawgrass is not a true grass, but a type of sedge. Sedges look like grasses, but the structure of their stems is different.

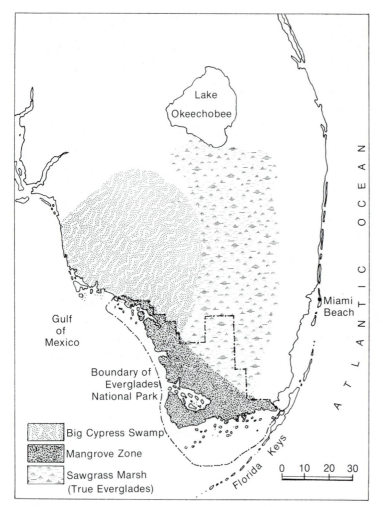

Figure 17–14 *Southern Florida and the Everglades.*

Map legend:
- Big Cypress Swamp
- Mangrove Zone
- Sawgrass Marsh (True Everglades)

Lake Okeechobee · Gulf of Mexico · Miami Beach · ATLANTIC OCEAN · Florida Keys · Boundary of Everglades National Park · 0 10 20 30

swamp. The roots of the plants are covered with water and do not burn. The tree islands are usually safe from early winter fires. At this time there is still enough water so that the plants are damp and resistant to fire. The grass fires move rapidly across the plain and burn around the tree islands but do not consume them.†

As the yearly drought continues, the waters recede still further. If not for another peculiar ecological adaptation, the swamp would dry up, and most of the animals in it would die. It is the alligators that save the swamp. These animals scoop out large depressions in the marsh with their tails. Water collects in these "gator holes."

When the plain dries up the fish seek the deep water of the holes to survive the drought. In fact, much of the aquatic life of the region becomes concentrated within the gator holes to survive until the next rainy season. Fish live and breed here. They also serve as food for the alligators as well as for predator birds and mammals.

The plants and animals of the Everglades have adapted to a complex cycle of seasonal growth, fire, and drought. Alligators alter their own physical environment. In doing so they have also ensured the continuation of other forms of animal life. The existence of all species is interconnected.

At the present time, civilization threatens this delicate cycle. Dairy farms, sugar cane fields, and orange groves lie just north of the Ev-

† Occasionally large fires do destroy the tree islands.

Figure 17–15 The Everglades. This river of grass, or Pay-hay-Okee as the Seminoles call it, is a strange mix of temperate and tropical zones. Forest and jungle, fresh and salt water ecosystems survive together in the 2000 square miles that comprise the national park. The glades are home to rare species of wildlife, such as the crocodile, manatee, and wood stork, which are found nowhere else in the United States. (Courtesy of National Park Service.)

erglades. When farmers fertilize their crops, some of the fertilizers spill into local streams and eventually enter the sawgrass marsh. Sewage runoff from the cities also flows into the Everglades and fertilizes it. Fertilizer promotes more rapid growth in the marsh. If the grasses and other plants grow too rapidly, the delicate relationships among plants and animals may be disturbed. Excess plant matter might choke the gator holes and fill the marsh with litter, algae, and weeds (Fig. 17–16). Then the alligators, fish, and birds might die during the dry season. The plant–animal balance could be altered and the glades could be changed forever.

Pesticides also leak into the Everglades killing plankton, fish, reptiles, and birds.

Perhaps the most serious threat to the marsh comes from water and flood control projects. The Everglades is dependent on a seasonal cycle of flood and drought. If the land did not flood during the wet season, the grasses and trees would not grow. If there were no drought, there would be no fire, and the marsh would slowly fill and die. But if the drought were too severe, the gator holes would dry up, and the animals would die. At the present time large quantities of water are pumped from Lake Okeechobee for irrigation and domestic use in the fields and cities of south Florida. Some of this water never returns to the Everglades. The United States Army Corp of Engineers has built a series of canals, levees, and water control gates in the marsh. They can reduce or even stop the flow of water to the southern Everglades. In fact, when a drought struck Florida from 1961 to 1965 most of the available water was pumped to nearby cities and farms, and millions of animals died. A public outcry forced the Water Control Bureau to reflood the 'Glades. The threat, however, still exists: In 1971 and again in 1981, seasonal rains were below normal, and the 'Glades became parched.

Figure 17–16 *Female alligator and young resting in a pool in the Everglades. (Courtesy of National Park Service, photo by Richard Frear.)*

The Everglades is a unique and fragile ecosystem. It is a product of millennia of coevolution of plants and animals with their physical environment. If excess pollution or improper control destroys the swamp, it will most probably be gone forever. It would be very unlikely that it could ever be re-created.

17.7 Tropical and Temperate Forests

If current trends continue . . . This phrase has been written time and time again throughout this book. It can be completed with a variety of ecological or economic catastrophes—fossil fuels will be exhausted; there will be too many people; soils will be depleted; there will not be enough food; there will not be enough water.

In modern times people are using a great many of the planet's resources faster than they are being replenished. Forests are no exception.

Four thousand years ago, a mere blink in geological time, much of the Earth was covered with forests. As civilization developed, most of the woodlands in Europe and large portions of those originally found in central Asia, the Mideast, and India were cleared. Later the same thing happened in North America, where in some regions in the midwest, 99.9 percent of the virgin timber has been cut. Today, most of the remaining woodlands in the developed countries are being replanted and reforested at the same rates at which they are being logged. But extensive deforestation is occurring in most tropical less-developed countries. The list is frightening. Between 1950 and 1980 Southeast Asia has lost nearly two thirds of its forests, Central America 40 percent, the Philippines 70 percent, Brazil 12 percent, and Africa 50 percent. If current trends continue many of these forests will be severely reduced or even obliterated by the year 2000.

There are many reasons why forests are being cut. Local residents need building materials, fuel, and paper. In addition, wood is a valuable export commodity, and virgin stands are often a valuable source of foreign exchange. Deforestation on a regional, national, and even continental scale is responsible for extensive ecological change.

Forests and Climate. A forest ecosystem controls the runoff of surface water, the trees maintain cooling shade, and the exchange of water and carbon dioxide in plant tissues maintains the balance of the Earth's atmosphere. No one knows exactly what will happen if continental-sized zones of tropical forests are destroyed. Replacement of forests with farmland in the United States and Europe has led to no obvious disaster. But these regions are temperate ecosystems, not tropical ones. If a small plot of tropical forest is cut, the temperature of the region will fluctuate from extreme highs during the day to cool temperatures (by tropical standards) during the night. In large areas that have been deforested, local weather patterns have been altered. For example, in Panama, rainfall in areas that were deforested 50 years ago has decreased by 1 cm every year (a total of 50 cm) compared to nearby unlogged regions. Many scientists fear that if a large segment of the globe is altered by deforestation, planetary climate could change. There is still no convincing evidence, however, either to support or contradict such forecasts.

Forests, Water Balance, and Erosion. In regions where the climate was favorable, forest systems once covered both hillsides and river valleys. Farmers cleared the lowland forests first because flat bottom land is best suited for agriculture. The wooded hillsides that remained were not used for food production directly but performed an important ecological function. The floor of a temperate forest is covered by a thick layer of partially decomposed litter. Beneath this litter the soil is interlaced with roots and richly endowed with humus. The soil and the litter are both effective in retaining and absorbing rainwater. As a result, the water from heavy spring rains does not run rapidly off the land but remains in the soil and percolates slowly downslope. This slowly moving water recharges creeks, streams, and groundwater supplies during the dry season. In addition, erosion is minimized when runoff rates are diminished.

As civilization and agriculture expanded during succeeding centuries, people have logged hillsides to acquire timber and to clear land for the expansion of agriculture. When forests are cut, the character of the land surface and the soil changes, reducing the ability of the system to retain water. Runoff is increased in the spring, and floods and landslides become more common. In addition, less water is available for natural plant growth or irrigation in the summer months. This series of events has occurred in certain regions all over the world, in both developed and less-developed countries, and has caused serious economic losses and increased human misery.

Forests and Agriculture. As discussed in previous chapters, nutrients in a tropical forest are locked mainly in the forest canopy, not in the soil. Therefore, when certain forest areas are cut and farmed, the thin topsoil cannot support the vast and continued agriculture that has been possible in temperate regions. Although some tropical soils are fertile after deforestation, most are not. In discussing these latter areas, one biologist stated, "We know of no sustainable use for the land after the tropical moist forests have been destroyed."

Forests and Species Destruction. Habitat destruction caused by logging of tropical rain forests is leading to the extinction of an uncounted number of species, perhaps as many as 1 million before the end of the century.

Forestry in the Developed Nations

Although the area of forested land in the developed countries is not changing appreciably, the quality of the timber is declining. Figure 17–17 shows the volume of wood produced by a tree as a function of time. In the early years of growth, the tree grows slowly and produces only a small amount of wood every year. Then, in its adolescence, the tree grows quite rapidly. Finally, in old age, growth rates diminish again. In a planned forest the maximum dollar yield per year is realized by cutting the tree just at the end of its period of maximum growth. In many commercial logging operations, trees are cut just after adolescence. But Figure 17–17 gives no information on lumber quality or, in a larger sense, the "quality" of a forest. As a tree ages, the wood becomes harder, denser, and stronger. Thus lumber from a mature tree is superior to that from an adolescent tree. To make up for this lack of quality, people have learned to substitute steel, concrete, synthetics, or glue-laminated beams made of small boards bonded together for applications that once required premium lumber.

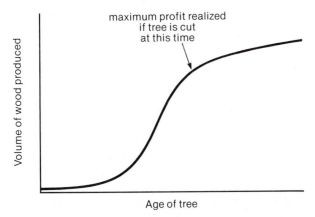

Figure 17–17 Lumber production as a function of the age of a tree. Note that the tree grows most rapidly during adolescence, and maximum profit can be realized by cutting it before it matures fully.

There are two different ways that an area can be logged. When a region is chosen for **selective cutting**, loggers harvest only those trees that produce the most valuable timber and leave the rest of the forest untouched. This approach minimizes ecological disruption, but in many cases it is more expensive than an alternate approach called **clearcutting** (Fig. 17–18). In a clearcutting operation, all the trees in an area are cut. Small logs are used to manufacture paper, and large ones are sawn into boards. The region is reduced to rubble, and then either replanted with new seedlings or allowed to regrow naturally.

To reduce environmental problems caused by clearcutting on federal lands, the U. S. Forest Service has limited the size of a clearcut to 16 hectares (40 acres). It is difficult to assess the overall impact of these logging practices. Patches of clearcut are not particularly disruptive to wildlife; in fact, many animals thrive on the lush new growth that sprouts up after the trees are cut. There is serious debate over how much soil depletion is caused when the indigenous vegetation is removed and logging roads are built. Some foresters claim that the overall soil loss is so great that the land will no longer be able to sustain forests after a few generations of cutting and regrowth. Others argue that, except when logging occurs on steep hillsides, grasses and brush grow back quickly, reduce erosion, and replenish lost nutrients after a year or two. Undoubtedly, the impact of clearcutting will vary considerably with local climate, type of soil, and many other factors.

Even in an area where soil is not destroyed by clearcutting, the question arises whether something has been lost by this operation. A noted environmentalist, Garrett Hardin, once asked the question, "How much is a redwood tree worth?" Suppose a person bought a seedling for $1 and planted it. It would take approximately 2000 years for the tree to grow to maturity. If the tree were then cut and sold for lumber, the investor would realize a profit of about ½ percent interest per year. That isn't very much, considering that money market accounts offer at least 10 percent interest per year or more. Should this calculation be used to argue that redwood trees have no value? Of course not! Since the beginning of time people have rec-

A

B

Figure 17–18 Clearcutting operations. *A,* Broad view of a clearcut logging operation in Willamette National Forest. (Courtesy of U.S.D.A. and the Forest Service.) *B,* Area in the foreground is a 15-year-old Douglas fir plantation that is well established. In the background is a recent clearcut, with mature timber on both sides. (Courtesy of U.S.D.A. and the Forest Service.)

ognized that places or things can have a value that cannot be expressed monetarily. People will lose a great part of their heritage if the beautiful places on this Earth are altered.

If you hike through a quiet stand of climax timber and then continue into a clearcut, stumbling over the rubble and the tire ruts, you can feel that something significant has been altered. In return, relatively inexpensive houses, newspapers, books, and packaging materials are made available. Jobs are created and our economic well-being is enhanced.

17.8 National Forests and Parks

Governments all over the world have realized that some ecosystems are too beautiful, too special, and too unique to be developed, industrialized, and destroyed. Therefore, different lands have been set aside to be preserved.

National Parks. In its purest concept, a national park is a piece of land that is set aside in its natural, "untouched" state to be enjoyed by all people forever. However, an inherent conflict exists between the concepts "untouched state" and "enjoyed by all." One argument contends that if people are to visit the areas and vacation in them, roads, gas stations, picnic grounds, hotels, and restaurants are needed (Fig. 17–19). The other view contends that this construction seriously alters the ecosystems. In recent years pressures on national parks have risen. Two short case histories, taken from the United States and Tanzania, illustrate the different types of problems that must be faced.

Grand Canyon Imagine yourself floating down the Colorado River through the Grand Canyon. For the moment the current is smooth and placid, but around the bend you hear the roar of white water. Soon you are committed to the rapids and dwarfed by standing waves that tower

Figure 17–19 A view of Yellowstone National Park. Even in winter, certain roads, restaurants, and hotels are maintained to simplify access. These facilities enable many to visit the park but detract from the wilderness quality. (Photo courtesy of D. Black.)

over your head. Today, many thousands of whitewater enthusiasts run the river every year. Some ride 40-foot motorized rafts while others seek challenge in tiny kayaks that can barely be seen amid the waves (Fig. 17–20). In many places the rocks of the canyon walls drop precipitously to the water's edge. In others, smooth sandy beaches offer comfortable camping sites. By the early 1970s it became obvious that most of the prime beaches were becoming polluted. Blackened fire sites and collections of human waste were despoiling the canyon. The National Park Service realized that something had to be done. There were simply too many people on the river. The Park Service placed severe restrictions on the handling of wastes and fires. They limited the traffic through the canyon to 89,000 user days per year. They also stipulated that 92 percent of the traffic should be carried by commercial guide services, and 8 percent of the permits could be used by private groups.

The problem is that private river runners feel cheated. Noncommercial groups are allowed only 8 percent of the user days, but there are many more people who would like to float the canyon on their own. A lottery is held every year. In 1977, there were 515 applications for private permits. Only 30 were approved.

Private boaters claim that the present system discriminates in favor of the rich. Just about anyone who cares to spend the money and travel with a guide can make the journey. But enthusiasts who wish to row or paddle their own boats and travel cheaply usually fail to obtain permits. Some environmentalists fear that a trend may develop. In the future, is it possible that hiking, backpacking, canoeing, scuba diving, rock climbing, and horseback riding will be similarly regulated? The problem is that some type of rules and regulations must be enforced when population pressures threaten an ecosystem. Yet it is hard to write laws that are universally just. Moreover, political lobbyists and commercial interests often sway legislative decisions. The challenge is to preserve natural ecosystems without discriminating unfairly against anyone.

The Serengeti The Serengeti plain in Tanzania (in East Africa) is the home of the largest herds of migrating ungulates left on Earth. As such, it is not only a national treasure but an interna-

Figure 17–20 *Grand Canyon National Park. A, The Grand Canyon of the Colorado. B, A large commercial outfitter preparing rafts for the journey downstream. (Courtesy of Hunt Worth.) C, An independent kayaker in the rapids of the Colorado. (Courtesy of Hunt Worth.)*

tional one as well. The Serengeti is the largest, and one of the most carefully administered parks in all of Africa (Fig. 17–21). Millions of

tourists visit the region, and tourism has become a significant source of foreign exchange and income to the country.

The concept of a park is to preserve a natural habitat as it has existed for centuries. For many thousands of years, native people have followed the game herds in Africa, hunted the animals, and lived within the Serengeti system. Some government officials say that these people must move away from the park because it is now to be preserved as a wildlife sanctuary. Others say that the tribal hunters are "natural" predators no different from lions or hyenas. The arguments go back and forth. Proponents of a peopleless park claim that times have changed. With the introduction of new medicines, tribal populations have skyrocketed so that they no longer live on a par with other predator species. Moreover, as guns have become increasingly available, the efficiency of hunting has improved so much that people could destroy the wild herds. On the other hand, if the native population were forcibly removed from the area, they would lose their land, their livelihood, and their heritage. At the present time, tribal hunters are being forced to live outside the park boundaries, but many sneak back in to poach illegally.

Wilderness. The United States government owns huge tracts of land, especially in the western states. Only a small portion of this area is set aside as parkland. The remainder is designated either as wilderness or as national forest.

A wilderness area is in many ways more protected than a national park. Wilderness areas are set aside to be preserved even if they cannot be "enjoyed by all." The original concept was that roads, buildings, motorized vehicles of any kind, and commercial activities would be prohibited from these places, and anyone who wanted to visit them must do so by foot or on horseback. The original Wilderness Act of 1964 set aside 54 of these special areas. In 1980, the total wilderness area was quadrupled by the Alaskan Lands Act, which set aside large tracts of wilderness in Alaska.

Not everyone approves of the concept of wilderness. Many argue that strategically valuable metals or fuels may exist in these areas, and our national interests demand that these be exploited. As a compromise, the Wilderness Act declared that mining would be allowed in the

Figure 17–21 Photo of Serengeti. (Photo courtesy of Stanley Wecker.)

wilderness areas on any claims filed before the end of 1983. As a result, exploration and preliminary development was intense in 1981 and 1982.

National Forests. National forests are administered according to a multiple use concept. Multiple use is established with the understanding that different people wish to use a forest in different ways. Some people like to hike or ski cross-country in solitude, others like to drive through the area in cars, motorcycles, or snowmobiles. Many would like to exploit the land for logging, mining, grazing, or commercial recreational development such as downhill skiing. Local residents often view a national forest as a potential site for gathering firewood or Christmas trees. The Forest Service has been given the nearly impossible task of managing the forests for use by all the various special interest groups simultaneously. Thus the Forest Service sells timber, issues grazing permits, leases land for ski areas, cuts and maintains hiking trails, and even contracts the construction of dams to conserve water and roads to ensure access. Of course, almost no one is pleased with all the decisions made on land use management. Some criticize the Forest Service for "selling out" to big logging interests and allowing too much clear-cutting. Others claim that resource development is being hampered and that logging and mining

should be further encouraged to provide jobs and accelerate economic development.

The questions and decisions about land use that must be made are not easy. It seems virtually certain that deep-rooted changes will occur in society within the next generation. Let us hope that in the process of making the necessary decisions, priceless qualities such as peace and solitude are not forfeited, and that short-term gains do not destroy long-range stability.

Summary

Demands on land use are varied and often conflicting. They include needs for food, water, waste disposal, living space, industrial and commercial zones, electrical energy, recreation, and transportation.

Urbanization in the developed countries has led to urban sprawl, which represents an inefficient use of land. Urbanization in the less developed countries is often associated with dismal slums and abject poverty.

Large areas of prime agricultural land are being permanently pre-empted by domestic and commercial development. The trend could be slowed by proper land use planning and a legal structure that recognizes the value of agricultural land.

Estuaries represent a small land area but are highly productive marine ecosystems. Unfortunately, they are often profitable locations for development. Development of coastal systems often runs counter to the natural movement of beaches.

Freshwater wetlands are valuable because they maintain groundwater levels, control flooding, purify natural waterways, and provide a habitat for many species of animals including migratory wildfowl.

Tropical rain forests are being depleted at an alarming rate by timber harvests and by clearing for agricultural land. Unfortunately, most tropical rain forest soils cannot support sustained agriculture.

National lands include wilderness areas where no roads or commercial development at all is allowed, parks where roads and tourist facilities but no commercial development is allowed, and forests that are designated as multiple use areas. All three regions are threatened by human impact.

Questions

Uses of Land

1. Describe the environment that you live in as urban, rural, or suburban. Obtain a map of your region and use different colored pencils to shade areas occupied by (1) factories and warehouses, (2) retail businesses, (3) low-density housing, (4) high density housing, (5) agricultural areas, and (6) open spaces. Try to design a more efficient system of land use. How would you alter the present system to make your region more efficient or more pleasant? Discuss.

2. Obtain a copy of your local zoning ordinance. Do the zoning laws promote efficient or inefficient land use? Would it be possible to improve the efficiency of land use in your neighborhood? Discuss with government officials any proposals you might have and report on their reaction.

3. What economic factors encourage construction of multistory buildings? Obtain a map of your county. Divide the map into sections rated according to the average building height in the area. What trends can be observed?

Land for Agriculture

4. Investigate local land and inheritance tax laws and report on how they affect farming in your area. Can you propose a more effective system to ensure maintenance of agricultural lands?

5. In the western United States, many farms are being subdivided into smaller lots, ranging from about 2 to 10 hectares. Owners of these lots frequently raise horses and a small garden. Would you classify this land as agricultural or suburban? Defend your answer.

Forest and Wilderness

6. Study an undeveloped forest, river, valley, estuary, or swamp in your area (if one exists). Discuss the present human impact on the area and the threat of increased pressure in the future. What is the value of the area in its present state? Discuss the tradeoffs involved in future development.

7. Many of the rich food-producing areas of the United States were once forested. Is it hypocritical for U. S. ecologists to recommend that people in tropical countries preserve their forests when North Americans have gotten rich destroying theirs? Defend your answer.

8. Obtain a list of state and national lands closest to your home. Are these parks, wilderness areas, or forests managed for multiple use? Describe the management of these areas.

Suggested Readings

General land use theory is discussed in:

Craig W. Allin: *The Politics of Wilderness Preservation.* Greenwood, Westport, CT, 1982. 304 pp.

G. Gordon Davis and Richard A. Liroff: *Protecting Open Space: Land Use Control in the Adirondack Park.*

Cambridge, MA, Ballinger Publishing Co., 1981. 302 pp.

Robert W. Burchell and Edward E. Duensing (Eds.): *Land Use Issues of the 1980's*. Piscataway, NJ, Center for Urban Policy Research, 1982. 220 pp.

Anthony J. Catanese and W. Paul Farmer (Eds.): *Personality, Politics, and Planning*. Beverly Hills, Sage Publications, 1978. 225 pp.

Ian McHarg: *Design with Nature*. New York, Doubleday and Company, 1969. 197 pp.

Brown Muller, Neil Pinney, and William Saislow: *Innovations in New Communities*. Cambridge, MA, MIT Press, 1972. 301 pp.

Development of agricultural land is documented in:

Lester R. Brown: *Building a Sustainable Society*. New York, W. W. Norton and Company, 1981. 433 pp.

Lester R. Brown: Vanishing croplands. *Environment*, 20(10): Dec. 1978.

Douglas M. Costle: Growth, land, and the future. *EPA Journal*, July/Aug. 1980. pp. 2 ff.

Robert G. Healy and James L. Short: The changing rural landscape. *Environment*, 23(10): 7, Dec. 1981.

Archibald M. Woodruff (Ed.): *The Farm and the City— Rivals or Allies*. Englewood Cliffs, NJ, Prentice-Hall, 1980. 184 pp.

Coastlines and shore erosion are discussed in:

Joseph M. Heikoff: *Politics of Shore Erosion: Westhampton Beach*. Ann Arbor, MI., Ann Arbor Science Publishers, 1976. 173 pp.

Thomas C. Jackson and Diana Reische (Eds.): *Coast Alert*. San Francisco, Friends of the Earth, 1981. 181 pp.

Bostwick H. Ketchum (Ed.): *The Water's Edge: Critical Problems of the Coastal Zone*. Cambridge, MA, MIT Press, 1972. 393 pp.

Wetlands are discussed in:

Elinor L. Horwitz: *Our Nation's Wetlands*. Washington, D.C., Council on Environmental Quality, U. S. Government Printing Office, 1978. 70 pp.

Two books that discuss the Everglades marsh system are:

Jean Craighead George: *Everglades Wild Guide*. U. S. Department of the Interior, National Park Service, 1972. 105 pp.

John Harte and Robert H. Socolow: *The Patient Earth*. New York, Holt, Rinehart and Winston, Inc., 1971. 364 pp.

Four articles concerned with the destruction of tropical rain forests are:

Stephen G. Bunker: Forces of destruction in Amazonia. *Environment*, 22(7): Sept. 1980.

Conservation Foundation Letter. Pressures on the world's forests keep mounting. Aug. 1981.

Norman Myers: The conversion of tropical forests. *Environment*, 22(6): July/Aug. 1980.

M. K. Ranjitsinh: Forest destruction in Asia and the South Pacific. *Ambio*, 8(5): 1979.

Urban land use is the subject of the following books:

John C. Bollens and Henry J. Schmandt: *The Metropolis—Its People, Politics, and Economic Life*. 3rd ed. New York, Harper & Row, 1975. 401 pp.

Melville C. Branch: *Planning Urban Environments*. Stroudsburg, PA, Dowden, Hutchinson, and Ross, 1974. 254 pp.

Donald A. Krueckeberg (Ed.): *Introduction to Planning History in the United States*. Piscataway, NJ, Center for Urban Policy Research, 1982. 235 pp.

W. Patrick Beaton (Ed.): *Municipal Expenditures, Revenues, and Services*. Piscataway, NJ, Center for Urban Policy Research, 1982. 166 pp.

David Listoken (Ed.): *Housing Rehabilitation: Economic, Social and Policy Perspectives*. Piscataway, NJ, Center for Urban Policy Research, 1982. 312 pp.

The economics, strategies, and politics of renewable resources are taken up in:

D. L. Little, R. E. Dils, and J. Gray: *Renewable Natural Resources: Our Next Crisis or Our Next Opportunity*. Boulder, CO, Westview Press, 1982. 316 pp.

Frank J. Popper: *The Politics of Land-Use Reform*. Madison, University of Wisconsin Press, 1981. 321 pp.

18

Nonrenewable Mineral Resources

Living organisms use a source of energy (ultimately, the Sun) to convert materials from the environment (nutrients) into body tissues. The time scale of these conversions is the time scale of the spans of life—months for plant fibers such as cotton, years for animal material such as bone and hide, and decades for wood. On the time scale of human lives, therefore, these materials are **renewable**; if we do not use them up faster than they are produced, they need never be exhausted.

Geological processes, like those of life, can also organize and concentrate materials, but here the time spans extend to millions and billions of years. Since humans cannot wait that long, mineral resources are said to be **nonrenewable**.

This chapter first reviews the structure of the Earth, the drift of continents, and the geological processes by which minerals and fossil fuels were formed. The pollution associated with mining operations is also discussed. Mining is but a first step in obtaining the materials used in modern technical societies. Metals, in particular, are needed because of their unique properties. The chapter closes with a consideration of the future prospects for nonrenewable mineral resources.

18.1 The Formation of Mineral Deposits

Structure of the Earth

Ancient philosophers theorized that the Earth was hollow, rather like a tennis ball, with a thin outer shell and a void in the center. This belief persisted in some circles even up to modern times. In the 1960s the Congress of the United States was considering legislation to finance the drilling of a test well under the ocean in an effort to penetrate the Earth's crust and sample the rock of the mantle. Several concerned people wrote letters telling their senators and representatives that if such a hole were bored it would pull the plug, so to speak, and all the oceans' water would drain away into the middle of the Earth. Such theories are not taken seriously. In fact, scientists have determined that the Earth is composed of several distinct layers of different kinds of solid or liquid matter.

Scientists believe that billions of years ago, natural decay of radioactive elements within the Earth released enough heat to melt vast volumes of primordial rock. Most of the dense elements

386

such as iron and nickel that were found in this mass of liquid rock gravitated toward the center, which we now call the **core**. These were then surrounded by a **mantle** of lighter rock and a surface **crust** that is generally of comparatively low density. The present structure of the Earth is shown in Figure 18–1.

The boundaries between the layers of rock inside the Earth are not rigid and impenetrable. Rather, rock flows upward and downward in slow but continuous and dynamic exchange. Thus the Earth is believed to have formed a hard solid crust about 3.5 to 4 billion years ago, but virtually none of that original crust remains. Almost all of the original rock has been pushed downward into the mantle, reabsorbed, and replaced by material coming from deep within the globe.

Careful studies have shown that the mantle itself is separated into several layers, each with its own unique physical characteristics. Some mantle rock is brittle. Other regions of the mantle are hot and semifluid and behave somewhat like heavy, viscous putty. Various forces within the mantle create large stresses, pushing the rock in one direction or another. When the brittle layers are stressed severely they may crack sharply, causing earthquakes. On the other hand, the puttylike "plastic" layers tend to flow gradually when stressed and do not break

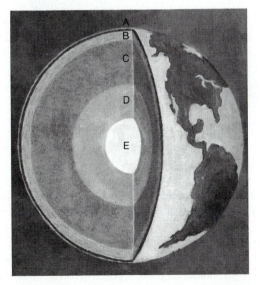

Figure 18–1 The structure of the Earth. *A*, Crust; *B*, mantle, upper layer; *C*, mantle, lower layer; *D*, outer core, probably liquid iron and nickel; *E*, inner core, probably solid iron and nickel.

sharply. There are certain regions within the plastic mantle layer that contain a much more fluid material. This fluid, called **magma**, consists of melted rock mixed with various gases such as steam. Magma flowing quickly to the surface of the Earth produces a volcano; the outpouring magma is called **lava**. If magma protrudes into the crust slowly through a crack or fissure in the rock but does not travel all the way to the surface, it will cool gradually deep within the crust.

For most practical purposes it may safely be assumed that the Earth's crust is a rigid mass of rock lying on the surface of our planet. Thus we expect that the distance between any two cities in the world will remain constant from year to year, and that the continents will always lie in the same relationship to each other as they now do. However, according to modern geological theory, the continents are not immobile and rigidly fixed in position. Geologists now believe that the Earth's crust is composed of several large continent-sized plates of solid rock and that each piece floats about on the semifluid plastic mantle. The continents and ocean basins float on the denser mantle fluid, much as tightly packed icebergs might float on water.

This concept of "floating" continents explains a great many observations. Look at a map of the world as shown in Figure 18–2A. If you were to cut out the continents and try to piece them together as part of a jigsaw puzzle, you would find that they fit together amazingly well, as shown in Figure 18–2B. From this evidence alone early scientists deduced that perhaps there once existed one or perhaps two large supercontinents. The supercontinent(s) then broke apart, and the pieces slowly drifted away from each other to their present position. This idea is called the theory of **continental drift**. Of course, as continents move, they carry their mineral deposits with them.

Formation of Mineral Deposits

The chemical composition of the rock and soil varies from place to place on the Earth's surface. These changes often occur abruptly, especially in mountainous regions. For example, outside of Boulder, Colorado, at the foothills of the Rocky Mountains, a series of uplifted sedimentary rock marks the landscape (Fig. 18–3A). Just a few kilometers away, however, many of the

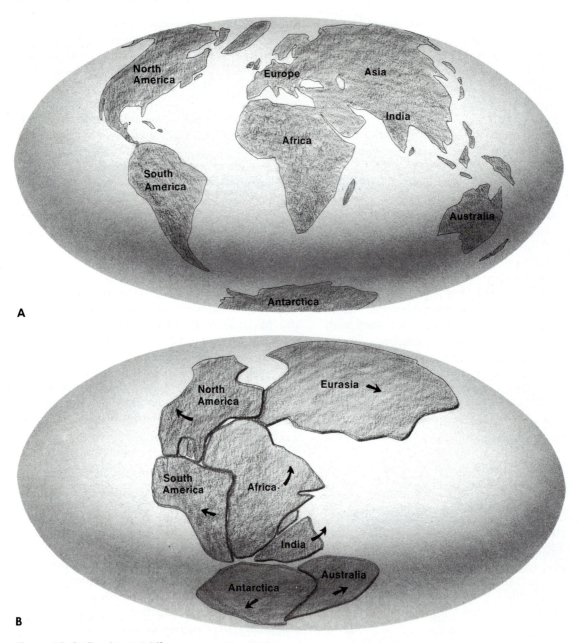

Figure 18–2 Continental drift.

imposing cliffs of Boulder Canyon are of granite (Fig. 18–3*B*). Rock chemistry may change even more abruptly, and two or more distinct types of minerals may appear in a single formation. Figure 18–4 shows small segments of one type of rock, called a **vein**, embedded in a dominant rock formation. One important aspect of geology is to try to understand the processes that cause local concentrations of one mineral or another.

Aside from theoretical interest, this problem has considerable practical importance, because geologists are always searching for concentrated deposits of metals, fuels, and fertilizers.

The magma lying in the Earth's mantle varies from place to place. The chemical composition of the lava coming from Mauna Loa in Hawaii is likely to be different from that coming from Vesuvius in Italy. But only a small fraction

A

The third Flatiron, a hard sandstone formation near Boulder, Colorado.

B

This rock outcrop, located only a few miles from the third flatiron, is composed primarily of granite.

Figure 18–3 The composition of rock often changes abruptly within short distances.

Figure 18–4 Veins of different minerals embedded in a dominant rock formation.

of the Earth's crust is composed of lava that shot rapidly out of a volcanic opening. Much of the igneous rock oozed up slowly through cracks in the crust, cooling gradually during its travel. If a layer of older rock is cracked apart and invaded by some upflowing magma, the final formation will contain veins of foreign rock.

Weathering and erosion also play an important role in developing heterogeneity. If several different rock types lie in the same region, the softer rock will erode away more quickly, leaving exposed layers of hard rock rising over a valley or plain of sedimentary material.

These and many other chemical and physical processes are responsible for separation of rock into distinct formations. It is these processes, in fact, that account for the development of economically significant deposits of ore, fuel, and fertilizer.

Separation by Gravity

Suppose that two minerals were mixed together within a molten lava. If the magma is agitated and pushed upward to the surface by a volcanic eruption and then frozen quickly, the minerals will be more or less evenly dispersed in the newly formed rock. But suppose, instead, that the magma had started to move up slowly through a fissure in the crust and cooled gradually while still kilometers under the surface. Say for example that one of the minerals solidified when cooled to 1500°C and the other solidified at 1200°C (Fig. 18–5). As the total mixture cooled, the mineral with the 1500°C melting point would start to solidify while the rest of the magma remained liquid. But now remember that the cooling process occurs slowly, sometimes over a period of many thousands of years. During this time, all the tiny crystals of solid minerals would settle downward through the lighter liquid melt until there existed a concentration of one mineral at the bottom of the uplifting magma. This deposit may lie deep in the crust, unreachable by modern mining techniques, or,

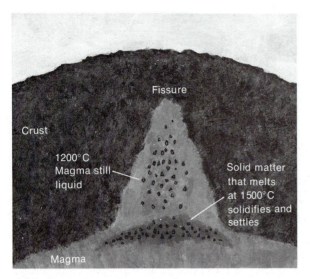

Figure 18–5 *Separation by gravity.*

alternatively, it may be uplifted by geological processes and exposed by erosion.

Separation by Differential Solubility

Minerals may also become concentrated inside the Earth by differential solubility. To understand how this process works, take a little bit of salt and mix it up with a lot of sand. It would be physically difficult to pick out the grains of salt from the grains of sand. But suppose you put the entire mixture in a glass of water and stirred it up. The salt would dissolve into the water, and the sand would settle to the bottom. You could then pour off the water and collect the sand. The salt, which remains in the water, could be retrieved by evaporating the water and collecting the residue that remained.

Similar processes can occur at the Earth's surface. Many valuable mineral deposits in Utah were formed by differential solution of rock in water. Rainwater traveling across rock and soil dissolves minerals such as salt (NaCl) and potash (K_2CO_3), whereas other minerals, such as quartz (SiO_2), are largely unaffected. In most natural ecosystems, these dissolved salts are either reabsorbed by plant roots or carried into the ocean. But occasionally, large land-locked lakes develop that have no outlet into open sea. Water flows into these lakes through streams and rivers but can escape only by evaporation. When salt water evaporates, the water escapes as va-

por and the salt remains behind. As time passes, the mineral concentration of land-locked lakes increases. There was once a large land-locked sea in North America, covering much of northwest Utah. The salinity of this sea increased from age to age. Then a major change of climate occurred. The region got hotter and drier, and streams flowing into this sea slowed down or dried up. The result was that more water escaped through evaporation than entered via streams, and the lake shrank to its present size. As the waters receded, mineral deposits were left on dry land, and these are now mined commercially.

Placer Deposits

There has always been a great deal of legend and intrigue about mining, especially in the pioneering days when Europeans colonized new continents of the Americas, Africa, and Australia and searched for mineral wealth. Prospectors probed the streams and hillsides for all kinds of metals, but most of all it was the search for gold that led them to face the dangers of unknown lands, frigid cold, and long periods of loneliness. Occasionally miners discovered veins of nearly solid gold, but much of the gold was taken out of streams or dug out of small **placer deposits**. Because gold is a particularly dense metal, if a mixture of gold dust, sand, and soil is swirled in a glass of water and the solids are allowed to settle out, the gold will fall to the bottom first. This action allows gold to be collected by **panning** (Fig. 18–6). A miner who suspects that gold may be found in a certain area places a small shovelful of soil in his pan, adds water, and swirls the mixture. Then gradually and carefully the miner dumps out the particles on the surface of the swirling liquid, adds more water, and continues the process. If there is any gold in the sediment, it will remain in the pan to the end.

Differential settling may occur by natural processes. Suppose, for example, that small amounts of gold from a mountainside are carried downstream by a river. Chances are that the concentration of gold in the river water and surface sediment will be too small to mine economically. But now imagine that the river is partially blocked downstream by a beaver dam. As the water reaches the pond behind the dam, it slows

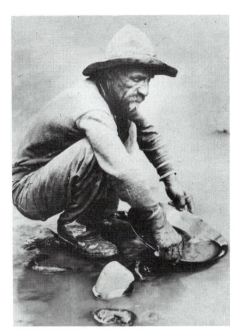

Figure 18–6 Panning for gold. (Courtesy of Colorado Historical Society.)

down but does not stop and continues over the dam and on toward the sea. When rains fall and snows melt in the early spring, the stream may carry a lot of soil—mostly fine sediment with an occasional speck of gold. When the water reaches the beaver pond and slows down, the sediment will settle. Gold, being much denser than rock, will settle first. If the stream flows at just the right speed, the water will carry most of the lighter sediments downstream. As the years go by, more and more gold specks collect in one location, until a small concentration, called a placer deposit, is formed.

BOX 18.1 PANNING FOR GOLD

"Looks good to me," he concluded, picking up his pick and shovel and gold pan.

He crossed the stream below the pool, stepping agilely from stone to stone. Where the sidehill touched the water he dug up a shovelful of dirt and put it into the gold pan. He squatted down, holding the pan in his two hands and partly immersing it in the stream. Then he imparted to the pan a deft circular motion that sent the water sluicing in and out through the dirt and gravel. The larger and the lighter particles worked to the surface, and these, by a skillful dipping movement of the pan, he spilled out and over the edge. Occasionally, to expedite matters, he rested the pan and with his fingers raked out the large pebbles and pieces of rock.

The contents of the pan diminished rapidly until only fine dirt and the smallest bits of gravel remained. At this stage he began to work very deliberately and carefully. It was fine washing, and he washed fine and finer, with a keen scrutiny and delicate and fastidious touch. At last the pan seemed empty of everything but water; but with a quick semicircular flirt that sent the water flying over the shallow rim into the stream, he disclosed a layer of black sand on the bottom of the pan. So thin was this layer that it was like a streak of paint. He examined it closely. In the midst of it was a tiny golden speck. He dribbled a little water in over the depressed edge of the pan. With a quick flirt he sent the water sluicing across the bottom, turing the grains of black sand over and over. A second tiny golden speck rewarded his effort.

(From) Jack London: All gold canyon. In *Short Stories of Jack London*. New York, Funk and Wagnalls, 1968, p. 222.)

Metal salts react with oxygen rich sea water. Minerals settle out.
Water with dissolved iron and manganese is heated and rises.
Hot rising magma

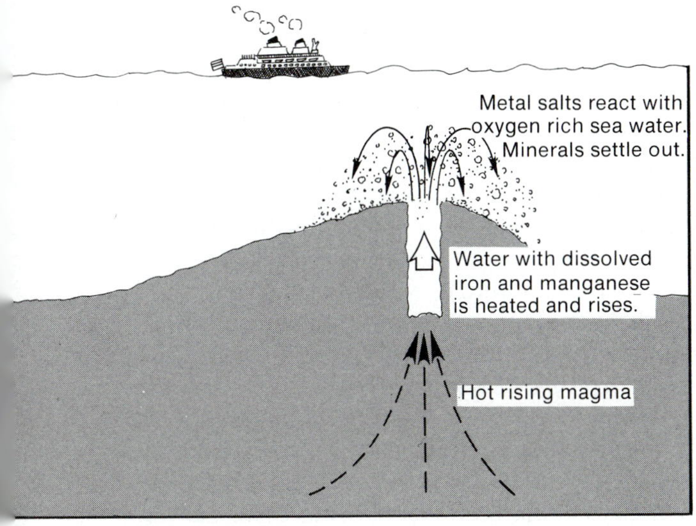

Figure 18–7 Formation of undersea mineral deposits.

Minerals Under the Sea

Many mineral deposits have formed on the ocean floor. Some of these were uplifted during the collisions of continental plates and became easily accessible (Fig. 18–7). The rich iron, copper, lead, and zinc deposits that contributed greatly to the wealth and power of the early Roman empire were formed in this manner. Other valuable deposits lie relatively inaccessible under several kilometers of sea water.

As continental plates separate and travel away from each other, new mineral matter from the mantle flows up through the gap and solidifies. When this occurs under the sea, large ridges rise up from the sea floor. Very deep cracks and crevices have been observed along these boundaries. Sea water seeps downward through the cracks and comes in contact with fresh, rising magma. Such valuable minerals as iron, manganese, copper, lead, and zinc may be dissolved or suspended in this water. As the water rises back up to the ocean floor, it mixes with the oxygen-rich waters above, and many metallic minerals can react with the oxygen to form metal oxides. These metal oxides are insoluble, and they fall to the sea floor near the ocean ridges in rich layers. As a result, valuable minerals are found near the midocean ridges.

18.2 Ore Deposits

As explained in Chapter 1, an **ore** is considered to be a rock mixture that contains enough valuable minerals to be mined profitably with currently available technology. The **mineral reserves** of a region are defined as the estimated supply of ore in the ground. Reserves are depleted when they are dug up, but our reserve supply may be augmented by either of two processes. First, new deposits may be discovered. In addition, there are many known deposits that are now uneconomical to mine. For example, a deposit containing 20 percent iron is not considered an ore because it is so expensive to extract the metal from the rock that it would not be profitable under current market conditions. If technology improves so that the materials can be refined cheaply, or if the market price of iron increases, the deposit will suddenly become an ore reserve.

Many of the very high-grade, concentrated, and easily accessible ores, such as the 50 percent iron deposit of the Mesabi Range in the north central United States, are being used up rapidly and either have been depleted or will be depleted in the near future. These mines are essentially nonrenewable, and once they are gone, our civilization will have suffered an irreplaceable loss. But our technological life will not end

with the exhaustion of these rich reserves, because less concentrated deposits are still available in great abundance. For example, there has been recent concern that phosphorus reserves needed for the manufacture of fertilizers will be depleted soon. At the present time it is economical to exploit a mine containing about 32 percent or more of phosphates. These reserves are being exploited heavily and will be depleted in the near future. There are, however, many deposits containg 29 percent phosphate, perhaps 100 times as much as the original supply of 32 percent ore. Much of this 29 percent phosphate lies under the sea and therefore would be expensive to extract, but it is available.

The situation is similar for a great many other minerals as well. For example, there are greater volumes of deposits containing 0.5 percent copper than 0.6 percent copper, and more ores containing 35 percent iron than 40 percent iron. Some of these lower grade deposits are easy to reach; others exist in harsh environments such as under the sea or in the Arctic. The question arises: Will the technology be developed to extract the lower grade ores at reasonable prices? Naturally, there are differences of opinion in this matter. Optimistic geologists say yes, the technology and machinery for refining ores are rapidly becoming more efficient, and progress will continue in the future. These people point to several phenomenal success stories. Shortly after World War II it became obvious that high-grade iron ore deposits in the United States were being depleted rapidly, and consequently there was a great deal of alarm about possible iron shortages. Fortunately metallurgical engineers discovered a new refining process to extract iron from low-grade ores at competitive prices. Thus the impending crisis was averted, and the price of iron actually decreased (see Section 1.4). As technology improves, many well-known mineral deposits become ores. In addition, new deposits continue to be discovered, with the result that the reserves of many important minerals have recently *increased* in some parts of the world.

Other geologists contend that such trends are not likely to continue for long. They point out that a finite supply of concentrated mineral deposits cannot last forever. Moreover, as increasingly lower grade ores are sought, the technological problems inherent in all aspects of mining and refining rise sharply. Dependence on technology to solve all problems may lead to grave disappointment. These scientists emphasize that the future availability of metal depends on many factors besides the quantity of ore in the ground and the state of refining technology. Some of these are discussed in the following paragraphs.

Availability of Energy

To extract metal from ore, the dirt and rock must be dug up and crushed, the ore itself must be separated and chemically reduced to the metal, and the metal must finally be refined to purify it. Each step, especially the chemical reduction, requires energy. Moreover, low-grade ores require much more energy to process than do high-grade ores. For example, if you are mining ore containing 29 percent phosphorus, more material must be dug, transported, and crushed to obtain a given yield of product than if you are mining ore containing 32 percent phosphorus. Some low-grade ores differ from high-grade ores not only in concentration but in chemical composition as well. Some chemicals are easier to purify than others. For example, it is energetically advantageous to extract lead from its sulfide ore (PbS), according to the generalized reaction:

$$PbS + O_2 \longrightarrow Pb + SO_2$$
lead sulfide + oxygen $\longrightarrow$ lead + sulfur dioxide

Lead also occurs in ores of other chemical compositions, but the overall energy requirements for processing them are much greater.

Pollution and Land Use

Most mining processes cause significant pollution of land, water, and air. For example, sulfur is found in large quantities in many ore deposits. This sulfur, bound in forms such as copper sulfide (CuS and Cu_2S), reacts with water in the presence of air to produce sulfuric acid (H_2SO_4), which runs off into the streams below the mine. This pollution, known as **acid mine drainage,** kills fish and disrupts normal aquatic life cycles. When sulfur accompanies other chemicals through refining processes, it is often converted to gaseous air pollutants such as hydrogen sulfide and sulfur dioxide. Sulfur, of course, is not the only polluting chemical from mining operations. Many other mine pollutants cause serious air and water pollution.

Just as more energy is required to handle low-grade ores than high-grade ones, more pollution generally results from processing these impure materials. The pollution can be controlled with highly specialized pollution-abatement equipment, but such measures are expensive and add to the total cost of refining ore.

The world is running short of food, energy, and recreational areas as well as of high-grade mineral deposits. What should our policy be if a valuable ore or fuel lies under fertile farmland or a beautiful mountain? Which resource takes precedence? At present this question is being raised principally with respect to exploitation of fuel reserves, for vast coal seams lie under the fertile wheat fields of Montana and the Dakotas in the United States, and southern Saskatchewan in Canada. If large areas of low-grade ore must be exploited, the problem will extend to metal reserves as well.

18.3 Key Metals and Minerals

Civilization has always been associated with metals. Ancient civilizations arose as the Stone Age was transformed to the Bronze Age. Modern civilization began with the Iron Age.

What is so special about metals? Most metals do not even occur in their metallic state in nature but have to be extracted from ores by processes that had to be invented, that require energy, and that produce pollution. On the other hand, there has always been a natural abundance of stone, clay, wood, bone, and fiber. So why are metals important? There are several reasons, of which three are listed here.

1. Metals have much greater strength than other materials (Table 18–1). Furthermore, metals are not brittle, whereas stone is. Heroic structures such as the pyramids (Fig. 18–8) can be built of stone, which resists compression. Long spans, however, such as those of modern bridges (Fig. 18–9), require the tensile strength that only metals can provide.

TABLE 18–1   Approximate Strengths of Materials

Material	Maximum Ultimate Strength (kg/cm^2)*
Metals:	
Wrought iron	3375
Steel**	17,600
Bronze†	9500
Aluminum	3900
Wood:	
Oak	310
Pine	360
Cypress	310
Stone:	
Granite	1750
Marble	1125
Limestone	1400
Sandstone	1340

*"Strength" of a material is its resistance to being deformed. Stone resists crushing (compressive strength) more than pulling (tensile strength) or slipping (shearing strength). The values given for stone are its compressive strength. The values for tensile strength are much lower. Metals and wood resist both crushing and being pulled apart. To convert kg/cm^2 to lb/in^2, multiply by 14.2.

**Steel is iron alloyed with other metals to impart strength and resistance to corrosion.

†Bronze is an alloy of copper and tin.

2. Metals can be melted and cast into any shape that a mold can provide. Metals can also be machined to precise dimensions (Fig. 18–10). Wood does not melt; it decomposes first. Some stones, such as limestone and marble, also decompose when heated. For stones that do melt, the high temperatures required and the brittleness of the product make casting in molds impractical.

3. Metals conduct electricity very well. No other materials come close. A society that needs electricity needs metals.

Table 18–2 summarizes the sources, uses, availability, and pollution problems associated with some key metals. Note some interesting aspects of these data. First, the supplies of metals listed refer to those in the Earth's crust, not in the entire Earth. (The Earth's core is iron and nickel, but there is no way we can get at it.) But abundance alone does not tell us all we need to know about availability. The key question is how concentrated the metal is in its ore. Note, however, that some metals that are very important to us (mercury and platinum, for example), are really very rare in the Earth.

Figure 18–8 One of the pyramids of Gizeh, near Cairo, Egypt.

18.4 Two Key Nonmetals: Helium and Phosphorus

Helium (He) is the second lightest (after hydrogen) of all gases. It is chemically inert—it does not react with anything. It does not even become radioactive with neutron bombardment. (The nucleus of the helium atom is an alpha particle, which is *extremely* stable; it is the *product* of nuclear fusion in the Sun and in the hydrogen bomb.) Helium has the lowest boiling point (−269°C) of any substance. When it does cool to a liquid it remains liquid almost down to absolute zero. Liquid helium (which is available commercially) is therefore the ultimate cooling agent.

Because of these unique properties, helium has uses for which there are no substitutes:

1. For lifting, with safety. Its lifting ability in balloons is almost as good as that of hydrogen (Box 18.2), but helium cannot burn.

2. To create an absolutely inert atmosphere, for welding and other processes, by excluding air.

3. For use as a cooling agent in some nuclear reactors and computers.

4. To test for potential leaks in anything. If helium doesn't leak, nothing else will.

5. For use as a substitute for nitrogen in divers' breathing gas. (The presence of nitrogen causes the painful "bends" when the pressure on the diver is reduced.)

Figure 18–9 Steel is much stronger than stone. *A,* The Pont-du-Gard aqueduct, erected in the last quarter of the first century B.C. as part of a canal that brought water to Nimes. Although this is an impressive structure, note that massive masonry is needed to support small spans. *B,* The New River Gorge Bridge, world's longest arch (520 meters) under construction. The delicate steel arch is designed to support a roadway above it.

Helium is present in the atmosphere (0.0005 percent) and in natural gas (about 0.4 percent). It is therefore $^{0.4}/_{0.0005}$, or 800 times as concentrated in natural gas as in the atmosphere. As billions of cubic feet of natural gas are burned each year, the helium present in it (which doesn't burn) simply enters the atmosphere and thus becomes 800 times more dilute. There is enough helium for our uses now, and there will always be enough (in the atmosphere), but it will take 800 times as much energy to extract it

when we run out of natural gas. What are we to do? Shall we extract and stockpile it now to save energy for future generations? Various recent reports express widely different views on such questions. The conservation and stockpiling position was taken in 1978 by the Helium Study Committee of the National Research Council and in 1979 by a House Subcommittee on Energy and Power. The opposite view, that the supply of helium is unlimited and that it would be unfair to penalize the present users of natural gas

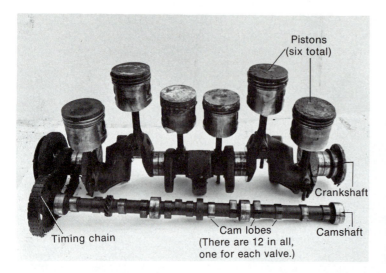

Figure 18–10 The crankshaft, camshaft, timing chain, and pistons of an automobile engine. Only metals can be so precisely machined and so strong.

(who would pay for the helium processing by a price rise) for an uncertain future benefit, was taken in 1978 by an Interagency Helium Committee.

Phosphorus (P) is not abundant in the Earth's crust. (At 0.13 percent, it ranks tenth among elements.) Furthermore, it is not easily replaceable in an ecosystem from which it is leached out or removed in the tissues of organisms. Fortunately, phosphorus is concentrated in the form of phosphate rock in various parts of the Earth, and these phosphates are mined and used as fertilizers that are essential to modern agriculture. Phosphate reserves are estimated to be sufficient for several hundred and perhaps even a thousand years, but that is not forever. It is not at all clear how global agriculture will be maintained when high-grade phosphate ores are depleted.

BOX 18.2 LIFTING ABILITIES OF HELIUM AND HYDROGEN

The density of a gas is proportional to its molecular weight. Helium (molecular weight 4) is therefore twice as dense as hydrogen (molecular weight 2). But the lifting ability of a gas depends on the difference between its molecular weight and that of air (molecular weight 29). The lifting ability of helium is therefore 29-4, or 25, compared with hydrogen's 29-2 or 27. Helium is therefore $\frac{25}{27}$, or 93 percent, as good as hydrogen in lifting ability.

18.5 Future Prospects

Are we or are we not in danger of running out of mineral resources? Consider iron, for example. In 1966 it was estimated* that the global resource reserve of iron was about 5 billion tonnes. At that time, the global annual consumption rate was about 280 million tonnes. If those figures were correct, and if consumption continued at a constant rate, then the iron reserves of the Earth would be consumed in 5 billion/280 million, or 18, years, bringing the date for the end of iron reserves to 1966 + 18 years, or 1984. There must be something wrong with such calculations. "Reserves" is not a constant factor but can change markedly with exploration and with the development of methods suitable for processing ores of lower grade. In the case of iron, the big change has been the development of improved methods of processing taconite rock for its iron content (see Table 18–2).

Would it be reasonable to expect similar dramatic improvements in the technology of mining other minerals? Such progress would make it profitable to mine less and less concentrated sources, and then we need never run out of anything. Such an approach ignores some serious difficulties. First, the concentrations of minerals

---

* B. Mason: *Principles of Geochemistry.* 3rd Ed., Appendix III. New York, John Wiley & Sons, 1966.

TABLE 18–2   Sources, Uses, and Environmental Problems of Some Important Metals

Metal and Its Abundance in the Earth's Crust	Sources and Reserves	Properties and Uses	Environmental Problems
Aluminum (Al) 8.13%	Bauxite, an impure form of Al_2O_3. Reserves are plentiful.	Only one-third as dense as iron; resists corrosion; good electrical conductor. Used in aircraft, automobiles, beverage cans, pots and pans.	Much electrical energy used in production (about 67,000 kilowatt hours per tonne of aluminum). Also some fluorine-containing gases and dusts are produced in manufacture.
Iron (Fe) 5.1%	Hematite (Fe_2O_3) and magnetite (Fe_3O_4). The highest grade hematite ores are largely exhausted. Magnetite is found in taconite rock; reserves are still vast, and extraction technology has improved.	Major metal for all structures and machinery.	Air, water, and land pollution occur both at the mine and in the chemical production of iron from the ore. Taconite deposits also contain asbestos.
Copper (Cu) 0.007%	Elemental copper, mined since ancient times, is largely exhausted. Major sources now are copper sulfides, CuS and Cu_2S.	Excellent conductor of heat and electricity. Used for electrical wire, water and steam pipes, and cooking utensils.	Acid mine drainage. Smelters that process the sulfides produce large amounts of sulfur dioxide, SO_2.
Lithium (Li) 0.003%	Widely distributed, but in small quantities, in various minerals.	Will be important for the fusion energy program because the ^6Li isotope is needed to make tritium (see Chapter 12).	No serious problems
Lead (Pb) 0.0016%	Major source is galena, PbS. Reserves are concentrated but not abundant.	Soft, dense metal that is fairly resistant to corrosion and has fairly low melting point (327°C). Used for pipes, solder, electrodes in batteries, and pigments in paint. Its use as an antiknock agent in gasoline is declining.	Acid mine drainage. SO_2 is produced in lead smelters. Lead compounds are cumulative poisons.
Gallium (Ga) 0.0005%	Found as an impurity in ores of zinc and aluminum but in small amounts.	Used in solar cells to convert solar energy to electricity. If the solar energy program expands, gallium reserves will be critical.	No special problems
Mercury (Hg) 0.00005%	Sometimes occurs as the native element in small amounts, but the important ore is cinnabar, HgS. Reserves are very limited.	The only metal that is liquid at all ordinary terrestrial temperatures. Used in electrical switches, in thermometers, and for many special chemical and medicinal purposes.	Mercury compounds are toxic.
Platinum (Pt) 0.00002%	Occurs as the native element in widely scattered ores.	Unsurpassed as a catalyst for oxidation reactions, for which it is used in catalytic converters to reduce pollution from automobile exhausts. Platinum is also widely used as a catalyst in industry.	No special problems

do not fall off gradually as the prime ores are mined. The facts are just the opposite: The boundaries between high and low concentrations are rather sharp, and the differences in concentration are great. (Note the similarity to helium.) Take the case of silver, for example. Rich sources occur in sharply defined veins that branch out between shallow rock layers like a tree. If you discovered a vein like the famous Comstock lode, you would be boundlessly rich.

This source yielded bonanzas of silver from 1860 to 1880 (Fig. 18–11). When the richest veins of such a lode are exhausted, production falls off, but later, as prices rise, it pays to mine lower grade ore in the same lode. When these in turn are exhausted, the mine is abandoned. The miners do not return until either the price rises to very much higher levels or an entirely new processing technology is developed. Note from Figure 18–11, however, that the *overall* production

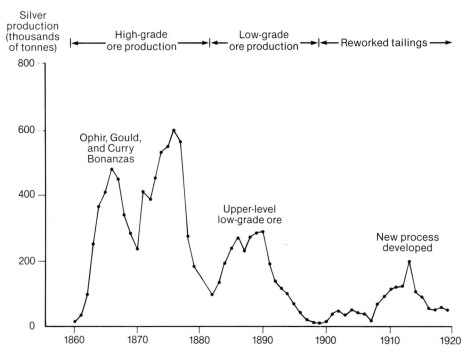

Figure 18–11 *The depletion of the Comstock silver lode.*

trend is down; the later peaks do not reach the heights of the earlier ones. If consumption continues to increase, the added time provided by new technology becomes less and less.

Suppose, now, we examine "average" rock—rock that can be found anywhere. All rock contains mineral matter. Isn't that inexhaustible? Again, it is a matter of concentrations, but here the differences are even greater. For many metals, the concentrations in ordinary rock may be as low as one ten-thousandth of those in commercial ores. The limiting factors, then, are energy consumption and the resulting pollution.

The technology needed for advanced mining methods is very elaborate. The ancient Greeks and Romans found copper under 10 cm or so of soil; today the depths of mines are measured in kilometers. We can never go back to picks and shovels. If deeper and less concentrated ores are mined, the disruption of the land becomes more and more extensive. Ultimately, if abundant energy (as from nuclear fusion) becomes available to extract minerals from ordinary rock, there is a final barrier—the limited ability of the biosphere to absorb the waste heat produced.

Assuming that our mineral reserves are not inexhaustible, what actions can be taken now?

One possibility would be to exhaust the reserves anyway and let future generations worry about the problem. Let us consider other options. First, we can look for entirely new areas to explore. What about the sea floor, or the Moon, or the planets? The sea floor was mentioned in Section 18.1. Various explorations have indicated that the mineral deposits on the sea floor are vast. Much of this material is concentrated in the form of round, flat, or odd-shaped pieces, typically weighing about a kilogram or so, that are rich in manganese. They are called **manganese nodules,** but they also contain copper, iron, nickel, aluminum, cobalt, and about 30 or 40 other metals. It is estimated that there are a trillion or more tonnes of these nodules on the sea floor. The sea floor does not have to be drilled or blasted, and explorations can be done with undersea TV cameras. Furthermore, no one "owns" the sea floor, but the question of just how the mining rights are to be allocated is still unclear. Nonetheless, the technology of collection is complicated, and costs are expected to be high. The sea is not the easiest environment in which to operate complex machinery. Possible methods of collection include scoops, dredges, and vacuum devices (Fig. 18–12). Various

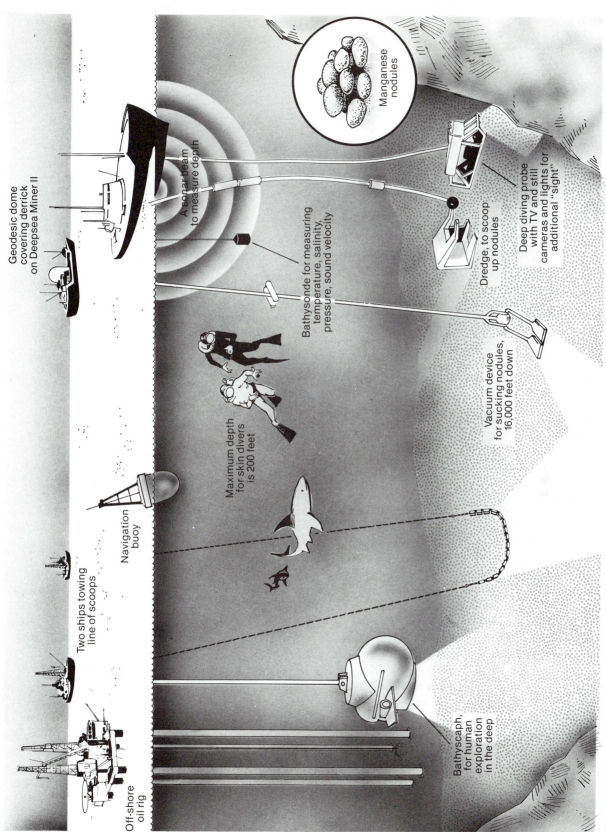

Geodesic dome
covering derrick
on Deepsea Miner II

Manganese
nodules

A sonar beam
to measure depth

Bathysonde for measuring
temperature, salinity,
pressure, sound velocity

Dredge, to scoop
up nodules

Deep diving probe
with TV and still
cameras and lights for
additional "sight"

Two ships towing
line of scoops

Navigation
buoy

Maximum depth
for skin divers
is 200 feet

Vacuum device
for sucking nodules,
16,000 feet down

Off-shore
oil rig

Bathyscaph,
for human
exploration
in the deep

Figure 18–12 Mining the sea floor for manganese nodules.

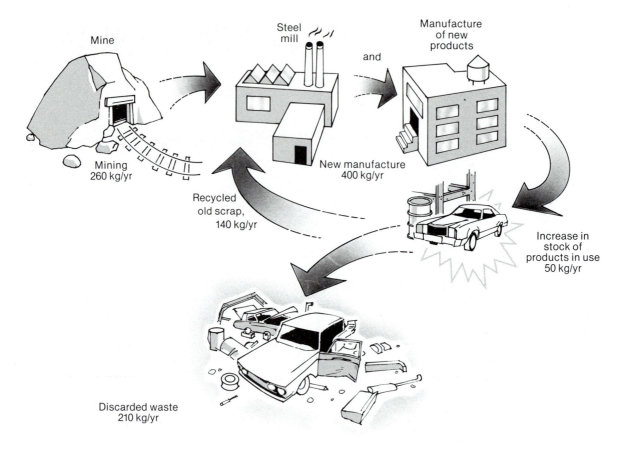

Figure 18–13 *Estimated annual manufacture and flow of iron and steel in the United States in kilograms per capita.*

groups of corporations are already involved in exploration and planning.

In the early years of the space age, some enthusiasts wrote optimistic speculations about minerals from the Moon and the planets. These voices are hardly ever heard anymore.

An alternative approach is to use less rather than to mine more. Consumption can be reduced by either conservation or recycling. Figure 18–13 shows the extent to which recycling in the United States could increase the life of iron reserves. The depletion of reserves is accounted for by the increase in stock plus waste, but waste accounts for over 80 percent (210 kg per year/260 kg per year). This means that if all the discarded iron and steel were recycled, reserves could last four times as long.

Such objectives could be realized by manufacturing products that are smaller and last longer and by taking better care of them. (It is much easier to keep on using a device that still works than it is to recycle it after it stops working.) Perhaps best of all is an improvement in technology or in methods that results in extensive conservation of materials. A good example is the modern pocket calculator replacing the old slow, mechanical, office calculating machine, which weighed about 20 or 25 kg.

Finally, what about substitutes for at least some nonrenewable mineral resources? Why not go back to the simpler styles of using wood, stone, and fiber, or go forward to using plastics that can be synthesized from coal, air, and water? To some extent, both paths are being followed. The use of sand and gravel (for concrete) and of stone is increasing (Fig. 18–14). Wood, too, is still a very desirable construction material. The use of many varieties of plastics is increasing much more rapidly than the use of metals. Improvements in the chemistry as well as in the mechanical makeup of plastics have made them competitive with metals in many applica-

Figure 18–14 *Concrete is used in the construction of highway bridges to save steel.*

tions, even where strength is an important factor. Some new technologies, however, can create a *greater* demand for metals. Note (Table 18–2) the examples of gallium for solar energy and lithium for fusion energy.

The factors involved are so complex that we cannot predict the future of mineral resources with much accuracy. However, as minerals continue to be extracted and dispersed around the globe, we can be sure of two things—the reserves of many important minerals are not inexhaustible, and the Second Law of Thermodynamics will not be denied.

Summary

The Earth consists of a core of iron and nickel surrounded by a mantle of lighter rock and surface crust, on which we live. Various forces within the mantle create large stresses, which produce motion. The entire crust is composed of several continent-sized plates of rock floating on a semifluid layer of mantle. The slow movement of these plates is called **continental drift.**

Crustal rock is far from homogeneous in composition; in fact, there are many sharp discontinuities. A small segment of one type of rock embedded in a dominant rock formation is called a **vein.** Some veins contain valuable minerals.

Discontinuities in crustal composition result from geological processes such as the invasion of older rock by upflowing **magma** (melted rock), differential weathering and erosion, slow melting of magma with gravitational settling of denser solids, differential solubility in water of various minerals, and differences in the rates of settling in streams that lead to **placer deposits,** notably of gold.

Mineral deposits are also found on the sea floor near midocean ridges formed from magma that rises up through gaps between the continental plates.

An **ore** is a rock mixture that contains enough valuable mineral matter to be mined profitably.

Mineral reserves are the estimated supplies of ore in the ground. This estimate can change with the discovery of new reserves, with improvements in extraction and refining, with new prices for minerals and for energy, and with increased pollution associated with the mining and processing of ores.

Metals are valuable because they are stronger than any other materials, they can be melted and cast into molds, they can be machined to precise dimensions, and they are by far the best conductors of electricity.

Helium and phosphorus are two key nonmetals. Helium is a light, inert gas. Phosphorus is an essential element in agricultural fertilizers. Concentrated sources of both elements are being depleted.

Future prospects for metals cannot be predicted accurately, because they depend on changes in reserves, on conservation and recycling, on changes in technology, and on the use of substitute materials. However, once the prime ores are depleted, the poor sources will provide *much* lower concentrations, and hence will require much more energy and will cause much more pollution.

One possible new source of minerals is the store of manganese nodules on the sea floor.

Questions

Structure of the Earth

1. Imagine that you were digging a hole to the center of the Earth. Starting with the surface on which you live, identify the layers in the order in which you would reach them. Which layer would be the thinnest?

Formation of Mineral Deposits

2. It is common for a single mine to contain fairly high concentrations of two or more minerals. Discuss how geological processes might favor the deposition of two similar minerals in a single location.

3. If one compound is to be separated from a complex mixture, it must somehow be transported away from the rest of the material. How are ores moved out of a mixture in each of the following processes—(1) separation by gravity, (2) separation by differential solubility, (3) formation of placer deposits.

Mineral Reserves

4. Explain why minerals on the sea floor are concentrated near midocean ridges.

5. What are mineral reserves? Mention three factors that can cause a change in the estimation of mineral reserves.

6. If minerals are widely distributed in ordinary rocks, why should we worry about ever running short?

Metals

7. Name three properties of metals that make them important in modern civilization.

8. Explain why it is very important to recycle aluminum.

9. What factors can make our metal reserves last longer? What factors can deplete them rapidly?

10. Name one synthetic and two natural classes of material that can be used as substitutes for metals. Which property of metals *cannot* be furnished by substitutes?

Suggested Readings

Books dealing with the mineral resources of the world include:

David N. Cargo and Bob F. Mallory: *Man and His Geologic Environment.* Reading, MA, Addison-Wesley, 1974. 548 pp.

Eugene N. Cameron (Ed): *The Mineral Position of the United States, 1975–2000.* Madison, The University of Wisconsin Press, 1973. 159 pp.

Gary D. McKenzie and Russell O. Utgart (Eds.): *Man and His Physical Environment.* Minneapolis, MN, Burgess Publishing Company, 1975. 387 pp.

Ronald Tank (Ed.): *Focus on Environmental Geology.* New York, Oxford University Press, 1973. 474 pp.

National Academy of Sciences Report: *Mineral Resources and the Environment.* Washington, D.C., National Academy of Sciences, 1975. 348 pp.

Brian J. Skinner: *Earth Resources.* 2nd ed. Englewood Cliffs, NJ, Prentice Hall, 1976.

The view that mineral resources are inexhaustible is expressed in the following references:

Wilfred Beckerman: The myth of "finite" resources. *Business and Society Review,* 12, Winter 1974–1975.

Julian L. Simon: *The Ultimate Resource.* Princeton, NJ, Princeton University Press, 1981. 363 pp.

The helium controversy is described in:

Earl Cook: The helium question. *Science,* 206:1141, Dec. 7, 1979.

The question of phosphorus reserves is reviewed in:

Phosphate: Debate over an essential resource. *Science,* 209:372, July 18, 1980.

The possibility of mining the ocean floor has stimulated much recent interest. Some popular articles on the subject are:

Peter Britton: Deep-sea mining. *Popular Science,* July, 1981. p. 64.

D. S. Cronan: Manganese nodules: Controversy upon controversy. *Endeavour* (new series), 2(2):80, 1978.

William Wertenbaker: Mining the wealth of the ocean deep. *New York Times Magazine,* July 17, 1977. p. 14.

David Sleeper: *Nations Eye Deepsea Minerals.* Washington, D.C., Conservation Foundation, April, 1981.

Economic and public policy aspects are emphasized in:

Dennis L. Little, Robert E. Dils, and John Gray (Eds.): *Renewable Natural Resources.* Boulder, CO, Westview Press, 1982. 316 pp.

John A. Butlin: *The Economics of Environmental and Natural Resources Policy.* Boulder, CO, Westview Press, 1982. 206 pp.

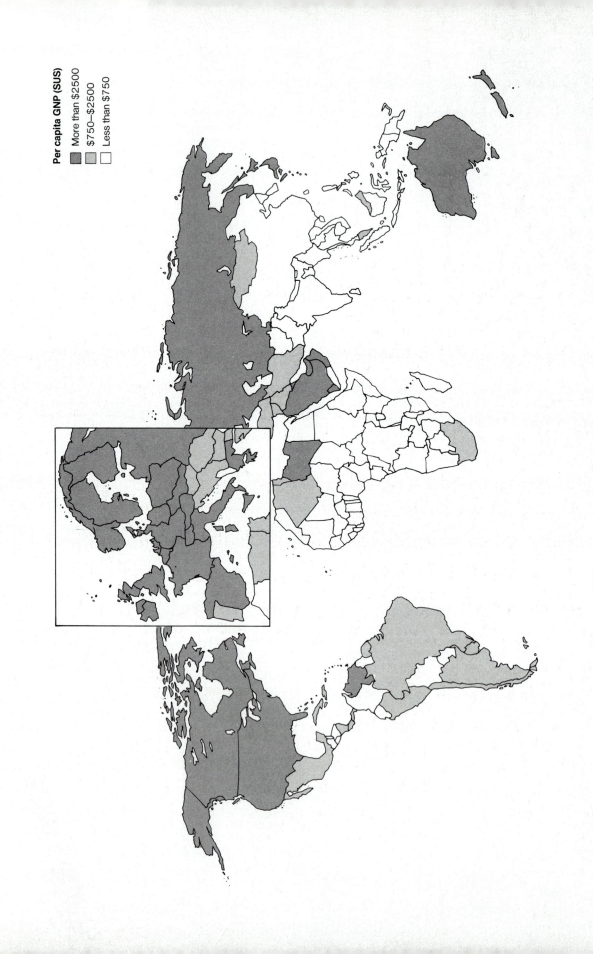

Per capita GNP ($US)

■ More than $2500
▨ $750–$2500
□ Less than $750

Air, Water, and Wastes

World per capita gross national product, in U.S. dollars. GNPs are largest in the United States, Canada, northern Europe, and Japan and smallest in parts of Africa, Asia, and South America. Oil-producing countries in other parts of the world, such as Saudi Arabia, Libya, and Venezuela, also show high GNPs. High economic productivity can lead to environmental disruption, depending on the extent to which pollution is regulated and controlled. (From *Our Magnificent Earth*. © 1979 by Mitchell Beazley as *Atlas of Earth Resources*. Published in the U.S.A. by Rand McNally & Company.)

19

Water Resources

Water is the most abundant substance on the Earth's surface. The oceans cover some 71 percent of the planet, glaciers and ice caps cover additional area, and water is also found in lakes and streams, in soils and underground reservoirs, in the atmosphere, and in the bodies of all living organisms. Thus, water in all its forms—ice, steam, and liquid—is very familiar to us. Compared with other substances, however, water has unique properties. Ice floats on liquid water; much energy is consumed when ice melts or when water evaporates; and water is an excellent solvent for many different kinds of substances. All of these properties of water affect its role in the biosphere. The movement of water on Earth (the hydrologic cycle) is closely related to the energy changes that take place when water changes its form between solid, liquid, and vapor.

Humans use water in the home, in industry, in agriculture, and for recreation. These applications differ widely in the quantities and quality of the water they require. In one way or another, we use all available sources—inland waters, groundwater, even ocean water. We pollute it, repurify it, and reuse it over and over again.

This chapter examines our practices and policies in regard to our water resources, with special attention to environmental effects and long-term consequences.

19.1 The Properties of Water

Chemical substances are generally classified into groups of compounds with similar properties such as salts, alcohols, or sugars. Water, however, is unique; there is nothing else like it. Thus, corn oil is more like cottonseed oil, kerosene is more like gasoline, grain alcohol is more like wood alcohol, and silver is more like gold than *anything* is like water.

To understand the properties of water, it is best to start with its molecules, which consist of two hydrogen atoms bonded to an oxygen atom. This composition is expressed by the familiar chemical formula H_2O. The molecule has a bent shape—the angle between the bonds is 105 degrees (Fig. 19–1A). The negative charges (electrons) in the H_2O molecule are crowded somewhat toward the oxygen atom. As a result, the

A

+H
 |
 + /
H—O−

B

        H
        |
H—O...
       ‥H
          |
    H—O
    ⋮
H—O
      \
       H

C

Figure 19–1 *A*, Structural formula for water, showing partial electrical charges. *B*, Water molecules bonded to each other (*dashed lines*). *C*, A drop of water hanging from a leaf. The forces that bond water molecules to each other pull the drop into a spherical shape.

oxygen end of the molecule has a partial negative charge. The other side of the molecule, where the hydrogen atoms are, is somewhat electron-deficient and therefore has a partial positive charge. The important consequence of all this electrical separation is that the oppositely

charged sides of different water molecules attract each other strongly, as shown in Figure 19–1*B*.

The shapes of water molecules and the strong bonding between them lead to some important and remarkable properties. As a result, water, more than any other substance, influences the physical environment of the biosphere.

1.   Ice has a high melting point and water has a high boiling point. It is true that melting ice feels cold to the touch, but think of it this way: A molecule of water is about the same size as a molecule of ammonia (NH_3) or a molecule of methane (natural gas, CH_4). The melting point of ice is 0°C. The melting point of solid methane is −183°C, and of solid ammonia it is −78°C (which is why they are seldom seen as solids). It is the strong bonding between the H_2O molecules that keeps ice solid up to 0°C. Similarly, the boiling point of water (100°C) is very high for a molecule as small as H_2O. Furthermore, it takes a lot of energy (80 cal per gram) to melt ice, almost twice as much as it takes to melt a gram of, say, beef fat.

2.   When ice does melt, the results, though familiar, are remarkable. Melting is a process in which the regularity of a crystal (Fig. 19–2) structure breaks apart, and the partially liberated molecules (liquid) move about more freely. For most substances, this freer movement causes ex-

Figure 19–2 Snowflake crystal.

pansion. As a result, most solids, being denser, sink in the liquids into which they melt. The behavior of water is just the opposite. When the crystal structure of ice breaks apart (melts), the partially liberated molecules attract each other so strongly that they pack together even closer than in the ice crystal. Consequently, ice is *less* dense than water and floats on it. If it did not, a body of water would freeze from the bottom up instead of from the top down, and aquatic life as we know it would not exist on Earth.

3.   Large amounts of energy are needed to melt ice or to boil water (not just to bring it up to the boiling point, but actually to boil it off into steam—that is, to vaporize it). When water vaporizes and becomes steam, or water vapor, the molecules are completely liberated—that is, all the bonds between the water molecules are broken. This requires even more energy. Thus, it takes about seven times as much energy to vaporize water as it does to melt ice. It also takes about seven times as much energy to vaporize water as it does to vaporize gasoline. When these processes (melting or vaporizing) are reversed, the energy flow is also reversed. Thus, when a gram of water freezes, 80 calories are released, and when a gram of water vapor liquefies, almost seven times that much energy, or 540 calories, is released. Since much of the surface of the globe is water, the Earth's climate is intimately related to the energy that is stored and released as the Earth's waters change from one form to another. Water is the essential fluid of all living organisms; it is the solvent of the blood and lymph of animals and of the sap of plants. Could any form of life exist without water? Not as far as we know; certainly if it did, it would not be life in any form we can now visualize.

19.2 The Hydrologic Cycle (Water Cycle)

If you travel widely on land, by sea, and in the air looking at the Earth's waters, three observations will be apparent. First, much of the water is *stored* in places that look rather permanent. The largest quantities, of course, are in the

oceans. But there are also the Arctic and Antarctic ice caps, as well as many lakes and glaciers. Second, much of the Earth's water is in motion: Snow and rain fall, clouds drift, and rivers flow toward the sea. Third, the water on land is very unevenly distributed. As you wander through tropical jungles, everything is wet, and water often drips on you throughout the day. But you had better not try to trek across Australia, Libya, or even southern California without taking all your water with you. These regions get very little rainfall; most of the year they get none.

The movement of water on Earth (the **hydrologic cycle**) is rather complex, involving various interrelated loops (Fig. 19–3A). However, it is convenient to break down the different transports of water into three simple categories—evaporation, precipitation, and runoff.

Evaporation or **vaporization** is the formation of water vapor. Dissolved solids such as salts remain behind when water evaporates. Most water vapor is produced by evaporation of liquid water from the surface of the oceans. Water can also vaporize *through* the tissues of plants, especially from leaf surfaces. This process is called **transpiration**. Ice can also vaporize without melting first. This process (which is called **sublimation**) is slower than vaporization of liquid water.

Precipitation means falling from a height. Referring to water, precipitation includes all forms in which atmospheric moisture descends to earth—rain, snow, hail, and sleet. The water that enters the atmosphere by vaporization must first condense into liquid (clouds and rain) or solid (snow, hail, and sleet) before it can fall. Recall that vaporization absorbs energy. (Water that evaporates from your skin absorbs heat, making you feel cold.) This energy is released in the form of heat when the water vapor condenses.

Runoff is the flow back to the oceans of the precipitation that falls on land. In this way the land returns the water that was carried to it by clouds that drifted in from the ocean. Runoff occurs both from the land surface (rivers) and from underground water.

Assuming that the water in the oceans and ice caps is fairly constant when averaged over a period of years, the water balance of the Earth's

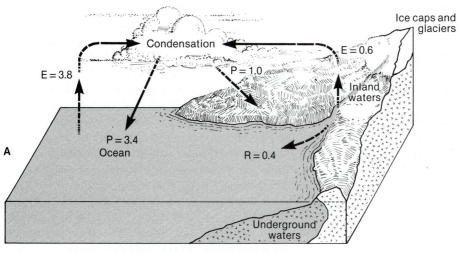

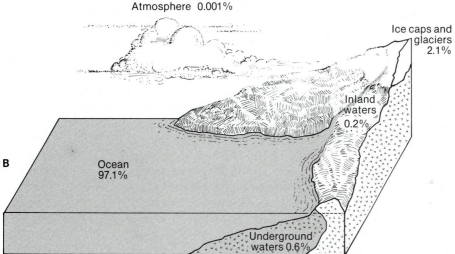

Figure 19–3 *A*, The hydrologic cycle. E = evaporation; P = precipitation; R = runoff. Numbers are in geograms (10^{20} grams) of water transferred per year. One geogram is about 100,000 cubic kilometers. *B*, Percentages of total global water in different portions of the Earth.

surface can be expressed by the simple relationship:

Water lost = Water gained

The oceans lose water by evaporation and gain water by precipitation and by runoff from the continents. The continents gain water by precipitation and lose water by runoff and by evaporation.

Therefore,

• *For the oceans*

$$\underbrace{\underset{E}{\text{Evaporation}}}_{\text{loss}} = \underbrace{\underset{P}{\text{precipitation}} + \underset{R}{\text{runoff}}}_{\text{gain}}$$

• *For the continents*

Evaporation + runoff = precipitation
$$\underbrace{E + R}_{\text{loss}} = \underbrace{P}_{\text{gain}}$$

With these relationships in mind, look at Figure 19–3A. The numbers (in geograms per year) for the oceans are

$3.8 = 3.4 + 0.4$

and for the continents

$0.6 + 0.4 = 1.0$

Now refer to Figure 19–3B, which is the same diagram with different numbers. Here the numbers tell you where the waters of the Earth *are* at any given time, not where they are going. The numbers are given in percentages of the total quantity of global water, which is about 1.35 billion cubic kilometers.

Note that of this vast amount, only 0.8 percent is in the form of inland and underground waters, and that most of this amount (0.6 percent) is underground. The remaining 0.2 percent is the inland surface water such as lakes and streams. The least amount of water is in the atmosphere (0.001 percent). Since all this water is in motion (Fig. 19–3A), it follows that all the waters of the Earth renew themselves—that is, they come and go.

An important question is, "How long does it take for water in a given part of the Earth to renew itself?" Consider, for example, two flows of pure water, one that goes into a small basin, the other into a large one (Fig. 19–4). Both basins are well stirred, so that the water in each is always uniform throughout. If the water in both basins is polluted, the time it takes for the fresh water to rinse out the pollutant depends on the flow rate of the water and the volume of the basin. The greater the flow rate and the smaller the basin, the faster the rinsing action. Under such circumstances, the *average* time that a water molecule spends in the basin is called the **residence time**. Some molecules, however, will remain for a longer or shorter time than the average. Any particular water molecule in a natural basin such as, say, Lake Erie, may be lost by evaporation or runoff the very next day, or it may still be there 1 year or 100 years from now. Table 19–1 gives average residence times for water in various parts of the hydrologic cycle.

Note that water spends the least time in the atmosphere and the longest time in the deepest ocean layers. Changes in global energy patterns can therefore readily affect atmospheric moisture, and hence rainfall and agricultural productivity.

The fresh water available for human use is the runoff from rivers and underground sources. Rivers renew themselves rapidly (in weeks), but groundwater is much slower (hundreds of years). Pollution of waters with long residence times is not easily reversed.

19.3 Human Use of Water

Water is used in the home or office, in industry, in agriculture, and for recreation. Both the quantities used and the water quality needed differ widely, depending on the application. In the home, for example, the amount of water used in one toilet flush would satisfy the drinking requirements of an adult (1 liter per day) for about three weeks; the water used for one load in a clothes washer would be enough for drinking for almost 6 months. The amounts used in industry and agriculture are far greater than those needed for any personal use. For example, the water used in industry to refine a tonne of petroleum would be enough to do about 200 loads in a clothes washer. When crops are irri-

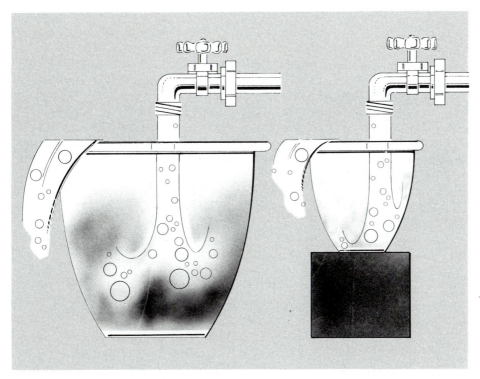

Figure 19–4 The same rate of water flow rinses pollutants out of a small basin faster than out of a large one.

gated, it takes much more water to grow a tonne of grain that it does to manufacture a tonne of most industrial materials (e.g., metals or plastics).

Table 19–1 Average Residence Times of Water Resources

Location	Average Residence Time
Atmosphere	9–10 days
Ocean	
Shallow layers	100–150 years
Deepest layers	30,000–40,000 years
World ocean average	3,000 years
Continents	
Rivers	2–3 weeks
Lakes	10–100 years
Ice caps and glaciers	10,000–15,000 years
Shallow groundwater	up to hundreds of years
Deep groundwater	up to thousands of years

The strictest requirements for quality apply to drinking water for humans. The least strict requirements probably apply to water used for cooling, where the prime concern is its temperature. Sea water is therefore adequate. For some industrial applications, the most important consideration is whether the water will corrode the equipment; control of acidity is often the only requirement in such cases.

For most human needs, however, including the large amounts used in agriculture and industry, water must be fresh, not salty. Figure 19–5 shows the pattern of consumption of fresh water in the United States. Note that the highest water consumption does not occur in the most densely populated areas. In general, rates of water consumption reflect the needs of agriculture much more than those of the home, of commerce, or of industry. One conclusion from this difference is that efforts to conserve water in the home, while locally helpful, cannot make a significant contribution to the demands of agriculture. A comparison of Figure 19–6 and Figure 19–5

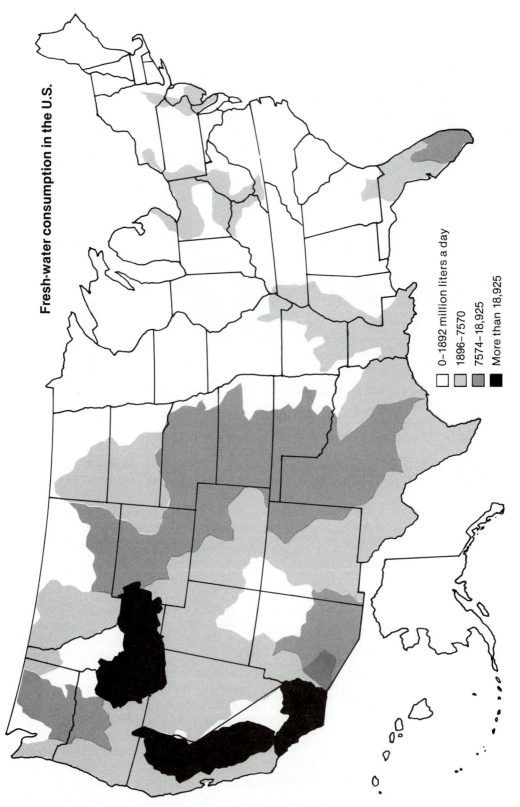

Figure 19–5 Freshwater consumption in the United States. (1 gallon = 3.785 L)

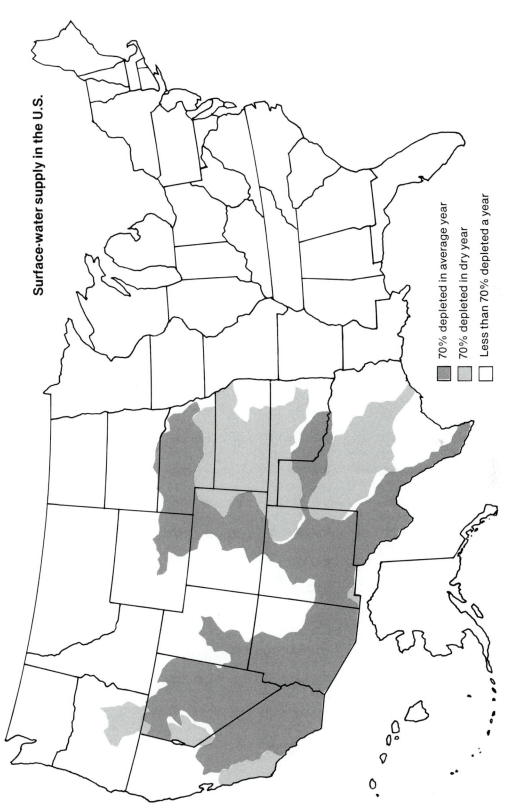

Figure 19–6 Surface water supply in the United States.

shows that some of the areas where the surface water supply is often depleted are also the areas where the largest quantities of fresh water are needed. Therefore, if the only fresh water available were the water on the Earth's surface, in lakes and rivers, it would hardly be enough. Even in years of average rainfall, much of the midwest and southwest regions of the United States deplete most of their surface waters. Therefore, we must turn to our groundwater, which makes up about 75 percent of the total inland waters.

Groundwater

When rain falls on the land, two forces act on it. One is the Earth's gravity, which pulls the water downward through any open path. The second force is the electrical attraction between water and other materials to which water tends to stick. (That is why water makes things wet. Not everything, however; water doesn't stick to oil, which is why ducks, whose feathers are oily, do not get wet and drown.) Thus, drops of rain-

water will stick to the vertical side of a window without falling. Water can also be pulled into tiny holes. If a corner of a paper towel is placed in a dish of water, the liquid travels upward, against the force of gravity. In this case, the electrical force attracting water molecules to paper molecules is stronger than the gravitational force pulling the water downward. The movement of water upward through small holes is called **capillary action**.

Now consider what happens when rain falls on dry soil. The first raindrops will simply wet the soil—they will not flow down or away. As the rain continues after all the land surface is wet, gravity will pull the excess water down through the spaces in the rock, sand, or gravel below. The larger the underground spaces, the faster the water will flow down. Eventually, the downward flow is stopped when the water meets rock that has no porosity. Since the water can go no further, it backs up, filling all the pores in the rock above the barrier. This completely wet section is called the **zone of saturation** (Fig. 19–7). The upper boundary of the zone of satu-

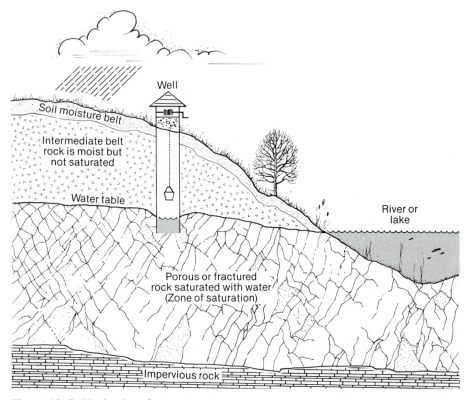

Figure 19–7 Distribution of groundwater.

ration is called the **water table**. Below the water table the rock is saturated; above it the rock may be moist but is not saturated. If rainfall recharges water faster than it is lost by flowing out to a lake or stream or by being pumped out, the water table rises. If the recharge is slower than the loss, the water table falls. A well dug below the water table will have water; a well that is not that deep will be dry.

When the rain stops, the water content of the topsoil decreases in three ways: (1) Water drains downward by gravity—most of this action is complete in a day or two. (2) Water evaporates from the surface. (3) Water is absorbed by plants and is then lost to the atmosphere by transpiration. To offset these losses, water can be drawn up toward the surface by capillary action but only within a narrow layer, called the **soil-moisture belt**.

These natural processes of gains and losses of water establish the essential features of the land's water balance—the moisture content of the soil and the level of the water table. If more groundwater is removed for irrigation, mining, or manufacturing than is replenished by runoff, the water table will drop. This problem exists in areas that are attractive because of their favorable climate and abundant natural resources but have low rainfall. The southwestern regions of the United States known as the "Sunbelt" constitute such an area.

When porous and impervious beds of rock alternate, there can be more than one layer of groundwater. Figure 19–8 shows inclined strata in which seepage introduces water into a lower layer of porous rock. Such a reservoir of groundwater is called an **aquifer**. The lower portion of an aquifer may be under considerable pressure from the weight of water above it. Under such circumstances, a deep well may allow the water to flow up without having to be drawn or pumped. A well of this kind is called an **artesian well**.

The total quantity of water in aquifers is very large. In fact, about half of all groundwater occurs in these deep layers. Note, however, from

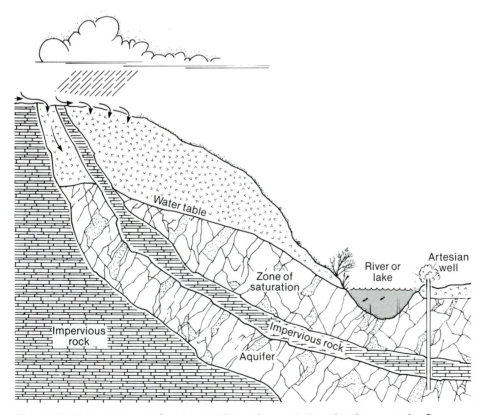

Figure 19–8 Sloped layers of porous and impervious rock that give rise to artesian flow.

Table 19–1 that these waters are replaced *very* slowly (up to thousands of years). Much of the water in some aquifers was accumulated many centuries ago in wetter climates than the present one. Under such conditions, deep groundwater may be considered to be, for all practical purposes, nonrenewable. Just as coal and petroleum are called fossil fuels, so is deep groundwater sometimes called ''fossil'' water. The removal of deep groundwater is therefore analogous to mining.

The Ogallala aquifer, which is one of the world's largest reservoirs of fresh groundwater, is a major source of water for farmers in parts of Kansas and its neighbors to the north, south, and west. The average annual rainfall in this region ranges from about 25 to 100 cm per year, in contrast to the more abundant 100 to 150 cm per year typical of the eastern United States. As a result, more water is taken out of the Ogallala aquifer then is replaced by rainfall.

It is estimated that some of the aquifer water tables in the United States are being lowered at rates of several centimeters to about half a meter a year. At such rates, serious depletion of the aquifers can occur early in the next century. It is important to understand that the viability of an agricultural-industrial-urbanized society that depends on fossil water can be at risk when that source is seriously depleted.

Two other problems besides depletion can arise from excessive removal of groundwater. One of these problems is **subsidence**, or settling, of the ground as deep groundwater is removed. This removal allows the rock particles to shift somewhat closer to each other, filling some of the space left by the departed water. As a result, the volume of the entire rock layer decreases, and the surface of the ground subsides. (Removal of oil from wells has the same effect.) Subsidence rates can reach 5 to 10 cm per year depending on the rate of water removal. These effects have been observed in such areas as the San Joaquin Valley of California, Houston (Texas), and Mexico City. Unfortunately, subsidence is not a readily reversible process. As a result, the water-holding capacity of a depleted aquifer may be permanently reduced so that it cannot be completely recharged even when water becomes abundant again.

The other problem is **salt water intrusion** (Fig. 19–9). As groundwater is removed from a coastal area, the zone of freshwater saturation is reduced from above and below. From above, the

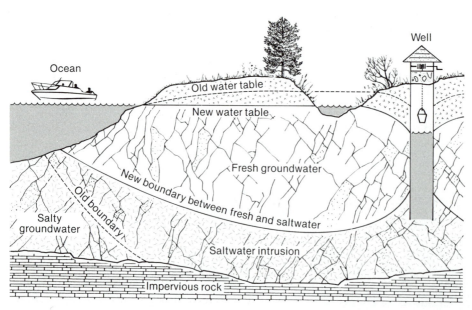

Figure 19–9 Salt water intrusion.

water table declines. From below, salt water seeps in. As a result, salt water may be drawn into wells, making the water unfit for drinking.

19.4 Human Control of Water

We cannot significantly affect the total quantity of water on Earth, but we can store it, change its flow patterns, and purify it. The choices that people make among such options depend heavily on the amount of water available, on sources of energy, on technological skills, and on prevailing patterns of living. In many arid less-developed areas, for example, water must be carried over distances of several kilometers from its source to where it is needed. Such tasks not only place a heavy burden on the carriers (generally women and children) but also create other problems. The precious water is used only for essentials, particularly for drinking and cooking. Unless it rains, there is not enough for sanitary uses. As a result, many people, especially children, suffer from eye infections and other diseases that could have been prevented by frequent washing. An interesting choice arises when a limited amount of capital becomes available that can be used either for a pumping and distribution system or for a water purification system but not both. One of us (A.T.) discussed these problems with a group of environmental engineers from the Cooperative for American Relief Everywhere (CARE) in a rural area of Kenya. The engineers pointed out that even though the water supply under consideration (a pond formed by damming a small river) was somewhat polluted by animal wastes, much of the water consumed by the people was first boiled, either in cooking or as tea. Furthermore, if the water were purified (by chlorination, see Chapter 20), its taste would be changed, and people might drink less of it, which would be undesirable. Finally, purification without distribution would not ease the burden of the water carriers. The choice was therefore for distribution over purification.

The following section will examine and evaluate the various options available for controlling water supplies. The reader should bear in mind,

however, that the choices are often made on a cultural as well as on a technical basis.

19.5 Storing Water by Means of Dams

As is apparent from the residence times listed in Table 19–1, water stays in lakes much longer than it does in rivers. Dams make lakes, or reservoirs. In fact, dams have been erected on many rivers, large and small, all over the world, from ancient times until the present. Let us look just at the physical events that start to happen when a river is dammed. First, the downstream flow is interrupted as the lake behind the dam begins to fill. The time needed for this step may be only a few weeks for a small lake fed by a fast-flowing stream. Or it can take years: For example, construction of Glen Canyon Dam in Arizona was started in the 1950s. Lake Powell, which is the reservoir created by the dam, was finally filled in 1980.

Once the lake is filled, water spills over the dam and the normal flow is re-established. Several benefits can now be realized from the dam and the lake it has created:

1. Hydroelectric Energy. The potential energy of the water has been increased by the rise in its level. The water can be made to fall through a turbine and generate electricity (Chapter 11). In effect, the dam has made it possible to convert the release of heat by friction in the old stream bed to electrical energy.

2. Control of Water Flow. Dams are provided with pipes and valves so that water can be drained from below. In a dry season, water from the reservoir can be released to feed the stream below the dam for irrigation or other uses. During heavy rains, the valves are closed, and the reservoir is allowed to fill up, thus providing protection against flooding.

3. Recreation. The lake or reservoir behind the dam can be used for fishing or for water sports. (But the flooding of canyons and whitewater play areas is a recreational and esthetic loss.)

There are also environmental problems associated with dams. These may be grouped into the following categories:

1. Loss of Water. The reservoir provides more surface for evaporation and more area for seepage from below compared with the stream that preceded it. For example, evaporation from Lake Powell removes about 270,000 cubic meters of water per year, enough to serve the water needs of a city with a population of half a million. Since salt does not evaporate, the remaining water becomes more saline, and its use for irrigation hastens the rate of salinization of the soil.

2. Silting. Consider the physical events that take place during the life of the dam and its reservoir. The water in the reservoir is calm and slow moving compared with the stream that it flooded. A rapid stream always carries some sandy particles in it, just as a high wind blows dust into the air. In still water a grain of sand might settle at a rate of about 2 cm per second (see Fig. 20–1), which is a bit less than 1 km per hour. Water in a stream can move faster than that and keep the sand in motion. But in a lake the sand and other sediment settle to the bottom. Most of the sediment that enters the reservoir from the runoff that feeds it, therefore, settles to the bottom before it can get a chance to spill over the dam. Thus the reservoir gradually fills up. Rates of buildup can reach 10 cm, or 0.1 meter, per year. At such a rate, the lakes behind high dams can last up to hundreds of years, but that is not forever.

3. Erosion. The water that flows over the dam is quite free of sediment. Even water that is discharged from outlets below the surface carries less sediment than would have been present before the dam was built. As the water now moves more rapidly in the stream below, it starts to scour out the stream bed because it has an unused *capacity* for carrying sediment. The water level below the dam is thus lowered. As a result, more energy must be used for pumping the water for irrigation or other needs. In addition, a deeper main canyon promotes more soil erosion in the side canyons.

4. Risk of Disaster. Areas near rivers (their flood plains) are attractive for farming, industry, and commerce, especially when flood control promises to make the region safe. But unusually heavy rainfall can fill and overflow a reservoir, which then can no longer limit the flow. Furthermore, dams have been known to break. Under such circumstances, the population in the flood path is vulnerable to disaster.

5. Ecological Disruptions. Disturbances of ecological factors are listed in Chapter 11.

19.6 Recharging

The water used for agriculture or industry is returned to the ground, and some is lost to the atmosphere. The situation may be pictured as shown in Figure 19–10. The percentage that is returned, or **recharged**, depends on the application and on the methods used to conserve water. In agriculture, a large loss is inevitable—water evaporates from the land and is transpired by plant matter. The percentage of loss depends on various agricultural factors. Much more water evaporates when plants are irrigated by spraying than when water is trickled at ground level. The choice of crop makes a big difference, too. For example, all the water needed by a crop of barley during its growing season would cover the land to a depth of about 0.5 meter. A crop of alfalfa, on the other hand, would need about three times as much—1.5 meters. A very rough estimate is that about 50 percent of the water used in agriculture is recharged and 50 percent is lost to the atmosphere.

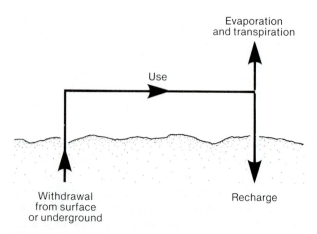

Figure 19–10 *Recharging.*

Industrial cooling water and wastewater can be fed back into the ground through recharge wells, in which water is pumped back into the wells rather than being withdrawn from them. The amount available from such sources is limited because it is sometimes chemically contaminated or is not available at the best locations for recharge.

Another method is to channel natural stream water into recharge wells during wet periods when the flow is high, thus augmenting the natural percolation through soil and rock. The problem of pollution of groundwater resulting from contamination during recharging will be discussed in the next chapter.

19.7 Diversion or Rechanneling of Water over Long Distances

It seems reasonable to divert water from an abundant source to a dry region where the soil would be fertile if irrigated. Likewise, fresh water can be taken from the countryside to a densely populated city that needs it. Southern California offers examples of both types of water diversion. The All American Canal channels water from the Colorado River to the farms and cattle ranges of the Imperial Valley (east of San Diego). The great Los Angeles Aqueduct, completed in 1913, brings water south from Owens Valley to Los Angeles. More recently, Californians have been considering additional diversion of water from rivers in the northern part of the state through a 42-mile Peripheral Canal, which, with its pumping stations and other facilities, would cost billions of dollars.

Such large water diversion projects can create a number of problems at both ends—at the source and at the area to which the water is supplied.

1.   Encouragement of Waste.   Planners often underestimate the ability and willingness of people to conserve when it is necessary to do so. (The unexpected decline in energy use in the early 1980s is one such example.) Most people use water wastefully when they know that it is abundant, but when it is scarce, conservation is

not seen as a serious burden. Thus, it is convenient to leave the water running while you are washing your hands or brushing your teeth. But if you must, you can use only about one-tenth as much water in a sink or bowl to accomplish the same purpose in about the same time. There are many other ways to conserve water in the home, on the farm, and in water distribution systems. In many cases, these measures would be an adequate or at least a partial substitution for water diversion projects.

2.   Salinity.   The problems caused by buildup of salts from irrigation water were discussed earlier in Chapter 14.

3.   Deterioration of Water Quality at the Source.   When water is diverted from an area, the *quality* of the remaining water can be affected, even if the amount is adequate. Fresh water from a stream or river becomes progressively saltier as it enters and mixes with ocean water. A fast-flowing river carries its fresh water well out to sea. A slow-flowing tidal basin (such as the Hudson River valley) is saline well inland. In a large water diversion project, the low natural flow in late summer may be further aggravated by peak demand for air conditioning and irrigation. Under such conditions saline water can back up into regions of a river delta that serve nearby farms. This phenomenon, called "reverse flow," can reduce agricultural production in the delta areas. Changes in water flow can also disrupt the life cycles of fish and wildlife in the area.

4.   Energy Consumption.   The great aqueducts of ancient Rome (Fig. 18–9) slope downward; there were no pumps. Ancient Egyptian farmers pumped the waters of the Nile a few feet up to their farms by human or animal power; many do the same to this day (Fig. 19–11). Modern water diversion projects, however, carrying water over hilly terrain, use electrically driven pumps. A cubic meter of water weighs 1000 kg, and it takes about 10 kW of power to lift this amount 1 meter every second. Diversion projects involve *large* quantities of water (hundreds of cubic meters per second) and therefore need much energy. One estimate for an expanded California state water project foresees a use of 10 billion

A

B

Figure 19–11 Waters from the Nile even today are pumped up to farmland (A) by human power, using a hand pump credited to Archimedes, or (B) by animal power.

kWh of electricity in the year 2000—about as much power as is used in 2 million homes.

19.8 Desalination

The lament of the Ancient Mariner—"Water, water everywhere, nor any drop to drink"—expresses what we all know about seawater: There is plenty of it, but you cannot use it for drinking or for irrigation of crops. The reason you cannot drink it is that the salt in seawater is more concentrated than the salts in your blood. **Osmosis** is the process by which a solvent (usually water) flows through a semipermeable membrane from a less concentrated into a more concentrated solution. (A semipermeable membrane does not allow dissolved matter to pass through.) Therefore, if you drink seawater, your blood will not gain water but will lose it by osmosis through the walls of your blood vessels. It would be better not to drink anything at all. For the same reason, salt water kills terrestrial plants; it does not water them.

Osmosis can be reversed. The pressure created by the flow of water during osmosis is called **osmotic pressure**. If enough back pressure (greater than the osmotic pressure) is applied, pure water may be forced out through a membrane, leaving the salts behind. This process, called **reverse osmosis**, is further described in Chapter 20. Another method of removing salt is **distillation**, in which water is boiled, and the resulting pure steam is then condensed to produce distilled water. Both reverse osmosis and distillation require energy and equipment, and the costs are high. Probably the best location for distillation would be a hot dry area near the sea, such as the Middle East, where solar energy could be used to boil the water.

19.9 Cloud Seeding

Water *vapor* is water in gaseous form; it is not visible. The moisture content of clear air is in the form of water vapor. A cloud consists of tiny droplets of water or tiny crystals of ice whose diameters may range from about 2 to 40 micrometers (μm). In still air, such particles settle very slowly (mostly slower than 1 cm per second, see Fig. 21–2) and are therefore kept aloft by the slightest air currents. This is the reason why clouds do not fall down. Thin clouds appear white as the sunlight is scattered going through them. Thick clouds look dark from below because they shade most of the sunlight; viewed from above, as from an airplane, they look white. For rain or snow to form, the water droplets or ice particles must become large—from about 0.5 mm (a drizzle) to about 5 or 6 mm (large raindrops). Often the precipitation starts as snow in the cloud and then melts to form rain on its way down.

The freezing point of water is 0°C. However, this statement does *not* mean that water will freeze when it is cooled to 0°C. The statement means only that when water and ice are in contact with each other, they come to a stable state, or equilibrium, at 0°C. Pure water can readily be cooled below 0°C if no ice is present; in fact, water droplets in clouds are often as cold as −10 or −12°C. Such water is said to be **supercooled**. If a crystal of ice (here called a **seed**) is added to supercooled water, freezing occurs rapidly as the cold water *warms up* to 0°C. The rapid formation of ice tends to produce large particles that can precipitate.

In the 1940s the scientists Vincent Schaefer and Irving Langmuir thought they might be able to seed clouds artificially to make rain. They tried it first in large freezers, and it worked. Next they did it from an airplane, pouring pellets of dry ice (frozen carbon dioxide, −78°C) into supercooled clouds. Sure enough, it worked again—ice particles grew rapidly. Later it was learned that the seed crystal need not be cold; it need only have a *shape* similar to that of an ice crystal. The most effective crystal was found to be silver iodide. Unfortunately, silver iodide is a poison and, in sufficient quantities, may produce toxic effects on plant and animal life.

Weather modification by cloud seeding methods has since been attempted from time to time, but the balance between costs and benefits is still uncertain. In fact, recent statistical analyses fail to confirm the successes claimed for most of the cloud seeding attempts carried out through 1981 (*Science*, 217:519, August 6, 1982). The technical and statistical problems are much more difficult than was appreciated in the earlier, more optimistic years. At best rainfall can be shifted from one location to another, with the object of watering an otherwise dry area such as the "rain shadow" (the dry side) of a mountain range. Precipitation can be spread out to a wider zone in an effort to reduce local intensive concentrations, such as the heavy winter snowfalls in the Buffalo, New York, area.

However, none of these methods creates water; they merely redistribute it. One region's gain is another region's loss. Furthermore, the control over the amount of rain that can be induced and the area in which it can be made to fall is not precise, and consequently, conflicts of interest and political problems can easily arise.

19.10 Water from Icebergs

When salt water is partially frozen, the salt remains in the liquid water, not in the ice. In this way the remaining water becomes saltier, but the ice is pure water. Therefore, if you are stranded on a ship in the Arctic ocean, and all your drinking water is gone, just grab a piece of ice, melt it, and drink the water.

It has been proposed that Antarctic icebergs, which are very large and rather flat, could be towed to dry southern coastal areas where they could be melted to provide fresh water. The potential supply is very large: A tablelike Antarctic iceberg may be 500 meters thick and cover an area of 250 square km or more (Fig. 19–12). The technology and economics, however, are uncertain. Some tests have been carried out with smaller icebergs, but the towing and handling of the Antarctic monsters may be impracticable.

Summary

Molecules of water, H_2O, attract each other strongly, and therefore water has high melting and boiling points compared with other substances of similar molecular size. It also takes a lot of energy to melt ice or to boil water.

The movement of water on Earth (the **hydrologic cycle**) involves evaporation, precipitation, and runoff. The oceans lose water by evaporation, and gain it by precipitation and runoff. The continents gain water by precipitation and lose it by evaporation and runoff. The Earth's waters spend the longest times (tens of thousands of years) in the oceans, ice caps, and gla-

Figure 19–12 Tabular Antarctic iceberg. (Official U. S. Coast Guard Photo.)

ciers, and the shortest times (days) in the atmosphere.

Humans use water in the home or office, in industry, in agriculture, and for recreation. The least amounts are needed for drinking, but this water must be of the highest quality. The largest amounts are used for industry and agriculture.

Of the water that falls on land, some percolates down until it is stopped by impervious rock. The overlying porous layers become saturated up to a level called the **water table**.

Water is drawn into narrow pores by **capillary attraction**, which is the result of the wetting action of water. The moist layer at the surface that is available for plant growth is called the **soil-moisture belt**.

Alternating strata of porous and impervious rock can provide lower layers of groundwater known as **aquifers**. These waters are replaced very slowly, and their rapid removal may be regarded as the mining of a nonrenewable resource. Other possible effects of removal of groundwater are **subsidence** of the land and **salt water intrusion**.

Water can be stored by means of dams, which create reservoirs behind them. The resulting benefits include hydroelectric energy, control of water flow, and the recreational use of the reservoir. Disadvantages include silting of the reservoir, increased loss of water by evaporation and seepage, erosion of the stream bed below the dam, the risk of disaster if a dam breaks or when heavy rains overflow the dam onto heavily populated areas, and ecological disruption related to changes of habitats and redistribution of nutrients.

Other methods of recycling or redistributing water for human use include diversion of fresh water by aqueducts or canals, recharging groundwater, desalination of seawater, cloud seeding to induce rain, and towing icebergs to dry areas.

Questions

Properties of Water

1. Speculate on what the Earth and its biosphere (if any) might be like if (1) ice were denser than water; (2) at normal atmospheric pressure, heat added to ice at 0°C caused it to evaporate directly to water vapor rather than to melt (just as "dry ice," which is solid CO_2, does), and liquid water could exist only under high pressure, such as under layers of rock; (3) the amount of energy needed to melt ice and to vaporize water were reduced by 90 percent but that no other properties of water in any of its forms were changed.

The Hydrologic Cycle

2. Describe the various ways in which (1) a rise and (2) a fall in the average global temperature could affect the hydrologic cycle.

3. (1) Under what conditions is the equation

$$\text{evaporation} + \text{runoff} = \text{precipitation}$$

true for the movement of water on the continents? (2) Write a more general equation that would hold true under *all* conditions for the continents. (3) Write a similar general equation for the oceans.

Human Uses of Water

4. Assume that you had water available from the following five sources—(1) rainwater drained from your roof; (2) good well water or tap water; (3) water from the wash cycle of your dishwasher or clothes washer; (4) water from the rinse cycle of your dishwasher or clothes washer; (5) water drained from your bath or shower. List all the applications in your home, in your garden, for your pets, or for other purposes for which each of these water supplies could be used.

5. List as many uses as you can think of for water in industry, in agriculture, and for recreation.

Groundwater

6. (1) Describe what happens to rainwater that starts to fall onto a very dry area and then continues heavily for several days. (2) Describe what happens to this water when there is no more rain for a month.

7. What is an aquifer, and how does water reach it? Under what conditions can a well that reaches an aquifer provide artesian flow?

8. Describe three problems that can arise from excessive use of groundwater.

9. Explain why land subsides when groundwater is depleted. If the removal of groundwater is stopped, will the land necessarily rise again to its original level? Defend your answer.

Dams

10. The following three statements refer to the costs and benefits to different groups of people that result from the building of a dam. For each statement, identify the different groups of people: (1) People in some locations are benefited; people in other locations are disadvantaged. (2) People in some years are benefited; people in other years are disadvantaged. (3) People with some occupations or hobbies are benefited; people with other occupations or hobbies are disadvantaged.

11. List the advantages and disadvantages of dams as (1) a means of providing electrical energy, (2) a source of water for irrigation, (3) a method of flood control.

Management of Water Resources

12. Imagine that you are the chief executive of a country that has been depleting its groundwater reserves. You call in your experts. One says, "Conserve and recharge our waste water!" The second says, "Desalinate our seawater!" The third says, "Seed the clouds!" and the fourth says, "Tow icebergs!" Your response is to ask them all for information to back up their advice. Make a list of the questions that you want each expert to answer for you.

13. Imagine that you live in an area with abundant water, and it is proposed that some of the excess water be diverted to another region that needs it. List the questions that an environmental impact study should consider before construction is started.

Suggested Readings

Some of the material in this chapter can be supplemented by the references given for Chapter 20 on water pollution. Additional valuable readings are given below.

A good text on environmental geology is:
Arthur N. Strahler and Alan H. Strahler: *Environmental Geoscience.* Santa Barbara, CA, Hamilton Publishing Co., 1973. 511 pp.

Data on the quality and use of water can be obtained from:
N. Wollman and G. Bonem: *The Outlook for Water.* Baltimore, Johns Hopkins University, 1971.
C. R. Murray and E. B. Reeves: *Estimated Use of Water in the United States.* Geological Survey Circular No. 765. Washington, D. C., U. S. Government Printing Office, 1977.

Two introductory treatments are:
L. B. Leopold: *Water, A Primer.* San Francisco, W. H. Freeman, 1974.
Cynthia Hunt and Robert Garrells: *Water, The Web of Life.* New York, W. W. Norton & Co., 1972.

The politics of water in California are discussed in:
Harry Dennis: *Water and Power.* San Francisco, Friends of the Earth Books, 1981. 167 pp.

Books that deal with public policy in water management:
Yacov Y. Haimes (Ed.): *Risk/Benefit Analysis in Water Resources Planning and Management.* New York, Plenum Publishing, 1981. 304 pp.
Laurence Pringle: *Water—The Next Great Resource Battle.* New York, Macmillan, 1982.
William Ashworth: *Nor Any Drop to Drink.* New York, Summit Books, 1982.

20

Water Pollution

Travelers from North America are advised not to drink the water in some less developed nations. In many regions of the world, water purification systems are either inadequate or nonexistent. Millions of infants die at an early age from drinking polluted water, and even adults suffer from stomach illness. North Americans have long prided themselves on the fine quality of their drinking water, but in recent years, many people have questioned whether the water is really so excellent. The tap water in several American cities has a strong smell and a bad taste. Local residents have turned to large-scale consumption of bottled spring water. Chemists analyzing water supplies have raised disturbing questions. Articles published both in technical journals and local newspapers have raised fears about water supplies. This chapter reviews some of the problems of water pollution and purification.

20.1 Types of Impurities in Water

Chemically pure water is a collection of H_2O molecules—nothing else. Such a substance is not found in nature—not in wild streams or lakes, not in clouds or rain, not in falling snow, nor in the polar ice caps. Very pure water can be prepared in the laboratory but only with considerable difficulty.

Water accepts and holds foreign matter in various ways:

1. Water is an unusually good solvent. It is especially good at dissolving mineral salts, which typically consist of negative and positive ions. The positive ions are those of metals, including many that are poisonous, such as the ions of copper, cadmium, mercury, and lead. Water is also a good solvent for many organic compounds that contain oxygen, such as alcohols, sugars, and organic acids. Furthermore, many materials that are normally considered insoluble in water are in fact very slightly soluble. Thus, hydrocarbons are said to be insoluble in water ("oil and water do not mix.") Yet benzene, for example, dissolves in water to the extent of almost 0.1 percent. That may not sound like much, but it would have a terrible taste, and besides, benzene is toxic to humans.

2. Insoluble particles, if they are small enough, may settle so slowly that for all practical pur-

poses they remain in water indefinitely (note the data in Fig. 20–1).

3. Some insoluble materials, especially certain metals, react with water to produce soluble products.

4. Nutrient matter is metabolized by living organisms in water, and the resulting waste products may be pollutants.

5. Living organisms themselves, if they are pathogens, may be considered pollutants. Their own energy of motion keeps them from settling out of solution.

6. A soluble substance may react with an insoluble contaminant and bring it into solution. For example, acids in water react with many minerals and thus dissolve them.

7. Finally, a contaminant may pollute water simply by floating on it. Water is denser than almost all hydrocarbons; therefore, petroleum floats on water. A floating oil spill is certainly a water pollutant.

It is useful to classify foreign substances in water according to the size of their particles because this size often determines the effectiveness of various methods of purification. Figure 20–1 shows a spectrum of particles arbitrarily divided into three classes—suspended, colloidal, and dissolved. Let us consider each in turn, referring to the figure.

Suspended particles, which have diameters of more than about 1 micrometer, are the largest. They are large enough to settle out of water reasonably quickly and to be retained by many common filters. They are also large enough to absorb light, thus making water containing them look cloudy or murky.

Colloidal particles are so small that their settling rate is insignificant, and they pass through the holes of most filter media; therefore, they cannot be removed from water by settling or by ordinary filtration. Water that contains colloidal particles appears cloudy when observed at right angles to a beam of light. (The same phenomenon occurs in air; colloidal dust particles can be seen best when observed at right angles to a sharply focused light beam in an otherwise dark room.) The colors of natural waters, such as the blues, greens, and reds of lakes or seas, are caused largely by colloidal particles.

Dissolved matter does not settle out, is not retained on filters, and does not make water cloudy, even when viewed at right angles to a beam of light. The particles of which such matter consists are no larger than about one-thousandth micrometer in diameter. If they are electrically neutral, they are called molecules. If they

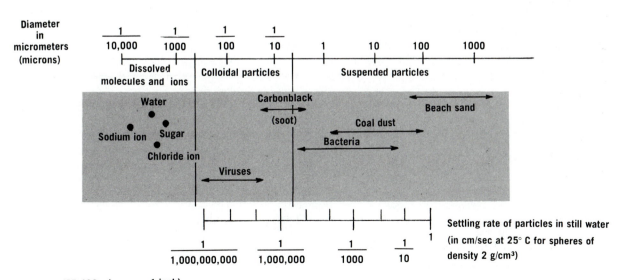

(25,400 microns = 1 inch)

Figure 20–1 Small particles in water.

bear an electrical charge, they are called ions. Cane sugar (sucrose), grain alcohol (ethanol), and "permanent antifreeze" (ethylene glycol) are substances that dissolve in water as electrically neutral molecules. Table salt (sodium chloride), on the other hand, dissolves as positive sodium and negative chloride ions.

Natural waters contain substances in all three categories, as shown in Table 20–1. Natural waters range from tastily potable to poisonous. In saltiness they range from fresh rainwater (not salty) to brackish (partly salty, where river water mixes with seawater), to ocean water, to the exceedingly concentrated salt solutions found in such places as the Dead Sea or Great Salt Lake, where water evaporation concentrates the salts.

20.2 Productivity in Aquatic Systems

The productivity of an ecosystem reflects the rate at which its producers photosynthesize (see Chapter 3). In most ecosystems, such as a field of crops, a forest, or an oceanic fishing ground, higher productivity is beneficial for humans because the system supplies useful products such as grains, wood, or fish. In lakes and streams, however, the opposite type of ecosystem is preferred. For example, picture a clear mountain lake full of trout and so pure that a person could drink the water. Such a system, called an **oligotrophic** lake, is characterized by *low* productiv-

TABLE 20–1   Impurities in Natural Water

Source	Particle Size Classification				
	Suspended	Colloidal	Dissolved		
Atmosphere	←Dusts→		*Molecules* Carbon dioxide, CO_2 Sulfur dioxide, SO_2 Oxygen, O_2 Nitrogen, N_2	*Positive ions* Hydrogen, H^+	*Negative ions* Bicarbonate, HCO_3^- Sulfate, SO_4^{2-}
Mineral soil and rock	←Sand→ ←Clays→ ←Mineral soil particles→		Carbon dioxide, CO_2	Sodium, Na^+ Potassium, K^+ Calcium, Ca^{2+} Magnesium, Mg^{2+} Iron, Fe^{2+} Manganese, Mn^{2+}	Chloride, Cl^- Fluoride, F^- Sulfate, SO_4^{2-} Carbonate, CO_3^{2-} Bicarbonate, HCO_3^- Nitrate, NO_3^- Various phosphates
Living organisms and their decomposition products	Algae Diatoms Bacteria ←Organic soil (topsoil)→ Fish and other organisms	Viruses Organic coloring matter	Carbon dioxide, CO_2 Oxygen, O_2 Nitrogen, N_2 Hydrogen sulfide, H_2S Methane, CH_4 Various organic wastes, some of which produce odor and color	Hydrogen, H^+ Sodium, Na^+ Ammonium, NH_4^+	Chloride, Cl^- Bicarbonate, HCO_3^- Nitrate, NO_3^-

ity. The water is clear because it contains comparatively few plankton or rooted plants. Ecologists have learned that productivity is low because the growth of producers is limited by a shortage of minerals. The low productivity of an oligotrophic lake or stream is beneficial for humans.

As time passes, nearby mountains or hillsides erode, and sediments are washed into all freshwater systems. These sediments carry nutrients, which fertilize the system and increase its productivity. A lake with high productivity, called a **eutrophic** lake, has a dense population of producers, often visible as a green scum on the surface of the murky water. In a natural system, many thousands of years may elapse before an oligotrophic waterway becomes visibly eutrophic. When this slow process is speeded up by human activity, such as by runoff from farms or sewage plants, it is called **cultural eutrophication**.

The inorganic nutrient whose absence most often limits freshwater productivity is phosphate. (Phosphates are a group of ions that contain the elements phosphorus and oxygen, such as PO_4^{3-}, $P_2O_7^{4-}$, and others.) The consequence of increasing the phosphate content of an oligotrophic lake was demonstrated in a dramatic experiment illustrated in Figure 20–2. A small lake in Manitoba was divided into two halves by suspending a plastic sheet across a narrow neck in the middle of the lake. The bottom of the sheet was secured to the lake's rocky bottom. A large amount of phosphate fertilizer was then added to one half of the lake, leaving the other half as a control. Within a few weeks the fertilized half had become opaque as a result of a massive bloom of plankton. Pathogens (disease-causing organisms) may be among the organisms whose populations grow with the influx of nutrients into water.

Figure 20–2 The effect of phosphorus on the productivity of a lake. This lake in Manitoba was divided in two by plastic sheeting across the narrow neck in the middle of the photograph. Phosphorus was added to the half of the lake in the upper part of the photograph. Several weeks later, the phosphorus-fertilized half of the lake was opaque as a result of massive plankton bloom; the lower part of the lake was as clear and oligotrophic as it was before the experiment. (Photograph courtesy of David Schindler.)

When phosphorus is added in a single dose, as in this experiment, the effect is short-lived because phosphorus is continuously removed from the system by deposition in bottom sediments or by stream outflow. On the other hand, if phosphorus is added continuously, as it often is in the form of sewage, phosphate-containing detergents, or fertilizer runoff from agricultural land, the productivity of a lake will remain high, and eutrophication will be hastened, changing the character of the lake, often irrevocably. Such changes have fundamentally altered Lake Erie, which is relatively shallow. Even deep oligotrophic lakes, such as Lake Tahoe on the California–Nevada border, have become noticeably more eutrophic in the last 20 years as a result of pollution.

Geologically young oligotrophic lakes are rare, especially in developed areas, where cultural eutrophication is common. When the American pioneers migrated to the west, they

BOX 20.1

Eutrophic originally meant "tending to promote nutrition." In this sense, a vitamin supplement would be a eutrophic medicine. The term was later applied to describe the nourishment of natural waters as a contributor to the process of succession.

found lakes that were so clear that the fish could be seen deep below the surface. Some 75 percent of large lakes in the United States today are eutrophic. In some, cultural eutrophication is advanced, and the problem is aggravated by the presence of an additional potential supply of nutrients stored in the sediments. Small concentrations of nutrients are essential to aquatic ecosystems, whereas large concentrations of the same materials are pollutants.

Severe eutrophication is sometimes considered to be almost irreversible. The situation is not always hopeless, however. Recent experience, for example with Lake Washington (Seattle), has shown that eutrophication caused by excessive fertilization can be reversed if the nutrient inflow is drastically reduced.

The Role of Oxygen

Most organisms require oxygen to metabolize their food and release the energy they need. This process is known as **aerobic respiration** ("aerobic" means *with oxygen*). Aquatic organisms use the oxygen dissolved in water for respiration. Oxygen is not very soluble in water. One liter of water in contact with air at 25°C contains 0.0084 g of oxygen. In contrast, 1 liter of air at 25°C contains 0.27 g of oxygen. Thus, aerated water contains only about one thirtieth as much oxygen as the same volume of air. Furthermore, when oxygen has been removed from water, it is not replaced rapidly except in places where turbulence mixes water and air, as it does in the "white water" of shallow rapids.

Because water does not contain much oxygen and because oxygen is not replaced rapidly when it is removed, the bottom mud of freshwater streams and lakes where the water does not move rapidly contains very little oxygen. This bottom mud provides the perfect environment for microorganisms that can respire **anaerobically** (without oxygen). These organisms live by releasing energy anaerobically from the dead organic matter, such as the bodies of phytoplankton, that falls from above. Many bacteria can respire either aerobically or anaerobically, depending on conditions. Since anaerobic respiration yields much less energy than aerobic respiration (Table 20–2), these bacteria use oxygen when it is available.

If slow-moving fresh water is rich in nutrients, however, the populations of various organisms will increase dramatically. In this situation, oxygen-consumers may use the oxygen in the water faster than it can be replaced by dissolving it from the air or by photosynthesis. Then most of the oxygen may disappear even from the surface layers of the water. When this occurs, phytoplankton that require oxygen die by the millions and are replaced by organisms, such as the bacteria from the bottom mud, that can survive anaerobic conditions. Many larger organisms such as trout need a lot of oxygen, and they die as well.

If the greenish scum of a bloom of photosynthetic plankton is unpleasant, a scum of dead plankton and fish, combined with a population explosion of anaerobic bacteria, is even more so.

Table 20–2 Energy Yields from Aerobic and Anaerobic Processes

Nutrient	Process	Products	Approximate Energy Yield per Gram of Nutrient (calories)
Carbohydrates	Respiration (aerobic)	$CO_2 + H_2O$	4000
	Fermentation (anaerobic)	Alcohol + H_2O	100
	Biogas production (anaerobic)	Methane + CO_2	220
Proteins	Respiration (aerobic)	$CO_2 + H_2O$ + nitrates and sulfates	4000
	Putrefaction (anaerobic)	CO_2 + ammonia, methane, and hydrogen sulfide	370

Anaerobic sulfur bacteria, for instance, produce hydrogen sulfide, a gas that smells like rotten eggs.

Because the addition of nutrients to fresh water initiates this cycle of plankton bloom, oxygen depletion, and anaerobic decay, nutrients must be considered pollutants. Whether they come from a chemical factory, a farmer's fields, the municipal sewage works, or inadequate septic systems, they render the water less fit for human use (Fig. 20–3).

Biochemical Oxygen Demand

Imagine that a **biodegradable** material (a material that can be utilized by living organisms) is added to a lake or stream. This material, by itself, may be harmless; it could be something as innocuous as waste food from a cannery. But once it is spilled into an aquatic ecosystem, it will be consumed and oxidized by a variety of organisms. If the organisms are aerobic, the process will also consume dissolved oxygen. If consumption is great enough, oxygen levels in the water will be depleted and aerobic organisms will die. Thus, the quality of the water is degraded and the water has been polluted.

The index of pollution by nutrients is called the **biochemical oxygen demand,** or **BOD.** The BOD is defined as the amount of oxygen that will be consumed when a biodegradable substance is oxidized in an aquatic system.

Some organic materials, such as chlorinated hydrocarbons, that are manufactured by industrial processes cannot be used as food by bacteria and therefore do not contribute to the BOD.

A

B

Figure 20–3 The choking of waters by weeds. A, The dam on the White Nile at Jebel Aulia near Khartoum, Sudan. The area was clean when photographed in October, 1958. B, The same area in October 1965, showing the accumulation of water hyacinth above the dam. (From Holm: Aquatic weeds. *Science* 166:699–709, Nov. 7, 1969. Copyright 1969 by the American Association for the Advancement of Science.)

BOD values are expressed in milligrams of oxygen per liter of water. Recall that pure water saturated with air at 25°C contains 0.0084 g, or 8.4 mg, of oxygen per liter. Compare this with some typical BOD values:

TYPE OF WATER	BOD (mg/L)
Pure water	0
Typical fresh natural water	2 to 5
Domestic sewage	hundreds
Sewage after primary and secondary purification (see Section 20.8)	10 to 20

BOX 20.2 MEASUREMENT OF BIOCHEMICAL OXYGEN DEMAND (BOD)

The rate of biochemical oxidation depends on the temperature of the environment and on the particular kinds of microorganisms and nutrients present. If the first two factors are controlled, then the rate of oxidation depends only on the amount of nutrient. After five days, under typical conditions, almost all the nutrient is gone and almost all the oxygen that is going to be used has been used. This length of time is considered to be a good compromise between completion of the oxidation and not having to wait forever. Therefore, a standard test is carried out by saturating a sample of the polluted water with oxygen at 20°C and determining how much oxygen has been used up after five days. The amount of oxygen thus consumed per liter of contaminated water is the practical measure of the BOD.

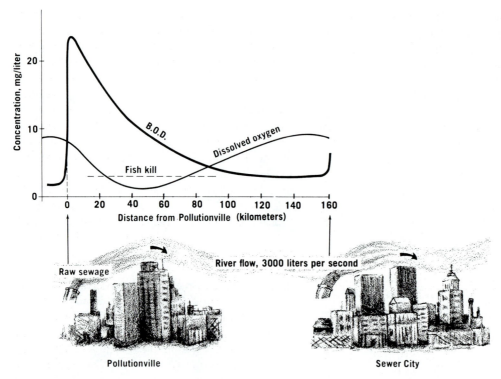

Figure 20–4 *River pollution from hypothetical cities.*

Thus domestic sewage requires more than 10 times as much oxygen as the 8.4 mg per liter available in pure aerated water, and even treated sewage demands up to twice as much oxygen as is available.

20.3 Pollution of Inland Waters by Nutrients

Streams and Rivers

When sewage is discharged into a freshwater stream, the stream becomes polluted. This does not mean that the oxygen content drops instantaneously. But the *potential* for oxygen depletion exists wherever there is sewage. The measure of this potential is the BOD, which rises as soon as the sewage goes in. Figure 20–4 shows this sudden increase in BOD in a hypothetical river. Now follow the water downstream from "Pollutionville." Three processes are going on, all at the same time.

● *Process 1.* The bacteria are feasting on the sewage. Because of this action, the amount of sewage in the water is decreasing, so the BOD is going down.

● *Process 2.* As the bacteria consume the sewage, they also use dissolved oxygen, so that concentration, too, starts to decrease.

● *Process 3.* Some of the lost oxygen is being replenished from the atmosphere and from photosynthesis by the vegetation in the stream.

For the first 50 km or so downstream, as shown in Figure 20–4, the natural ability of the river to recover its oxygen (process 3) simply cannot keep up with the feasting bacteria (process 2), so the dissolved oxygen concentration goes down. The fish begin to die (Fig. 20–5), but it is not the sewage that is killing them. (In fact, the sewage provides food.) Instead, the fish die from lack of oxygen, beginning when the dissolved oxygen concentration falls below about 4 mg per liter, depending on the particular species.

Figure 20–4 shows that the fish kills start about 15 km downstream from the introduction of the raw sewage. In time, as the sewage is used up by bacteria, the BOD goes down (process 1), the consumption of oxygen also slows

Figure 20–5 A fish kill caused by water pollution.

down, and the natural ability of the river to recover (process 3) becomes predominant. The river then begins to repurify itself. About 90 km downstream the fish begin to survive again, and at about 140 km the oxygen content has increased to its former, unpolluted level.

Of course, if additional sewage is discharged before recovery is complete, as shown in the illustration at 160 km, the river becomes polluted again. When sources of pollution are closely spaced, pollution becomes practically continuous. Rivers in such a condition, which unfortunately can be found near densely populated areas all over the world, support no fish, are high in bacterial content (usually including pathogenic organisms), appear muddily blue-green from choking algae, and, in extreme cases, stink from putrefaction and fermentation.

Lakes

Water flows more slowly in lakes than in rivers, and therefore the residence time of water is much longer in a lake. Figure 20–6 shows typical conditions in winter and in summer in a small lake in a temperate climate, such as New En-

gland. In the summer, the upper waters, called the **epilimnion** (the "surface lake") are warmed by the Sun. These warmer waters, being lighter than the colder ones below, remain on top and maintain their own circulation and oxygen-rich conditions. The lower lake waters (the **hypolimnion**) are cold and relatively airless. Between the two lies a transition layer, the **thermocline**, in which both temperature and oxygen content fall off rapidly with depth. As winter comes on, the surface layers cool and become denser. When they become as dense as the lower layers, the entire lake water circulates as a unit and becomes oxygenated. This enrichment is, in fact, enhanced by the greater solubility of oxygen in colder waters. Furthermore, the reduced metabolic rates of all organisms at lower temperatures result in a lesser demand for oxygen. When the lake freezes, then, the waters below support the aquatic life through the winter.

With the spring warmth, the ice melts, the surface water becomes denser,* and again the lake "turns over," replenishing its oxygen supply.

Now, what are the effects of oxygen-demanding pollutants on these processes? During the summer, increased supplies of organic matter serve as nutrients in the oxygenated upper waters; the oxygen is replaced as needed by physical contact with the air and from photosynthesis by algae and other water plants. But some organic debris rains down to the lower depths, which are reached neither by air nor by sunlight. Therefore, in an organically rich or eutrophic lake, the bottom suffers first. Fish that live best at low temperatures are therefore the first to disappear from lakes as the cold depths they seek become depleted of oxygen by the increased inflow of nutrients. These fish are frequently the ones most attractive to human diets, such as trout, bass, and sturgeon.

Soap and Detergents

Soap was produced in ancient times by heating animal fat with wood ashes. Soap is biodegradable but does not contain nitrogen or phosphorus. The mineral matter in most groundwater (calcium, magnesium, and iron)

---

* Water reaches its maximum density at 4°C, so any approach to this temperature, from above or below, is accompanied by an increasing density.

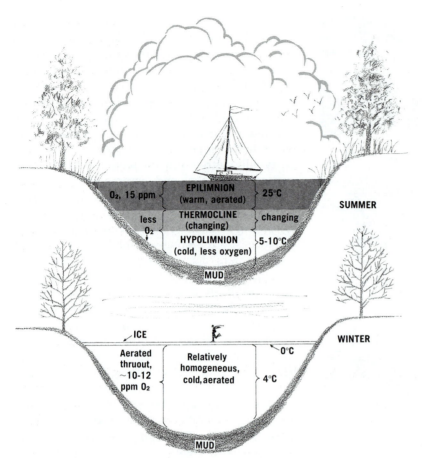

Figure 20–6 Thermal stratification in a small, temperate lake.

makes soap insoluble, leaving a "ring around the collar" or around the bathtub. Such water is said to be "hard."

Starting in the 1940s, the chemical industry developed and marketed a series of synthetic substitutes for soap called **synthetic detergents**, which could be used in hard water. Other ingredients, such as brighteners, fabric softeners, and bleaching agents were also added. These new detergents, however, contained phosphates and therefore contributed to the cultural eutrophication of the receiving waters. In recognition of this problem, several states and cities in the United States have banned or limited phosphates in detergents. Some of the synthetic substitutes for phosphates, however, may have other detrimental effects, such as irritation of the tender skin of infants.

What, then, can the individual do? One possibility is to return to soap, if "soft " water is available. Soap and rainwater (which is soft) makes a very effective combination. If only hard water is available, a combination of soap and washing soda (also called sal soda), in a ratio of about 5 to 1, is effective. Another possibility is simply to use less detergent. Manufacturers often recommend quantities appropriate (or even excessive) for extreme conditions. Frequently ½ of the recommended amount of detergent gives very good cleaning action, and as little as ⅛ gives fairly good cleaning. As far as personal health is concerned, soap sanitizes clothes very effectively. Therefore, the use of bleaches makes clothes whiter, but not more sanitary. As far as "brightening" effects are concerned, you must decide for yourself how important it is to have "dazzling" underwear.

20.4 Industrial Wastes in Water

Several years ago we (A.T. and J.T.) toured a factory that recovered felt from old fur scraps.

Figure 20-7 Industrial wastes in water.

The company purchased smelly scraps of animal hides and then boiled them down in a huge vat of dilute sulfuric acid. The acid separated the valuable fur fibers from the bits of skin and flesh. The fur was then collected and used to manufacture felt products such as hats or the insoles of winter boots. The hot sulfuric acid solution of decomposed skin and animal fat was then dumped directly into the river. This small company was eventually forced out of business by pollution control laws and a changing economy. But the pollution of waterways continues to this day. Large quantities of water pollutants are currently released into streams and rivers from paper production, food processing, chemical manufacturing, steel production, and petroleum refining (Fig. 20-7).

Some of these wastes are known to be poisonous. The effects of others are unknown. Some have existed in drinking water since ancient times. Many are quite recent, and new types of wastes continue to appear as new technology develops.

Many large American cities obtain their drinking water from nearby rivers. Rivers are also convenient repositories for the discharge of industrial waste waters. In the United States and most other developed countries, there are legal requirements for cleanup of such wastes before they leave the factory boundaries. Many factories and chemical plants include elaborate water

purification systems as part of their manufacturing operations. However, the purification process is never 100 percent complete, and "purified" industrial wastewater cannot be assumed to be fit to drink. As a result, the major rivers of the industrialized countries, such as the Mississippi, the Rhine, and the Volga, contain small concentrations of thousands of industrial chemicals. Cities and towns along the river draw this water and purify it again before it is piped to homes for drinking. This purification process, however, is also incomplete; sometimes the major objective is only to kill microorganisms and to make the water look clear. Trace concentrations of heavy metals, pesticides, and industrial organic chemicals, some of them suspected carcinogens, are therefore found in the drinking water of many cities that are located near major rivers.

20.5 Pollution of Groundwater

The accumulation, use, and recharging of groundwater were discussed in Chapter 19. Pollution as well as purification can occur during these transfers. In recent years, however, the pollution of groundwater in some areas has increased to the point where its quality as drinking water is seriously threatened.

If raw sewage is dumped onto the soil (Fig. 19-7), the liquid percolates into the ground. The soil acts as a filter, blocking large solid particles while allowing the liquid to pass through. Smaller particles and even molecules of contaminants, although not physically blocked, adhere to the soil particles, and they, too, are removed. At the same time, however, the percolating water dissolves mineral matter out of the soil and rock. Therefore, all groundwater contains some dissolved inorganic salts. In rare instances, such dissolved matter may be toxic, as for example when the water seeps through areas containing lead or arsenic minerals. Some groundwaters contain natural concentrations of fluorides, which reduce the incidence of dental cavities in children but which in larger concentrations mottle the teeth.

The pollution of groundwater from human sources presents special problems that are different from the pollution of surface waters such as lakes or streams. There are two main reasons for this difference:

1. Most groundwater moves quite slowly through its zones (see Table 19–1); a typical rate might be about 30 cm (1 foot) per day. Furthermore, the water does not mix as much during its motion through porous rock as it would, say, in a river (Fig. 20–8). Instead, the flow of water advances more like a column of marchers in a parade, who do not mingle much with the crowds on either side. Consequently, contaminants introduced into groundwater are not readily diluted. Figure 20–8B shows a schematic view of a wastewater pond that leaks into the ground. The contaminated "plume" follows the flow of underground water until it is discharged into a lake or stream. Water taken from above or below this plume (wells A or B) may be uncontaminated, but the unfortunate user of well C will get polluted water.

2. Groundwater does not have the access to air that is available to surface waters. Therefore, the

Figure 20–8 A, Wastewater is diluted in a turbulent stream. B, Wastewater leaks into the ground from an improperly lined storage basin. The "plume" of contaminated water follows the prevailing groundwater flow but does not mix much with it and is therefore not diluted. Water taken from well A (below the contaminated plume) or from well B (above it) is good, but water from well C is polluted.

oxidation that can purify or decontaminate surface water does not occur in deep aquifers.

The sources of groundwater pollution from human activities include the following:

1. Wastewater is often stored in a basin, pit, pond, lagoon, or other such facility. The purpose of such storage is to hold the wastewater prior to treatment or simply to allow oxidation by air to decontaminate it. There are probably between 100,000 and 150,000 such sites in the United States; about 75 percent of them are industrial, the remainder are agricultural and municipal. Many of them are unlined, and the soils beneath them are permeable (as in Fig. 20–8), so the polluted water can seep down. The same problem is often associated with solid wastes, which are stored in landfills (Chapter 22). These, too, often leak contaminants into the ground.

2. Bioresistant pesticides, which are sometimes heavily applied (or misapplied) to deal with large infestations of insects, can percolate down to groundwater.

3. Mine tailings can be very troublesome because they may contain toxic mineral matter. When these wastes are exposed to the outside, rainwater (which is slightly acid naturally but more so if it is polluted) can dissolve some of these minerals and carry them down to an aquifer.

4. Liquid chemical wastes that are injected into deep wells below the aquifers (Fig. 19–8) may sometimes leak or migrate up into groundwater sources. Pipes can burst under high pressure or can be corroded by acidic wastes.

5. Viruses, which can multiply in sewage and in landfills, are very difficult to remove from water. Viruses pass through ordinary filters, and are resistant to chlorination. Seepage from improperly designed landfills therefore imposes a danger of groundwater contamination by viruses.

Most groundwaters are still safe, but once a source is contaminated, there is no practicable way to clean it up. Furthermore, natural processes of self-purification are far too slow. Pollution of groundwater must therefore be prevented. It is for this reason that regulations such as the Safe Drinking Water Act of 1974 (Section

20.9), as well as various state and local laws, have been enacted.

20.6 Pollution of the Oceans

For many years people were relatively unconcerned about pollution of the oceans because the great mass of the sea can dilute a huge volume of foreign matter to the point where it has little effect. In recent years this attitude has changed for several reasons. For one, pollutants are often added in such relatively high concentrations in local areas that environmental disruption does occur. In addition, the quantity of pollutants dumped into the ocean has grown so large that some scientists fear that global effects may be significant.

Oil

On March 16, 1978, the oil tanker *Amoco Cadiz* lost her steering off the coast of Brittany in France. High winds blew the ship ashore, and within the next few days she broke apart, spilling approximately 1.6 million barrels (220 thousand tonnes) of crude oil onto the beaches and into the water (Fig. 20–9). It is virtually impossible to estimate the economic damage done by this spill. Brittany had long been a vacation area, and tourists avoided the area for years after the spill (Fig. 20–10). Over a million sea birds were also killed within a few days of the disaster. The oil clogged their feathers and respiratory tracts so that they drowned or died from inhaling the oil. A huge oyster fishery was destroyed. Plankton, fish, and other sea animals were killed.

Between 1969 and 1974, there were 500 tanker accidents that involved oil spills. More than 1 million tonnes of oil were released altogether. In 1976, five major accidents off the coasts of the United States combined to spill another 35,000 tonnes of oil. But accidents are not the only source of oil in the sea. Tanker captains often clean the oily holds of their ships by washing them out with seawater despite the fact that this is strictly illegal in many parts of the world. It has been estimated that up to 90 percent of the oil in the oceans comes from these small discharges. No matter where you may be, on the coast of Africa or a South Seas island, beach-

A

B

Figure 20–9 *A,* Big waves are rolling over the wrecked tanker Amoco Cadiz while large amounts of oil are still pouring out. (Wide World Photos.) *B,* Location of the wreck.

combers will probably agree that there is hardly an unoily beach left in the world.

Figure 20–10 Cleaning oil-soaked beaches in northern France after the wreck of the Amoco Cadiz. (Wide World Photos.)

Offshore drilling operations also contribute their share of oil to the sea. The largest spill of all started on June 3, 1979, at an offshore oil well owned by Pemex, the national Mexican oil company. Mud, rapidly followed by oil and gas, started to gush through an unsealed drill pipe. The fumes ignited on contact with the pump motors, the drilling tower collapsed, and the spill was out of control. It took nine months to stop it, by which time 3.1 million barrels (440,000 tonnes) of oil—about twice the amount lost by the Amoco Cadiz—had been spilled into the Gulf of Mexico.

The total quantity of oil that finds its way into the sea each year is very large. It has been estimated that close to one million tonnes of oil are spilled into the ocean each year from ships and oil drilling operations alone. But there are also many "mini-spills." Some examples are sludges from automobile crankcases that are dumped into sewers, routine oil-handling losses at seaports, leaks from pipes, and the like. Some oil aerosols also settle into the sea from the atmosphere. The grand total from all these sources is difficult to estimate, but it could well reach 10 million tonnes per year or more.

Shipwrecks have occurred ever since people first went to sea in ships. In ancient times these accidents were disastrous for the sailors, their families, and the ship owners. Today tanker wrecks and discharges represent an even greater hazard because they are threatening the very life of the sea. Can anything be done? The sea has always been an international domain. No nations owns it, no nation can impose laws concerning it. Therefore, individual shipping companies can make decisions concerning the seaworthiness of their vessels. Often these decisions are based more on profit than on safety. For example, it is possible to build a ship with two rudders, two propellers, and two independent steering mechanisms. Then if a steering gear breaks, the ship will not flounder helplessly and be smashed to bits on the shore. But such careful design is expensive. A ship owner can realize higher profits if ships are built cheaply without such safety devices.

Governments cannot regulate foreign ships when they are far out to sea. But they can impose laws on vessels that sail into port. After the disastrous winter of 1976 when five tankers sank off the coast of the United States, many lawmakers suggested strict controls for all tankers landing in the country. But strict laws have not been passed. Meanwhile, oil continues to accumulate in the oceans. Ship captains report sighting large oil slicks daily in the North Atlantic Ocean.

Crude oil is crude indeed, in the sense that it consists of many thousands of components of widely differing molecular weights. It is usually a dark brown, smelly liquid, about as thick as engine oil. It is largely composed of hydrocarbons, but there is an appreciable proportion of sulfur, and there are trace concentrations of metals such as vanadium and nickel.

Most hydrocarbons are less dense than water, and therefore the major portion of a mixture of hydrocarbons such as crude oil floats. However, some hydrocarbons are dense enough to sink even in seawater, and these materials, together with a portion of the metallic components, may settle to the bottom, where they have the potential to disrupt generations of aquatic organisms. Furthermore, the oxidation of floating oil also yields some products that are denser than seawater.

If a typical crude oil is heated to 100°C, some 12 percent of its volume boils off; if it is heated to 200°C, an additional 13 percent boils off. The total (25 percent) may be considered to be the volatile fraction that will evaporate from the floating oil surface within a few days. The remaining oil is slowly metabolized by bacteria, and some of it slowly evaporates. After about three months, practically all the material that can evaporate has evaporated, and all that can be eaten has been eaten. The persistent remainder is an asphaltic residue, representing about 15 percent of the original oil. These leftovers occur as small tarry lumps all over the Earth's seas.

Plastics

Plastics are generally less biodegradable than oil. Sailors in small boats report that the world's oceans are awash with polystyrene cups in a continuous trail across the Atlantic and Pacific Oceans. Polystyrene breaks down into microscopic globules. Water samples from the depths of the Antarctic and the Arctic oceans contain polystyrene particles; they are obviously everywhere in the oceans. We have no idea what long-term effects these plastics may have on marine organisms that ingest them. These plastics have mainly been thrown overboard from boats. Most boating associations now urge their members never even to take anything disposable and plastic to sea and encourage ocean liners to follow suit. But much of the damage has, obviously, already been done.

Other Chemical Wastes

There is no known inexpensive and guaranteed safe method of disposing of highly poisonous chemical wastes, such as byproducts from chemical manufacturing, chemical warfare agents, and pesticide residues. It is cheap and therefore tempting to seal such material in a drum and dump it in the sea. But drums rust, and outbound freighters do not always wait to unload until they reach the waters above the sea's depths. As a result, many such drums are found in the fisheries on continental shelves or are even washed ashore. It is estimated that tens of thousands of such drums have been dropped into the sea.

Of course, all the river pollutants enter the same sink—the world ocean. The organic nutrients are recycled in the aqueous food web. But the chemical wastes from factories and the seepages from mines are all carried by the streams and rivers of the world into the sea.

And where do air pollutants go? Airborne lead and other metals from automobile exhaust, mercury vapor, and the fine particles of agricultural sprays ride the winds and fall into the ocean.

Is There an Overall Threat to Life in the Sea?

In ocean regions near large cities such as New York, pollution has killed most marine life in wide areas. These places have come to be known as "dead seas." Is it possible that the entire ocean may die?

It is, in general, very difficult to predict how complex ecosystems react to environmental pollution. Sometimes natural systems are amazingly stable and seem to be barely affected by pollution. Thousands of tonnes of crude oil have been dumped in midocean, and in a few months the residues seem to disappear quietly. In other situations ecosystems seem to be extremely fragile. Scientists are not sure what goes out of adjustment. But sometimes small concentrations of pollutants disturb entire ecosystems.

These concerns, expressed in 1972 at the United Nations Conference on the Human Environment in Stockholm, led to a study of "The Health of the Oceans" that was released in October, 1982. The findings of the study were optimistic. The world's oceans actually seemed healthier in 1982 than they were in 1972. Some of this improvement is a result of environmental laws that now restrict the production and distribution of many toxic substances such as pesticides and harmful metals in the most industrialized countries. In addition, various natural biological and chemical processes serve to assimilate or degrade oil spills and other toxic materials enough to render them harmless. These conclusions apply specifically to the open oceans, hundreds of miles from any shoreline. Even in intertidal and subtidal areas, however, where

Figure 20–11 Oil-soaked gannet (a gull-like sea bird), Jones Beach, Long Island. (© Komorowski, from National Audubon Society.)

the impact of oil can be severe, the report said that there was no evidence "that oil alone could threaten the survival of a species."

20.7 The Effects of Water Pollution on Human Health

On a worldwide scale, the pollution of water supplies is probably responsible for more human illness than any other environmental influence. The diseases so transmitted are chiefly due to microorganisms and parasites. Two examples will illustrate the dimensions of the problem. Cholera, an illness caused by ingestion of the bacterium *Vibrio cholerae,* is characterized by intense diarrhea, which results rapidly in massive fluid depletion and death in a very large percentage of untreated patients. Though its distribution in the past was virtually worldwide, it has been largely restricted during the twentieth century to Asia, particularly the area of the Ganges River in India (Fig. 20–12). During the nine years from 1898 to 1907, about 370,000 people died from this disease, and thousands of Indians continue to die each year even at present.

Most Americans have never heard of schistosomiasis. This is actually a group of diseases caused by infection with one of three related types of worms. (Which worm you get depends on where in the world you live.) Current estimates are that over 100 million people are infected with schistosomiasis; these cases are distributed throughout the African continent, in parts of Asia, and in areas of Latin America. Estimating the amount of human suffering caused by schistosomiasis is much more difficult than for a disease like cholera, because unlike cholera it is a cause of much chronic as well as acute disease. For both these illnesses the main mode of transmission is water supplies contaminated with the feces of infected individuals. Other bacterial illnesses, such as the salmonelloses (of which typhoid fever is a leading example), and viral infections such as poliomyelitis and hepatitis may also be disseminated in this way. In bacterial and viral illnesses, the organisms themselves are shed in the stool and must be ingested by others to cause disease.

In the case of schistosomal infections, however, the eggs of the organisms are shed. They then hatch into forms that must find a certain type of snail to complete their life cycle. Once safely in the snail, the worm develops into a free-living form that leaves the snail and may infect people if ingested in drinking water. Alternatively, it may penetrate human skin on con-

Figure 20–12 In India, wastes are discharged directly into the Ganges and other rivers. People use the same water for washing and cooking. As a result, waterborne diseases are widespread.

tact and enter the bloodstream. The lesson is never to bathe in fresh water in the tropics where schistosomiasis occurs.

In the United States, however, the picture is very different; in fact, nowhere is the contrast between developed and underdeveloped countries starker than in the comparison between the health effects of water pollution on the respective populations. During the decade 1961–1970, there were 130 reported outbreaks of disease attributable to contaminated water supplies in the United States; of these, all but a very few were probably due to the presence of microorganisms rather than chemicals. A total of 46,000 people became ill, but only 20 died. Although the existence of such outbreaks in a technological society such as ours is deplorable, it is immediately evident that water pollution is a very minor source of acute fatal illness in the United States. In this age of extreme mobility made possible by international travel, the possibility always exists that a disease such as cholera could spread to the United States and attain epidemic proportions here. That this could happen on a large scale, however, seems unlikely since about three quarters of the American population derive their water from sources that are monitored by state and federal agencies.

The usual measure of microbiologic purity of a water supply is the so-called coliform count (coliforms are the class of bacteria present in the human intestine); therefore, the concentration of coliforms in a water supply is a measure of the amount of human fecal contamination, not a direct measure of the number of disease-causing microorganisms. Water is generally considered safe if it contains fewer than 10 coliforms per liter. Though this method generally serves to safeguard the purity of water, its major pitfall is that some steps in water purification, notably chlorination, may destroy bacteria without killing viruses; hence viral disease may be transmitted by water that satisfies rigid bacteriologic standards.

As noted earlier, water supplies may become contaminated with a wide variety of chemical substances. It is surprising, therefore, to realize that although the potential for the production of disease from this source exists, actual accounts of major illness due to chemically contaminated water are, fortunately, quite few. However, the simple fact that acute illness is uncommon does

SPECIAL TOPIC A
Legionnaire's Disease

In the summer of 1976 the Pennsylvania Division of the American Legion held its convention in Philadelphia at the Bellevue-Stratford Hotel. An epidemic of disease broke out among the attendees. The initial symptoms of fever, chills, and chest pains were followed by a type of pneumonia. Similar outbreaks have since occurred in other locations, and examination of old records showed that the same disease had also been noted but not identified before. About 15 percent of all known cases have been fatal.

The cause was a mystery that took six months of intensive, careful study to solve. Finally, an unfamiliar, rod-shaped bacterium, which was named *Legionella pneumophila,* was found to be responsible. The human body is not the natural home of this organism. Therefore the disease, unlike most pneumonias, is not contagious. Instead, the organism's natural habitat is water. It has been isolated from many lakes and streams and occasionally from mud. It thrives in warm water, and its concentration can therefore be increased by thermal pollution. As a result, the organism is widely distributed in the aqueous environment—including lakes and ponds as well as cooling towers and shower heads.

L. pneumophila is transmitted through the air on water droplets. It enters the human body through the lungs, not through the stomach. The danger therefore comes from contaminated water that is sprayed into the air in fine droplets—from cooling towers, decorative fountains, and water-cooled air conditioning systems. The air conditioning industry has pointed out that transmission of the disease in air conditioning systems can be controlled by rigorous compliance with standard maintenance procedures, which are now followed in only about 50 percent of existing installations.

not rule out the possibility of chronic illness, about which there is very little definite information. Over the years the U.S. Public Health Service has suggested standards for drinking water in the form of maximal allowable concentrations of various substances, particularly metals and some classes of organic pollutants. Some of these may be acutely toxic; others produce chronic illness.

20.8 Water Purification

Water molecules have no memory, and therefore it is silly to talk about the number of times that the water you drink has been polluted and repurified, as if the molecules gradually wore out. All that is important is how pure it is when you drink it.

The purification of water has developed into an elaborate and sophisticated technology. However, the general approaches to purification should be comprehensible, and in some cases even obvious, from a general understanding of the nature of water pollution.

In Section 20.1 impurities in water were classified as *suspended, colloidal,* or *dissolved.* Suspended particles are large enough to settle out or to be filtered. Colloidal and dissolved impurities are more difficult to remove. One possibility is somehow to make these small particles join together to become larger ones, which can then be treated as suspended matter. Another possibility is to convert them to a gas that escapes from the water into the atmosphere. Whatever the approach, it must be remembered that energy is required to lift water or to pump it through a filter.

With these principles in mind, consider the procedures used in purifying municipal waste waters. The first step is the collection system. Waterborne wastes from sources such as homes, hospitals, and schools contain food residues, human excrement, paper, soap, detergents, dirt, cloth, other miscellaneous debris, and, of course, microorganisms. This mixture is called **sanitary** or **domestic sewage.** (The adjective "sanitary" is rather inappropriate because it hardly describes the condition of the sewage; it presumably refers to that of the premises whose wastes have been carried away.) These waters, which are sometimes joined by wastes from commercial buildings, by industrial wastes, and by the runoff from rain, flow through a network of sewer pipes, as shown in Figure 20–13. Some systems separate sewage from rainwater, others combine them. The combined piping is cheaper and is adequate in dry weather, but during a storm the total volume is apt to exceed the capacity of the treatment plant, so some is allowed to overflow and pass directly into the receiving stream or river.

Bacterial and microbial actions occur during the flow of wastes through the sewer pipes; high-energy food chemicals are degraded to low-energy compounds, with consumption of oxygen. The more such action occurs before the sewage is discharged into open waters, the less occurs afterward; therefore, this process must be regarded as the beginning of purification.

Primary Treatment

When the sewage reaches the treatment plant (see schematic diagram of Fig. 20–14), it first passes through a series of screens that remove large objects such as rats or grapefruits, and then through a grinding mechanism that reduces any remaining objects to a size small enough to be handled effectively during the remaining treatment period. The next stage is a series of settling chambers designed to remove first the heavy grit, such as sand that rainwater brings in from road surfaces, and then, more slowly, any other suspended solids—including organic nutrients—that can settle out in an hour or so. Up to this point the entire process, which is called primary treatment, has been relatively inexpensive but has not accomplished much. If the sewage is now discharged into a stream (as, unfortunately, is often the case), it does not look so bad because it bears no visible solids, but it is still a potent pollutant carrying a heavy load of microorganisms, many of them pathogenic, and considerable quantities of organic nutrients that will demand more oxygen as their decomposition continues.

Secondary Treatment

The next series of steps is designed to reduce greatly the dissolved or finely suspended organic matter by some form of accelerated biological action. What is needed for such decomposition is oxygen and organisms and an environment in which both have ready access to the nutrients. One device for accomplishing this objective is the **trickling filter,** shown in Figure 20–15. In this device, long pipes rotate slowly over a bed of stones, distributing the polluted water in continuous sprays. As the water trickles over and around the stones, it offers its nutrients in the presence of air to an abundance of rather unappetizing forms of life. A fast-moving food chain is set in operation. Bacteria consume molecules of protein, fat, and carbohydrate. Protists consume bacteria. Farther up the chain are worms, snails, flies, and spiders. Each form of life plays its part in converting high-energy

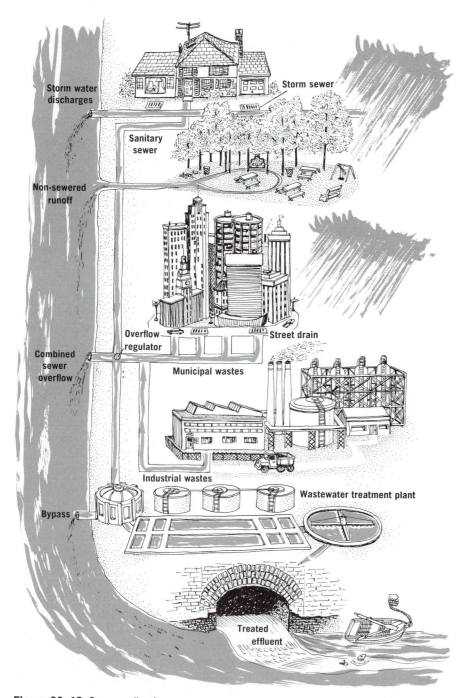

Figure 20–13 *Sewer collection system.*

chemicals to low-energy ones. All the oxygen consumed at this stage represents oxygen that will not be needed later when the sewage is discharged to open water. Therefore, this process constitutes a very significant purification.

An alternative technique is the **activated sludge** process, shown schematically in Figure 20–16. Here the sewage, after primary treatment, is pumped into an aeration tank, where it is mixed for several hours with air and bacteria-

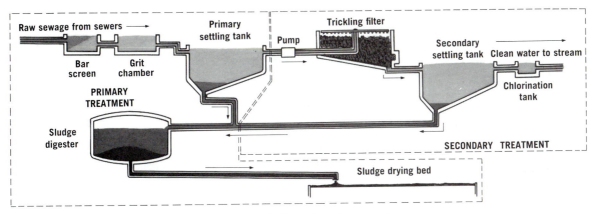

Figure 20–14 Sewage plant schematic, showing facilities for primary and secondary treatment. (From *The Living Waters,* U.S. Public Health Service Publication No. 382.)

laden sludge. The biological action is similar to that which occurs in the trickling filter. The sludge bacteria metabolize the organic nutrients; the protozoa, as secondary consumers, feed on the bacteria. The treated waters then flow to a sedimentation tank, where the bacteria-laden solids settle out and are returned to the aerator. Some of the sludge must be removed to maintain steady-state conditions. The activated sludge process requires less land space than the trickling filters, and, since it exposes less area to the atmosphere, it does not stink so much. Furthermore, since the food chain is largely confined to microorganisms, there are not so many insects flying around. However, the activated sludge process is a bit trickier to operate and can

be more easily overwhelmed and lose its effectiveness when faced with a sudden overload.

The effluent from the biological action is still laden with bacteria and is not fit for discharge into open waters, let alone for drinking. Since the microorganisms have done their work, they may now be killed. The final step is therefore a disinfection process, usually chlorination. Chlorine gas, injected into the effluent 15 to 30 minutes before its final discharge, can kill more than 99 percent of the harmful bacteria.

Let us now return to the sludge. Each step in the biological consumption of this waterborne waste, from sewage nutrients to bacteria to protozoa and continuing to consumers of higher orders (such as worms), represents a degradation

Figure 20–15 A trickling filter with a section removed to show construction details. (From Warren: *Biology and Water Pollution Control.* Philadelphia, W. B. Saunders Co., 1971.)

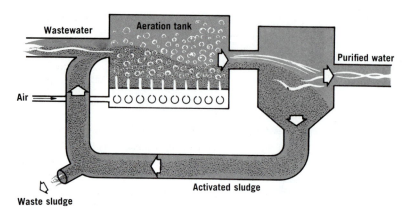

Wastewater

Aeration tank

Purified water

Air

Activated sludge

Waste sludge

Figure 20–16 Activated sludge process (schematic view.)

of energy, a consumption of oxygen, and a reduction in the mass of pollutant matter. Also, and perhaps most important from a practical point of view, the process brings about an increase in the average size of the pollutant particles. Look at Figure 20–1 to see how dramatic this change can be. Sugar is dissolved in water in the form of molecules that never settle out. Partially degraded starch and protein occur as colloidal particles in approximately the same size range as viruses. Bacteria are much larger, growing up to about 10 micrometers. Protists are gigantic by comparison; some amoeba reach diameters of 500 micrometers and thus are comparable in size to fine grains of beach sand. Some agglomeration also occurs in the metabolic processes of the protozoa, so that their excreta are usually larger than the particles of food they

ingest. Finally, when the microorganisms die, their bodies stick together to form aggregates large enough to settle out in a reasonably short time. This entire process of making big particles out of little ones is of prime importance in any system of waste water treatment. The mushy mixture of living and dead organisms and their waste products at the bottom of a treatment tank constitutes the biologically active sludge (Fig. 20–17). Typical sewage contains about 0.6 g of solid matter per liter, or about 0.06 percent by weight, whereas a liter of raw waste sludge contains about 40 to 80 g of solid matter, corresponding to a concentration of 4 to 8 percent. Even after this magnification, however, the raw sludge is still a watery, slimy, malodorous mixture of cellular protoplasm and other offensive residues. The organic matter can undergo still further decomposition, but its high concentration engenders anaerobic conditions. Recall that anaerobic processes generate methane, CH_4, and carbon dioxide, CO_2, among other gases. Such conversions thus decrease the content of solid carbonaceous matter still further, although the process is necessarily accompanied by offensive nitrogenous and sulfidic odors. The final disposal of the sludge residue, whether by incineration, landfill (Fig. 20–18), or other means, becomes a problem in the handling of solid wastes (see Chapter 22).

Tertiary or "Advanced" Treatments

Although considerable purification is accomplished by the time wastewaters have passed through the primary and secondary stages, these treatments are still inadequate to deal with

Figure 20–17 Sewage sludge.

Figure 20–18 The Metropolitan Sanitary District of Chicago transports its sewage and sludge by barge and pipeline(A) to a 6000-hectare site in Fulton County, Illinois, where it is spread on agricultural areas that had been left in poor condition by strip-mining (B).

some complex aspects of water pollution. First, many pollutants in sanitary sewage are not removed. Inorganic ions, such as nitrates and phosphates, remain in the treated waters; these materials, as we have seen, serve as plant nutrients and are therefore agents of eutrophication. If chlorination is incomplete, microorganisms will remain; in any case, chlorine will remain in some form or other, frequently as chlorinated organic matter that seriously impairs the taste of the water and that can even introduce new toxins.

Additionally, many pollutants originating from sources such as factories, mines, and agricultural runoffs cannot be handled by municipal sewage treatment plants at all. Some synthetic organic chemicals from industrial wastes are foreign to natural food webs (that is, they are non-biodegradable); they not only resist the bacteria of the purification system but may also poison them, and thereby nullify the biological oxidation that the bacteria would otherwise provide. There are also inorganic pollutants, including acids and metallic salts, as well as suspended soil particles from chemical and mining operations and from natural sources. Some of these materials occur as very fine particles from roadways, construction sites, or irrigation runoffs. These sediments are troublesome before they settle, because they reduce the penetration of sunlight, and afterwards, because they fill reservoirs, harbors, and stream channels with their silt.

The treatment methods available to cope with these troublesome wastes are necessarily specific to the type of pollutant to be removed, and they are generally expensive. A few of these techniques are described below.

Coagulation and Sedimentation

As mentioned earlier in the discussion of biological treatment, it is advantageous to change little particles into big ones that settle faster. So it is also with inorganic pollutants. Various inorganic colloidal particles are waterloving (hydrophilic) and therefore rather adhesive; in their stickiness they sweep together many other colloidal particles that would otherwise fail to settle out in a reasonable time. This process is called **flocculation.** Lime, alum, and some salts of iron are among these so-called flocculating agents.

Adsorption

Adsorption is the process by which molecules of a gas or liquid adhere to the surface of a solid. The process is selective—different kinds of molecules adhere differently to any given solid. To purify water, a solid that has a large surface area and binds preferentially to organic pollutants is needed. The material of choice is activated carbon, which is particularly effective in removing chemicals that produce offensive tastes and odors. These include the biologically resistant chlorinated hydrocarbons. (See also Chapter 21.)

Other Oxidizing Agents

Potassium permanganate, $KMnO_4$, and ozone, O_3, have been used to oxidize water-borne wastes that resist oxidation by air in the presence of microorganisms. Ozone has the important advantage that its only byproduct is oxygen.

$$2O_3 \longrightarrow 3O_2$$

Reverse Osmosis

Osmosis (see Chapter 19) is the process by which water passes through a membrane that is impermeable to dissolved ions. In the normal course of osmosis, as illustrated in Figure 20–19A, the system tends toward an equilibrium in which the concentrations on both sides of the membrane are equal. This means that the water flows from the pure side to the concentrated "polluted" side. This is just what we don't want, because it increases the quantity of polluted water. However, if excess pressure is applied on the concentrated side (Fig. 20–19B), the process can be reversed, and the pure water is squeezed through the membrane and thus freed of its dissolved ionic or other soluble pollutants.

20.9 Economics, Social Choices, and Legal Strategies in Water Pollution Control

Domestic wastes were once collected in pits called cesspools, from which they were periodically shoveled out and carted away. As cities grew denser, the task became more onerous, and toward the end of the last century it became customary to connect series of cesspools with conduits so that they could all be flushed out with water in a single operation. The next obvious step was to eliminate the cesspools and use the piping system alone with flushing water

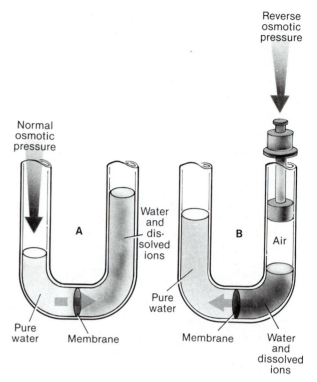

Figure 20–19 Reverse osmosis.

continuously available. Thus were sewer systems born. According to some environmentalists, this was the point at which sanitary engineers and public health officers took civilization down the wrong road. What we are doing now is discharging our wastes into the public waters and then spending billions of dollars to restore the water to a quality that is fit for drinking. The wastefulness of the process is illustrated by the fact that the average toilet flush uses about 20 liters of water to carry away about ¼ liter of body wastes and that the average user of a flush toilet flushes it some seven times a day.

Few recommend a return to the outhouse or cesspool. Before examining reasonable alternatives, it is important to decide on our objectives. In the United States national goals are proclaimed in federal legislation. The major national law is the federal Water Pollution Control Act of 1972 and its subsequent amendments starting in 1977. The amended law is now known as the Clean Water Act. Its stated goals are:

1. Wherever attainable, to achieve, by July 1, 1983, an interim goal of water quality that pro-

vides protection and encourages propagation of fish, shellfish, and wildlife and provides for recreation in and on the water.

2. To eliminate, by 1985, the discharge of pollutants into navigable waters. (Note: Goals 1 and 2 have not yet been achieved, as of 1983.)

3. To prohibit the discharge of toxic pollutants in toxic amounts.

4. To provide federal financial assistance to construct publicly owned waste treatment works.

5. To make a major research and demonstration effort to develop the technology necessary to eliminate the discharge of pollutants into navigable waters and the oceans.

It is important to recognize that these goals can be attained only by establishing standards and enforcing them. Under the 1977 amendments, the conventional types of water pollution, such as BOD, suspended solids, fecal coliforms, and acidity, are to be treated by the "best conventional technology" (BCT). The assessment of BCT includes a reasonable consideration of the costs of control and the benefits of the reduction in pollution. In other words, BCT is to be viewed as a cost/benefit compromise rather than as a strict requirement of pollution control at any cost. For toxic pollutants, the law requires the "best available technology" (BAT) that is economically achievable and that will result in "reasonable further progress toward the national goal of eliminating the discharge of all pollutants." Thus the BAT requirement is stricter than the BCT, but it is still not an absolute prohibition against all toxic pollutants.

The 1983 goal of achieving "fishable-swimmable" waters "wherever attainable" has been hampered by rising costs and by changing political priorities. Federal funds for sewage treatment plants have been cut, and deadlines for cleaning up have been extended. Many coastal communities have applied for waivers to dump raw or partially treated sewage into the ocean or tidal rivers. These changes represent a shift from a regulatory policy of total environmental protection to one of cost/benefit assessment and strategy.

In 1974, Congress passed the Safe Drinking Water Act, which was designed to ensure that water supply systems serving the public meet minimum national standards for protection of public health. The Act gave the Environmental Protection Agency (EPA) responsibility for setting minimum national drinking water regulations throughout the United States. Interim regulations, which were published on December 24, 1975, and became effective June 24, 1977, set maximum levels permitted for bacteria, cloudiness, and concentrations for a number of organic and inorganic chemicals.

The central factor that must be confronted in trying to reconcile national goals, industrial objections, and recommendations for alternate strategies is the *quantity* of water that is involved. The average total volume of water supplied to the United States per day by rain and snow is close to 4 trillion liters. About 1½ trillion liters, more than a third of the total supply, is used daily by the manufacturing and power industries, by agriculture, and by cities and towns. If all of this water were to be purified to the highest standards of quality for drinking, the costs might well become prohibitive. Figure 20–20 shows how purification costs rise as the purity rises. The key to achieving national goals for water quality at bearable costs must therefore lie

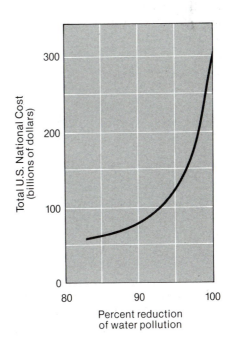

Figure 20–20 Cost of reducing water pollution. (Estimates from the Environmental Protection Agency, 1972.)

A

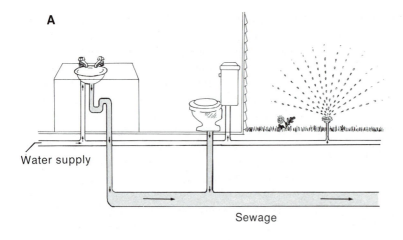

Water supply

Sewage

B

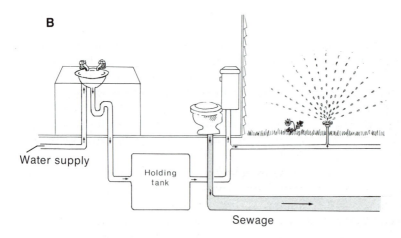

Water supply

Holding
tank

Sewage

Figure 20–21 Water use. *A*, Present, wasteful system—pure water is used for sinks, toilets, and lawns. *B*, More conservative system—waste water from sinks is stored and used for toilets and lawn irrigation.

in conservation. For example, think of water consumption in the home. Considerable quantities could be saved simply by wasting less. Or, water could be reused. Under present practices, pure, clean, drinking water is used to flush toilets and irrigate lawns (Fig. 20–21*A*). If waste water from bathtubs and sinks were recycled through toilets and outdoor sprinklers (Fig. 20–21*B*), there would be considerably less demand on the purification systems. As another possibility, waterless toilets (Fig. 20–22) have been developed and could easily be used. Some of these burn wastes, others compost them for use as fertilizer.

There have been many suggestions over the years to return to traditional agricultural practices by piping liquid sewage directly to farms and woodlands, where it may be sprayed as fertilizer. As an added benefit, the water, filtered by the soil, replenishes the water table. Such practices are especially beneficial in the arid areas of the Southwest and have operated in some hundreds of communities in those regions. However, they are hardly applicable to large cities, where the requisite farm areas simply do not exist. Furthermore, the use of domestic sewage as a fertilizer must not be considered an unmixed blessing, because it is no longer a simple mixture of biodegradable organic matter. Instead, such sewage is always mixed with some industrial wastes that contain metallic compounds and other nonbiodegradable chemicals that may accumulate in the soil. The agricultural use of sludge or compost from household bio-

Figure 20–22 Chemical toilet: A new style in an old setting.

logical toilets would be a much sounder practice, especially if people could be discouraged from dumping pesticides or paints in them.

Summary

Water can be polluted by impurities that dissolve in it, are suspended in it, live in it, float on it, or enter it as a result of some chemical or biochemical change. Particles that are too small to be filtered out or to settle out but are large enough to make the water look cloudy are called **colloidal particles.**

The oxygen dissolved in water can be depleted faster than it is replaced from the atmosphere. Oxygen therefore becomes the limiting factor when organic and inorganic nutrients are plentiful. Under such conditions microorganisms become more numerous while larger organisms, such as fish, decline in population. Excess nutrients thus pollute water. Many bacteria can survive even when no oxygen is present. The **anaerobic** decomposition of carbohydrates is called **fermentation,** and that of proteins is called **putrefaction.** Anaerobic processes yield much less energy than **aerobic** ones and produce foul-smelling pollutants such as hydrogen sulfide. **Biochemical oxygen demand** (BOD) is the measure of water pollution by nutrients.

The BOD of a stream or river increases instantaneously when raw sewage is added to it, but the concentration of dissolved oxygen decreases gradually. As the river continues to flow, it recovers oxygen from the atmosphere and from photosynthesis.

In lakes, the oxygen content and temperature of the various layers—**epilimnion** (surface), **hypolimnion** (lower), and **thermocline** (intermediate)—change with the seasons. The pollution of a lake by nutrients is called **eutrophication.** Phosphorus, which is the usual limiting nutrient in natural lakes, is often introduced from sewage, detergents, or agricultural fertilizers.

Groundwater can be polluted by contamination from pesticides, mine tailings, viruses, wastewater storage basins, and liquid chemical injection wells. Natural purification of groundwater is very slow because it is not readily diluted and it does not have access to air.

Many water pollutants, such as heavy metals and organic chemicals, come from industrial sources. A major source of pollution in the oceans is petroleum oil from tankers. The ultimate effect of pollution on aquatic life is uncertain, although the immediate effects can be devastating.

Worldwide water pollution is probably the major environmental cause of human illness. Waterborne diseases are due chiefly to microorganisms and parasites. Reports of chemical poisoning from polluted water are still relatively rare, although the potential for such illness may be serious and the possible long-term effects are not well known.

Water purification systems comprise three stages. **Primary treatment** consists of mechanical processes such as screening and settling. **Secondary treatment** is a biological process in which microorganisms, in the presence of an adequate supply of oxygen, consume the organic pollutants. The effluent from this stage is then disinfected, usually by chlorination. The resulting mushy mixture of organisms and waste products, called **sludge,** is disposed of by incineration, landfill, or ocean dumping. **Tertiary** or **advanced treatment** may be one of a series of chemical or physical processes such as adsorption, oxidation, or reverse osmosis that removes additional pollutants.

An ideal objective for improving water quality is the elimination of *all* discharges of pollutants. An alternative viewpoint, favored by industry, is that the goal of zero discharge should be replaced by standards for water quality. The central factor that must be dealt with is the quantity of water involved. If the quantities can be reduced by conservation and reuse of water, the goal of pure water can be realized at a much lower cost.

Questions

Water Pollution

1. When you are healthy you live in harmony with bacteria in your digestive system. Why, then, should water that contains digestive bacteria be considered to be polluted?

2. Suppose that a manufacturing plant in your area starts to generate liquid chemical wastes before its waste water treatment plant, now under construction, is ready for use. Meanwhile, there are three possible ways to handle the wastes: (1) Store them temporarily in a holding basin. (2) Dump them in a local river. (3) Pump them into a deep injection well below the local aquifer level. Describe the conditions, if any, under which you would favor each of the above choices.

3. List the various categories of groundwater pollution. Explain why polluted groundwater does not repurify itself naturally as fast as surface water does.

4. What are the major sources that contribute to the pollution of the oceans?

5. The contents of our stomachs are acidic, and we drink acidic fruit juices without doing ourselves any harm. Why, then, are acids considered to be pollutants in drinking water?

6. The half-life of carbon-14, which is produced in the atmosphere by cosmic rays, is about 5700 years. As a result, recently produced organic matter has practically its original concentration of carbon-14, whereas "old" organic matter, such as fossil fuels, has practically none. Explain how you could differentiate between sewage and oil pollution in a stream, based on observations of carbon-14 levels.

7. In its article on "Sewerage," the eleventh edition of the *Encyclopaedia Britannica*, published in 1910, states, "Nearly every town upon the coast turns its sewage into the sea. That the sea has a purifying effect is obvious. . . . It has been urged by competent authorities that this system is not wasteful, since the organic matter forms the food of lower organisms, which in turn are devoured by fish. Thus the sea is richer, if the land is the poorer, by the adoption of this cleanly method of disposal." Was this statement wrong when it was made? Defend your answer. Comment on its appropriateness today.

Productivity in Aqueous Systems

8. Explain how a nontoxic organic substance, such as chicken soup, can be a water pollutant.

9. What is eutrophication? Explain how it occurs and why it is hastened by the addition of inorganic matter such as phosphates.

10. It has been suggested that the world food shortage could be alleviated if we cultivated algae in sewage to produce a new food in the form of "algaeburgers." (1) Could such production be carried out on a 24-hour basis? Only during the daytime? Only at night? Explain. (2) If the sewage were used as the food in a "fish farm," would the product be able to feed more people or fewer people? Explain.

11. The BOD curve of Figure 20–4 shows that the rise and the fall occur sharply but not

instantaneously. How would the curve look if both rise and fall did start instantaneously? Which of the following is the more reasonable explanation for the noninstantaneous character of the changes—(1) some smaller discharges, such as those from individual homes or small farms, occur both before and after the main sewer effluent; (2) the sewage does not react instantaneously with oxygen. Defend your answer.

Health Effects

12. What are the criteria for water that is considered fit for drinking? Is such water always safe? Is water that does not meet these criteria always harmful? Explain.

Water Purification

13. Imagine that you had a sample of water containing all the impurities listed in Table 20–1, and that you purified it in the following successive stages: (1) Filter it through insect screening. (2) Filter it through filter paper that removes suspended but not colloidal particles. (3) Boil it so that dissolved gases are expelled. (4) Distill it so that inorganic compounds are left behind.

    List typical substances that would be removed in each step.

14. An alternate method of waste water treatment is the **stabilization** or **oxidation pond,** which is a large shallow basin in which the combined action of sunlight, algae, bacteria, and oxygen purifies the water. It may be said that the stabilization pond trades time, space, esthetics, and flexibility for savings in capital and operating costs. Explain this statement.

15. Biological treatment of wastewater reduces the mass of pollutant. Where does the lost matter go?

16. Distinguish among primary, secondary, and tertiary types of waste water treatment.

17. Is the speed of settling of particles in water directly proportional to their diameters? If the diameter is multiplied by 10, is the settling speed 10 times faster? (Justify your answer with data from Figure 20–1.) Is a set-

tling pond a good general method of water pollution control? Explain.

18. In a combined piping system, some untreated sewage is dumped into the receiving watercourse during rainstorms. Is this procedure more acceptable than it would be in dry weather? Defend your answer.

19. List and explain four methods of "advanced" water treatment.

Suggested Readings

The following two companion books are the standard engineering texts on water pollution and its control:
Metcalf and Eddy, Inc. (George Tchobanoglous, Ed.): *Wastewater Engineering: Treatment, Disposal, Reuse.* New York, McGraw-Hill Book Co., 1979. 890 pp.; and *Wastewater Engineering: Collection and Pumping of Wastewater.* New York, McGraw-Hill Book Co., 1981. 432 pp.

An older but less technical book is:
Charles E. Warren: *Biology and Water Pollution Control.* Philadelphia, W. B. Saunders Co., 1971. 434 pp.

For a detailed study of eutrophication, refer to:
National Academy of Sciences: *Eutrophication: Causes, Consequences, Correctives.* Washington, D.C., National Academy of Sciences Press, 1969. 661 pp.

The biological aspects of water treatment are described in the following texts:
H. A. Hawkes: *The Ecology of Wastewater Treatment.* London, Pergamon Press, 1963. 203 pp.
Larry D. Benefield and Clifford W. Randall: *Biological Process Design for Wastewater Treatment.* Englewood Cliffs, NJ, Prentice Hall, 1980. 526 pp.
Anthony F. Gaudy, Jr., and Elizabeth T. Gaudy: *Microbiology for Environmental Scientists and Engineers.* New York, McGraw-Hill Book Co., 1980. 736 pp.

Two advanced books on water chemistry are:
Samuel D. Faust and Osman M. Aly: *Chemistry of Natural Waters.* Woburn, MA, Butterworths, Inc., 1981. 400 pp.
David Jenkins: *Water Chemistry.* New York, John Wiley and Sons, 1980. 463 pp.

There are also various popular books that deal with the crisis of water pollution. Most of them were published in the

1960s and are out of print. However, they may be available from your school or local library:

Wesley Marx: *The Frail Ocean.* New York, Ballantine Books, 1967.

D. E. Carr: *Death of the Sweet Waters.* New York, W. W. Norton & Co., 1966. 257 pp.

F. E. Moss: *The Water Crisis.* New York, Encyclopaedia Britannica, Praeger Publishers, 1967. 305 pp.

G. A. Nikolaieff (Ed.): *The Water Crisis.* New York, W. H. Wilson Co., 1969. 192 pp.

Wesley Marx: *The Oceans: Our Last Resource.* Sierra Club Books, 1981.

Air Pollution

Most of us have had direct and personal experiences with air pollution. Perhaps you were standing behind the smoky exhaust of a car, bus, or truck. Or maybe you were driving from the hilly countryside down into a city and suddenly felt a slight irritation in your eyes or throat as you entered the yellow-brown haze that hung close to the ground. You might have been in an airplane, or on a hilltop, or even on a tall building looking down on that haze and wondering how hazardous it was to the health of the people who lived and worked in it. The city might have been any of a number of commercial-industrial centers. It might have been Denver, Tokyo, Los Angeles, Milan, or Montreal.

Air pollution is not confined to cities, nor is it produced only by human activity. Volcanic eruptions, naturally occurring forest fires, and dusts stirred up by storm winds can also contaminate the atmosphere.

This chapter will discuss the sources of air pollution and its effects on human health and welfare. It will also describe how air pollution can be controlled by technology and by public policy.

21.1 The Atmosphere

Before life evolved, the primary constituents of Earth's atmosphere were nitrogen, ammonia, hydrogen, carbon monoxide, methane, and water vapor. Oxygen was present only in trace quantities. Although some scientists believe that geological processes altered atmospheric composition, most feel that the excess oxygen released by the first autotrophs built up slowly over the millennia until its concentration reached about 0.6 percent of the atmosphere. Multicellular organisms could have evolved only at this point, because aerobic respiration is a prerequisite for their development. The emergence of various organisms about 600 million years ago triggered an accelerated biological production of oxygen. The present oxygen level of about 21 percent of the atmosphere was reached some 450 million years ago. Although there have been several more or less severe oscillations since that time, an overall oxygen balance has always been maintained.

If the oxygen concentration in the atmosphere were to increase even by a few percentage points, fires would burn uncontrollably across the planet; if the carbon dioxide concentration were to rise by a small amount, plant production would increase drastically. Since these apocalyptic events have not occurred, the atmospheric oxygen must have been balanced to the needs of the biosphere during the long span of life on Earth. By what mechanism has this gaseous atmospheric balance been maintained? The answer appears to be that it is maintained by the living systems themselves. The existence of an effective homeostatic mechanism of the whole biosphere has led J. E. Lovelock, an English chemist who has done important work on the analysis of trace gases and atmospheric pollutants, to liken the biosphere to a living creature. He calls that creature by the Greek name for Earth, *Gaia*. He believes that not only is the delicate balance of oxygen and carbon dioxide biologically maintained but also that the very presence of oxygen in large quantities in our atmosphere can be explained only by biological maintenance. If all life on Earth were to cease and the chemistry of our planet were to rely on inorganic processes alone, oxygen would once again become a trace gas.

The concept of biological control over the physical environment warrants careful consideration. It holds that, just as an organism is more than an independent collection of its organs and an ecosystem is more than an independent collection of its organisms, the biosphere is more than an independent collection of its ecosystems. The body chemistry of a human being can function only at or within a few degrees of 37°C (98.6°F), but normally we are not in mortal danger of very high or low body temperatures because there are mechanisms by which our bodies maintain the proper temperature. Similarly, Lovelock believes that modern life can exist only in an atmosphere at or close to 21 percent oxygen, but we are not normally in mortal danger of conflagration or starvation because there are mechanisms by which we, the species of Earth, maintain the proper oxygen concentration.

An alternative theory claims that our physical environment has evolved through a series of inorganic reactions, and that biological and physical evolution were independent. The difference between these two beliefs is not trivial.

If Lovelock is correct, then a large biological catastrophe such as the death of the oceans or the destruction of the rain forest in the Amazon Basin could cause reverberations throughout our physical world that might create an inhospitable environment for the rest of the biosphere. Alternatively, if the physical world did evolve independently of the biological world and is currently controlled by inorganic processes, such a doomsday prediction concerning oxygen balance might be considered unnecessarily alarming.

The Earth's atmosphere today is a mixture of gases, water vapor, and a variety of solid particles and liquid droplets. In some respects, air differs from place to place around the globe. The air in a tropical rain forest is hot and steamy. People travel to the seaside to enjoy the "salt air." Visitors to the Smoky Mountains in Tennessee view the bluish hazy air. On a cold night in the Arctic the air feels particularly dry and "pure." Dry, filtered air is roughly 78 percent nitrogen, 21 percent oxygen, and 1 percent other gases (Fig. 21–1). A more detailed breakdown is given in Table 21–1. Most samples of natural air contain some water vapor as well. In a hot, steamy jungle, air may contain 5 percent water vapor, whereas in a dry desert or a cold polar region there may be almost none at all.

If you sit in a house on a sunny day you may see a sunbeam passing through the window. Look at right angles to the sunbeam and

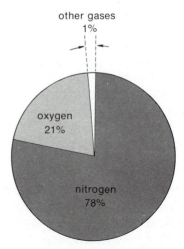

Figure 21–1 *Approximate gaseous composition of natural dry air.*

TABLE 21–1   Gaseous Composition of Natural Dry Air

Gas	Concentration (by volume)*	
	ppm	percent
"Pure" air		
Nitrogen, N_2	780,900	78.09
Oxygen, O_2	209,400	20.94
Inert gases, mostly argon, (9300 ppm) with much smaller concentrations of neon (18 ppm), helium (5 ppm), krypton and xenon (1 ppm each)	9,325	0.93
Carbon dioxide, CO_2	335	0.03
Methane, CH_4, a natural part of the carbon cycle of the biosphere; therefore, not a pollutant although sometimes confused with other hydrocarbons in estimating total pollution	1	
Hydrogen, H_2	0.5	
Natural pollutants		
Oxides of nitrogen, mostly N_2O (0.5 ppm) and NO_2 (0.02 ppm), both produced by solar radiation and by lightning	0.52	
Carbon monoxide, CO, from oxidation of methane and other natural sources	0.3	
Ozone, O_3, produced by solar radiation and by lightning	0.02	

*See Appendix A for expressions of concentration. Concentration of gases "by volume" is the same as concentration by number of molecules. Thus, a concentration of nitrogen of 78 percent by volume means that there are 78 molecules of nitrogen in every 100 molecules of air. The initials ppm mean "parts per million." To change percent to ppm, multiply by 10,000.

you will see tiny specks of dust suspended in the air. There are many sources of airborne particles. Sand, dirt, pollen, bits of cloth, hair, and skin are all suspended in air. There are living particles as well, such as bacteria and viruses. All these materials are natural components of air.

What, then, is an air pollutant? There are two ways to answer the question. One is to say that an **air pollutant** is any substance that adversely affects the quality of the air—its fitness to support life on Earth. Another answer is simply to define "pure air" as the composition of the first six components listed in Table 21–1—anything else is a pollutant. Any significant variation in the compositions shown in the table could also be harmful. For example, air containing, say, 10 percent carbon dioxide would be toxic; therefore carbon dioxide in high concentrations is a pollutant.

21.2 Sources of Air Pollution

Air pollution from human activity is probably as old as our ability to start a fire. But large-scale air pollution from industry is a relatively recent development. Furthermore, the resulting contaminants, by the nature of the activities that

BOX 21.1 GASES AND SMOKES

The properties of gases and smokes arise in large measure from the sizes of the particles that compose them. Gases consist of molecules in constant motion, moving in straight lines between collisions with each other or with the walls, floor, or ceiling. The molecules are so small, so speedy, so numerous, and their collisions so frequent, that nothing in our experience (not even, say, a swarm of gnats) can be called upon to help us visualize them in a quantitative way. Most of the space in a volume of gas is empty. At ordinary pressures, the molecules themselves occupy less than 1 percent of the total volume. Thus "1 liter of gas" consists of many molecules darting about in almost 1 liter of empty space.

Smokes consist of particles. A particle is a very small portion of matter, and therefore a molecule is a particle, and so is an atom or even an electron. But in air pollution usage the word "particle" has come to have a more restricted meaning: It refers to portions of matter that, although small, are much larger than molecules, large enough, in fact, to settle out, or at least to reflect or scatter a beam of light that shines on them. (Hence smoke is visible.) Grains of sand are particles, and so are droplets of mist and tiny organisms like protozoa or bacteria. Figure 21–2 shows their relative sizes.

Matter consisting of particles is called "particulate matter." This term is often shortened to "particulates," which means simply "particles."

Just as Eskimos have many words for different kinds of snow, so air pollution engineers have many words for different kinds of airborne particles. All the common ones are listed in Table 21–2.

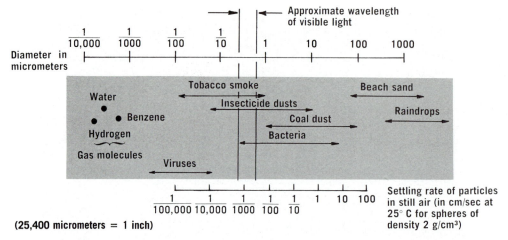

Figure 21-2 Small particles in air. Note that the size scale, shown in micrometers (μm), is not linear but progresses by factors of 10. The largest diameter shown, 1000 μm, is equal to 1 mm; it is the size of large grains of sand or small raindrops. The lower scale tells how rapidly or slowly particles settle in still air. Very small particles can be kept aloft indefinitely by wind currents. The smallest particles are not removed from air by gravitational settling. Usually these small particles have or acquire some electrical charge, and the consequent electrostatic forces are much greater than gravitational ones. As a result, the particles are attracted to surfaces with an opposite charge or to other particles and stick to them.

produce them, are likely to be emitted to the air in regions where many people live. Therefore the effects, even if small on a global scale, may be locally very severe. The following paragraphs summarize the major categories of air pollution from human sources.

Stationary Combustion Sources

Since the Industrial Revolution, the major fuels in the developed areas of the world have been coal and petroleum. Coal is largely carbon, which, when it burns completely, produces car-

TABLE 21-2 Vocabulary of Particles

Term	Meaning	Examples
Aerosol	General term for particles suspended in air	
Mist	Aerosol consisting of liquid droplets	Sulfuric acid mist
Dust	Aerosol consisting of solid particles that are blown into the air or are produced from larger particles by grinding them down	A dust storm; dust from the grinding of grain or metal and so on
Smoke	Aerosol consisting of solid particles or a mixture of solid and liquid particles produced by chemical reactions such as fires	Cigarette smoke, smoke from burning garbage
Fume	Generally means the same as smoke, but often applies specifically to aerosols produced by condensation of hot vapors, especially of metals	Zinc fumes, lead fumes, smelter fumes
Plume	The geometrical shape or form of the smoke coming out of a chimney.	
Fog	Aerosol consisting of water droplets	
Haze	Any aerosol, other than fog, that obscures the view through the atmosphere	Summer haze around a large pine forest
Smog	Popular term originating in England to describe a mixture of smoke and fog. Now subdivided into two different categories: "London smog" and "Los Angeles smog"	
Smaze	Popular term originating in New York to describe a mixture of smoke and haze	

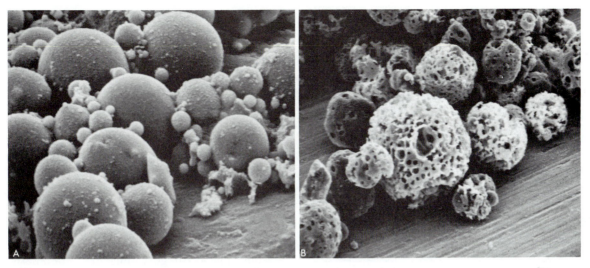

Figure 21–3 Particulate matter sampled from the effluent of (A) coal-fired and (B) oil-fired power plants. (Illustrations obtained with scanning electron microscope, courtesy of Atmospheric Sciences Research Center of the State University of New York at Albany.)

bon dioxide, CO_2. Petroleum consists largely of hydrocarbons (compounds of hydrogen and carbon), which, when they burn completely, form CO_2 and water. Fossil fuels contain other chemicals, and combustion produces other products. Coal is always mixed with some of the incombustible mineral matter of the Earth's crust; when the coal burns, some of this mineral ash flies out the chimney. The smoke it produces is called **fly ash** (Fig. 21–3).

Sulfur is an essential element of life, and since coal and oil are derived from living organisms, these fuels always contain some sulfur. When the fuels burn, so does the sulfur, producing a mixture of oxides, mainly **sulfur dioxide,** SO_2, and **sulfur trioxide,** SO_3. High SO_2 concentrations have been associated with major air pollution disasters of the type that have occurred in large cities, such as London, and that were responsible for numerous deaths.

Nitrogen, like sulfur, is common to living tissue and therefore is found in all fossil fuels. This nitrogen, together with a small amount of atmospheric nitrogen, also oxidizes when coal or oil is burned. The products are mostly nitrogen oxide, NO, and nitrogen dioxide, NO_2. Chemists frequently group NO and NO_2 together under the general formula NO_x. NO_2 is a reddish-

brown gas whose pungent odor can be detected at concentrations above about 0.1 ppm, and it therefore contributes to the "browning" and the smell of some polluted urban atmospheres.

Finally, even if no other elements are present, carbon and hydrocarbons produce pollutants because they usually burn incompletely. Some of these pollutants are gaseous, others are particulate. One such gas is **carbon monoxide,** CO, which is colorless, odorless, and nonirritating, yet very toxic. Particles that consist mostly of carbons are called, collectively, **soot,** which is known to be carcinogenic.

Mobile Combustion Sources

With the demise of the steam locomotive, the prime sources of energy for mobile engines are gasoline, diesel fuel, and jet fuel. The internal energy of these materials, like that of coal and oil, lies in the ability of their carbon and hydrogen atoms to oxidize, respectively, to CO_2 and H_2O. The resulting air pollutants, therefore, do have features in common with those from stationary combustion sources, but there are important differences.

The fuels used by mobile power plants are more highly refined than those in large station-

Figure 21-4 Automobile pollution along a freeway in Denver, Colorado. (Photo by Mel Schieltz; courtesy of *Rocky Mountain News.*)

ary power plants, and therefore trucks, automobiles, and airplanes are not a major source of fly ash. However, they do produce gases, smoke, and smog (see Fig. 21-4 and the Case History of Section 21.7).

Manufacturing Sources

Many thousands of products are manufactured for industry, for commerce, and for domestic use. Some of the manufacturing processes produce unwanted particulate matter or gases that can escape into the atmosphere. Following are three important classes of air polluting processes:

(a) Any Process that Uses Air. Air is used in industry as an oxidizing agent (to support combustion), as a coolant, and as a carrier. The use of air as an oxidizing agent invariably produces some unwanted impurities that find their way into the atmosphere. When air is used as a cooling agent (as in wet cooling towers), any pollutants that can evaporate into the air stream are carried into the atmosphere. Air is also widely used as a "carrier" gas, especially in drying operations in which hot air carries away moisture. Air is also used to blow dust particles, as in sand blasting. The air always takes some of the contaminants along with it to the outside atmosphere.

(b) Any Process that Uses High Temperatures. Heat promotes evaporation, which produces gases. Metals, too, can evaporate. For example, lead evaporates rapidly at the temperature of an ordinary gas flame. The resulting lead vapor cools in air to produce the aerosol known as a lead fume. Heat also causes compounds to break down, or decompose. The decomposition products may be gaseous or particulate matter that escapes into the atmosphere. When air is heated to a high temperature, as in flames, some of its nitrogen and oxygen combine chemically to produce oxides, mostly NO and NO_2, which are pollutants.

(c) Any Mechanical Process that Breaks Down Materials. This category includes operations such as blasting, drilling, crushing, or grinding. These mechanical processes are carried out extensively in industries such as mining, agriculture, construction, and metallurgy. In fact, the manufacture or handling of any product that consists of small particles such as sand, cement, soil, fertilizer, or flour can give rise to particulate air pollution.

21.3 The Meteorology of Air Pollution

To assess the extent to which your health is affected by, say, sulfur dioxide, you must be concerned with the quantity or concentration that you inhale. The total quantity of SO_2 in the Earth's atmosphere, or the concentration in the exhaust gas of some particular copper smelter, does not affect you directly because you do not breathe all the world's air nor do you stick your head into the smokestack. Therefore, you must consider how pollutants are transported in the atmosphere and how atmospheric conditions affect their concentrations. The science that deals with these and other atmospheric phenomena is **meteorology.**

The atmosphere is heated from above by direct sunlight and from below by radiation from the Earth's surface. If you were to ascend vertically from the Earth in a balloon, under average meteorological conditions, you would find that the air temperature drops steadily with altitude.

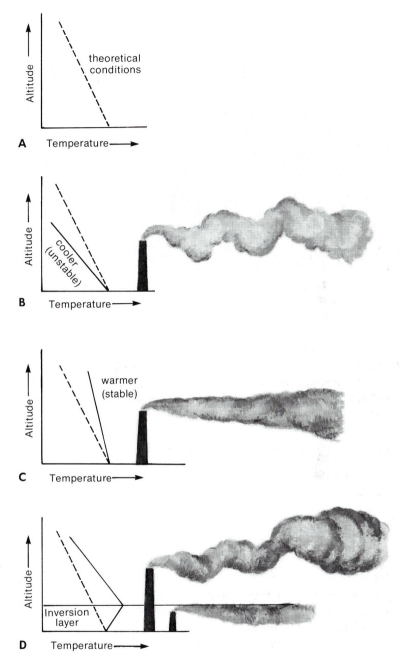

Figure 21–5 Air temperature versus altitude. *A,* Theoretical condition when gains and losses of energy are in balance. *B,* Actual temperatures *(solid line)* are cooler than theoretical conditions. *C,* Actual temperatures *(solid line)* are warmer than theoretical conditions. *D,* Atmospheric inversion layer.

This cooling, which is shown in Figure 21–5*A,* occurs because the heat from the Earth's surface is dissipated at high altitudes. Thus mountain tops are generally colder than valley floors.

Atmospheric conditions do not always conform to this theoretical model. At any particular time, for example, the actual temperature at some elevation above ground level may be cooler than the equilibrium conditions. This situation is unstable because cool air, being denser than warm air, tends to fall. The result is an unstable, turbulent atmosphere with gusty winds blowing this way and that. If polluted air from a chimney enters such an atmosphere, it mixes well with the turbulent air, and this mixing helps to dilute the pollutants. The resulting smoke pattern would resemble that shown in Figure 21–5B.

Now consider the situation in which the air at some given altitude is *warmer* than the equilibrium conditions (Fig. 21–5C). Such a warm layer rests in a stable pattern over the cooler, denser air beneath it. This condition is called an **atmospheric inversion** (Fig. 21–6). Such a situation may start to develop an hour or two before sunset after a sunny day, when the ground starts to lose heat and the air near the ground also begins to cool rapidly. This inverted condition often continues through the cool night and is usually broken up by turbulent mixing. Sometimes, however, atmospheric stagnation allows the inversion to persist for several days. Imagine that such an inversion layer existed between the ground and, say, 200 meters but not at higher altitudes. Even if the lower air were turbulently stirred by warm currents, the total volume available for mixing would be limited by the 200-meter ceiling, and any pollutants discharged at lower levels would concentrate beneath the ceiling. The longer the inversion lasted, the greater the buildup. However, a chimney tall enough to penetrate the 200-meter barrier would discharge its effluents into the upper atmospheric layers where they would be diluted in a much larger volume. Such a condition is shown in Figure 21–5D.

The world's tallest chimney (as of 1983), built for the Copper Cliff smelter in the Sudbury District of Ontario, Canada, is as tall as the Empire State Building. It is illustrated in Figure 21–7, together with three smaller chimneys. Such heroic structures can be quite effective in reducing ground-level concentrations of pollutants. For example, during a 10-year period studied by the Central Electric Generating Board in Great Britain, the SO_2 emissions from power stations increased by 35 percent, but, because of the construction of tall stacks, the ground-level concentrations decreased by as much as 30 percent. However, tall chimneys do not collect or destroy anything, and therefore they do not reduce the total quantity of pollutants in the Earth's atmosphere.

Atmospheric inversions have been associated with air pollution disasters that have caused human injury and death. In all such cases, the inversion layer traps the pollutants in

Figure 21–6 Three views of downtown Los Angeles. *Top:* A clear day. *Middle:* Pollution trapped beneath an inversion layer at 75 meters. *Bottom:* Pollution distribution under an inversion layer at 450 meters. (Photos from Los Angeles Air Pollution Control District.)

a limited volume of air, where their concentrations can build up to harmful levels. Examples are given in Table 21–3, which appears in the next Section.

21.4 The Effects of Air Pollution

Methods of appraising the effects of air pollution are now much more sophisticated than they used to be. Malaria is no longer blamed on swampy odors, and there is no longer any doubt

Figure 21–7 World's tallest chimney. Built in Ontario, Canada, at a cost of $5.5 million, this chimney stands 380 meters high. (Photo courtesy of M. W. Kellogg Company, division of Pullman Inc.)

that cigarette smoke and coal dust are harmful to health. However, there is still much that is not known, especially with regard to chronic or long-term effects. Furthermore, much is suspected but is not easily proved, since many pollutants have been tested only on animals. These uncertainties, taken together with the differences in interests among various sectors of the population, lead to a variety of recommendations for public policy.

It is important, therefore, to learn about air pollution effects and to be careful to distinguish among what is "known" with a high degree of confidence, what is suspected, and what is merely possible (that is, not yet proved to be untrue).

Air pollution effects may be classified into five categories: (1) modification of climate, (2) harm to human health, (3) injury to vegetation

and animals, (4) deterioration of materials, and (5) aesthetic insults. The following sections will discuss each in turn.

Effects on Atmosphere and Climate

The possibility that the Earth's climate, particularly its temperature, may be drifting away from its previous range of conditions is alarming. A warming portends a melting of glaciers and a flooding of the populous coastal plains. A cooling presages a new ice age. Changes in precipitation may not be quite so cataclysmic, but, nonetheless, a greater or lesser rainfall, or its redistribution, can have profound effects on agriculture, and so on people.

The major factor that determines global climate is solar radiation. Therefore to predict the possible effects of air pollution on climate, it is necessary first to consider what happens to the solar radiation that the Earth receives. Only 21 percent of the incident solar radiation strikes the Earth directly. The other 79 percent is intercepted by the atmosphere—the clouds, gases, and aerosol particles (Fig. 21–8). Some of this intercepted radiation is reflected back to space, some is absorbed as heat, and some is rescattered down to Earth. On a global average, just about half of the heat energy received from the Sun reaches the surface of the Earth. Some of this heat is reflected back into the atmosphere, while some is absorbed, thereby warming the Earth. Heat energy is then carried back into the atmosphere by several processes, including evaporation and recondensation of water, conduction, and reradiation. The warm atmosphere radiates some heat back to Earth and some to outer space. Thus a complex set of interactions occurs, in which radiation bounces back and forth between the surface and the atmosphere until it is ultimately lost to space.

Reflection of Radiation

Energy is scattered into space by reflection. Dust reflects light. Pollution makes dust. Therefore, one would expect that the effect of dust pollution would be to cool the Earth. So far, major volcanic eruptions far outstrip human activities in producing dust. The most spectacular eruption in modern times was that of Krakatoa (near Java) in 1883; its dust particles stayed in

Figure 21–8 Energy balance of the Earth. The sets of numbers in the dashed areas total 100 percent.

the atmosphere for five years. Summers seemed to be cooler in the Northern Hemisphere during this period, although the extent of temperature fluctuations makes even this observation somewhat doubtful. More recently, the violent eruptions of Mt. St. Helens in the state of Washington released great quantities of volcanic dust (Fig. 21–9). The best evidence seems to be that such particles do account for some back-scattering of solar radiation.

There is no convincing evidence that pollutant dusts in the lower atmosphere have any important effect on the Earth's temperature. However, the dust that settles to the ground is another matter. Most of us have seen how snow in the city can become dirty after a few days. Dust that falls on snow and ice in mountainous and polar regions depresses their reflectivities. Snow and ice that retain heat may melt, and if

this occurs extensively, it may produce a rise in temperatures, with significant global consequences.

Absorption of Radiation: The Greenhouse Effect

If there were no atmosphere, the view from the Earth would be much like what the astronauts see from the Moon—a terrain where starkly bright surfaces contrast with deep shadows, and a black sky from which the Sun glares and the stars shine but do not twinkle. The atmosphere protects us by serving as a light-scattering and heat-mediating blanket. A large portion of the heat emitted from the Earth is reabsorbed by the atmosphere and is, in effect, conserved, with the result that the surface of the Earth is warmer than it would otherwise be. This warming is called the **greenhouse effect,** by analogy

Figure 21–9 *The upper portion of Mt. St. Helens, Washington, protruding through the cloud layer during one of its eruptions in 1980. Large quantities of smoke and ash are discharged 15,000 m into the atmosphere. The peak to the right (actually 80 km northeast and about 1450 m higher) is Mt. Rainier, a dormant volcano. (Wide World Photo.)*

with the ability of a greenhouse to keep its inside warmer than the outside during the daytime. The energy emitted from the Earth is, of course, invisible (the Earth does not shine); it is largely infrared radiation, sometimes called heat rays.

Some molecules in the atmosphere absorb infrared radiation, and others do not. Oxygen and nitrogen, which together comprise almost 99 percent of the total composition of dry air at ground level, do not absorb infrared. On the other hand, molecules of water, carbon dioxide, and ozone do absorb infrared. Water plays the major role in absorbing infrared because it is so abundant. Ozone is the least important because there is so little of it. Carbon dioxide is also important, particularly because our combustion of fossil fuels produces more and more of it. Taking all factors into account, our best estimate is that the worldwide carbon dioxide concentration has increased from about 290 ppm in 1870 to 335 ppm in 1980.

Some scientists estimate that the carbon dioxide concentration will increase enough in the next 40 years to warm the Earth significantly, perhaps by as much as 0.5°C. It is diffi-

cult to predict the effects of such a temperature change on world climate.

Precipitation

The aerosol particles in cities whose atmospheres are polluted serve as nuclei for the condensation of moisture. Fogginess, cloudiness, and perhaps rainfall are usually increased considerably in contrast to the less polluted countrysides. However, the rise of warm air from cities, and the circulation of cooler surrounding air that is thereby induced, can sometimes reverse this situation.

The worldwide effects of pollutants on precipitation are even more difficult to assess. The concentration of small particles in the lower atmosphere is increasing over wide areas. The lead compounds from automobile exhaust are of particular concern because some of them, particularly lead chloride and lead bromide, are similar to cloud-seeding compounds. That is to say, they serve particularly well as nuclei for the condensation of water droplets or ice crystals. This action could lead to more precipitation, if the droplets coalesce, or less precipitation, if the nuclei are so small that the droplets stabilize and coalescence is inhibited.

Another effect that has been evident in recent years is **acid rain** (Fig. 21–10). Under natural conditions, rainwater is slightly acid because of the carbon dioxide, CO_2, dissolved in it. A small fraction—less than 1 percent—of the CO_2 reacts with water to produce the weak, unstable carbonic acid, H_2CO_3.

$$CO_2 + H_2O \rightarrow H_2CO_3$$

The pH of the resulting solution is about 6 or a little lower. (The pH scale is explained in Appendix C.) But in recent years some rainfall in many parts of the world has become much more strongly acid. Much of our acid rainfall has been between pH 4 and 5, but more severely acidic episodes occur from time to time. For example, a rainstorm in Pitlochry, Scotland, on April 10, 1974, had a pH of 2.4—as acidic as vinegar! Even that record has since been broken several times!

The source of this acidity is the oxides of sulfur and of nitrogen, especially SO_2, NO, and NO_2, as well as some acidic dusts. In the atmo-

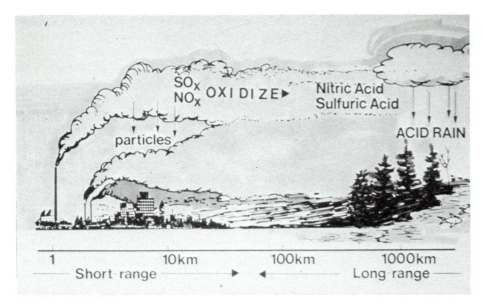

Figure 21–10 Acid rain.

sphere, SO_2 oxidizes to SO_3, which then reacts with water to produce sulfuric acid:

$$2SO_2 + O_2 \rightarrow 2SO_3$$

$$SO_3 + H_2O \rightarrow \quad H_2SO_4$$
$$\text{sulfuric acid}$$

The oxides of nitrogen eventually produce nitric acid, HNO_3. Both H_2SO_4 and HNO_3 are very strong acids.

The acidity of rain in North America increases from west to east because prevailing winds are westerly, and sulfur-containing pollutants accumulate in the atmosphere as an air mass travels across the country. Acid rain dissolves paint and stone from buildings, causing damage estimated at several billion dollars a year, and it also affects natural ecosystems. A 1981 survey in the Adirondack Mountains of upstate New York showed that the pH of hundreds of ponds and lakes was less than 5 and that 600 of these had completely lost their fish populations. Plants are also affected. A mysterious blight that has killed increasing numbers of oak and maple trees in recent years has now been traced to acid rain.

Acid rain is an increasing problem in industrial areas in America, Europe, and Asia where large amounts of fossil fuel are burned. But in 1981, American ecologists were surprised to find rain and streams with a pH of 2 in the agricultural plains of central China, an area that is not downwind of industrial areas and where the underlying rock has a pH of almost 7. The explanation turned out to be that the fuel used for heat and cooking in this area is yak dung, an organic fuel that also releases sulfur and nitrogen compounds into the air. Therefore, it seems that as long as human beings burn organic compounds—whether wood, animal dung, oil, or coal—for fuel, the acidity of rain all over the world will steadily increase as the human population of the world increases. If acid rain is not controlled, the damage it does to buildings, animal life, crops, and other plants will continue.

Depletion of the Ozone Layer

Ozone, O_3, is very different from oxygen, O_2. Pure ozone is a blue, explosive, poisonous gas. Very small amounts of ozone are produced naturally in the atmosphere by the action of sunlight on oxygen. The natural concentration of O_3 in the lower atmosphere is about 0.02 ppm; in the stratosphere it is about 0.1 ppm. At one time you could buy home "air purifiers" that were supposed to make your air fresher by producing ozone. These devices did not purify the air; they polluted it.

Ozone in the stratosphere, however, is not a pollutant. On the contrary, it protects life on Earth from excess solar ultraviolet (UV) radiation. UV radiation from the Sun tans our skin. Heavier doses of UV can cause burns and can increase the chances of skin cancer. If more of the UV light that reaches our upper atmosphere were to penetrate to the surface of the Earth, the risks of such damages would be increased.

Plants, too, might be adversely affected, and preliminary data suggest that the growth of some food crops, such as tomatoes and peas, is retarded by high doses of ultraviolet light. Fortunately, the high-energy UV radiation is removed by a series of reactions in the upper atmosphere before it reaches the Earth. In these reactions, stratospheric ozone acts as a catalyst for converting UV radiation into a warming effect in the upper atmosphere, which is harmless.

The most serious threat to the stratospheric ozone layer is thought to come from chlorine atoms, which remove ozone by catalyzing its conversion into oxygen. The important stratospheric sources of atomic chlorine are the chlorofluoromethanes, often referred to as **Freons,** which are DuPont trade names. Freons are used as propellants in aerosol cans, and as refrigerants. These compounds are stable in the lower atmosphere. As a result, they persist long enough to diffuse into the stratosphere. There they become exposed to solar ultraviolet radiation that is energetic enough to break the C—Cl bonds and release Cl atoms.

Aerosol cans using Freons have been banned in the United States but are still produced in many parts of the world.

Jet Aircraft

There has been concern that jet aircraft, particularly supersonic airplanes which fly at 18,000 to 20,000 meters, could engender stratospheric air pollution with consequent changes in climate. Jet exhaust contains water, CO_2, oxides of nitrogen, and particulate matter. For example, it is estimated that a fleet of 500 supersonic aircraft over a period of years could increase the water content of the stratosphere by 50 to 100 percent, which could result in a rise in average temperature of the surface of the Earth of perhaps 0.2°C and could cause destruction of some of the stratospheric ozone that protects the Earth from

ultraviolet radiation. On the other hand, the particulate matter would nucleate clouds of ice crystals and would increase stratospheric reflection to some extent. This could lead to a slight cooling of the Earth. One visible effect of such continued high-altitude pollution would be that the skies would gradually become hazier and lose some of their blueness. Since water vapor is predicted to warm the atmosphere and particulate matter may cool it, the net effect of supersonic aircraft is uncertain.

Bacterial Decomposition and Fertilizers

Bacterial decomposition of organic matter produces nitrogen-containing gases—NH_3, N_2O, and NO. A recent, fairly large-scale addition to this natural atmospheric load of nitrogen compounds comes from the extensive use of nitrogen fertilizers, specifically ammonium nitrate, NH_4NO_3. This compound decomposes to produce the gas N_2O.

$$NH_4NO_3 \rightarrow N_2O + 2H_2O$$

These gases then add to the other nitrogen oxides produced by stationary and mobile combustion sources.

Effects of Air Pollution on Human Health

Acute Effects

Much of the public attention to air pollution has been focused on episodes in which deaths occurred during periods of pollution and were clearly caused by its toxic effects. These disasters were generally caused by three conditions, all operating at the same time—(1) a severe atmospheric inversion, (2) the presence of airborne particulate matter such as fog, and (3) the continued generation of pollutants during the inversion; Table 21–3 outlines several such episodes. The combination of particulate and gaseous pollutants can be very harmful because both are inhaled into human lungs together (Fig. 21–11).

The fate of a particle, once it is inhaled, depends largely on its diameter. If it is greater than about 2 micrometers, it is usually trapped in the nasal passage or in the mucus of the bronchi and is generally coughed up and perhaps swallowed. If the diameter is less than 2 micrometers, the particle may be carried all the way

TABLE 21–3   Some Early Examples of Air Pollution Disasters During Atmospheric Inversions

Location and Meteorology	Sources of Pollution	Effects
Meuse Valley, Belgium, 25 km long, with 100-m high hills on either side. Atmospheric inversion, Dec. 1 to 5, 1930	Coke ovens, steel mills, glass factories, zinc smelters, sulfuric acid plants	60 people died. Many illnesses of respiratory tract—chest pains, shortness of breath, and so on. Also deaths of cattle
Donora, Pennsylvania, on inside of a horseshoe bend of the Monongahela River. Steep hills on either side of the valley. Very severe inversion and fog, Oct. 26 to 31, 1948	Steel mills, sulfuric acid plant, zinc production plant, and other industries	20 deaths and about 6000 illnesses. Severe irritations of throat and lungs
Poza Rica, Mexico, low inversion and fog during the night of Nov. 23–24, 1950	A single pollutant, hydrogen sulfide gas, H_2S, released between 4:45 and 5:10 A.M. on Nov. 24	320 people hospitalized; 22 died. H_2S first stinks, then causes loss of sense of smell, then severe respiratory irritation and death
London, fog and atmospheric inversion from Dec. 5 through Dec. 9, 1952	Coal smoke from individual home heating units and fireplaces	Between 3500 and 4000 excess deaths during the month of December, mostly in older age groups

through the air passages into the air sacs (alveoli) of the lung, where it may be trapped by specialized cells, or alternatively, it may be absorbed into the bloodstream.

The fate of gases is determined largely by the solubility of the gas in water. Since biological tissues are rich in water, a water-soluble gas such as SO_2 will rapidly dissolve in the soft tissues of the mouth, nose, throat, bronchi, and eyes where it produces the characteristic dry mouth, scratchy throat, and smarting eyes that most city dwellers have sometimes experienced.

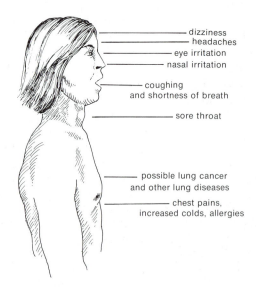

- dizziness
- headaches
- eye irritation
- nasal irritation
- coughing and shortness of breath
- sore throat
- possible lung cancer and other lung diseases
- chest pains, increased colds, allergies

Figure 21–11 Effects of air pollution on the human body.

By contrast, NO_2, which is relatively insoluble, may bypass this part of the respiratory tract and be carried to the alveoli where in very high doses it may cause gross accumulation of fluid in the air spaces, and thus make effective lung function impossible.

The net toxic effect when various pollutants are inhaled together, however, may be different from the sum of the effects of these same pollutants if they are inhaled separately. An interaction that produces *more* than a merely additive effect is called **synergism.** It is known, for example, that SO_2 may be adsorbed onto particles; if these particles are smaller than 2 micrometers, molecules of SO_2 may thus gain access to the alveoli in greater concentrations than they would otherwise. The retention of carcinogenic hydrocarbons in the human body has been shown to be greatly enhanced if they are first adsorbed onto soot particles. Also, oxygen and water can react with SO_2 to form sulfuric acid and with NO_2 to form nitric acid.

Personal Air Pollution—Cigarettes

The air whose quality is important to your health is the air *you* breathe, not the entire global atmosphere. If the outdoor air in any community resembled the self-polluted air that a smoker inhales, it would be considered a national disaster. Smokers are, on the average, approximately 10 times more likely to develop and die of lung cancer as nonsmokers. They are also

6 times as likely to die from pulmonary disease and nearly twice as likely to die from coronary heart disease. A host of other illnesses are also significantly associated with smoking. The more you smoke, the greater are your chances of getting lung cancer.

Chemical analysis of tobacco and cigarette smoke has revealed the presence of at least seven distinct hydrocarbons that have been shown to be able to produce cancer in animals. Cigarette smoke also contains polonium-210, a radioactive substance that may be carcinogenic.

Chronic Effects of Community Air Pollution

Another question of concern to medical scientists for many years has been the issue of whether exposure to mildly polluted air results in a higher rate of illness and death than is experienced by those who breathe relatively clean air. Many illnesses have been examined, but the ones that have prompted more research than any others have been diseases of the respiratory tract—lung cancer, chronic bronchitis, and emphysema.

Various studies have shown that lung disease is up to four times as prevalent in cities as in rural areas. It is possible that the differences in any disease from city to country may be due to factors other than air pollution. Many of the better studies of this problem take some of the differences between city and country living into account. The preponderance of the evidence indicates that, after most of these factors have been taken into consideration, there still remains an excess incidence of lung disease in urban environments. The most likely element of the "ur-ban factor" seems to be air pollution, though conclusive proof is lacking.

Moreover, cigarettes and the urban environment seem to have a more than additive effect on the incidence of lung cancer. That is, if one compares urban nonsmokers with rural nonsmokers, the difference in incidence of lung disease is small; but urban smokers have a much greater incidence of the disease than rural smokers. These studies indicate that the combination of smoking and living in large cities is particularly pernicious.

Those with pre-existing cardiac disease, particularly the type caused by narrowing of the blood vessels to the heart (so-called coronary heart disease) risk serious illness when exposed to very high levels of air pollution. Air polluted by ordinary automobile exhaust can have deleterious effects on the cardiovascular function of people whose heart is already compromised by other disease. One wonders how many people have died (or killed others) on crowded highways from heart attacks triggered by dirty air.

Injury to Vegetation and Animals

Air pollution has caused widespread damage to trees, fruits, vegetables, and ornamental flowers (Fig. 21–12). The total annual cost of plant damage caused by air pollution in the United States has been estimated to be in the range of $1 to $2 billion. Such figures are obtained in part from experiments carried out in chambers into which analyzed concentrations of pollutants are introduced. For example, California naval orange trees exposed to 0.0625 ppm of NO_2 for 290 days yielded 57 percent less fruit than those in filtered air. The results of such tests, combined with analyses of air pollution in the areas of orange groves, permit calculations of the total crop loss and its economic cost. Some plant scientists point out that the figures obtained in this way may be too low because they do not take various factors into account such as losses due to unregulated pollutants, to synergistic effects of pollutant mixtures, to predisposition to plant diseases caused by pollutants, and so on.

The most dramatic early instances of plant damage were seen in the total destruction of vegetation by sulfur dioxide in the areas sur-

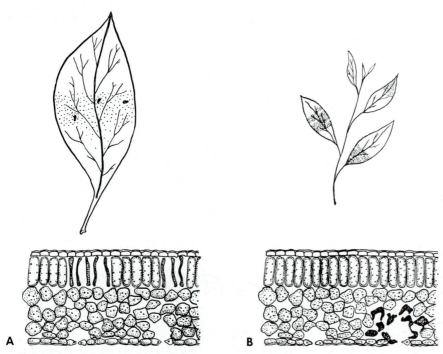

Figure 21–12 Effects of air pollution on plants. *A,* Ozone injury. Note the flecking or stippled effect on the leaf. On sectioning, only the palisade layer of cells is affected. *B,* Smog-type injury. Note the change in position of the effect with the age of the leaf. On sectioning, initial collapse is found in the region of a stoma. (From A. Stern: *Air Pollution,* New York: Academic Press, 1962.)

rounding smelters (see Fig. 1–4), where this gas is produced by the "roasting" of sulfide ores. There is a wide variety of patterns of plant damage by air pollutants. For example, all fluorides appear to act as cumulative poisons to plants, causing collapse of the leaf tissue. Photochemical (oxidant) smog bleaches and glazes spinach, lettuce, chard, alfalfa, tobacco, and other leafy plants. Ethylene, a hydrocarbon that occurs in automobile and diesel exhaust, makes carnation petals curl inward and ruins orchids by drying and discoloring their sepals.

Countless numbers of North American livestock have been poisoned by fluorides and by arsenic. The fluoride effect, which has been the more important, arises from the fallout of various fluorine compounds on forage. The ingestion of these pollutants by cattle causes an abnormal calcification of bones and teeth called **fluorosis,** resulting in loss of weight and lameness (see Fig. 21–13). Arsenic poisoning, which is less common, has been transmitted by contaminated gases near smelters.

Deterioration of Materials

Acidic pollutants are responsible for many damaging effects, such as the corrosion of metals and the weakening or disintegration of textiles, paper, and marble. Hydrogen sulfide, H_2S,

Figure 21–13 A cow afflicted with fluorosis.

tarnishes silver and blackens leaded house paints. Ozone produces cracks in rubber.

Particulate pollutants driven at high speeds by the wind cause destructive erosion of building surfaces. The deposition of dirt on an office building, as on a piece of apparel, leads to the expense of cleaning and to the wear that results from the cleaning action (see Fig. 21–14). The total annual cost in the United States of these effects is very difficult to assess but has been estimated at several billion dollars.

Aesthetic Insults

A view of distant mountains through clear, fresh air is aesthetically satisfying, and an interfering acrid haze is therefore a detriment. Unpleasant aesthetic effects cannot be neatly separated from the other disruptions caused by air pollution. The acrid haze is sensed not only as an annoyance but also as a harbinger of more direct harm, somewhat as the smell of leaking gas forebodes an explosion. Thus the pollution engenders anxiety, and anxiety may depress our appetites, or rob us of sleep, and these effects, in turn, can be directly harmful.

21.5 Standards of Air Quality

Outdoor Air

The wind dies down, and the air is still, but traffic continues to move, homes are heated, and industry operates. The air becomes hazy, murky, then uncomfortable. How dangerous is it? At what point should factories be shut down and traffic stopped? Clearly, to answer such questions something must be measured. Should it be SO_2, particulate matter, carbon monoxide, or all such pollutants together? After the measurements have been made, what is the most reasonable basis for action? The usual approach is to establish a set of air pollution standards that can serve as guidelines for governmental policies or regulations. Since any action will depend on the concentrations of pollutants, the conditions under which the analyses are carried out must be specified.

Figure 21–14 Old post office building being cleaned in St. Louis, Missouri, 1963. (Photo by H. Neff Jenkins. From A. Stern: *Air Pollution.* 2nd ed. New York, Academic Press, 1969.)

An air pollutant can be measured at the point where it is discharged to the atmosphere, such as at the chimney top, or in the surrounding (ambient) atmosphere where people live. The concentration at the source is not *directly* related to effects on human health, because people do not live in chimneys. Nonetheless, such measurements are valuable aids to the enforcement of air pollution control regulations. When permissible limits are established for such sources, they are called **emission standards.** The ambient concentrations are those in the air that people breathe, and the recommended limits are called **ambient air quality standards.**

TABLE 21–4   U.S. Primary* Ambient Air Quality Standards (as of August 1981)

Air Pollutant	Averaging Time**	Concentration†	
		µg/m³	ppm by Volume
Sulfur oxides	24 hr	365	0.14
	1 yr	80	0.03
Particulate matter	24 hr	260	—
Carbon monoxide	1 hr	40,000	35
	8 hr	10,000	9
Ozone	1 hr	235	0.12
Hydrocarbons (excluding methane)	3 hr (6 to 9 AM)	160	0.24
Nitrogen dioxide	1 yr	100	0.05
Lead	3 mo	1.5	—

*Ambient air quality standards are *maximum* allowable outdoor concentrations of air pollutants. Primary standards are those necessary to protect the public health. Secondary standards (not shown here) are those necessary to protect against other effects, such as damage to plants or materials. They differ somewhat from the primary standards only for sulfur oxides and particulate matter.

**The 1-hr, 3-hr, and 24-hr standards are not to be exceeded more than once per year.

†For gases at 25°C and 1 atm pressure, the relationship is µg/m³ = ppm × mol. wt./0.0245. Concentration in ppm does not apply to particulate matter.

Table 21–4 displays air quality standards set up by the U.S. Environmental Protection Agency (EPA). Note the following:

• The concentrations refer to levels that must not be exceeded, that is, to maximum levels. In discussions of public policy in relation to pollution standards, it is best not to talk about "raising" or "lowering" them because such expressions are ambiguous. It is better to refer to making the standards more lenient (that is, relaxing them) or making them more stringent or strict (tightening them).

• For any one pollutant, the longer the averaging time, the lower the permissible concentration. Thus, for example, the standard allows exposure to 35 ppm of carbon monoxide for one hour, but if the exposure lasts for eight hours, the concentration must be reduced to 9 ppm.

• The hydrocarbon entry is interesting; note that it applies only to morning hours. The reason is that the standard recognizes only that hydrocarbons are converted to smog by the action of sunlight. Therefore the hydrocarbons exhausted from automobiles during the morning

traffic rush can be irradiated throughout the day—for this reason they are called **smog precursors**. Hydrocarbons from evening traffic, on the other hand, will usually be dissipated during the night.

To advise the public about levels of air pollution, the EPA and other agencies have developed a Pollution Standards Index (PSI), as shown in Table 21–5. A score of 100 means that the short-term (24-hour or less) ambient air quality standards are just met. If the concentrations are higher than the standards, the PSI goes up proportionately. If the concentrations are lower, the PSI goes down. Health effects and warnings to the public are also shown in the table.

Indoor Air

Most people in developed areas of the world spend most of their lives indoors. Many in fact spend very little time outdoors at all—from home to car perhaps, and from car to factory or office. It is therefore important to consider the quality of indoor air and the methods of purifying it.

Indoor air carries pollutants from outdoors that are drawn in by fans or blown in by the wind as well as pollutants that are generated indoors. Because of this double burden, indoor air is generally more polluted than outdoor air. The largest burden of total indoor suspended particulate matter comes from cigarette smoke. (This statement assumes a proportion of smokers corresponding to the United States national average, which is about one smoker for every three persons over the age of 16.) Particulate matter also seeps out of wood stoves, which are becoming very common as the price of heating oil rises. Users of wood stoves should learn how to operate them properly to minimize the amount of smoke they produce.

Other indoor sources include gas cooking stoves, aerosol spraying, and pollutants that come from the building itself. The latter category is an important one. Some poorly manufactured grades of insulation decompose in moist air, generating formaldehyde, a toxic gas. In addition, most stone, brick, and concrete structures, as well as some soils, contain tiny concentrations of radioactive materials that release radon, a ra-

TABLE 21–5   Pollutant Standards Index—Health effects information and cautionary statements.

PSI	Air Quality	Health Effects	Warnings
0–50	Good		
51–100	Moderate		
101–200	Unhealthful	Mild aggravation of symptoms in susceptible persons, irritation symptoms in healthy population	Persons with existing heart or respiratory ailments should reduce physical exertion, outdoor activity.
201–300	Very unhealthful	**First stage alert** Significant aggravation of symptoms, decreased exercise tolerance in persons with heart or lung disease, widespread symptoms in healthy population	Elderly persons with existing heart or lung disease should stay indoors, reduce physical activity.
301–400	Hazardous	**Second stage alert** Premature onset of certain diseases, significant aggravation of symptoms, decreased exercise tolerance in healthy persons	Elderly persons with existing heart or lung disease should stay indoors, avoid physical exertion; general population should avoid outdoor activity.
401–500	Hazardous	**Third stage alert** Premature death of ill and elderly, healthy people experience adverse symptoms that affect normal activity	All persons should remain indoors, windows and doors closed; all persons should minimize physical exertion, avoid traffic.

dioactive gas. In very tightly sealed masonry houses, radon gas can build up to concentrations that are considered to be hazardous.

Measures that conserve heat by reducing leaks of air make matters worse by slowing down the escape of indoor air pollutants to the outdoors. The control of indoor air pollution, therefore, is a significant public health problem. Nevertheless, it is very unlikely that indoor air quality standards will be established in the United States because any such regulation would be an invasion of privacy. However, there are various measures that you as an individual can take to protect the quality of the air you breathe indoors, as shown in Table 21–6. Air conditioning (which is not listed in the table) cools air but is not in itself a method of purification.

21.6 Control of Air Pollution

It is easy to think of air pollution control as something one does to the source of the pollution—the engine or tailpipe of an automobile, or the smokestack of a power plant. That is, of course, a valid concept, but it is not the only one. One could also make cultural choices of many kinds that serve to reduce air pollution. Some of these are discussed in the sections of Chapter 11 that deal with energy sources.

The emission from a polluting source such as an industrial smokestack contains many harmless gases, such as nitrogen or oxygen, mixed with a much smaller concentration of pollutants. Typically, the volume of polluted matter in a contaminated air stream is in the range of 1/1000 to 1/10,000 of the total volume.

There are two general approaches for controlling air pollution:

1. The pollutants can be separated from the harmless gases and disposed of in some way, or

2. the pollutants can be somehow converted to harmless products that can then be released to the atmosphere.

Particulate matter can be separated from a moving gas stream in a variety of ways that depend on the physical and chemical properties of the pollutants. Commercial air pollution control devices such as the ones shown in Figure 21–15 consist of large, semiporous bags that filter dust from a polluted air stream.

There are various mechanical collection devices that depend on the fact that particles are *heavier* than gas molecules. If a gas stream that contains particulate pollutants is whirled around

TABLE 21–6   Measures to Protect Your Own Indoor Air Quality

What You Can Do	Advantages	Disadvantages
Don't stop *all* the air leaks—allow for a reasonable amount of ventilation.	Outdoor air replaces some indoor air, forcing pollutants outside.	Either (a) your fuel bills will be higher or (b) you will have to put up with cooler temperatures in winter and less air conditioning in summer.
Do not allow smoking indoors	Less indoor pollution	Inconvenient for smokers
Insulate only with high-grade material such as fibrous glass.	Does not decompose readily	Cannot be blown behind walls in existing homes
After using paint, lacquers, glues, and so on, ventilate thoroughly until all odors are gone.	Removes large amounts of temporary indoor pollution	Fuel bills will be temporarily higher
After clothes are dry cleaned, air them outdoors for a day before bringing them inside.	Much of the residual dry-cleaning solvent (trichlorethylene or perchlorethylene) evaporates	Inconvenient. Risk of rain
If you use a wood stove, learn how to operate it so that it produces a minimum amount of smoke.	Reduces indoor pollution by particulate matter	None
If your building is so tightly sealed that some forced ventilation is needed, bring air in through an air-to-air heat exchanger	Saves heating and air-conditioning costs	None except for initial cost
Use an indoor air purifier.	Saves energy compared with outdoor ventilation	Some electrostatic indoor air purifiers generate ozone, which is toxic. Air cleaners based on adsorbers are effective, but the adsorber must be replaced when it becomes saturated.

in a vortex (Fig. 21–16), the more massive pollutant particles may be spun out to the outer walls. These particles will then settle toward the bottom while clean gases will move upward and out the top of the collector. Such a device is called a **cyclone.**

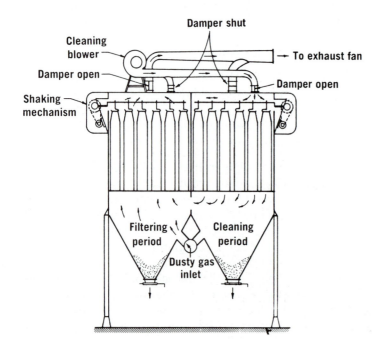

Figure 21–15 Typical bag filter employing reverse flow and mechanical shaking for cleaning. The figure shows the dusty gas being blown from the inlet toward the left and up. When the dusty gas reaches the six bags on the left, the gas goes into the bags while the dust remains on the outside (just like a vacuum cleaner running backward). The cleaned gas then comes out the tops of the bags and is discharged via the exhaust fan to the atmosphere. Meanwhile, the bags on the right, which have collected dust on their outer surfaces from the previous cycle, are being shaken and blown so that their dust falls to the bottom, where it can be removed. When the bags on the right are clean and those on the left are dusty, the air flow pattern is reversed. (From Stern: *Air Pollution.* 2nd ed. New York, Academic Press, 1968.)

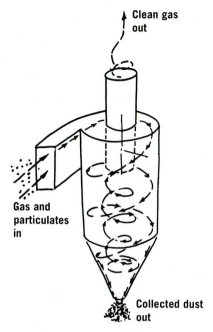

Particles may also be removed from a gas stream by virtue of their behavior in an electrical field. Recall that if a balloon is given an electrical charge by rubbing it against your hair, it will stick to a wall. Similarly, if dust particles are charged, they will adhere to a chamber wall or to each other. An **electrostatic precipitator** (Fig. 21–17) operates on just this principle. Electrostatic precipitators operate at efficiencies of 99 percent or higher and can convert a smoky exhaust to a visually clear one (Fig. 21–18). Furthermore, the energy requirements are modest compared with those for other air cleaning devices, because work is done only on the particles to be removed, not on the entire air stream. However, the installations are expensive and require much space. Most electrostatic precipitators are used in power plants that burn fossil fuels.

Figure 21–16 Basic cyclone collector. (From Walker: *Operating Principles of Air Pollution Control Equipment.* Bound Brook, N.J., Research-Cottrell, Inc., 1968.)

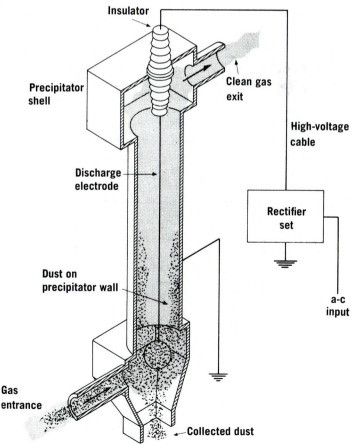

Figure 21–17 Basic elements of an electrostatic precipitator. A dusty gas is blown upward through a tube that contains a fine wire (the discharge electrode) running through the center. A high voltage on the wire transfers an electrical charge to nearby dust particles. This charge forces the dust particles to drift to the outer tube, where they stick to the walls. The accumulated dusty layer can then be removed while the clean gases exit out the top of the chamber. (From White: *Industrial Electrostatic Precipitation*, Reading, Mass., Addison-Wesley Publishing Company, 1963.)

Figure 21–18 Top view shows how fumes would pour from the smokestack of a steel furnace if pollution control devices were not installed. Bottom view shows operation with an electrostatic precipitator. (Photo courtesy of Beth-lehem Steel.)

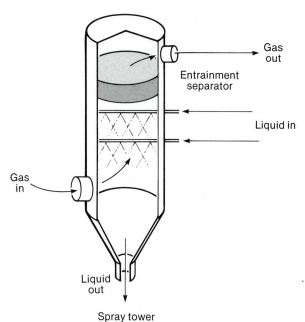

Figure 21–19 Schematic drawing of a spray collector, or scrubber. (After Stern: *Air Pollution.* 2nd ed. New York, Academic Press, 1968.)

Pollutant gases cannot feasibly be collected by mechanical means, because their molecules are not sufficiently larger or heavier than those of air. However, some pollutant gases may be more soluble in a particular liquid (usually water) than air is. For example, ammonia, NH_3, is a common gas that is very soluble in water. If an air stream is polluted with ammonia, the mixed gases can simply be bubbled through a water bath or sprayed with a water mist. The ammonia will dissolve and be removed with the water, whereas the air molecules will pass through un-affected. Devices that effect such separation are called **scrubbers** (Fig. 21–19).

Gas molecules adhere to solid surfaces. Even an apparently clean surface, such as that of a bright piece of silver, is covered with a layer of molecules of any gas with which it is in contact. The gas is said to be **adsorbed** on the solid. If a solid is perforated with a network of fine pores, its total surface area (which includes the inner surfaces of the pores) may be increased so much that its capacity for gas collection becomes significant. Such a solid is **activated carbon,** which can have many thousands of square meters of surface area per kilogram.

Activated carbon is made from natural carbon-containing sources, preferably hard ones, such as coconut shells, peach pits, dense woods, or coal, by charring them and then causing them to react with steam at very high temperatures. The resulting material can retain over 10 percent of its weight of adsorbed matter in many air purification applications. Furthermore, the adsorbed matter can be recovered from the carbon and, if it is valuable, can be recycled.

The adsorbent bed consists of carbon granules about 3 mm in diameter. It may seem strange that gases, which pass through the finest filter, can be effectively purified by a bed of loose granular material. But gas molecules move very rapidly between collisions. This rapid darting motion, when superimposed on the much slower speed of the entire current of gas, enables the molecules to reach some surface of the granular adsorbent particles in a very short time.

Sometimes it is difficult to remove chemical pollutants from a rapidly moving air stream, and it is easier to convert them to some relatively harmless compounds without separating them at all. By far the most important *conversion* of pollutants is by oxidation in air. If a gas stream is passed through a hot incinerator, the pollutants will burn and be converted to less objectionable products. When organic substances containing only carbon, hydrogen, and oxygen are completely oxidized, the resulting products are carbon dioxide and water, both innocuous. However, the process is often very expensive because considerable energy must be used to keep the entire gas stream hot enough (about 700°C) for complete oxidation to occur.

One way to reduce this energy requirement is with the aid of a catalyst. The application of this method to automobile exhaust is described in the Case History that follows.

21.7 Case History: Pollution by Automobile Exhaust

By the early 1900s, many industrial cities were heavily polluted. The major sources of pollution were no mystery. The burning of coal was number one. Other specific sources, such as a steel mill or a copper smelter, were readily identifiable. The major air pollutants were mixtures of soot and oxides of sulfur, together with various kinds of mineral matter that make up fly ash. When the pollution was heavy, the air was dark. Black dust collected on window sills and shirt collars, and new-fallen snow did not stay white very long.

Imagine now that your great grandfather had decided to go into the exciting new business of making moving pictures. Old-time photographic film was "slow" and required lots of sunlight, so he would hardly have moved to Pittsburgh. Southern California, with its warm,

sunny climate and little need for coal, was more like it. A region of Los Angeles called Hollywood thus became the center of the movie industry. Population boomed, and after World War II, automobiles became almost as numerous as people. Then the quality of the atmosphere began to deteriorate in a strange way. It was certainly air pollution. But it was somehow different from the smog in London or Pittsburgh. The differences may be summarized as follows:

LOS ANGELES (PHOTOCHEMICAL) SMOG	LONDON SMOG
Begins only during daylight	Begins mostly at night
Smells something like ozone; can also irritate the nose	Smells smoky
Looks yellow to brown	Looks gray to black
Damages certain crops such as lettuce and spinach	Damages stone buildings, especially marble
Irritates the eyes, causes blinking	
Makes rubber crack	

Most puzzling of all, people did not burn coal in Los Angeles! In 1951 A. J. Haagen-Smit, a chemist, reported on his experiments with automobiles. He piped automobile exhaust into a sealed room equipped with sun lamps (ultraviolet, UV) (Fig. 21–20). The room contained various plants and pieces of rubber. The room was also provided with little masklike windows that permitted people to stick their faces in and smell the inside air.

His results may be summarized as follows:

EXPERIMENT	CONDITION	RESULTS
1	Auto exhaust piped into room. UV lamps turned on	Smog! Plants were damaged, rubber cracked. If you stuck your head in the window, your eyes became irritated.
2	Auto exhaust piped into room. UV lamps turned off	Smelled like auto exhaust but not like smog. Smog effects not evident
3	Auto exhaust piped into room. UV lamps turned off. Ozone added to room	Smog again, as in Experiment 1

Figure 21–20 Smog is produced when automobile exhaust is exposed to sunlight.

These were the crucial experiments that pointed the accusing finger at the combination of automobile exhaust and sunlight. The results explained why Los Angeles smog begins in the daytime, not at night. The role of ozone was also understood. Ozone is an essential chemical agent in the production of smog, as shown by the result of Experiment 3. In the simplest terms, the chemistry of Los Angeles smog can be expressed in two equations:

(1) oxygen + ultraviolet light $\longrightarrow$ ozone

(2) auto exhaust + ozone $\longrightarrow$ smog

Of course, these equations are an extreme simplification. The detailed chemistry is devastatingly complex. Many separate chemical steps are involved. The rates of these various reactions change during the course of the day as the intensity of the sunlight changes. In fact, the whole story—all the reactions and their rates—is not yet known. One thing, however, is known: If hydrocarbons and other organic compounds were not introduced into the atmosphere, the polluting process would not take place. The remedy, then, is to drive the combustion of gasoline to completion, to produce only CO_2 and H_2O:

$$\text{hydrocarbon} + O_2 \longrightarrow CO_2 + H_2O$$

There are fundamentally two ways to reduce hydrocarbon emission from automobile exhaust. One is to improve the design of the engine itself so that the gasoline is burned more completely. Such a design approach also improves engine efficiency and results in substantial fuel savings. The other approach is to oxidize unburnt fuel before it is released into the air.

The second objective is achieved by using the catalytic converter (Fig. 21–21). The best catalysts for speeding up the oxidation of organic molecules to CO_2 and H_2O are certain heavy metals, especially platinum and palladium.

However, two major problems arise. For many years, a lead compound, tetraethyl lead,

Figure 21–21 Cutaway view of catalytic converter, showing catalyst pellets.

exhaust from engine

purified exhaust

$Pb(C_2H_5)_4$, was added to gasoline to improve engine performance. But the lead poisons the catalyst, destroying its effectiveness. It is for this reason that automobiles equipped with catalytic converters must use unleaded gasoline. The second problem is that the catalytic oxidation of the gasoline hydrocarbons generates heat within the catalytic converter that promotes other, environmentally unfavorable oxidations. Probably the most harmful of these is the increased conversion of N_2 to NO and NO_2.* The production of oxides of nitrogen is just what we want to avoid, since these compounds are also involved in the photochemical smog sequence.

The technical development of antipollution systems for automobiles is far from complete, and the next step may well be directed to the objective of minimizing the production of oxides of nitrogen. Our experience with the problems of atmospheric pollution from automobiles shows that the environmental aspects of this single process are extremely complex. Of course, decisions of public policy are also involved; pollution from the use of gasoline can be effectively reduced by using less gasoline—by driving fewer miles in smaller cars with more efficient gasoline consumption.

21.8 Legislation, Public Policy, and Strategies in Air Pollution Control

In Western society, the significant beginnings of legislative approaches to air pollution control occurred during the early use of coal in Great Britain, starting in the fourteenth century. During the reign of Edward II (1307–1327), for example, a man was put to torture for filling the air with a "pestilential odor" through the burning of coal. Later, less brutal and (one hopes) more effective methods of regulation took the form of taxation, restriction of the movement of coal into congested areas such as the City of London, and application of the common law of nuisance (see Chapter 2).

Polluters are no longer tortured, although it is possible under United States federal law for responsible company executives to be sentenced to prison for violation of air pollution regulations. Many communities establish regulations that are based on the nature of the localities whose air is polluted. For example, higher concentrations are allowed in areas zoned for industry than in residential areas. There are also some zones of pure air, such as in wilderness areas, where very little pollution is allowed. Such discrimination may be regarded as a form of regulation of land use. Although the legislative protection of pure air no longer depends entirely on statutes declaring air pollution to be a public nuisance, much of the tradition that pollution is only a minor crime remains. In particular, it is still customary to be "flexible" in administering air pollution regulations, particularly when an injunction would close down an otherwise lawful business where the immediate economic well-being of the community is seen to be involved.

In the United States, the first federal legislation exclusively concerned with air pollution was enacted in July, 1955. A very modest beginning, it authorized the Public Health Service to perform research, gather data, and provide technical assistance to state and local governments. The major air pollution legislation in the United States is now the Clean Air Act of 1963, together with a series of amendments added in later

---

* The N_2 is part of the air drawn in to the cylinder. Since only a little of it reacts there, most of it goes out with the exhaust gases and enters the catalytic converter.

years. This legislation recognized at the start that polluted air crosses state boundaries, and that in such instances, if individual states did not act to correct the problems, the federal government could do so. The Act further called for the publication of documents on air quality criteria and control techniques. The states were then to develop ambient air quality standards and plans for implementing them.

It turned out that these measures were not sufficient to promote rapid progress in air cleanup, and in 1970 amendments to the Clean Air Act called for the development of national ambient air quality standards. These are the standards discussed in Section 21.5 and displayed in Table 21–3. The law also limits the quantities of air contaminants that may be emitted by any new factory. These standards are distinct from the ambient air quality standards, because they are intended to control directly the pollution given off by a specific source, such as a power plant or foundry.

Furthermore, the legislation is not static. Debates on environmental policy occur in every session of Congress, and new amendments are added from time to time.

In view of all these legislative safeguards, how is it that you can still see and smell air pollution if you travel through industrial areas of the United States? This deficiency may be blamed, in part, on the usual problems of enforcement—administrative complexities, inadequate staffing, judicial delays. But there is more to it than that. Recall first the traditional reluctance to close down an otherwise lawful business for the "minor" crime of committing an air pollution nuisance. Currently, this reluctance takes the form of legal or administrative extensions of compliance dates for any of a variety of reasons. For example, a given company or even an entire industry may claim that it will take more time to develop, test, and install the control systems needed to reduce their air pollution emissions than the law allows. Representatives of the government may or may not agree. Lengthy negotiations usually follow, and extensions are often granted, especially if a plant is experimenting with "innovative" technology that promises to be more efficient than existing control methods. In some cases there may be a penalty for delayed compliance, such as a monthly payment, which may be refunded to the company when compliance is finally achieved. Temporary measures, such as using cleaner fuels or shutting down the entire plant during temperature inversions, may be accepted as a substitute during the extension period.

Such negotiations are sensitive to economic conditions. In the years before an "energy crisis" was recognized, some environmentalists took the position that *any* air pollution should be opposed. During times of high energy costs and economic slowdown, however, public policy tends to turn in the opposite direction. The research studies on which air quality standards were based are reexamined, and statistical uncertainties are pointed out. Arguments are made in favor of relaxing the standards. Sometimes the argument takes another form, which might be expressed as follows: "Let us maintain our pollution standards, but we must relax the time schedule for achieving them. The technical and economic problems can be reviewed each year to see how much further progress we can make toward our air quality goals." This viewpoint may be more palatable to the public, but the actual effect of an indefinite delay in reaching a standard is not really different from relaxing the standard.

Summary

Dry air is roughly 78 percent nitrogen, 21 percent oxygen, and 1 percent other gases. Natural air also contains water vapor and various suspended particles such as soot, pollen, bacteria, and viruses. An **air pollutant** is any substance that adversely affects air quality.

Air pollution from human activity originates mainly from stationary combustion sources, mobile combustion sources, and manufacturing sources. The latter category includes any process that uses air or heat or that breaks down materials mechanically.

Under average meteorological conditions, air temperature drops steadily with altitude. However, if the upper air is *warmer* than the air beneath it, the atmosphere becomes very stable, and pollutants are trapped under the warm air.

This condition is called an **atmospheric inversion.**

The effects of air pollution are (1) alteration of the atmosphere and climate, including reflection or absorption of radiation, acid rain, and depletion of the ozone layer; (2) acute or chronic damage to human health; (3) injury to other animals or to vegetation; (4) deterioration of materials; and (5) aesthetic insults.

Legally mandated limits to outdoor air pollution are called **ambient air quality standards.** Indoor air carries additional burdens from such sources as tobacco smoke, heating and cooking stoves, aerosol sprays, and pollutants that come from the building itself, including formaldehyde from insulation and radon from masonry.

Particulate air pollution can be controlled by bag filters, cyclones, and electrostatic precipitators. Gaseous air pollution can be controlled by activated carbon adsorbers, scrubbers, and incinerators with or without the use of catalysts.

The combination of automobile exhaust and sunlight is primarily responsible for photochemical ("Los Angeles") smog. In contrast, "London" smog is caused primarily by pollutants such as those from the burning of coal and does not require sunlight.

Air pollution legislation provides for the support of research, the dissemination of information, and the enforcement of controls. The degree of control to be mandated, however, is often seen as a compromise between the conflicting goals of air quality and those of economic development.

Questions

Air Pollutants and Their Sources

1. Identify each of the following substances as either a gas, smoke, dust, or mist. (1) A gray material stays suspended in the air without settling. A flashlight beam that shines through it is clearly visible, even when viewed from the side. (2) A brown transparent material is in a closed container. When the container is opened, the brown color becomes lighter, first near the top, then throughout the container. Finally, the material disappears entirely from the container. (3) Black particles slowly settle from the air to the ground. The settled material feels gritty. (4) Transparent particles slowly settle from the air to the ground. The settled material feels wet.

2. Table 21–1 refers to 14 gases. List them in descending order of their concentrations in natural dry air. (Careful—you must separate some gases that are grouped together in the table.) What is the sum of all their concentrations, expressed in ppm? Divide this number by 10,000 to get the sum expressed in percent. Is the total 100 percent? If not, how can you account for the discrepancy?

3. A concentration of 1 ppb of ethylene (C_2H_4) gas in the air cannot be smelled and has no demonstrable effect on people, animals, or materials. However, it does produce dried sepal injury in growing orchids—to such a degree that they become unfit for sale. Furthermore, ethylene at a concentration of 0.1 percent is used by food wholesalers to ripen bananas. Care must be exercised in this process, for if the concentration rises to 2.7 percent, the ethylene-air mixture becomes explosive.

Some ethylene is produced by fruit as part of the natural ripening process during its growth, and some ethylene is also discharged from the exhausts of automobiles and trucks, especially from those with diesel engines.

(1) Taking all these facts into consideration, would you classify ethylene as an air pollutant? Would your answer depend on its concentration? On its source? On your occupation?

(2) Would you invest your savings in a greenhouse for growing orchids that was located near a banana-ripening building? Near a busy highway?

4. State which of the following processes are likely to be sources of gaseous air pollutants, particulate air pollutants, both, or neither: (1) Gravel is screened to separate sand, small stones, and large stones into different piles. (2) A factory stores drums of liquid chemicals outdoors. Some of the drums are not

tightly closed, and others have rusted and are leaking. The exposed liquids evaporate. (3) A waterfall drives a turbine, which makes electricity. (4) Coal is heated in a large oven at 1000°C to drive off volatile matter, which is piped away to be refined into various chemicals. After 18 hours, the oven door is opened and a great ram ejects the glowing, sizzling residue, which is **coke,** onto a rail-car, where it is quenched with water.

5. Is the speed of settling of particles in air directly proportional to their diameters? (If the diameter is multiplied by 10, is the settling speed 10 times faster?) Justify your answer with data from Figure 21–2. Is a settling chamber a good general method of air pollution control? Explain.

Meteorology

6. Under some conditions, two inversion layers may exist at the same time in the same vertical atmospheric structure. Draw a diagram of temperature vs. height that shows one inversion layer between the ground and 200 meters, and another aloft, between 1000 and 1200 meters, while the temperature variations between them approximate equilibrium conditions.

7. "Since tall chimneys do not collect or destroy anything, all they do is protect the nearby areas at the expense of more distant places, which will eventually get all the pollutants anyway." Argue for or against this statement.

8. Describe the meteorological conditions most conducive to the rapid dispersal of pollutants. Describe those that are least conducive.

9. Refer to Figure 21–8. (1) What percent of the incident solar energy is received by the Earth? (2) Does the Earth's surface receive any additional energy? If so, from what source? (3) Is the amount of energy emitted by the Earth greater, less, or the same as that which it receives from incident solar radiation? Explain.

10. Imagine that you must determine whether some particular climatic effect, such as increased fog or rainfall in a given area, is

caused by human activity. Which of the following experimental method(s) would you rely on? Defend your choices. (1) Compare current data with that of previous years, when population and industrial activity were less. (2) Compare the effects during weekdays, when industrial activity is higher, with those on weekends, when it is low. (3) Compare effects during different seasons of the year. (4) Compare effects just before and after the switch from Daylight Saving Time to see whether there is a sharp 1-hour shift in the data. (5) Compare effects in different areas where population and industrial activities differ.

Health Effects

11. Cigarettes have been implicated as an extreme menace to the health of smokers, and hazardous even to nonsmokers who live or work in smoky areas. (1) Discuss the proposition that all smoking should be banned. (2) What recommendations, if any, would you favor regarding regulation of smoking in schools, in public buildings, in work places?

12. Which of the following groups of subjects would you study to learn the separate effects of cigarettes and air pollution on human health? Explain. (1) Urban smokers vs. urban nonsmokers; (2) urban smokers vs. rural smokers; (3) urban smokers vs. rural nonsmokers; (4) urban nonsmokers vs. rural smokers; (5) urban nonsmokers vs. rural nonsmokers; (6) rural smokers vs. rural nonsmokers.

Control Methods

13. Suppose that you keep some animals in a cage in your room and you are disturbed by their odor. Comment on each of the following possible remedies, or some combination of them, for controlling the odor: (1) Spray a disinfectant into the air to kill germs. (2) Install a device that recirculates the room air through a bed of activated carbon. (3) Clean the cage every day. (4) Install an exhaust fan in the window to blow the bad air out. (5) Install a window air conditioning unit that recirculates and cools the room air. (6) Install an ozone-producing device. (7) Spray a

pleasant scent into the room to make it smell better. (8) Light a gas burner in the room to incinerate the odors. (9) Keep an open tub of water in the room so that the odors will dissolve in the water.

14. Distinguish between separation methods and conversion methods for source control of air pollution. What is the general principle of each type of method?

15. Explain the air pollution control action of a cyclone; a settling chamber; a scrubber; activated carbon; an electrostatic precipitator; an incinerator.

Legislation and Public Policy

16. A report of air pollutant concentrations shows 24-hour average values of 1.0 ppm for sulfur dioxide, 100 $\mu g/m^3$ of particulate matter, and 0.5 ppm of nitrogen dioxide. However, this analysis was carried out at the top of a 200-meter stack of a power plant, the only factory in town that is a potential source of air pollution. If you were the health officer or mayor, what action, if any, would you recommend? Would you require any additional information? If so, describe the data you would request.

17. Explain the legal concept of a nuisance. What implications does this concept have with regard to air pollution regulations?

18. Describe some of the important features of the federal Clean Air Act of 1963 and its amendments. (If you or the entire class wish to study this in more detail, get a copy of the Act from your U.S. representative or senator.)

Suggested Readings

The basic text on air pollution is a five-volume work:
Arthur C. Stern: *Air Pollution.* 3rd ed. New York, Academic Press. 1976–1977. Volume 1, 715 pp.; Volume II, 656 pp.; Volume III, 799 pp.; Volume IV, 946 pp.; Volume V, 672 pp.

Three good one-volume texts are:
Arthur C. Stern, Henry C. Wohlers, Richard W. Boubel and William P. Lowry: *Fundamentals of Air Pollution.* New York, Academic Press, 1973, 492 pp.
Samuel J. Williamson: *Fundamentals of Air Pollution.* Reading, MA, Addison-Wesley Publishing Co., 1973. 473 pp.
Samuel S. Butcher and Robert J. Charlson: *An Introduction to Air Chemistry.* New York, Academic Press, 1972. 241 pp.

A textbook specifically devoted to control methods is:
Howard E. Hesketh: *Air Pollution Control.* Ann Arbor, MI, Ann Arbor Science Publishers, 1979. 362 pp.

A less technical introductory book is:
Virginia Brodine: *Air Pollution.* New York, Harcourt Brace Jovanovich, 1973.

Various introductory pamphlets as well as more technical documents on the effects of specific air pollutants and on methods for controlling them are available from the U.S. Environmental Protection Agency, Washington, D.C., or from its various regional offices.

OF TRENCH

- NO -
DEAD ANIMALS
OR ANIMAL PARTS
TO BE DUMPED IN
THIS LANDFILL

22

Solid Wastes

Today's landscape is not littered with huge mounds of dinosaur bones or ancient ferns. Waste products from living things are broken down by decomposer organisms, and the chemicals of one organism's refuse are reused by others. Occasionally, chemical elements are locked for long periods of time inside glaciers or geological deposits such as coal, but these deposits are eventually returned into the world's ecosystems.

In modern industrial societies, waste products are not reused efficiently. An apple grown in an orchard in the state of Washington may be shipped to a city on the Atlantic seaboard. After someone eats it, the core is usually not left out to be consumed by decay organisms and returned to the soil. Rather, it is stored in a trash pail, picked up by a truck, and transported to a large garbage dump. Similarly, the feces of a person who has eaten the apple may ultimately be washed into a sewer, and the farmer in Washington must purchase manufactured fertilizers to grow more food.

This chapter will discuss the sources of solid waste, the extent to which they are recycled, and the problems involved in their disposal. Radioactive wastes were discussed in Chapter 12 and will not be included here.

22.1 Sources and Quantities

Perhaps the most noteworthy characteristic of solid wastes is their variety. Household garbage includes food scraps, old newspapers, a variety of plastic bottles and utensils, discarded paper, wood, lawn trimmings, glass, cans, furnace ashes, old appliances, tires, worn-out furniture, broken toys, and a host of other items. The total quantity of solid waste is large and increasing. In the United States municipal solid wastes averaged 1.2 kg per person per day in 1920. The quantity rose to 2.3 kg in 1970, and to about 3.6 kg in 1980. In other words, we throw away three times as much trash as our grandparents did. The waste disposal system of an average city must accommodate about 85 kg of refuse per week for every family of four, and that is a lot of trash. In the year 1980 alone, about 300 million tonnes of municipal refuse accumulated in the United States. If all this trash were dumped in a single landfill, it would cover an area of 91 square km to a depth of nearly 5 meters.

The amount of solid waste generated is directly related to societal affluence. In contrast to

482

the 3.6 kg per person per day in the United States, residents of Australia produce 0.8 kg per person per day. The average person in India produces only about 0.2 kg per day. In the United States packaging materials amount to between 30 and 40 percent of municipal refuse (Fig. 22–1). Over 50 percent of the cost of soft drinks is for the bottle, and almost half of the cost of many other items lies in the package. Yet Americans have the money to buy these products.

There is another way to look at the problem. In the United States, 115 million tonnes of household and consumer goods were produced in 1978. (Durable goods such as appliances and automobiles are not counted in this figure.) Of this total, approximately 100 million tonnes were thrown in the trash. About 11 million tonnes were recycled, and the remainder, about 4 million tonnes, was saved and reused in subsequent years.

Municipal sources contribute only a fraction of the types and amounts of solid wastes discarded in the United States. Agricultural activities, for example, produce over 1.8 billion tonnes of wastes each year (Fig. 22–2). About three quarters of this is manure. Much of this manure is piled in dumps, where it pollutes streams and waterways. Yet at the same time, farmers across the continent are suffering from worn-out and depleted soils that could be enriched with manure. The balance of agricultural waste includes a variety of items. Some, like branches and slash left over from logging, are not particularly harmful to the environment. In fact, if left alone, these logging wastes eventually rot and enrich

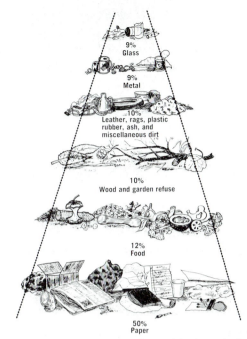

Figure 22–1 Composition of municipal trash in the United States.

the soils. Many byproducts of food processing operations, such as the discarded parts of fruits and vegetables, presently place a strain on local waste disposal facilities, but they could be shredded and returned to the land as fertilizers. Animal wastes from slaughter houses are a more difficult problem and require special processing (see Section 22.5). Pesticide residues are perhaps the most serious of all, because they are usually toxic.

Mining operations produce about 1.35 billion tonnes of debris per year. Most of this material

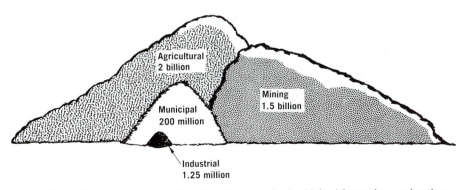

Figure 22–2 Sources and quantities of solid wastes in the United States (approximations expressed in tonnes per year).

is rock, dirt, sand, and slag that remain behind when metals are extracted from the Earth. These piles do not represent a loss of valuable raw materials, but they are ugly and are a source of erosion and acid mine drainage as well.

Look about the room you are in and note the number of manufactured objects. Each different kind of object was produced by a series of industrial operations, and some solid wastes were generated at each stage of production. Whenever raw materials such as metals, fossil fuels, or agricultural products are used to manufacture airplanes, shoes, beer cans, or even balloons, solid wastes are always generated. It should not be surprising, then, that the categories of indus-

trial solid wastes are more varied than those of municipal wastes.

When most people see an old soup can or pizza wrapper lying in the street, their first reaction is, "What an eyesore!" The environmental problems of solid wastes include, but go far beyond, the aesthetic insult of litter.

Pick anything out of the trash can and consider the total impact of producing that item. Perhaps you find a soup can made primarily of iron. That iron was dug out of an open pit mine that defaces the surface of the land. Air and water pollution are closely linked with the mining and refining operations. Large quantities of energy are consumed as well, and of course a valuable and limited resource is being depleted. What about that paper pizza wrapper? That was once part of a tree in a forest. Logging operations disrupt natural habitats, deplete soils, and encourage pesticide use in natural forests.

The total environmental impact of packaging practices in the United States is severe. Figure 22–3 shows the percentage of the total production of some important resources that are currently used for packaging in the United States. An automobile made of steel and aluminum may be used for 10 years before it is discarded, but a steel or aluminum can is used only once before it is dropped in a garbage can.

Table 22–1 lists some of the environmental problems involved in the production of consumer goods, most of which end up in the garbage can. You can see that the list is, in a way, a summary of most of the environmental problems discussed so far in this text.

22.2 Disposal on Land and in the Ocean

The least acceptable disposal method is the **open dump** (Fig. 22–4). Waste is collected and, to save space and transportation costs, is compacted. The compacted waste is hauled to a suitable site and simply dumped on the ground. Organic matter rots or is consumed by insects, by rats, or, if permitted, by hogs. In some communities the pile is set afire in the evening to reduce the total volume and the odor. There are

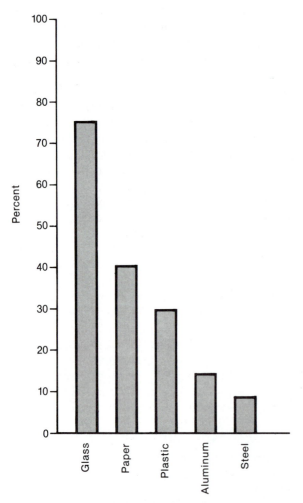

Figure 22–3 Percentage of several resources used in packaging in the United States.

TABLE 22–1   **Some Environmental Effects of the Production of Consumer Goods**

Activity	Typical Environmental Effects
Mining and oil drilling	Surface land disruption, acid mine drainage, mine tailings, sludge ponds, oil spills, fuel consumption
Agriculture	Disruption of natural habitats, soil erosion, fertilizer runoff, water consumption, poisonous pesticides
Forestry	Habitat disruption, soil erosion, pesticides, disruption of natural streams
Raw materials processing	Air pollution, acid rain, water pollution, dust, mine tailings, noise, fuel consumption.
Energy conversion and transmission	Depletion of resources, economic disruption, air pollution, water pollution, weather modification, thermal pollution, radioactive wastes, disruption of land for transmission rights of way, oil and chemical spills
Disposal	Litter, land use disruption, air pollution, water pollution, fuel consumption, release of hazardous wastes, disruption of natural waterways, dust, noise

serious ecological problems with the open dump. The dump itself is a potential source of disease. The fires are uncontrolled and therefore are always smoky and polluting. Rain erodes the

Figure 22–4 This untended landfill in southern Colorado has gradually turned into an open dump.

dump, and the polluted water flows into nearby rivers and groundwater reserves. And of course, the dumps are ugly. In urban regions open dumping is illegal, but it is practiced in many rural counties.

Ocean dumping is practiced by many coastal cities. Barges carrying the refuse travel some distance from the harbor and dump their loads into a natural trench or canyon on the ocean floor. In this way most of the trash is removed from sight, though not from the biosphere. Not surprisingly, ocean dumping upsets the ecological balance of regions of the sea. Many organisms are killed outright. Although certain plankton and fish survive in these areas, they are affected by the unusual environment. For example, flounder caught in the former New York City dump region have had an off-taste. Biologists have found old adhesive bandages and cigarette butts in fish stomachs. Therefore, it is not surprising that the flesh had a foul flavor.

The **sanitary landfill** is far less disruptive to the environment than uncontrolled dumping on land or into the ocean. A properly engineered landfill should be located on a site where rainwater will not flow through the trash and pollute nearby ecosystems. After waste is brought to a landfill, it is further compacted with bulldozers or other heavy machinery. Each day 15 to 30 cm of soil is pushed over the trash to exclude air, rodents, or vermin (see Fig. 22–5). In practice, however, the distinction between a sanitary landfill and the open dump is not always sharp. For example, a thin layer of earth may be an in-

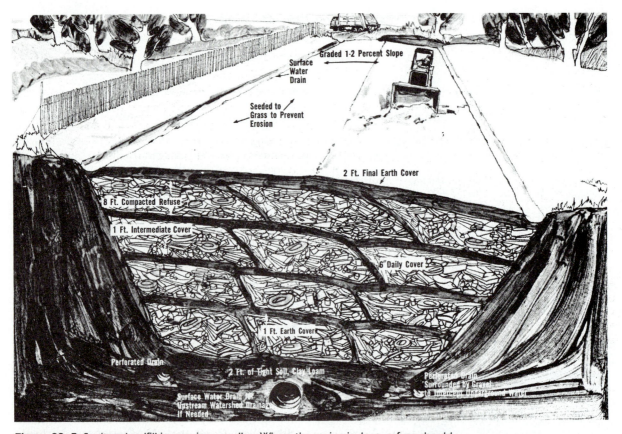

Figure 22–5 Sanitary landfill in a ravine or valley. Where the ravine is deep, refuse should be placed in lifts of 6 to 10 feet deep. Cover material may be obtained from the sides of the ravine. To minimize settlement problems, it is desirable to allow the first lift to settle about a year. This is not always necessary, however, if the refuse has been adequately compacted. Succeeding lifts are constructed by trucking refuse over the first one to the head of the ravine. Surface and groundwater pollution can be avoided by intercepting and diverting water away from the fill area through diversion trenches or pipes, or by placing a layer of highly permeable soil beneath the refuse to intercept water before it reaches the refuse. It is important to maintain the surface of completed lifts to prevent ponding and water seepage. (Courtesy of New York State Department of Environmental Conservation.)

effective barrier against burrowing rats, flies, or gases evolving from decomposition.

The sanitary landfill is an effective way to dispose of trash. It does not produce much pollution, disease or unsightliness. With proper ingenuity, landfills may even reclaim land unfit for industrial development. In some cases swamps and marshes have been filled with garbage and used as building sites for apartment complexes or parks. In other cases, landfill mountains have been constructed and used as ski slopes, and in one case as a "soapbox derby" raceway.

On the other hand, several serious problems are associated with sanitary landfills. First, land conversion is not always desirable. As we saw in Chapter 17, marshes and swamps house many valuable species of plants and animals. To destroy them is to destroy valuable ecosystems. Second, many large metropolitan areas have used up their available sites for landfills. They are now being forced to transport their trash further into the countryside. In these instances, transportation costs are high. Third, and perhaps most serious, such disposal represents a depletion of resources. Food wastes and sewage sludge that could be used as fertilizers are buried deep underground. Paper and wood scraps that could be recycled are lost, and nonrenewable supplies of metals are dissipated.

22.3 Energy from Refuse

Many metropolitan areas burn their garbage rather than dump it. The process, as applied to waste disposal, is more complex than simply setting fire to a mass of garbage in an open dump. In a modern incinerator unit the trash is burned in a carefully engineered furnace. The heat from the fire is used to boil water and produce steam. Then the steam is sold for industrial use. Thus, the trash is used as an energy source.

In some incineration plants, the money received from the sale of steam does not pay the operating costs. Economic problems arise because trash is not an ideal fuel. Municipal wastes contain food scraps and wet garbage that are difficult to burn efficiently. Also, the incineration of certain waste products produces acidic gases that corrode furnace walls. Particularly notorious is polyvinyl chloride (PVC), a plastic used in the manufacture of rainwear, toys, containers, garden hoses, and records. The burning of PVC produces hydrogen chloride gas. This gas reacts with water to produce a strongly corrosive liquid, hydrochloric acid. Even more threatening is the fact that some of the PVC decomposes before it burns and releases vinyl chloride, a known carcinogen.

In spite of the difficulties, incineration may become profitable in the future because:

1. Increasingly large quantities of dry paper and cardboard have appeared in refuse, thereby increasing the fuel content of trash.
2. The price of fuel has skyrocketed since 1973, and therefore the value of steam has also skyrocketed. On the other hand, costs of handling and burning trash have increased only moderately.
3. The rising cost of land has made it harder to find adequate sites for landfills. If valuable land is used for dumping, or trash is hauled long distances to less expensive sites, landfills become more expensive. A high cost of discarding trash makes alternative solutions more desirable.

At the present time, many industries and municipalities across the globe are burning trash as fuel. As mentioned in Chapter 11, large-scale incineration is practiced in France. Other countries in Europe are also active in this type of recycling. In 1980, municipalities in Denmark were using 60 percent of their wastes to produce energy, and those in Switzerland, the Netherlands, and Sweden were not far behind. In contrast, less than 1 percent of the trash in the United States was burned for fuel.

22.4 Recycling—An Introduction

Most refuse contains a wealth of raw materials that can be easily reused or recycled. There are many kinds of recycling paths. Consider, for example, the element aluminum. Large quantities of this metal go into nondurable goods that are discarded soon after they are purchased. "So what," you may say, "aluminum metal does not disappear. Can't the old dumps eventually be mined?" To answer this question, one must consider the total environmental costs of waste recovery. Dumps contain aluminum mixed with many other different materials. The aluminum could be extracted, but it is not very concentrated. Therefore, mining old dumps is so expensive and uses so much fuel that it is imprac-

---

BOX 22.1

One evenin' he decided t'go out an' still hunt; try t'kill 'im a bear or somethin'. Sittin' at th' head a'th' swamp an' there's three bear come walkin' out. A small little bear in front, and they's a big he bear in th' center, an' they's a little cub behind this'um. An' he waited 'til this big bear got betwixt him an' a tree t'shoot it wi' his hog rifle so he could save his bullet—go cut it out of th' tree. And he shot this big bear. . . . An' he went home an' took a axe an' cut th' bullet out. He'd take it back an' remold it. Lead was hard t'get, so that's th' way they'd try t'save their bullets.

From Eliot Wigginton (Ed.): *The Foxfire Book.* Garden City, NY, Anchor Books, Doubleday & Co., 1972, 380 pp.

Figure 22–6 Much energy could be conserved if glass jars were refilled rather than discarded. In this store, cooking oil is dispensed from drums into reusable jars. (Photo by Marion Mackay.)

tical. Resources would be conserved most efficiently if metals were recycled directly.

Think about the articles that people normally throw away. What would be the best way to minimize such waste? Obviously, there would be less junk if fewer items were discarded in the first place. That peanut butter jar in your garbage could be used over and over again if there were a large tub of peanut butter at the grocery store. Then you could simply take your jar to the store and refill it. The paper towel that you threw away need never have been purchased, for a cloth towel would have worked just as well. Or what about that old automobile that was carted off to the dump? If an automobile consumes too much oil and runs inefficiently, the engine can probably be rebuilt; perhaps it could be used for years to come. Similarly, when a refrigerator no longer cools efficiently, the fault generally lies with the compressor or the thermostat, not with the frame or the food compartment. If new parts were installed, many valuable raw materials would be conserved.

Not all items are repairable or reusable. A car that has been in a bad accident often cannot be rebuilt. A tire can be recapped only once with safety. Week-old newspapers and spoiled meat are useless to most people. When an item cannot be used in its present condition, it must be destroyed and treated somehow to extract its useful raw materials. For example, used tires can be shredded and converted to raw rubber. Old newspapers can be repulped and converted to new paper. Spoiled meat can be rendered and converted to tallow and animal feed. Discarded metals can be melted and reused. In general, recycling conserves not only material resources but fuel reserves as well. For example, nearly 20 times as much fuel is needed to produce aluminum from virgin ore as from scrap aluminum. Over twice as much energy is needed to manufacture steel from ore as from scrap. It is also twice as costly in energy to make paper from trees as it is to recycle used paper.

In many cases, recycling operations also emit less pollution than the original processes. Significant quantities of pollutants are released when paper is manufactured from wood pulp or when metal is refined from ore. The Environmental Protection Agency (EPA) has estimated that recycling all the metal and papers in municipal trash in the United States would prevent the release of over 2000 tonnes of air pollutants and 700 tonnes of water pollutants every year.

One can easily envision an ideal sequence in which durable goods are used for a long time, then repaired or patched to prolong their lives still further. Finally they could be recycled for use as raw materials. Unfortunately, no such plan has operated effectively in modern society. Complex social and economic factors have interfered.

BOX 22.2 RECYCLING PAPER

In 1979, about 21 percent of the paper in the United States was recycled, compared with about 50 percent in Japan. If the rate of recycling were increased to 50 percent in the United States, approximately 100 million trees would be saved, and the energy saved would amount to enough to supply 750,000 homes with electricity.

22.5 Recycling Techniques

Much municipal, industrial, and agricultural trash can be neither reused nor repaired and consequently must be reduced to raw materials suitable for remanufacture. Several techniques are available for this type of recycling.

Many materials such as metal, glass, and some plastics can be **melted,** purified, and reused.

Any material containing natural cellulose fiber such as wood, cloth, paper, sugar cane stalks, and marsh reeds can be beaten, **pulped,** and made into paper. Thus, for example, scrap newspaper can easily be repulped to manufacture recycled paper (Fig. 22–7). Agricultural wastes can be reused as well. When sugar is extracted from cane, the remaining fibrous stalks are well suited to paper production. These stalks currently contribute 60 million tonnes of solid waste annually and could easily be converted into paper.

If organic wastes are partially decomposed by bacteria, worms, and other living organisms, a valuable fertilizer and soil conditioner can be produced. This process is called **composting.** Almost any plant or animal matter, such as food scraps, old newspaper, straw, sawdust, leaves, or grass clippings, will form a satisfactory base for a composting operation.

Composted sewage sludge represents a valuable soil conditioner that can be used to increase the humus content of soils. Problems arise, however, because most domestic sewage contains small amounts of toxic metals, and if sludge is used as a fertilizer, these metals enter into our food chain. As a result, some food processors will not accept produce fertilized with sludge.

When sewage sludge and animal manure are composted, large quantities of methane gas are released. Methane is an excellent fuel. Some farmers have collected methane from cow manure, used the fuel to drive their tractors, and then recycled the compost as fertilizer. Although many such small-scale methane generators have

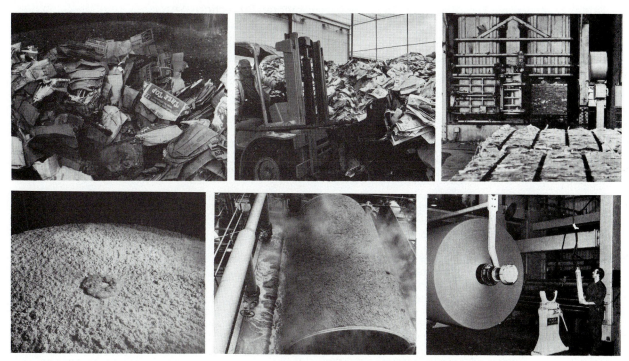

Figure 22–7 Wastepaper cycle: Scrap paper is collected, pressed into bales, pulped into a slurry, and reformed into new paper. (Courtesy of Container Corporation.)

been used throughout the world, few large-scale units are in operation.

If animal wastes such as fat, bones, feathers, or blood are cooked (**rendered**), several valuable products can be produced. These include a fatty product called tallow, which is the raw material for soap, and a nonfatty product that is high in protein and can be used as an ingredient in animal feed. The raw material for a rendering plant contains wastes from farms, slaughterhouses, retail butcher shops, fish processing plants, poultry processors, and canneries. If there were no rendering plants, these wastes would impose a heavy burden on sewage treatment plants. They would add pollutants to streams and lakes and nourish disease organisms. At the rendering plant, the waste materials are sterilized and converted to useful products such as tallow and chicken feed. But the rendering process generates odors. Although these odors can be controlled most of the time, they do get away now and then.

Municipal Trash Recycling

If people would separate their household trash into piles of aluminum, steel, paper, plastics, and food scraps, then perhaps all these materials could be recycled. But homemakers have not been willing to do this. Although some people have suggested that governments pass laws requiring residents to separate their trash, such a law could hardly be enforced. Policemen would have to examine garbage cans and try to discover who put the chicken bone in with the beer bottle, and that just wouldn't work.

It is possible to separate municipal garbage automatically in large recycling plants. Several garbage separating factories have been designed. The trash is dumped into a huge bin and carried into the separator by means of a conveyor belt. Useful materials are removed and recycled, as shown schematically in Figure 22–8. In 1979, about a dozen of these plants were in operation in the United States. It sounded like a wonderful idea. Homeowners pay to have their trash removed, and then the trash is separated and sold as scrap. Unfortunately, the cost of maintaining these plants has, in practice, been prohibitive. Of the plants in operation, all have lost money or at best have come close to breaking even. Several have shut down after severe financial losses (Fig. 22–9). What happened?

TABLE 22–2   Various Recycling Routes for Some Common Wastes

Waste	Recycling Possibilities
Paper	Use the backs of business letters for scrap paper or personal stationery; lend magazines and newspapers to friends. Repulp to reclaim fiber Compost Incinerate for heat
Glass	Purchase drinks in deposit bottles and return them; use other bottles as storage bins in the home. Crush and remelt for glass manufacture Crush and use as aggregate for building material or antiskid additive for road surface
Tire	Recap usable casings Use for swings, crash guards, boat bumpers, etc. Shred and use for manufacture of new tires Grind and use as additive in road construction
Manure	Compost or spread directly on fields Ferment to yield methane, use residue as compost Convert to oil by chemical treatment Treat chemically and reuse as animal feed
Food scraps	Save for meals of leftovers Sterilize and use as hog food Compost Use as culture for yeast for food production
Slaughterhouse and butcher shop wastes	Sterilize and use as animal feed Render Compost

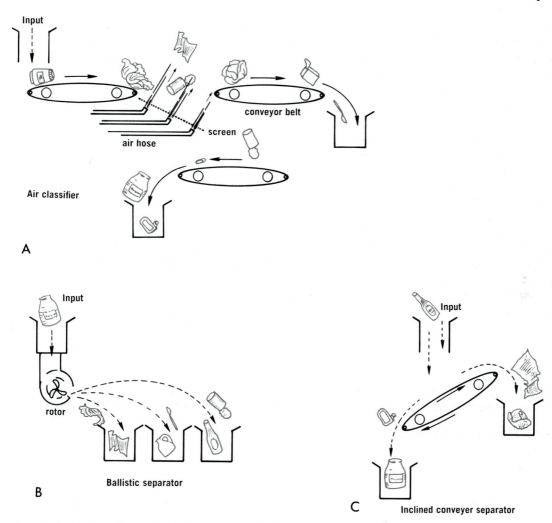

Figure 22–8 Types of waste material separators. *A,* Air classifier. *B,* Ballistic separator. *C,* Inclined conveyor separator.

The problem is that at the present time high construction costs, high interest rates, and high labor costs make it cheaper to simply throw the trash away. Yet, as will be discussed in the next section, this accounting does not include economic externalities.

22.6 The Future of Recycling

Techniques are available for recycling most solid wastes. Recall from the earlier discussion, however, that less than 10 percent of household and consumer goods are ever reused. Why is there such a large gap between available techniques and actual practice?

Even though the cost of fuel has risen markedly in North America since 1973, energy is still relatively cheap compared with the price of labor. In general, recycling is labor-intensive, whereas mining and logging use less labor and larger amounts of energy. Therefore, at the present time recycling is often not economical. The economic externalities of waste, however, are not counted in this balance sheet. In the United States, an aluminum beer can is worth about 1¢. For a penny it is often easier to throw the can in the trash than to store it in a separate bin and later take it to the recycling center. But think of the *real* cost of that can. Remember that 20 times as much energy is needed to manufacture a can from raw ore as from recycled scrap. In the United States that increased energy consump-

Figure 22–9 The Franklin, Ohio, Recycling Center. By 1979 this plant was losing $300,000 per year, and it closed down, leaving the town with a debt of $65,000 per year until 1996.

tion causes increased dependence on foreign oil, which in turn fuels inflation and even world-wide political instability. Furthermore, it de-pletes domestic reserves, which will lead to a rise in fuel prices within our lifetimes. In addi-tion, mining and logging add to global pollution. If these costs were added to the price of a beer can, then the cost of discarding one would be much greater than 1¢. In Europe and South America, fuel prices are much higher in relation to the cost of living than they are in North America. Recycling is more profitable and consequently it is more popular. For example, in the late 1970s, a bottle of beer in Ecuador cost approximately 40¢ (measured in U.S. currency). The deposit on the bottle, however, was three times the price of the beer. Thus, empty bottles were worth $1.20 each, and people simply did not discard them.

In the United States today, external economic incentives actually favor processing virgin materials over recycling. Tax laws in the United States favor overuse of resources. Tax structures give special deductions for mineral exploration and development of mines. Depletion allowances offer tax incentives for every tonne of ore removed from the ground. Likewise, timber companies are favored by complex tax laws that offer economic incentives for cutting trees. In addition, railroads charge more to transport scrap materials than logs and virgin ores. It is cheaper to ship a tonne of iron ore than a tonne of recycled iron, and cheaper to ship a tonne of logs than a tonne of waste paper. Railroads not only charge more, they also realize a higher profit from transportation of recycled goods. This rate structure is approved by the federal government. In a true free-market system, recycling could be more economically attractive than it is today.

Is it possible to increase the use of recycling techniques now? One way to do this is to pass laws that encourage reuse. It would be possible, of course, to enact legislation to reduce the tax and transportation advantages currently enjoyed by mining and logging operations. Although this has not been done in recent years, there are some encouraging signs. Within the last decade, some laws have been passed that encourage household recycling. For example, the states of Vermont, Oregon, Maine, Connecticut, Michigan, Iowa, Massachusetts, Delaware, and New York have banned most nonreturnable beverage containers. However, the usual 5¢ deposit required may be too low. The total environmental cost of a discarded bottle includes the energy and raw materials of manufacture as well as the litter and waste disposal problem. Therefore, a bottle's true cost in 1980 was close to 25¢. Per-

haps a law requiring a 25¢ deposit on glass bottles would be an effective inducement for returning them.

Industry responds to a variety of pressures, not only economic but also social, political, and legal. For example, the General Services Administration of the United States government and many corporations have ordered large quantities of recycled paper for stationery or annual reports. At present there is no economic advantage to the use of recycled paper, but these companies have responded to public pressure to reuse scrap.

22.7 Case History: Recycling of Beverage Containers

A century ago, milk, beer, cooking oil, and other liquids were shipped to general stores in large barrels. Customers then filled their own reusable jars. Many solid items such as coffee and nuts that are now available in glass containers were scooped from kegs into cloth sacks. These sacks were used over and over again. In time, glass packaging became increasingly popular, especially for foods. Later the soft drink, dairy, and beer industries added a new approach to glass packaging—the deposit bottle. These generally well-constructed bottles cost more to manufacture than the 2¢ or 5¢ deposit values assigned to them. But the small deposit was sufficient incentive for the customer to return the bottles. This system worked so well that at first a deposit bottle averaged about 30 round trips. Some were broken, lost, neglected, or used as flower vases and therefore never returned. These were, of course, replaced with new bottles. Gradually, consumers became so indifferent to the deposit that by 1960 the aver-

Figure 22–10 An old advertisement for bottle deposits.

age number of round trips had declined to four. Moreover, the cost of washing bottles increased, so bottling companies initiated a shift to no-deposit, no-return containers.

This shift to throwaway glass has led to a tripling of energy consumption in the bottling industry. Glass from a no-deposit, no-return bottle can be recycled, but it must be collected, transported to the recycling plant, crushed, sorted according to color, remelted, and reprocessed. Once again, a key problem here is collection. In a recycling-conscious society people would ideally sort their waste glass and deposit it in convenient centers. Under these conditions, recycling of glass would be profitable and would conserve energy. In our present society, the collection process is so inefficient that overall, more energy is needed to recycle glass from a no-deposit bottle than is needed to manufacture one from virgin material. It is, therefore, obvious that it would be environmentally more sound to sell only returnable bottles. The present use of throwaways is wasteful of resources.

In the 1940s some beer and soft drink manufacturers shifted partially to the use of steel cans as containers. Cans are lighter, easier to ship, and less susceptible to breakage than glass. Steel cans are hard to open, however, so in the late 1950s aluminum cans appeared on the market. Soon thereafter "flip-tops" were added. Aluminum cans require even more energy to manufacture than glass bottles do. But they are cheaper to handle for the beverage bottlers and shippers. Therefore, they have been marketed extensively during the past decade. In 1980, more than 60 percent of packaged beer and soft drinks in North America were sold in aluminum cans.

Seven hundred watt hours of electricity, enough to light a 100-watt bulb for seven hours, are needed to manufacture *one* 12-ounce aluminum beverage can. In 1980 some 40 billion aluminum beverage cans were produced in the United States. If they were all made from raw materials, approximately 28 billion kilowatt hours of electricity would be required to manufacture them. Electricity is produced with 33 percent efficiency, so 50 million barrels of oil are needed to manufacture this number of beverage cans.

Fortunately, recycling of aluminum cans has grown significantly in the United States in recent years. Figure 22–12 shows the growth of alumi-

Figure 22–11 Energy is conserved when aluminum is recycled.

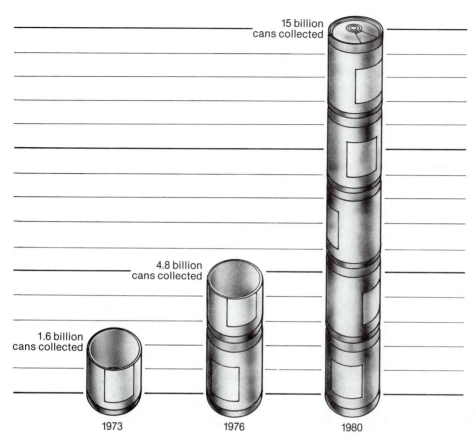

15 billion
cans collected

4.8 billion
cans collected

1.6 billion
cans collected

1973   1976   1980

Figure 22–12 *More and more Americans are collecting used aluminum cans. In 1980 alone over 15 billion cans were turned in for recycling. That's almost nine times as many cans as were collected in 1973. (Courtesy of Aluminum Corporation of America.)*

num recycling in this country. In 1980, approximately one third of all aluminum beverage cans sold were recycled. This is important when you recall that recycling uses only 5 percent of the energy needed to process raw ore. The energy savings from aluminum recycling in 1980 conserved enough electricity to power a city the size of Pittsburgh, Pennsylvania, for about 18 months. This is encouraging—unless you look at it the other way. The energy wasted by throwing away aluminum beverage cans in 1980 was enough to power two cities the size of Pittsburgh for 18 months.

Summary

Municipal refuse is composed largely of unnecessary packaging materials and of items that have been discarded because they weren't built to last in the first place. Depletion of resources and environmental disruptions are caused when these goods are produced.

Much of the solid waste in the United States is deposited in unsightly, uncontrolled, smelly, polluting **open dumps.** Some wastes are dumped untreated into the ocean. A **sanitary landfill** is an area where trash is deposited, compacted, and covered with soil daily to reduce pollution.

Many metropolitan areas **incinerate** their garbage and use the heat to produce steam. In recent years the fuel content of trash, the value of the steam, and the operating costs of alternative methods of disposal have all increased, and thus incineration has become more desirable.

The real objective of recycling is to guarantee the greatest possible use of a material in the most efficient manner. If an item cannot be reused or repaired, it may be practical to recover and reuse the materials of which it is made. In general, recycling conserves not only material resources but fuel reserves as well.

Some recycling techniques include melting, repulping, composting, rendering, and industrial salvage.

Recycling has stagnated in our society mainly because the economic externalities of unecological waste practices are generally ignored. At the present time, some laws actually favor the production of goods from virgin materials over recycling.

Questions

Solid Wastes

1. Plastics, made from coal and oil, have a high heat content. If garbage incinerators were commonly used in the United States, and the resulting heat of combustion were used industrially, would you feel that plastic packaging would be an advantageous way to use fossil fuels twice? Defend your answer.

2. An old-timer complains that years ago a man could store canned milk in the creek for three years before the can would rust through, but now a can will only last one year in the creek. Would you agree with the old man that cans should be made to be more durable? What about automobiles? Explain any differences.

Recycling

3. As mentioned in the text, broken or obsolete items can be repaired, broken down for the extraction of materials, or discarded. Which route is more conservative of raw materials and energy for each of the following items? (1) a 1948-model passenger car that doesn't run; (2) a 1980-model passenger car that doesn't run; (3) an ocean liner grounded on a sandbar and broken in two; (4) an ocean liner sunk in the central ocean; (5) last year's telephone directory; (6) an automobile battery that won't produce current because the owner of the car left the lights on all day; (7) an empty ink cartridge from a fountain pen.

4. Sand and bauxite, which are the raw materials for glass and aluminum, respectively, are plentiful in the Earth's crust. Since we are in no danger of depleting these resources in the near future, why should we concern ourselves with recycling glass bottles and aluminum cans?

5. List the most efficient recycling technique and the resultant products for each of the following: (1) steer manure; (2) old clothes; (3) scrap lumber; (4) aluminum foil used to wrap your lunch; (5) a broken piece of pottery; (6) old bottle caps; (7) stale beer; (8) old eggshells; (9) a burnt-out power saw; (10) worn-out furniture; (11) old garden tools; (12) tin cans; (13) disposable diapers. How does your choice of technique depend on your location?

6. At the present time, recycling of paper is a marginally profitable business, and many repulping mills have gone bankrupt in recent years. List some of the economic externalities associated with the production of paper from raw materials. Who bears these costs? How could the burden be shifted to encourage paper recycling?

7. A family lives in a sparsely populated canyon in the northern Rocky Mountains. Their household trash is disposed of in the following manner: Papers are used to start the morning fire in the potbelly stove; food wastes are either fed to livestock or composted. Ashes are incorporated into the compost mixture; metal cans are cleaned, cut open, and used to line storage bins to make them rodent-resistant; glass bottles are saved to store food; miscellaneous refuse is hauled to a sanitary landfill. Comment on this system. Can you think of situations where this system would be undesirable? Do you think that it is likely that many people will adopt this system?

8. Rendering plants recycle various slaughterhouse and cannery wastes. Despite careful controls these factories sometimes emit foul odors. Comment on the overall environmental impact of a rendering plant.

9. Environmental organizations have been active in establishing collection centers for old newspapers, cans, and so on. Can you think

of other activities that these groups might engage in that would produce increased recycling?

10. Consider the possible reasons why the average number of round trips of a returnable bottle, originally about 30, fell to 4 by 1960. Can you think of other reasons besides the fact that a 2¢ or 5¢ deposit is not worth as much now as it used to be? Ask your parents or grandparents to recall what was actually involved in returning bottles for deposit. How would you feel about depositing bottles today? Also, interview some store managers, including some older ones, and get their answers to these questions.

11. In 1976 voters in Colorado were asked to decide whether or not to levy a mandatory tax on nonreturnable cans and bottles. The beverage industry strongly opposed the law. In one brochure published by a major beer manufacturer it was stated:

> Claim: Amendment #8 [the proposed bottle law] would conserve energy and resources.
> Fact: Any savings in coal consumption resulting from the law would be offset by an increase in the consumption of gasoline, natural gas and water.
> Returnable bottles are heavier than nonreturnable cans. Manufacturing their heavy, durable carrying cases would require an increase in energy consumption.
> Bottles would also require twice as much space as cans. Trucks would have to make at least twice as many trips to haul refillable containers, to say nothing of the extra trips to pick up empties, resulting in increased gasoline consumption.
> Washing re-usable bottles requires five times more water than cans. Heating the water for sterilization and removing the detergents that are used means increased energy consumption.

   Examine these statements critically. Have all the facts been presented fully and

accurately, or do you feel that the brochure states the case incompletely? Is it reasonable or misleading? Defend your position.

12. In arguing against a mandatory bottle deposit bill, the president of a supermarket chain recommended instead a bill for "total litter control." Such a law would provide public funds to pay jobless youth at minimum wage to remove all litter (not just beverage containers) from streets and highways and to establish community recycling centers and a public education campaign. Prepare a class debate in which one side favors the bottle deposit bill, the other side the "total litter control" approach.

Suggested Readings

General references on solid wastes include:
Solid Wastes. Environmental Science and Technology Reprint Book. Washington, D.C., American Chemical Society, 1971.
Choices for Conservation. Research Conservation Committee Report to the President and Congress, July 1979.
N. Y. Kirov: *Solid Waste Treatment and Disposal.* Ann Arbor, MI, Ann Arbor Science Publishers, 1972.

Information on recycling can be found in
Bruce Hanon: *System Energy and Recycling, A Study of the Beverage Industry.* Center for Advanced Computation Document #23, University of Illinois Press, 1973.
Dennis Hayes: *Repairs, Reuse, Recycling—First Steps Toward a Sustainable Society. New York, Unipub (World-Watch Institute), 1978.*
Arthur H. Purcell: *The Waste Watchers: A Citizen's Handbook for Conserving Energy and Resources.* New York, Anchor Press/Doubleday, 1980.

Epilogue—Planning for a Sustainable Society

The desert in western Egypt is a harsh land noted for low rainfall, intense heat, and periodic windstorms. As a result, the human population in the area has always been low, and human existence has remained at subsistence level even in modern times. Early in the 1980s, an interdisciplinary research team discovered a new source of wealth in the area—vast underground reservoirs of water. Engineering and economic studies showed that if deep wells were drilled, enough water could be pumped to the surface to irrigate farmlands and raise the local standard of living considerably. When the villagers in the area received news of this plan, they asked if the water reserves were renewable like those of a river or likely to be depleted like those of an oil well. The developmental team explained that the water had been trapped in the rock structure millions of years ago and was therefore more analogous to an oil well than to a river, but the pools were so large that they would not be depleted for 100 to 200 years, so there was no need to worry. Upon hearing this news, the local farmers explained that their people had lived in the desert for many thousands of years, not

hundreds. Therefore it would be unwise to build a new prosperity that was likely to collapse within the time span of a few generations.

The difference in outlook between the developers and the villagers is typical of conflicts that underlie many environmental issues. The developers' view is based on the concept that technology has improved the level of living of the human race immensely over the past few centuries and will continue to offer an ever-brightening prospect in the future. Their argument continues: Accept the technological solutions today and enjoy the wealth and ease they bring, and then trust that scientists and engineers will find new cures when the resources are depleted in the future. The other viewpoint, represented by the villagers, is that the technological fix today will allow people to change their lifestyles and increase the human population but that these changes are not sustainable. When the resources, in this case water, run out, the legacy will be overpopulation of the ecosystem, followed by starvation and misery.

In 1982, a short pamphlet entitled *Six Steps to a Sustainable Society* was published by the

Worldwatch Institute. The primary thesis of this work was that irrevocable depletion of the resource base for technological development has already begun (Fig. E–1) but that a sustainable society can be built if the danger is recognized and appropriate action is taken immediately. The six steps necessary for a smooth transition into the twenty-first century provide a summary review of the environmental issues already discussed in this book.

1. Stabilize World Population

Figure E–1 shows that the per capita production of key commodities has been decreasing in recent years. During this time period, the *total* production of these commodities has been increasing, but the gains in output have been more than offset by the increase in population. Certainly, the Earth cannot support a continuously growing population indefinitely.

In the 1970s the Swiss demographer Pierre Prodervand visited a village in southern Senegal and noted that the people expressed little interest in family planning and contraception. When he returned a few years later, attitudes had changed, and people were eager for information on how to reduce births. This change in outlook was attributed to inflation. In the time period between the two visits, the price of basic necessities such as food, kerosene, and clothing had increased dramatically, and people had realized that a decent existence was possible only if the sizes of their families were reduced. This attitude is becoming more and more common throughout the world, but change is slow. In many European countries, population growth has slowed or even stopped, but in most less-developed countries rates of natural increase remain high.

The population of the world in 1982 was approximately 4.5 billion. The United Nations demographic council predicted that world population would stabilize at about 9.8 billion, but many people feel that the Earth cannot adequately support twice as many people as there are today. The first step toward building a sustainable society is achieving zero population growth.

2. Protecting Cropland

As more and more people populate the Earth, increasing quantities of food are needed, but at the same time, prime cropland is being eroded or converted to nonagricultural uses. A two-pronged program is necessary to reverse these trends.

1. Use croplands in the most efficient manner possible.

(a) *Developed countries.* In the developed world, the rapid destruction of farmland for commercial development can be slowed in several ways. One approach involves economic incentives such as the use of a tax structure that values agricultural land differently from land used for other purposes. Alternatively, direct subsidies can be used to maintain land for agriculture. A third possibility involves the passage of zoning laws that prohibit development in prime agricultural areas.

(b) *Less-developed countries.* In many cases, fertile river-bottom land held by the wealthy minority is used for grazing, whereas the peasants must farm the less productive hillsides. In these situations, land redistribution programs would promote the most efficient use of soils.

2. Reduce erosion. World food output has more than doubled since 1950, but between one fifth and one third of the global cropland is being eroded so severely that long-range productivity is threatened. A study of land erosion in Iowa has shown that the short-term costs of soil conservation are three times as great as the short-term financial returns. Similar unfavorable balance sheets are observed in the less-developed countries. When farmers must meet expensive mortgage payments or when people are hungry, it is often difficult to set money and time aside for conservation. In order to achieve the goal of a sustainable society, economic incentives such as tax relief or even direct aid for programs that reduce erosion are essential.

3. Reforesting the Earth

Forty percent of the people of the world use wood as a primary fuel. In addition, wood is used as a building material and as a raw material in the manufacture of paper and a variety of other products. On the other hand, living trees are valuable as well, because forests hold rainwater much more tightly than bare soil does.

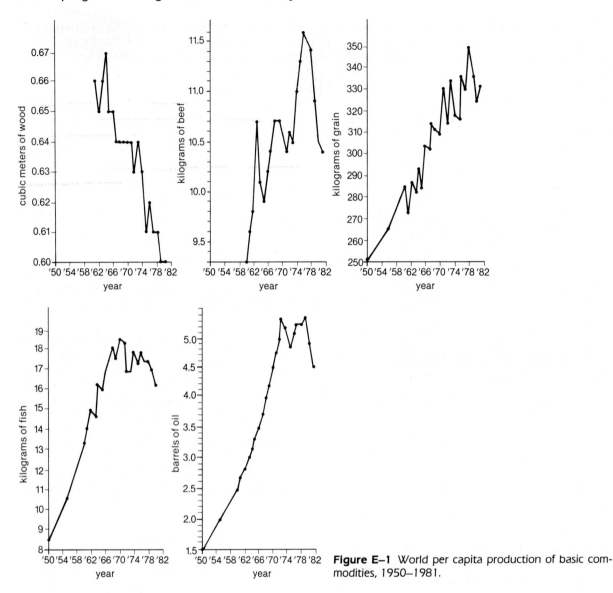

Figure E–1 World per capita production of basic commodities, 1950–1981.

When rainwater is retained, river runoff and erosion are reduced, groundwater supplies are maintained, and reserves of irrigation water are ensured. Therefore rapid deforestation has serious economic and ecological consequences. The solutions to the problem of deforestation are obvious: Plant trees in barren areas and control cutting in existing woodlands. However, the problem is also one of short-term economics versus long-range stability. Today, profits can be made by clearcutting forests and jungles, and many are eager to farm deforested regions even if the soil quality will be lost in the near future.

4. Moving Beyond the Throwaway Society

In the developed countries, the throwaway society originated in an era when energy and raw materials were cheap and abundant. Today, people know that if current trends continue, scarcity is not far off. Waste can be reduced if manufactured goods are more durable in the first place, and if discarded goods are recycled. In general, recycling not only conserves materials, it also saves energy and reduces pollution. Some encouraging steps taken in various countries around the world are listed on page 501.

India. In a land where resources are expensive relative to the cost of labor, garbage is carefully sorted by hand, and most discarded products are reused.

U.S.S.R. An automated resource recovery plant in Leningrad sorts municipal trash mechanically, processing 575,000 tonnes of garbage every year.

Denmark. A program under study proposes that *all* beverage containers must be marketed in one of five standard sizes and colors. All bottlers would draw from a common pool of containers that can easily be collected, sorted, and recycled.

Japan. Recycling jumped from 16 percent of all raw materials in 1974 to 48 percent in 1978. (In contrast, recycling in the United States and the United Kingdom is closer to 10 percent.)

Norway. Laws have been enacted that require homeowners, merchants, and manufacturers to separate waste paper from other garbage. The paper is then collected and recycled.

East Germany. In 1980, the government-owned recycling centers increased the price paid for recycled newspapers by a factor of three, while payment for glass containers was multiplied by six. This economic incentive dramatically increased the recycling of wastes.

Although the programs outlined above are an encouraging start, a continuing acceleration of this effort is essential.

5. Conserving Energy

In the 1950s and 1960s one common economic viewpoint held that economic growth was dependent on high energy consumption. Today people have realized that energy is expensive not only in direct costs but also in the form of economic externalities related to the drain of financial reserves needed to pay for imported fuels. Moreover, it has been shown that economic growth can be maintained in a conservation-minded society as well as in a wasteful one.

Conservation is attractive because some of its effects can be felt immediately. There is no need to make capital investments or to wait for new technology to form car pools, drive fewer miles, turn the thermostat down, or recycle beverage containers. Other approaches to conservation employ technologies that are readily available but require some time and capital to

implement. Examples include use of more efficient appliances and automobiles, installation of more insulation and weather stripping in buildings, and relocation of industries to reduce commuting distances.

People in less-developed countries use far less energy than those in developed regions, but appropriate technology can conserve energy here, too. For example, although fuel is woefully scarce in many regions, inefficient open fireplaces waste as much as 90 percent of the energy used for cooking. Construction of small cookstoves could save half the firewood or dung now used.

6. Developing Renewable Energy

Eventually a sustainable society must be based on renewable energy sources or nuclear power because fossil fuel supplies are limited. No one expects this transition to be rapid, because far-reaching changes in the structure of society are required. The world consumption of energy, by sources, is listed in Table E–1. Predictions of future energy use made by the Worldwatch Institute are summarized in Table E–2. This table shows that the use of renewable energy should rise from 16 percent of the total in 1980 to 24 percent by the year 2000. At the same time, this report predicts that the impact of nuclear energy will continue to be small relative to energy from renewable sources. In 1980, fission

Table E-1 World Energy Consumption in 1980

Source	Percentage of Total
Coal	36
Petroleum	29
Natural gas	16
Wood	9.2
Hydroelectric power	5.5
Nuclear power	2.2
Crop residues	0.9
Cow dung	0.5
Geothermal power	0.1
Waste steam	0.1
Methane generated from wastes	0.04
Wind	0.03
Energy crops	0.03
Solar collectors	0.01

Table E-2   World Energy Consumption in 1980 with Projections to the Year 2000

Year	Fossil Fuels (%)	Nuclear (%)	Renewable (%)	Total Consumption (millions of tonnes of coal equivalent)
1980	81	2.2	16	10,914
1990	77	4.8	18	13,321
2000	71	4.8	24	15,217

reactors accounted for only 2.2 percent of the global energy consumption, and this proportion is expected to rise only to 4.8 percent at the turn of the century.

The mode of transition to a sustainable energy supply will vary from region to region depending on the resources available. Norway and Sweden are expected to develop hydroelectric and wind resources; China has already been active in the construction of small-scale hydroelectric dams; energy crops and logging in Brazil will be important; and people in the United States are expected to use ever-increasing quantities of energy from wind, solar sources, and firewood.

In conclusion, the authors of *Six Steps to a Sustainable Society* note:

> The transition to a sustainable society will challenge the capacity of countries everywhere to change and adapt. Some adjustments will occur in response to market forces, some in response to public policy changes, and still others as a result of voluntary changes in lifestyles. In order to take the necessary steps, all nations will have to make major financial commitments as soon as possible. . . .
>
> Taking part in the creation of a sustainable society will be an extraordinarily challenging and satisfying experience, enriched by a sense of excitement that our immediate forebearers who built fossil-fuel-based societies did not have. The excitement comes from both the vast scale of the undertaking and the full knowledge of the consequences of failure.

Suggested Readings

Lester R. Brown and Pamela Shaw: *Six Steps to a Sustainable Society*. Worldwatch Paper 48. Washington D.C., Worldwatch Institute, March, 1982. 63 pp.

A book that relates environmentalist movements to patterns of social change is:
Stephen Cotgrove: *Catastrophe or Cornucopia: The Environment, Politics and the Future*. Somerset, NJ, John Wiley & Sons, 1982. 176 pp.

Another book that deals with social policies in environmental issues is:
Albert H. Teich and Ray Thornton (Eds.): *Science, Technology, and the Issues of the Eighties: Policy Outlook*. Boulder, CO, Westview Press, 1982. 290 pp.

The following books and article take the position that the environmental movement has exaggerated the dangers of pollution, population, and economic development:
William Tucker: *Progress and Privilege: America in the Age of Environmentalism*. New York, Doubleday & Co., 1982. 314 pp.
Mary Douglas and Aaron Wildavsky: *Risk and Culture: An Essay on the Selection of Technical and Environmental Dangers*. Berkeley, University of California Press, 1982. 221 pp.
Alvin M. Weinberg: Avoiding the Entropy Trap. *Bulletin of Atomic Scientists*, October, 1982, pp. 32–35.

Physical Concepts and Units of Measurement

FACTOR	PREFIX	SYMBOL	EXAMPLE
10^{20}	geo*		geogram = 10^{20} grams
10^{9}	giga	G	gigajoule, GJ = 1 billion joules
10^{6}	mega	M	megawatt, MW = 1 million watts
10^{3}	kilo	k	kilogram, kg = 1000 grams
10^{2}	hecto	h	hectare, ha = 100 ares (measure of area)
10^{-1}	deci	d	decibel, dB = $\frac{1}{10}$ Bel (measure of sound)
10^{-2}	centi	c	centimeter, cm = $\frac{1}{100}$ meter
10^{-3}	milli	m	millimeter, mm = $\frac{1}{1000}$ meter
10^{-6}	micro	μ	micrometer μm = a millionth of a meter

*Not an official SI prefix.

The International System of Units (SI) defines various units of measurement as well as prefixes for multiplying or dividing the units by decimal factors. The prefixes used in this book are as shown in the table above.

- **Time.** The SI unit is the **second, s** or sec, which used to be based on the rotation of the Earth but is now related to the vibration of atoms of cesium-133. SI prefixes are used for fractions of a second (such as milliseconds or microseconds), but the common words **minutes, hours,** and **days** are still used to express multiples of seconds.

- **Length.** The SI unit is the **meter,** m, which used to be based on a standard platinum bar but is now defined in terms of wavelengths of light. The closest English equivalent is the **yard** (0.914 m). A **mile** is 1.61 kilometers (km). An inch is exactly 2.54 centimeters (cm)* A **nautical mile** is 1.15 miles, or 1.85 km. A **knot** is the speed of a nautical mile per hour, or 1.15 miles per hour.

The SI rules specify that its symbols are not followed by periods nor are they changed in the

---

* All relationships between units of the same system are exact. Those between different systems are approximate, except when otherwise noted.

503

640 acres = 1 sq. mile

plural. Thus, it is correct to write "The tree is 10 m high" not "10 m. high" or "10 ms high."

- **Area** is length squared, as in **square meter, square foot,** and so on. The SI unit of area is the **are,** a, which is 100 square m. More commonly used is the **hectare, ha,** which is 100 ares, or a square that is 100 m on each side. (The length of a U.S. football field plus one end zone is just about 100 m.) A hectare is 2.47 acres. An **acre** is 43,560 square feet, which is a plot of, say, 220 ft by 198 ft.

- Volume is length cubed, as in **cubic centimeter,** cm^3, **cubic foot,** ft^3, and so on. The SI unit is the **liter,** L, which is 1000 cm^3. A **quart** is 0.946 L; a U.S. liquid **gallon** (gal) is 3.785 L. A **barrel** of petroleum (U.S.) is 42 gal, or 159 L. A **drum** (not a standard unit of volume) of the type generally used by the chemical industry and often found in chemical waste dumps is 55 gal, or 208 L.

- **Mass** is the quantity that is popularly called **weight.** (Weight depends on gravity; an astronaut in space has no weight but still has mass. On Earth the two terms are often used interchangeably.) The SI unit is the **kilogram,** kg, which is based on a standard platinum mass. A **pound** (avdp), lb, is 0.454 kg. A **metric ton,** also written as **tonne,** is 1000 kg, or about 2205 lb. In the English system, a **short ton** is 2000 lb and a **long ton** is 2240 lb. A tonne is therefore between the two English tons but closer to the long ton. This book always uses the term "tonne" because it is the metric unit.

- **Temperature.** The SI unit is the **kelvin** (K). In measuring differences in temperature, such as the rise in temperature from the melting point of ice to the boiling point of water, one kelvin is the same as 1 degree Celsius (°C). However, Celsius temperature (not temperature difference) is related to kelvin temperature as follows:

Celsius temperature (°C)
= kelvin temperature (K) − 273 K

Freezing point of water	Boiling point of water
0°C or 273 K	100°C or 373 K

difference = 100°C or 100 K

In describing very high temperatures, such as the millions of degrees in stars or nuclear reactions, the 273-degree difference between the two scales is too small to matter, so either the kelvin or the Celsius scale can be used. Fahrenheit temperature (°F) is not used in scientific writing, although it is still popular in English-speaking countries.

- **Concentration** is the quantity of a substance in a given volume of space or in a given quantity of some other substance. "Quantity" can be expressed in units of mass or volume, or even in molecules.

percent = parts per 100 parts

ppm = parts per million parts

ppb = parts per billion parts

Concentration is therefore always a ratio:

$$\frac{\text{quantity of substance X}}{\text{volume of space}}, \text{ or}$$

$$\frac{\text{quantity of substance X}}{\text{quantity of substance in which X is dispersed}}$$

For example, 1 kg of ocean water contains about 65 mg of the element bromine. The concentration of bromine in the ocean is therefore

$$\frac{65 \text{ mg bromine}}{\text{kg of ocean water}}$$

But a kilogram is 1 million milligrams, so we can write

$$\text{concentration} = \frac{65 \text{ mg bromine}}{1 \text{ million mg ocean water}}$$

When both quantities are expressed in the same units, the units can be dropped (they cancel out), and we can say simply that the concentration of bromine is 65 parts per million (ppm) by weight (or mass).

Quantities of gases are usually expressed in units of volume. For example, the concentration of carbon dioxide, CO_2, in air is 335 ppm by volume. Because the volume of a gas is proportional to the number of molecules it contains, the concentration of a gas "by volume" really means "by number of molecules." Thus, there are 335 molecules of CO_2 per million molecules of air. The concentration of dusts in air cannot be expressed in volume units, however, because dusts are not gases. The concentration of particulate matter in air is expressed in mass per unit volume. For example, the maximum exposure to the pesticide 2,4-D allowed by OSHA for an eight-hour work shift is 10 mg per cubic meter (10 mg/m^3) of air.

Concentrations expressed in parts per million or per billion by volume or mass seem quite small. In terms of molecules, however, such concentrations are very large. For example, if there is 10 ppb by weight of the pesticide DDD in water, each gram of water contains 20×10^{12}, or 20 trillion molecules of DDD. That concentration is high enough to kill some species of trout.

● **Energy** is a measure of work or heat, which were once thought to be different quantities. Hence two different sets of units were adopted and still persist, although we now know that work and heat are both forms of energy.

The SI unit of energy is the **joule,** J, the work needed to accelerate a 2-kg mass from a dead start to a speed of 1 m per second. In human terms, that is not much—it is about the amount of work required to lift a 100-g weight to a height of 1 m. Therefore, joule units are too small for discussions of machines, power plants, or energy policy. Larger units are

megajoule, MJ =
10^6 J (a day's work by one person)
gigajoule, GJ =
10^9 J (energy in half a tank of gasoline)

Another unit of energy, used for electrical work, is the **watt hour,** Wh, which is 3600 J. A **kilowatt hour,** kWh, is 3.6 MJ.

The energy unit used for heat is the **calorie,** cal, which is exactly 4.184 J. One calorie is just enough energy to warm 1 g of water 1°C. The more common unit used in measuring food energy is the **kilocalorie,** kcal, which is 1000 cal. When **Calorie** is spelled with a capital C, it means kcal. If a cookbook says that a jelly doughnut has 185 calories, that is an error—it should say 185 Calories (capital C), or 185 kcal.

A value of 185 calories (small c) would be the energy in about one quarter of a thin slice of cucumber.

The unit of energy in the British system is the **British thermal unit,** Btu, which is the energy needed to warm 1 lb of water 1°F.

1 Btu = 1054 J = 1.054 kJ = 252 cal

The unit often referred to in discussions of national energy policies is the **quad,** which is a quadrillion Btu, or 10^{15} Btu.

Some approximate energy values are

1 barrel (42 gal) of petroleum = 5900 MJ

1 tonne of coal = 29,000 MJ

1 quad = 170 million barrels of oil,
or 34 million tonnes of coal

● **Efficiency of energy use**

Imagine that you want to heat 1 kg of water from, say, 10°C to 90°C by using the heat from a fuel such as gasoline. The heat that must be absorbed by the water to reach the required temperature is 80 kcal, but let us say that you actually have to use 100 kcal to do the job. Since energy cannot be destroyed, where did the other 20 kcal go? Not into the water, obviously. The pot that contained the water absorbed some heat and so did the air in the room. The efficiency of your operation, therefore, was

Efficiency
(1st law) =

$$\frac{\text{energy absorbed by the desired process}}{\text{total energy actually supplied}} \times 100\%$$

$$= \frac{80 \text{ kcal}}{100 \text{ kcal}} \times 100\% = 80\%$$

The assumption that the energy not absorbed by the water was merely wasted, but not destroyed, is really a statement of the first law of thermodynamics. This is the reason why the efficiency described by the above equation is called the **first law efficiency.**

Now imagine that you take the same amount of energy (100 kcal) and use it to run a heat engine, such as a gasoline engine, to make the engine work. The engine extracts the heat from the fuel and converts it to work. But heat can be extracted from a body only by flowing from a higher temperature to a lower temperature. (That is a statement of the second law of thermodynamics.) If the engine ran in an environment in which the outside temperature was absolute zero (zero kelvin, or $-273°C$), then the engine theoretically could convert all of the heat into work. But suppose, to pick convenient numbers, that the engine operated at 1000 K and the outside temperature was 500 K. The outside temperature would then be only halfway down from the operating temperature (1000 K) to absolute zero, and the engine could convert only half of its heat to work, yielding only 50 kcal of work. Anything better than that would be impossible; anything less would be inefficient. Therefore the measure of efficiency is

$$\text{Efficiency (2nd law)} = \frac{\text{the minimum amount of useful energy (work) needed at a given temperature to do a desired task}}{\text{the amount of useful energy or work actually supplied}} \times 100\%$$

The efficiency calculated in this way is called the **second law efficiency.**

- **Power** is a measure of energy per unit time. The SI unit is the **watt,** W, which is a joule per second. Other common units are the **kilowatt,** kW, which is 1000 watts, and the **megawatt,** MW, which is 1 million watts. The older English unit is the **horsepower,** which is about three quarters of a kilowatt. The watt rating of a light bulb is its power rating. A 100-watt bulb generates 100 watts of power when it is lit. Your electric bill is a charge for energy, not power—you pay for the wattage multiplied by the time. If the bulb is lit for a day, you pay for 100 watts $\times$ 24 hours, which is 2400 watt hours or 2.4 kilowatt hours. A factory that generates electricity is called a power plant, not an energy plant, because it is rated according to the power it can produce (usually expressed in megawatts).

Chemical Symbols, Formulas, and Equations

Atoms or elements are denoted by symbols of one or two letters, like H, U, W, Ba, and Zn.

Compounds or molecules are represented by formulas that consist of symbols and subscripts, sometimes with parentheses. The subscript denotes the number of atoms of the element represented by the symbol to which it is attached. Thus H_2SO_4 is a formula that represents a molecule of sulfuric acid, or the substance sulfuric acid. The molecule consists of two atoms of hydrogen, one atom of sulfur, and four atoms of oxygen. The substance consists of matter that is an aggregate of such molecules. The formula for oxygen gas is O_2; this tells us that the molecules consist of two atoms each.

Chemical transformations are represented by chemical equations, which tell us the molecules or substances that react and the ones that are produced, and the molecular ratios of these reactions. The equation for the burning of methane in oxygen to produce carbon dioxide and water is

$$CH_4 + 2O_2 \rightarrow CO_2 + 2H_2O$$

Each coefficient applies to the entire formula that follows it. Thus $2H_2O$ means $2(H_2O)$.

The atoms in a molecule are held together by chemical bonds. Chemical bonds can be characterized by their length, the angles they make with other bonds, and their strength (that is, how much energy would be needed to break them apart).

In general, substances whose molecules have strong chemical bonds are stable, because it is energetically unprofitable to break strong bonds apart and rearrange the atoms to form other, weaker bonds. Therefore, stable substances may be regarded as chemically self-satisfied; they have little energy to offer and are said to be energy-poor. Thus, water, with its strong H—O bonds, is not a fuel or a food. The bonds between carbon and oxygen in carbon dioxide, CO_2, are also strong (about 1.5 times as strong as the H—O bonds of water), and CO_2 is therefore also an energy-poor substance.

In contrast, the C—H bonds in methane, CH_4, are weaker than the H—O bonds of water. It is energetically profitable to break these bonds and produce the more stable ones in H_2O and CO_2. Methane is therefore an energy-rich substance and can be burned to heat houses and drive engines.

Acidity and the pH Scale

Hydrogen ions render water acidic. The original meaning of acid is "sour," referring to the taste of substances such as vinegar, lemon juice, unripe apples, and old milk. It has long been observed that all sour or acidic substances have some properties in common, notably their ability to corrode (rust, or oxidize) metals. When the attack on a metal by an acidic solution is vigorous, hydrogen gas (H_2) is evolved in the form of visible bubbles. Acidic solutions also conduct electricity, with evolution of hydrogen gas at the negative electrode (cathode). These circumstances imply that acid solutions are characterized by the presence of positive ions bearing hydrogen. A hydrogen ion, or proton, designated H^+, cannot exist as an independent entity in water because it is strongly attracted (in fact, chemically bonded) to the oxygen atom of the water molecule. The resulting hydrated proton is formulated as $H(H_2O)^+$, or H_3O^+. The simpler designation H^+ may therefore be regarded as an abbreviation.

Some slight transfer of protons occurs even in pure water.

$$H_2O + H_2O \leftrightarrows H_3O^+ + OH^-$$

Hydroxyl ions, OH^-, can neutralize H_3O^+ ions by reacting with them to produce water, as indicated by the arrow pointing left. The concentration of H_3O^+ and of OH^- in pure water at 25°C is 1.0×10^{-7} moles/liter.* This solution is said to be *neutral*, because the concentrations of the two ions are equal. When the hydrogen ion concentration is greater than 1.0×10^{-7} moles/ liter at 25°C, the solution is *acidic*. Hydrogen ion concentrations are usually expressed logarithmically as pH values, where pH = $- \log_{10}$ (hydrogen ion concentration).

---

* One mole of hydrogen atoms is the amount that weighs one gram.

Recall that the logarithm of a number "to the base 10" is simply the *number of times 10 is multiplied by itself* to give the number. When the number is a multiple of 10, its log is simply the number of zeros it contains. When the number is 1 divided by a multiple of 10, its log is *minus* the number of zeros in the denominator. Therefore, when the number expresses hydrogen ion concentration, the pH is *plus* the number of zeros in the denominator:

CONCENTRATION OF H$^+$ (MOLES/LITER)	pH
1/10	1 (acidic)
1/10,000,000	7 (neutral)
1/1,000,000,000	9 (basic)

Any pH less than 7 connotes acidity. The lower the pH of a body of water, the more prone it is to be corrosive and, thereby, to become polluted with metallic compounds. Any pH greater than 7 is basic.

Chemical Formulas of Various Substances

Organochlorides

Dichlorodiphenyltrichloroethane (DDT)

Dichlorodiphenyldichloroethane (DDD)

Aldrin

Lindane (Containing 99% γ-Isomer)

2,4-Dichlorophenoxyacetic Acid (2,4-D)

2,4,5-Trichlorophenoxyacetic Acid (2,4,5-T)

Organochlorides

Dioxin

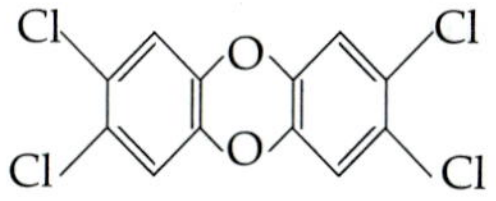

Polychlorinated Biphenyls
(PCBs)

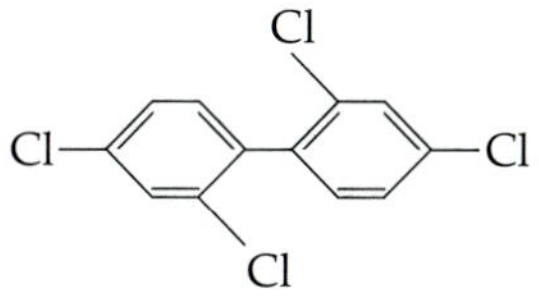

And Other Isomers

Organophosphate Pesticides

Parathion

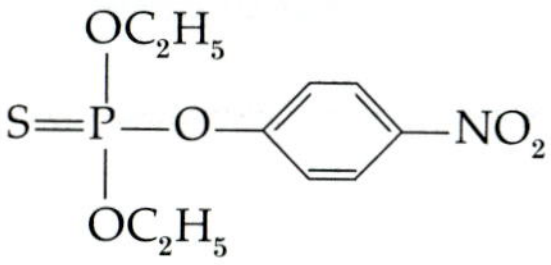

Organophosphate Pesticides

Malathion

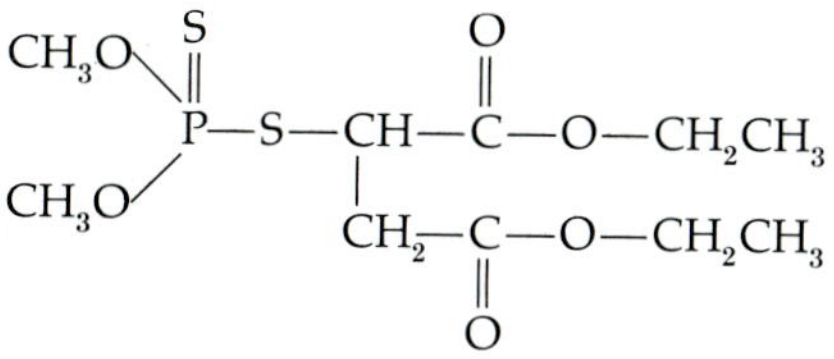

A Carbamate Pesticide

Sevin®
(1-Naphthyl-N-Methylcarbamate)

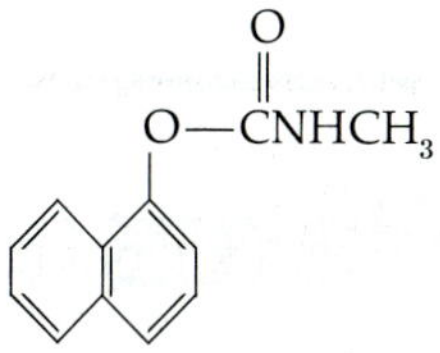

Growth Rates and Doubling Time

A **geometric** (or **exponential**) **growth rate** is calculated as follows: If there are x individuals in the zeroth generation and ax individuals in the first generation (where a is greater than 1), then there will be $a(ax)$ or a^2x in the second generation, a^3x in the third, and so on to a^nx in the nth generation.

GENERATION, n	NUMBER OF INDIVIDUALS ($x = 10$)	
	$a = 2$ (doubling)	$a = 3$ (tripling)
0	10	10
1	20	30
2	40	90
$\vdots$		
5	$2^5 \times 10 = 320$	$3^5 \times 10 = 2430$
$\vdots$		
n	$2^n \times 10$	$3^n \times 10$

In an **arithmetic** (or **linear**) **growth rate,** each generation increases by a constant amount, and there will be $x + na$ individuals in the nth generation.

GENERATION, n	NUMBER OF INDIVIDUALS ($x = 10$; $a = 2$)
0	10
1	12
2	14
$\vdots$	
n	$10 + 2n$

Geometric growth is always eventually much faster than arithmetic growth.

To compute doubling time from rate of growth, think of an analogy with compound interest at the bank. If a rate of growth (interest rate) is applied once a year to a population of size P_o (capital in the bank), the population (capital) at the end of one year is

$$P_1 = P_o + P_o r = P_o (1 + r)$$

More generally, if population growth is compounded n times each year, then the population in year t is

$$P_t = P_o(1 + r/n)^{nt}$$

512

It is reasonable to suppose that populations grow continuously or that $n \to \infty$. From elementary calculus,

$$\lim_{n \to \infty} (1 + r/n)^{nt} = e^{rt}$$

The doubling time, t_d, is then the solution of the equation

$$2P_o = P_o e^{rt_d}$$

Or, taking logarithms on both sides of the equation,

$$t_d = 0.693/r.$$

The Decibel Scale

Physicists have created a unit that defines a tenfold increase in sound intensity and names it a **Bel,** after Alexander Graham Bell. If the sound of a garbage disposal unit is 10 times as intense as that of a vacuum cleaner, it is one Bel more intense. A rocket whose sound is a million, or 10^6, times as intense as the vacuum cleaner is therefore 6 Bels more intense. The Bel is a rather large unit, so that it is convenient to divide it into tenths, or **decibels,** dB. A decibel is one tenth of a Bel. These definitions lead to the following relationship:

Difference in intensity
between two sounds,
X and Y, expressed in
decibels, dB

$$= 10 \log_{10} \left(\frac{\text{sound intensity of X}}{\text{sound intensity of Y}} \right)$$

Finally, it is convenient to start a scale somewhere that can be designated as zero. The most convenient zero decibel level is the softest sound level that is audible to the human ear. Then,

Intensity in decibels of any given sound
$= 10 \log_{10}$

$$\left(\frac{\text{intensity of the given sound}}{\text{intensity of a barely audible sound}} \right)$$

Example: The sound of a vacuum cleaner in a room has 10 million times the intensity of the faintest audible sound. What is the intensity of the sound in decibels?
   Answer:

$$\left(\frac{\text{sound intensity of vacuum cleaner}}{\text{faintest audible sound}} \right)$$
$$= 10,000,000$$

$\log 10,000,000 = 7$

Sound intensity $= 10 \times \log 10,000,000$
$= 10 \times 7$
$= 70$ decibels

Table of Atomic Weights
*(Based on Carbon-12)

	Symbol	Atomic No.	Atomic Weight		Symbol	Atomic No.	Atomic Weight
Actinium	Ac	89	227.0278	Europium	Eu	63	151.96
Aluminum	Al	13	26.98154	Fermium	Fm	100	[257]
Americium	Am	95	[243]**	Fluorine	F	9	18.998403
Antimony	Sb	51	121.75	Francium	Fr	87	[223]
Argon	Ar	18	39.948	Gadolinium	Gd	64	157.25
Arsenic	As	33	74.9216	Gallium	Ga	31	69.72
Astatine	At	85	[210]	Germanium	Ge	32	72.59
Barium	Ba	56	137.33	Gold	Au	79	196.9665
Berkelium	Bk	97	[247]	Hafnium	Hf	72	178.49
Beryllium	Be	4	9.01218	Helium	He	2	4.00260
Bismuth	Bi	83	208.9804	Holmium	Ho	67	164.9304
Boron	B	5	10.81	Hydrogen	H	1	1.0079
Bromine	Br	35	79.904	Indium	In	49	114.82
Cadmium	Cd	48	112.41	Iodine	I	53	126.9045
Calcium	Ca	20	40.08	Iridium	Ir	77	192.22
Californium	Cf	98	[251]	Iron	Fe	26	55.847
Carbon	C	6	12.011	Krypton	Kr	36	83.80
Cerium	Ce	58	140.12	Lanthanum	La	57	138.9055
Cesium	Cs	55	132.9054	Lawrencium	Lr	103	[260]
Chlorine	Cl	17	35.453	Lead	Pb	82	207.2
Chromium	Cr	24	51.996	Lithium	Li	3	6.941
Cobalt	Co	27	58.9332	Lutetium	Lu	71	174.967
Copper	Cu	29	63.546	Magnesium	Mg	12	24.305
Curium	Cm	96	[247]	Manganese	Mn	25	54.9380
Dysprosium	Dy	66	162.50	Mendelevium	Md	101	[258]
Einsteinium	Es	99	[252]	Mercury	Hg	80	200.59
Erbium	Er	68	167.26	Molybdenum	Mo	42	95.94

Table continued on next page.

	Symbol	Atomic No.	Atomic Weight		Symbol	Atomic No.	Atomic Weight
Neodymium	Nd	60	144.24	Scandium	Sc	21	44.9559
Neon	Ne	10	20.179	Selenium	Se	34	78.96
Neptunium	Np	93	237.0482	Silicon	Si	14	28.0855
Nickel	Ni	28	58.70	Silver	Ag	47	107.868
Niobium	Nb	41	92.9064	Sodium	Na	11	22.98977
Nitrogen	N	7	14.0067	Strontium	Sr	38	87.62
Nobelium	No	102	[259]	Sulfur	S	16	32.06
Osmium	Os	76	190.2	Tantalum	Ta	73	180.9479
Oxygen	O	8	15.9994	Technetium	Tc	43	[98]
Palladium	Pd	46	106.4	Tellurium	Te	52	127.60
Phosphorus	P	15	30.97376	Terbium	Tb	65	158.9254
Platinum	Pt	78	195.09	Thallium	Tl	81	204.37
Plutonium	Pu	94	[244]	Thorium	Th	90	232.0381
Polonium	Po	84	[209]	Thulium	Tm	69	168.9342
Potassium	K	19	39.0983	Tin	Sn	50	118.69
Praseodymium	Pr	59	140.9077	Titanium	Ti	22	47.90
Promethium	Pm	61	[145]	Tungsten	W	74	183.85
Protactinium	Pa	91	231.0359	Uranium	U	92	238.029
Radium	Ra	88	226.0254	Vanadium	V	23	50.9415
Radon	Rn	86	[222]	Xenon	Xe	54	131.30
Rhenium	Re	75	186.207	Ytterbium	Yb	70	173.04
Rhodium	Rh	45	102.9055	Yttrium	Y	39	88.9059
Rubidium	Rb	37	85.4678	Zinc	Zn	30	65.38
Ruthenium	Ru	44	101.07	Zirconium	Zr	40	91.22
Samarium	Sm	62	150.4				

*Atomic weights given here are 1977 IUPAC values.
**A value given in brackets denotes the mass number of the longest-lived or best-known isotope.

GLOSSARY

acid mine drainage—Mineral acids that may leach from mining operations to pollute ground and surface water supplies.

acid rain—A condition in which natural precipitation becomes acidic after reacting chemically with pollutants in the air.

activated sludge—See *Sludge*.

adaptation—The process of accommodation to change; the process by which the characteristics of an organism become suited to the environment in which the organism lives.

adsorption—The process by which molecules from a liquid or gaseous phase become concentrated on the surface of a solid.

aerobiosis (*adj.*, aerobic)—Bacterial decomposition in the presence of air.

aerosol—A substance consisting of small particles, typically having diameters that range from 1/100 µm to 1 µm. See also *Fume; Smoke*.

air pollution—The deterioration of the quality of air that results from the addition of impurities.

air quality standards—Maximum allowable outdoor concentrations of air pollutants.

albedo—A measure of the reflectivity of a surface, measured as the ratio of light reflected to light received. A mirror or bright snowy surface has a high albedo, whereas a rough, flat road surface has a low albedo.

altruism—Devotion to the interests of others.

anaerobiosis (*adj.*, anaerobic)—The biological utilization of nutrients in the absence of air.

aquaculture—The science and practice of raising fish in artificially controlled ponds or pools.

aquifer—An underground layer of rock that is porous and permeable enough to store significant quantities of water.

arithmetic growth—In population studies, growth characterized by the addition of a constant number of individuals during a unit interval of time. For instance, if there are x individuals in year 0 and x + *a* in year 1, arithmetic growth implies x + n*a* in year n.

artesian well—A well in which the groundwater has sufficient pressure to rise above the level of its aquifer.

atmosphere—The predominantly gaseous envelope that surrounds the Earth.

atom—The fundamental unit of the element.

atomic nucleus—The small positive central portion of the atom that contains its protons and neutrons.

atomic number—The number of protons in an atomic nucleus.

autotroph—An organism that obtains its energy from the Sun, as opposed to a *heterotroph*, which is an organism that obtains its energy from the tissue of other organisms. Most plants are autotrophs.

background radiation—The level of radiation on Earth from natural sources.

baleen—A set of elastic, horny plates that form a seivelike region in the mouths of certain whales.

517

Plankton-eating whales have no true teeth, only baleen.

Bel—Ten decibels.

benthic organism—A plant or animal that lives at or near the bottom of a body of water.

biochemical oxygen demand (BOD)—A measure of pollution of water by organic nutrients that recognizes the rate at which the nutrient matter uses up oxygen as well as the total quantity that can be consumed.

biodegradable—Refers to substances that can readily be decomposed by living organisms.

biogas—Gas, consisting mostly of methane, that is produced by fermentation of organic waste.

biomass—The total weight of all the living organisms in a given system.

biome—A group of ecosystems characterized by similar vegetation and climate and that are collectively recognizable as a single large community unit. Examples include the arctic tundra, the North American prairie, and the tropical rain forest.

biosphere—The part of the Earth and its atmosphere that can support life.

birth cohort—A group of individuals born in a given period of time, such as in a particular year.

birth rate—The number of individuals born during some time period, usually a year, divided by an appropriate population. For example, the crude birth rate in human populations is the number of live children born during a given year divided by the midyear population of that year.

black lung disease—A series of debilitating and often fatal diseases that affect the lungs of miners who work in underground coal mines.

bloom—A rapid and often unpredictable growth of a single species in an ecosystem.

boreal forest—See *Taiga*.

branching chain reaction—A chain reaction in which each step produces more than one succeeding step.

breakwater—See *Groin*.

breeder reactor—A nuclear reactor that produces more fissionable material than it consumes.

calorie—A unit of energy used to express quantities of heat. When calorie is spelled with a small c, it refers to the quantity of heat required to heat 1 g of water 1°C. (This definition is not precise, because the quantity depends slightly on the particular temperature range chosen.) When Calorie is spelled with a capital C, it means 1000 small calories, or 1 kilocalorie (kcal), the quantity of heat required to heat 1000 g (1 kg) of water 1°C. Food energies in nutrition are always expressed in Calories. The exact conversions are: 1 cal = 4.184 joules; 1 kcal = 4184 joules.

cancer—An abnormal, unorganized, and unregulated growth of cells in an organism.

capillary action—The movement of water upward against the force of gravity, through narrow tubes. The water is pulled upward by attractive forces between the water molecules and the sides of the tubes.

carnivore—An animal that eats other animals.

carrying capacity—The maximum number of individuals of a given species that can be supported by a particular environment.

census—A count of a population.

chain reaction—A reaction that proceeds in a series of steps, each step being made possible by the preceding one. See also *Branching chain reaction.*

chaparral—A biome characterized by a dry climate with little or no summer rain. Vegetation is dominated by shrubs that have adapted to regrow rapidly after fires, which occur frequently during the dry season.

chelating agent—A molecule that can offer two or more different chemical bonding sites to hold a metal ion in a clawlike linkage. The bonds between chelating agent and metal ion can be broken and reestablished reversibly.

chemical bond—A linkage that holds atoms together to form molecules.

chemical change—A transfer that results from making or breaking of chemical bonds.

chemical energy—The energy that is absorbed when chemical bonds are broken or released when they are formed.

China Syndrome—A facetious expression referring to a nuclear meltdown in which the hot radioactive mass melts its way into the ground toward China. Although a meltdown through the Earth to China is, of course, impossible, an accident in a nuclear power station may potentially lead to a situation where a hot radioactive mass melts its way through the containment structure into the Earth, contaminating neighboring environments and groundwater supplies.

clearcutting—The practice of cutting all the trees in a designated area. Clearcutting leaves behind barren, open regions but is sometimes the cheapest way to harvest timber if only short-term economic evaluation is used. See also *Selective cutting.*

climate—The composite pattern of weather conditions that can be expected in a given region. Climate refers to yearly cycles of temperature, wind, rainfall, and so on, not to daily variations.

climax community—A natural system that represents the end, or apex, of an ecological succession.

cogeneration—A tandem operation in which waste heat from one industrial process, such as the generation of electricity, is used in another process, such as oil refining. In general, such a system uses fuel more efficiently than would two facilities operating separately.

colloid—Material composed of minute particles, generally within a size range too small for gravitational settling but large enough to scatter light.

commensalism—A relationship in which one species benefits from an unaffected host.

common mode failure—An accident in which a single event causes multiple failures and thereby knocks out redundant systems. See *Redundancy*.

competition—An interaction in which two or more organisms try to gain control of a limited resource.

composting—The controlled, accelerated biodegradation of moist organic matter to form a humus-like product that can be used as a fertilizer or soil conditioner.

conservation of energy, law of—See *Thermodynamics*.

containment structure—A large vapor-proof, steel-lined, reinforced concrete structure built around a nuclear reactor and designed to contain all matter released inside it even if the largest primary piping system in the reactor were to rupture.

continental drift—The theory that the Earth's crust and the continents that lie on it are slowly moving in relation to one another.

control rod—A neutron-absorbing medium that controls the reaction rate in a nuclear reactor.

cooling pond—A lake or pond used to cool water from an electric generating station or any other industrial facility.

cooling tower—A large towerlike structure used to cool water from an electric generating station or any other industrial facility.

core (of the Earth)—The central portion of the Earth, believed to be composed mainly of iron and nickel. See also *Crust, Mantle*.

cost/benefit analysis—A system of analysis that attempts to weigh the benefit of a given level of pollution control against its economic cost.

critical condition—In nuclear science, a critical condition is said to exist when a chain reaction continues at a steady rate, neither accelerating nor slowing down. See also *Critical minimum size*.

critical mass (in a nuclear reaction)—The quantity of fissionable material just sufficient to maintain a nuclear chain reaction.

critical minimum size—In general, a critical condition relates to a point at which some property changes very abruptly in response to a small change in some other property of the system. In ecology, a population is said to be reduced to its critical minimum size when its numbers are so few that it is in acute danger of extinction.

crude rate—A vital rate with the entire population of some area as the denominator.

crust (of the Earth)—The solid outer layer of the Earth; the portion on which we live. See also *Mantle; Core*.

cyclone (for air pollution control)—An air cleaning device that removes dust particles by throwing them out of an air stream in a cyclonic motion.

death rate—The number of individuals dying during some time period, usually a year, divided by an appropriate population. For example, the crude death rate in human populations is the number of deaths during a given year divided by the mid-year population of that year.

decibel (dB)—A unit of sound intensity equal to one tenth of a Bel. The decibel scale is a logarithmic scale used in measuring sound intensities relative to the intensity of the faintest audible sound.

demographic transition—The pattern of change in vital rates typical of a developing society. The process can be outlined briefly as follows: Birth and death rates in preindustrial society are typically very high; consequently, population growth is very slow. The introduction, or development, of modern medicine causes a decline in death rates and hence a rapid increase in population growth. Finally, birth rates fall, and the population grows slowly once more.

demography—The branch of sociology or anthropology that deals with the statistical characteristics of human populations, with reference to total size, density, number of deaths, births, migrations, marriages, prevalence of disease, and so forth.

deoxyribonucleic acid—See *DNA*.

desert—The biome in which rainfall is less than 25 cm per year. These systems support relatively little plant or animal life.

detergent—A cleaning agent that acts by binding water molecules to molecules of grease or other soiling substances.

doldrums—A region of the Earth near the Equator in which hot, humid air is moving vertically upward, forming a vast low-pressure region. Local squalls and rainstorms are common, and steady winds are rare.

DNA—A substance consisting of large molecules that determines the synthesis of proteins and accounts for the continuity of species.

doubling time—The time a population takes to double in size, or the time it would take to double if its annual growth rate were to remain constant.

dust—An airborne substance that consists of solid particles typically having diameters greater than about 1 μm.

ecological niche—The description of the unique functions and habitats of an organism in an ecosystem.

ecological succession—See *Succession*.

ecology—The study of the interrelationships among plants and animals and the interactions between living organisms and their physical environment.

economic externality—That portion of the cost of a product that is not accounted for by the manufacturer but is borne by some other sector of society. An example is the cost of environmental degradation that results from a manufacturing operation.

ecosystem—A group of plants and animals occurring together plus their physical environment.

ecosystem homeostasis—The condition of an ecosystem by which negative feedback systems act automatically to maintain constancy.

electron—The fundamental atomic unit of negative electricity.

electrostatic precipitator—A device that electrically charges particulate air pollutants so that they drift to an electrically grounded wall from which they can be removed easily.

element—A substance in which all atoms have the same atomic number.

emission standards—Legal limits on the quantities of pollutants that are permitted to be discharged to the atmosphere from specific sources or processes.

energy—The capacity to perform work or to transfer heat. See also specific types of energy (e.g., *Geothermal energy, Heat, Solar energy*).

entropy—A thermodynamic measure of disorder. It has been observed that the entropy of an undisturbed system always increases during any spontaneous process—that is, the degree of disorder always increases.

environmental resistance—The sum of various pressures, such as predation, competition, adverse weather, and so on, that collectively inhibit the potential growth of every species.

epidemiology—The study of the distribution and determination of health and its disorders.

epilimnion—Upper waters of a lake.

estuary—A partially enclosed shallow body of water with access to the open sea and usually a supply of fresh water from the land. Estuaries are less salty than the open ocean but are affected by tides and, to a lesser extent, by wave action of the sea.

euphotic zone—The surface volume of water in the ocean or a deep lake that receives sufficient light to support photosynthesis.

eutrophication—The enrichment of a body of water with nutrients, with the consequent increase in growth of organisms.

evapotranspiration—The movement of liquid (usually water) by a combination of evaporation and transpiration.

expectation of life—The number of years an infant can be expected to live according to a specified schedule of death rates.

exponential growth—See *Geometric growth.*

externality—See *Economic externality.*

extrapolation—The prediction of points on a graph outside the range of observation.

evolution, theory of—A theory that states that species are not unchangeable but arise by descent and modification from pre-existing species.

fermentation—An anaerobic process by which certain microorganisms consume sugars, starches, or cellulose to produce various organic byproducts, particularly alcohols. In this manner, some low-quality organic wastes can be converted to useful fuels or animal feeds.

fire climax—An ecosystem whose continuance depends on periodic fires.

First Law of Thermodynamics—See *Thermodynamics.*

fission (of atomic nuclei)—The splitting of atomic nuclei into approximately equal fragments.

flocculation—The process by which colloidal particles are bound into larger aggregates by chemical agents.

food chain—An idealized pattern of flow of energy in a natural ecosystem. In the classic food chain, plants are eaten only by primary consumers, primary consumers are eaten only by secondary consumers, secondary consumers only by tertiary consumers, and so forth.

food web—The actual pattern of food consumption in a natural ecosystem. A given organism may obtain nourishment from many different trophic levels and thus give rise to a complex interwoven series of energy transfers.

Freon—A trade name of the Dupont Company that refers to the class of chlorofluorocarbons. The compounds that may be implicated in stratospheric pollution are Freon-11 ($CFCl_3$) and Freon-12 (CCl_2F_2).

fume—An aerosol that is usually produced by condensation of hot vapors, especially of metals.

fusion (of atomic nuclei)—The combination of nuclei of light elements (particularly hydrogen) to form heavier nuclei.

Gaia—The ancient Greek goddess of the Earth. This word has recently been used to describe the biosphere and to emphasize the interdependence of the Earth's ecosystems by likening the entire biosphere to a single living organism.

gas—A state of matter that consists of molecules that are moving independently of each other in random patterns.

gaseous diffusion—The movement of a gas in space by the random motions of its molecules. Lighter gases diffuse faster than heavier ones.

gasohol—A motor fuel consisting of approximately 90 percent gasoline and 10 percent ethyl alcohol.

gene pool—The aggregate of all genes in an interbreeding community.

genetic engineering—The process of artificially removing specific genes from one organism and replacing them with genetic material from another.

geometric growth—In population studies, growth such that in each unit of time, the population increases by a constant proportion. Also called *Exponential growth.*

geothermal energy—Energy derived from the heat of the Earth's interior.

germ cells—Cells involved in reproduction. See also *Somatic cells.*

GRAS list—A list compiled by the FDA of food additives that are Generally Recognized As Safe.

Green Revolution—The realization of increased crop yields resulting from the development of new high-yielding strains of wheat, rice, and other grains in the 1960s.

greenhouse effect—The effect produced by certain gases, such as carbon dioxide or water vapor, that cause a warming of the Earth's atmosphere by absorption of infrared radiation. The term is an inappropriate analogy to greenhouses, which were once thought to keep themselves warm by admitting sunlight but retaining infrared radiation. However, it has been shown that most of the heat retention in greenhouses results from the conservation of warm air. The atmosphere, of course, is open, so the mechanism is not the same.

groin—A stone or concrete structure built perpendicular to a beach to interrupt coastal currents and trap sand in a local area. Groins impede the normal flow of sand along a shore.

gross national product (GNP)—The total value of all goods and services produced by the economy in a given year.

half-life (of a radioactive substance)—The time required for half of a sample of radioactive matter to decompose.

hard energy system—An energy system based on large centralized electric generating facilities. See *Soft energy system.*

heat—A form of energy. Every object contains heat energy in an amount that depends on its mass, its temperature, and the specific heat of the materials of which it consists. Heat can also be generated by chemical reactions such as combustion.

heat engine—A mechanical device that converts heat to work.

heat pump—A mechanical device that uses an outside source of energy to force a separation of heat into a cool reservoir and a warmer one.

heavy oil—Natural petroleum deposits that are too thick to be extracted by conventional pumping techniques. Heavy oils are found in "dry" conventional wells as well as in oil shales and tar sands.

hectare—A metric measure of surface area. One hectare is equal to 10,000 square meters or 2.47 acres.

herbicide—A chemical used to kill unwanted plants.

herbivore—An animal that eats only plant matter.

heterotroph—An organism that obtains its energy by consuming the tissue of other organisms.

home range—The area in which an animal generally travels and gathers its foods.

hormone inhibitors—A class of compounds that block the action of hormones. These compounds can be used as insecticides, because if sprayed properly, they will disrupt the life cycles of specific insect larvae and kill them.

humus—The complex mixture of decayed organic matter that is an integral part of healthy soil.

hunter-gatherers—People who obtain their food by collecting it from the wild rather than by cultivating plants or domesticating animals.

hydroelectric power—Power derived from the energy of falling water.

hydrologic cycle (water cycle)—The cycling of water, in all its forms, on the Earth.

hypolimnion—The lower levels of water in a lake or pond that remain at a constant temperature during the summer months.

imbibition—A process by which water is tightly bound to soil particles by chemical attractions.

industrial melanism—The shift in color from light to dark of moths inhabiting areas in which the surfaces of trees and other objects have been darkened by industrial pollution.

injunction—A legal order requiring a defendant to stop doing a wrongful act.

innate capacity for increase—The maximum rate of growth of a population as determined by the reproductive capacity of the organism.

intertidal zone—The region along the ocean seacoast that lies between the high and low tide marks. See also *Sublittoral zone.*

inversion—A meteorological condition in which the lower layers of air are cooler than those at higher altitudes. This cool air remains relatively stagnant and causes a concentration of air pollutants and unhealthy conditions in congested urban regions.

ion—An electrically charged atom or group of atoms.

isotopes—Atoms of the same element that have different mass numbers.

joule—A fundamental unit of energy; 4.184 joules = 1 calorie.

juvenile hormone—A chemical naturally secreted by an insect while it is a larva. When the flow of juvenile hormone stops, the insect metamorphoses to become an adult. These compounds can be used as insecticides because, if sprayed at critical times, they will interrupt the natural metamorphoses and eventually kill specific insect pests.

kinetic energy—Energy of motion.

krill—Small, shrimplike crustaceans that grow in large numbers in the cold waters of the southern oceans. Krill represent a primary food supply for many species of whales.

laterite—A soil type found in humid tropical regions that contains a large proportion of aluminum and iron oxides and only a thin surface layer of organic matter. Lateritic soils cannot support sustained agriculture.

lava—The material produced when magma pours onto the surface of the Earth rapidly through fissures in the crust. A site where lava appears is called a volcano.

leaching—The extraction by water of the soluble components of a mass of material. In soil chemistry, leaching refers to the loss of surface nutrients by their percolation downward below the root zone.

legal standing—The set of requirements that must be met before a plaintiff can pursue a case in court.

legume—Any plant of the family *Leguminosae,* such as peas, beans, or alfalfa. Bacteria living on the roots of legumes change atmospheric nitrogen, N_2, to nitrogen-containing salts, which can be readily assimilated by most plants.

littoral zone—Intertidal zone.

macroeconomic effect—An economic force on the entire society, such as change in total production, general price level, employment, or economic growth.

magma—A fluid material lying in the upper layers of the Earth's mantle, consisting of melted rock mixed with various gases such as steam and hydrogen sulfide.

magnetohydrodynamic generator (MHD)—A type of electrical generator that operates by passing ions through a magnetic field. MHD systems are more efficient than conventional mechanical generators because there are fewer moving parts and hence fewer frictional losses.

mantle—The solid but partly semiplastic portion of the Earth that surrounds the central core and lies under the crustal surface.

mass number—The sum of the number of protons and neutrons in an atomic nucleus.

megalopolis—An agglomeration of several smaller cities and towns into a single, expanded, urban area.

meteorology—The science of the Earth's atmosphere.

metric system—See *Système International d'Unitès.*

MHD generator—See *Magnetohydrodynamic generator.*

microeconomic effect—An economic effect on an individual family, farm, or industry.

mine spoil—The earth and rock removed from a mine and discarded because the mineral or fuel content is too low to warrant extraction.

mineral reserves—The estimated supply of ore in the ground.

mist—An airborne substance that consists of liquid droplets typically having diameters greater than about 1 μm.

moderator—A medium used in a nuclear reactor to slow down neutrons.

molecule—The fundamental particle that characterizes a compound. It consists of a group of atoms held together by chemical bonds.

mortality factors—Factors that lead to mortality of a population. *Density-dependent mortality factors* are those that kill a larger proportion of individuals as the population density increases. *Density-independent mortality factors* are those that kill a constant proportion of the population, regardless of its density.

mutagen—A substance that causes mutations.

mutation—An inheritable change in the genetic material of an individual.

mutualism—An interaction beneficial to both interacting species.

natural selection—A series of events occurring in natural ecosystems that eliminates some members of a population and spares individuals endowed with certain characteristics that are favorable for reproduction.

nekton—The collective name for larger aquatic animals that are powerful enough to swim independently of water currents. Fish, sharks, and marine mammals are all types of nekton.

neutron—A fundamental particle of the atom that is electrically neutral.

niche—See *Ecological niche.*

nodules—Small pieces of rock, typically having a mass of about 1 kg, that are found on the floor of the ocean. These nodules are rich in manganese and also contain several other minerals such as copper, iron, aluminum, and cobalt.

noise—Unwanted sound.

nuisance—In law, the "substantial, unreasonable interference with the reasonable use and enjoyment of property."

nuclear reactor—A device that utilizes nuclear reactions to produce useful energy.

nucleus—See *Atomic nucleus.*

ocean thermal power—Electrical power generated from the temperature difference between the surface of the ocean and the cooler underwater layers.

oligotrophic—Pertaining to a body of fresh water that contains few nutrients and few living organisms.

omnivore—An organism that eats both plant and animal tissue. Common omnivores include bears, pigs, rats, chickens, and humans.

open dump—A site where solid waste is deposited on a land surface with little or no treatment.

ore—A rock mixture that contains enough valuable minerals to be mined profitably with currently available technology.

organic farming—A set of farming practices in which no synthetic fertilizers or chemicals are used.

organochlorides—A class of organic chemicals that contain chlorine bonded within the molecule. Some organochlorides, such as DDT, are effective pesticides. They are generally broad-spectrum and are long-lived in the environment.

organophosphates—A class of organic compounds that contain phosphorus and oxygen bonded within the molecule. Some organophosphates, used as pesticides, are broad-spectrum and extremely poisonous, although they are not long-lasting in the environment.

osmosis—The process by which water passes through a membrane that is impermeable to dissolved matter. The water passes from the less concentrated to the more concentrated solution. See also *Reverse osmosis.*

oxidant—An oxidizing agent. Oxidants in polluted air typically contain O–O chemical linkages.

oxidation—The addition of oxygen to a substance. More generally, oxidation is a loss of electrons.

ozone—Triatomic oxygen, O_3.

parasitism—A special case of predation in which the predator is much smaller than the victim and obtains its nourishment by consuming the tissue or food supply of a larger living organism.

PCB—See *Polychlorinated biphenyl.*

permafrost—A layer of permanent ice or frozen soil lying beneath the surface of the land in the arctic or tundra.

perpetual motion machine—A machine that will run forever and perform work without the use of an external energy supply. Such a machine is impossible to build.

pheromone—A substance secreted to the environment by one individual and received by a second individual of the same species. Reception of the signal releases specific behavioral reactions such as alarm, sex attraction, and trailing. When pheromones are discharged artificially into the environment, they lead to inappropriate behavior in the organisms that respond to them.

photosynthesis—The process by which chlorophyll-bearing plants use energy from the Sun to convert carbon dioxide and water to sugars.

phytoplankton—Any microscopic, or nearly microscopic, free-floating autotrophic plant in a body of water. A great many different species usually exist in a community of phytoplankton; these plants occur in large numbers and account for most of the primary production in deep bodies of water.

plankton—Any small, free-floating organism living in a body of water. See *Phytoplankton* and *Zooplankton.*

plasma—A gas at such a high temperature or pressure that the electrons have been stripped from their atoms, resulting in a mixture of nuclei surrounded by rapidly moving electrons.

pollution—The impairment of the quality of some portion of the environment by the addition of harmful impurities.

pollution tax—A tax on a polluter that is determined by the quantity of pollutants emitted. Also called *Residual charge.*

polychlorinated biphenyl (PCB)—A class of organochloride chemicals that are structurally related to DDT. They were used widely in the plastics and electrical industries until they were found to be potent environmental poisons.

population—The breeding group to which an organism belongs in practice. A population is generally very much smaller than an entire species, because all the members of a species are seldom in close proximity to each other.

population bloom—See *Bloom.*

population distribution—The composition of a population categorized by several variables, often age and sex.

population ecology—The branch of ecology dealing with the size, growth, and distribution of populations of organisms.

power—The amount of energy delivered in a given time interval.

$$\text{Power} = \frac{\text{energy}}{\text{time}}$$

prairie—Name given to grassland biome in North America. Rainfall is too low to support shrubs or trees.

predation—An interaction in which some individuals eat others.

predator—An animal that attacks, kills, and eats other animals. More broadly, an organism that eats other organisms.

pressure—Force per unit area.

primary consumer—An animal that eats plants.

primary treatment (of sewage)—The first stage of removal of impurities from water, generally by simple physical methods such as screening and settling.

productivity—The rate at which energy is stored as organic matter by photosynthesis. **Gross primary productivity** is the total organic matter fixed during photosynthesis. **Net primary productivity** is the plant matter produced that is not used during the plant's own respiration. **Secondary productivity** is the rate of formation of new organic matter by heterotrophs.

proton—A fundamental particle of the atom that bears a unit positive charge.

putrefaction—The anaerobic decomposition of proteins.

pyrolysis—The process by which a material is decomposed by heating it in the absence of air, sometimes yielding valuable products.

rad—a unit of radiation dosage, equivalent to the absorption of 100 ergs of energy per gram of biological tissue.

radioactivity—The emission of radiation by atomic nuclei.

radioisotope—A radioactive isotope.

rate of natural increase—The difference between the crude birth and crude death rates. Also called the *crude reproductive rate.*

rate of population growth—The increase in population during a given time period divided by the initial population.

recycling—The process whereby waste materials are reused for the manufacture of new materials and goods.

reduction—The removal of oxygen from a substance. More generally, reduction is a gain of electrons.

redundancy—Superabundance. In the context of safety systems, redundancy refers to the provision of a series of devices that duplicate each other's functions and that are programmed to go into operation in sequence if a preceding device in the series fails.

rendering—The cooking of animal wastes such as fat, bones, feathers, and blood to yield tallow (used in the manufacture of soap) and high-protein animal feed.

replacement level—The level of the total fertility rate, which, if continued unchanged for at least a generation, would result in an eventual population growth of zero.

residual charge—See *Pollution tax*.

respiration—The process by which plants and animals release energy by the stepwise oxidation of organic molecules.

resource—Any source of raw materials or means of producing raw materials.

reverse osmosis—The application of pressure to direct the flow of water through a membrane from the more concentrated to the less concentrated phase, which is the opposite of the normal osmotic flow.

saline seep—A process by which excess water leaches through the subsurface layers of soil, carrying dissolved salts. When this solution collects in a local depression in the land, the water may evaporate and the accumulated salts may destroy the fertility of the soil.

salt water intrusion—The movement of salt water from the ocean to terrestrial groundwater supplies that occurs when the water table in coastal areas is reduced.

sanitary landfill—A site where solid waste is deposited on a land surface, compacted, and covered with dirt to reduce odors and prevent disease and fire.

saprobe—An organism that absorbs food across its body surface, unlike animals that ingest food by mouth.

saprophyte—An organism, usually a plant such as a mold or a fungus, that consumes the tissue of dead plants or animals.

savanna—Name given in Africa to grassland biome with enough rainfall to support sparse shrubs and trees.

scrubber (in air pollution engineering)—A device that removes air pollutants by passing the polluted exhaust through a suitable solvent (usually water).

Second Law of Thermodynamics—See *Thermodynamics*.

secondary consumer—A predator that eats an animal that eats plants.

secondary treatment (of sewage)—The removal of impurities from water by the digestive action of various small organisms in the presence of air or oxygen.

selective cutting—The process of cutting selected trees in a stand of timber. See also *Clearcutting*.

SI system—See *Système International d'Unités*.

sigmoid curve—A mathematical function that is roughly S-shaped, characterized by an initially slow rate of increase, followed by a rapid growth and then by a second stage of slow, near zero rate of increase.

sludge—Wet residues removed from polluted water. When the sludge is laden with microorganisms that promote rapid decomposition, it is said to be activated.

smog—Smoky fog. The word is used loosely to describe visible air pollution.

smog precursor—An organic gaseous air pollutant that can undergo chemical reaction in the presence of sunlight to produce smog.

smoke—An aerosol that is usually produced by combustion or decomposition processes.

soft energy system—An energy system based on the use of small, home-sized, decentralized devices. See also *Hard energy system*.

soil-moisture belt—The layer of soil from which water can be drawn to the surface by capillary action.

solar cell—A semiconductor device that converts sunlight directly into electrical energy.

solar collector—A device designed to concentrate solar energy for a useful purpose.

solar design—active systems—Heating systems that use a controlled flow of some substance, usually air or water, to collect, store, and transmit heat from the Sun. **passive systems**—Construction systems that use the structure of a building itself to collect, store, and transmit heat from the Sun.

solar energy—Energy derived from the Sun.

somatic cells—All the cells in the body that are not involved in reproduction.

species—A group of organisms that interbreed with other members of the group but not with individuals outside the group.

strip mining—Any mining operation that operates by removing the surface layers of soil and rock, thereby exposing the deposits of ore to be removed.

sublittoral zone—The region along an ocean seacoast extending from the low tide outward to a depth of over 200 meters. These regions, lying above the continental shelves, are highly productive.

subsidence—The settling of the surface of the ground as coal, oil, or deep groundwater is removed.

subsidies—Economic incentives to persuade polluters not to despoil the environment.

succession—The sequence of changes through which an ecosystem passes during the course of time. **Primary succession** is a sequence that occurs when the terrain is initially lifeless, or almost so. **Secondary succession** is the series of community changes that takes place in disturbed areas that have not been totally stripped of their soil and vegetation.

survivorship curve—A function showing the patterns of mortality for a birth cohort of a given species.

synergism—A condition in which a whole effect is greater than the sum of its parts.

synfuels—An abbreviation for synthetic fuels—liquid fuels produced from oil shales, tar sands, or coal.

synthetic chemicals—Compounds produced by chemical reactions performed in a laboratory as opposed to those formed naturally.

Système International d'Unités (SI)—Commonly called the metric system. A system of measurement used in all scientific circles and by lay people in most nations of the world. The standard units in the SI are length, meter; mass, kilogram; time, second; electric current, ampere; temperature, kelvin or degree Celsius; luminous intensity, candela; amount of substance, mole.

taiga—The northern forest of coniferous trees that lies just south of the arctic tundra.

technological fix—An approach to solving a social problem, such as environmental degradation, by technical or engineering methods.

temperate forest—A biome that occurs in temperate regions with abundant rainfall.

tertiary, or "advanced," treatment (of sewage)—Any of a variety of special methods of water purification, such as adsorption or reverse osmosis, that are more effective than simple physical or biological processes for special pollutants.

thermal pollution—A change in the quality of an environment (usually an aquatic environment) caused by raising its temperature.

thermocline—Middle waters of a lake, where temperature and oxygen content fall off rapidly with depth.

thermodynamics—The science concerned with heat and work and the relationships between them. **First Law of Thermodynamics**—Energy cannot be created or destroyed. **Second Law of Thermodynamics**—It is impossible to derive mechanical work from any portion of matter by cooling it below the temperature of the coldest surrounding object.

thermonuclear reaction—A nuclear reaction, specifically fusion, initiated by a very high temperature.

threshold level—The minimum dose of a toxic substance that causes harmful effects.

tidal energy—Energy derived from the movement of the tides.

tort—A noncriminal action that results in personal injury or damage to property.

total dependency ratio—The ratio of the elderly plus the young to the total number of working-age people.

total fertility rate (TFR)—The total number of infants a woman can be expected to bear during the course of her life if birth rates remain constant for at least one generation.

toxic substance—Any substance whose physiological action is harmful to health.

transpiration—The evaporation of water vapor from the surface of leaf tissue.

trespass—An illegal intrusion or invasion of other people's property.

trophic levels—Levels of nourishment. A plant that obtains its energy directly from the Sun occupies the first trophic level and is called an autotroph. An organism that consumes the tissue of an autotroph occupies the second trophic level, and an organism that eats the organism that had eaten autotrophs occupies the third trophic level.

tropical rain forest—A biome that occurs when high, fairly constant rainfall and temperature permit plants to grow rapidly throughout the year.

tumor—An abnormal growth of unregulated cells.

tundra—Arctic or mountainous areas that are too cold to support trees and are characterized by low mosses and grasses.

turbine—A mechanical device consisting of fanlike blades mounted on a shaft. When water, steam, or air rushes past the blades, the shaft turns, and thus mechanical energy can be used to generate electricity.

upwelling—An ocean current that moves vertically upward, bringing nutrients to the surface. Such oceanic regions are characterized by high productivity.

urban sprawl—The growth of suburban areas and the expansion of cities to pre-empt increasingly large areas of land.

urbanization—A demographic process characterized by movement of people from rural to urban settlements, from small towns to large cities, and from large cities to their suburbs.

vein (of rock)—A thin layer of one type of rock embedded in a dominant rock formation.

vital event—In demography, a birth, a death, a marriage, a termination of marriage, or a migration.

vital rate—The number of vital events occurring in a population during a specified period of time divided by the size of the population.

water cycle—See *Hydrologic cycle*.

water pollution—The deterioration of the quality of water that results from the addition of impurities.

water table—The upper level of water in the zone of saturated subsurface soil and rock.

work—The energy expended when something is forced to move.

zero population growth—A condition in which the birth and death rates in a population are equal. Therefore the size of the population remains constant.

zooplankton—Microscopic or nearly microscopic free-floating aquatic animals that feed on other forms of plankton. Some zooplankton are larvae of larger animals, whereas others remain zooplankton during their entire life cycle.

Laboratoratory Manual
of Take-Home Experiments

The experiments in this section are designed to illustrate various principles in environmental science. The equipment and supplies needed for most of the experiments are available in the household. In some cases, additional material may have to be obtained from a hardware or hobby shop. Some of the experiments may also be used as classroom projects.

Most measuring devices in the United States are nonmetric. The following conversions (to the nearest g or mL) will therefore be helpful.

Volume

1 teaspoon	= 1/3 tablespoon	= 5 mL
1 tablespoon	= 1/16 cup	= 15 mL
1 cup	= 1/4 qt	= 237 mL
1 fluid ounce	= 30 mL	
1 mL of water	= 15 to 20 drops	

Mass

1 lb (avdp)	= 454 g
1 oz (avdp)	= 28 g
1 oz (troy)	= 31 g

Temperature (at normal atmospheric pressure)

	°C	°F
Arctic winter	−40	−40
Ordinary winter temperature	−10	14
Melting point of ice	0	32
Brisk autumn temperature	10	50
Normal room temperature	20	68
Summer day	30	86
Body temperature	37	98.6
Boiling point of water	100	212
Baking a cake	177	350
Baking fish	204	400

Some operations required by various experiments are described below.

Filtration. Laboratory filter paper is used as shown in Figure L–1. If such supplies are not available, a coffee filter can be used. As a last resort, a paper towel can be folded into a filter. Paper towels are more porous than laboratory filter paper and therefore will not retain fine particles. However, if the filtered liquid is recycled through the paper several

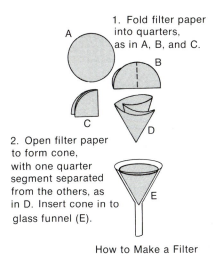

1. Fold filter paper into quarters, as in A, B, and C.

2. Open filter paper to form cone, with one quarter segment separated from the others, as in D. Insert cone in to glass funnel (E).

How to Make a Filter

Figure L–1 *How to make a filter.*

times, the pores will gradually fill up, and the filtering action will improve.

Evaporation. When water that contains dissolved solids, such as salts, is evaporated, the solids remain behind. Some of the experiments call for such a procedure. The easiest method is to place the liquid in a large, flat, open pan and set the assembly outdoors if the weather is sunny, hot, dry, and not too windy. Failing this, the pan can be set on a warm surface, such as the top of a radiator or wood stove, or on a hot plate or electric stove turned to the lowest point on the dial. If you use a glass dish, it must be heat-resistant. Even so, turn off the heat when evaporation is almost complete and let the liquid dry gradually on the residual heat. In no case should you continue to heat the dry residue, because the high temperature can decompose the solids you are isolating and also damage or break the dish.

Weighing. If a laboratory balance is not available, you will have to improvise one. You can use a small postal balance, a medicine dropper, and two small plastic coffee measuring cups (see Fig. L–2). Place one cup on each side of the balance. If they are not equal in weight, cut off a little piece of handle from the heavier one until they are equal. Now put a new penny in one cup and counterbalance it with water from your dropper in the other cup, counting the drops. (It will take approximately 50 drops, depending on the size of your dropper.) A new penny weighs about 3060 mg. Therefore, each drop of water from your dropper weighs

$$\frac{3060 \text{ mg/penny}}{\underline{\hspace{1cm}} \text{drops/penny}} = \underline{\hspace{1cm}} \text{mg/drop}$$

Now pour out the water, dry the wet cup, and remove the penny. To use your balance, put the material to be weighed in one cup, and counterbalance it with pennies in the other cup. When you have added one penny too many, remove the last penny and add drops of water until the weights are balanced. Then,

$$\begin{aligned} \text{weight of} \\ \text{sample} \end{aligned} = \left(\begin{array}{c} \text{number of} \\ \text{pennies} \end{array} \times 3060 \text{ mg} \right) + \left(\begin{array}{c} \text{number} \\ \text{of drops} \end{array} \times \underline{\hspace{1cm}} \text{mg} \right)$$

If you do not have any kind of balance available, you can make a crude one out of a 12-inch plastic ruler that has a small hole in the center, at the 6-inch mark. Tape or glue one small coffee measuring cup at each end of the ruler. Support the assembly with a sharpened pencil held vertically, with the eraser end down and its point in the center hole of the ruler. If the cups do not balance, cut enough of the handle from the heavier end until they do.

Ecosystems

EXPERIMENT 1   Soil Organisms

Obtain about a kilogram of fertile soil from a farm, woodland, or garden shop. Spread the soil carefully on a smooth piece of paper, and, using a magnifying glass, search for any living organisms. How many do you find? Draw pictures of them. If a micro-

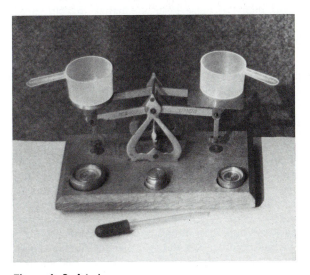

Figure L–2 *A balance.*

scope is available, place a small quantity of the soil on a slide and examine it. Can you see more organisms? What are they doing? How can the quantity and variety of life in your sample influence the quality of the soil?

EXPERIMENT 2 Interactions between Species

In this experiment you will observe and record one or more cases of competition and predation. The observations can be made in the field or in the laboratory. There is no limitation to the type of study that can be conducted. Two examples are given below, but you are encouraged to use your own imagination.

1. Set a bird feeder in a convenient location and keep the feeder well stocked with bread and seeds. Observe the behavior of the birds. Do some individuals chase others away? Do some species of birds dominate the feeder? Describe your observations. If possible, take photographs, and bring them to class.
2. Take a slow walk in the woods or in a park. Can you observe any instances where two plants appear to be competing for light, space, or water? Can you prove that competition is occurring, or would a further experiment be necessary? Can you observe the growth of any plant parasites? Are any insects present on the plants? Can you see the insects eating the plants or eating each other? Describe your observations. If possible, take photographs, and bring them to class.

EXPERIMENT 3 Foreign Species

Go to local farmers or nursery owners and ask them to list the major insect pests in your region. How many of these pests are native to your area? How many have been imported from foreign lands?

EXPERIMENT 4 Nutrient Cycling

Fill two small flower pots of equal size with sterile potting soil. Plant some rapid-growing rye grass seed in one of the pots, and water it as needed. Do not plant anything in the other pot, but keep it moist just the same. After the grass has grown, test each pot as follows: Seal the bottom opening with a piece of tape. Place the pot in a larger pan or bucket and set the assembly under the cold water faucet of your sink or bathtub. Run the water into the flower pot until about a liter overflows into the bucket. Now pour all the overflow liquid, including any soil it contains, from the bucket into a pan and evaporate the liquid to dryness. Weigh or estimate the amount of solid matter. Compare the solid matter from the two pots. What do

the results imply about the role of plant matter in protecting soil and recycling minerals in an ecosystem?

EXPERIMENT 5 Insect Pests

Go to your local garden supply store and make a list of all the insect control products that are for sale. How many products are chemical pesticides? How many utilize other methods? (A baited trap is not classified as a chemical pesticide.)

Demography and Patterns of Growth

EXPERIMENT 6 Population Growth

(a) Yeast.   Prepare and knead a bread dough according to the recipe in any standard cookbook. Remove ½ cup of the dough, and place it in a glass measuring cup. (A graduated cylinder would be even better; fill it one-third to one-half full with the dough.) Measure and record the volume of the dough. Now set the cup in a warm place (follow the directions in the cookbook), and record the volume of the dough at 10-minute intervals for two hours, then at half-hour intervals for an additional three hours. When the dough has almost doubled its volume, punch it down again and continue to measure the volume change. Now draw a graph of volume versus time. However, do not count the loss of volume that resulted from punching down the dough. The rising of the dough results from the carbon dioxide released by the yeast during its growth and is therefore a measure of growth. What is the shape of your graph? Interpret your results.

(b) Mold.   Allow the bulk of the dough to rise, then bake it in a greased pan at 350°F (177°C) for approximately 45 minutes. You now have a warm loaf of bread with no preservatives added. Eat most of the bread, but save one slice. Place this slice in an open dish in a quiet place in the room. Within a few days, mold will start to grow on the bread. In this part of the experiment you will measure the growth of the mold as a function of time.

Take a piece of wax paper and draw a series of horizontal and vertical lines on it about a centimeter apart. (Use a crayon.) You now have a grid with a series of squares that are each 1 cm on a side. When the mold first starts to form, place the grid over the bread and use it to estimate (1) the total surface area of the bread and (2) the surface area covered by mold. Remove the paper. Repeat this measurement once a day for 10 days. Draw a graph showing the growth of the mold population as a function of time. What type of function do you observe?

EXPERIMENT 7 Arithmetic vs Geometric (Exponential) Growth

This experiment can be done with your pocket calculator. To make it more interesting, two people, each with a calculator, can play against each other. Imagine that Player A (arithmetic growth) is given $1000 and can continue to gain as many thousands of dollars as can be added, $1000 at a time, to the calculator. Player B (geometric growth) is given only 1¢ (enter 0.01 in the calculator) but can double it repeatedly as fast as can be done on the calculator. Which player will have more money in 10 seconds? 20 seconds? 30 seconds?

EXPERIMENT 8 Demography

Using whatever reference sources are available, try to build a demographic history of your community for the past 50 years. Briefly outline the history of population changes in the community. What ethnic group or groups have migrated into or out of the area? How have population movements been related to patterns of employment?

Health

EXPERIMENT 9 Efficiency of a Cigarette Filter*

Find the smoked butt of a filter cigarette. Carefully slice off the filter, using a razor blade or sharp knife. Now slice the filter across in half, marking the tobacco end T and the mouth end M. Unwrap each half and estimate the relative shades of the brown colors as follows: Assume that the brown shade of the tobacco end is 100. Relative to this, estimate the shade of the mouth end. If you think it is three quarters as dark, call it 75, if it is 90 percent as dark, call it 90; if it is half as dark, call it 50, and so on. Better yet, ask a few of your friends to estimate the shades, but don't tell them what the T and M stand for. Now average the results. Calculate the efficiency of the filter, E, as follows:

$$E = \left(1 - \frac{M}{100}\right) \times 100\%$$

---

*The filter must be one of the type that is white before the cigarette is smoked. The usual material is cellulose.

Note that this calculation applies only to the efficiency of filtration of tarry particles whose color is visible. It does not apply to gases in cigarette smoke, such as carbon monoxide, which are not filtered at all.

EXPERIMENT 10 Food Additives

Make lists of ingredients of various packaged foods in your kitchen. Classify the function of each ingredient as preservation, nutritional supplement, colorant, flavorant, emulsifier (to disperse oil and water in each other so that they do not separate), thickener, or some other purpose. If the function is not as obvious as, say, vitamins for nutritional supplements, or is not stated on the label, see whether you can get information from some other source. The best single reference volume is the Merck Index, which lists some 10,000 chemicals alphabetically and gives their uses. It should be in your school or public library.

Energy

EXPERIMENT 11 Work and Heat

Take two rubber bands—each about 1 cm or more in width—and slip one inside the other to make a double thickness. Now, hold the double band so that it touches your lips lightly and you can feel its temperature. While the rubber band still brushes your lips, stretch it suddenly. Can you sense a rise in its temperature? Release it. Can you feel it cool down? When you stretch rubber, some of the work is converted to the potential energy of the stretched rubber and some is converted to heat. Feel a tire on an automobile or truck that has just been driven for several miles or a nail just as it is pried out of a board. Why are they warm?

EXPERIMENT 12 Electric Meter

Find the electric meter in your house or apartment. If you live in a private house, the meter is probably located on the outside wall, whereas if you live in an apartment, it is probably in the basement. As shown in Figure L–3, a meter has several small dials and a horizontal disc. The disc spins when electric energy is being used. The rate of spinning is proportional to the amount of electricity consumed. Thus, if many appliances are plugged in, the disc will spin rapidly, and if little electricity is being used in the house, the disc will barely move. Turn off all the electrical appliances in your house, and unplug the refrigerator and freezer. The disc should now be stationary. Plug in one 100-watt light bulb. How many seconds

Figure L–3 Household electric meter.

are required for the disc to make one complete revolution? Plug in a 50-watt light bulb and measure the time required for one revolution. Now plug in an appliance whose power requirement you do not know. Record the time required for one revolution of the disc. How much power is consumed by the appliance? (Don't forget to plug in your refrigerator and freezer when you finish this experiment.)

EXPERIMENT 13 Fuel for Cooking

Turn on your stove and wait until the burner reaches constant heat output. Pour 2 cups of cold water into a pan, cover it, and measure the time needed for the water to boil. Empty the pan, cool it, add another 2 cups of cold water, and measure the time needed for the water to boil with no cover on it. Compare your results.

EXPERIMENT 14 Conserving Electricity

Look at your last month's electric bill and record the number of kilowatt hours of electricity used at that time. Now initiate an energy conservation program in your home. Report on the conservation measures used. Look at the electric bill one month later, and record the amount of electricity saved. (Caution—if your house is heated or cooled electrically, there may be a large monthly variation in electric bills due to change of seasons. For example, less heat is needed

in March than in February. If the records are available, compare electric consumption for a month this year with the consumption for the same month last year.) Did your program save money? Discuss any inconveniences raised by the conservation practices.

EXPERIMENT 15 Conserving Energy

Using a hot air dryer of the type available in some public restrooms, measure the time needed to dry your hands without first shaking off the excess water, and then repeat the test after shaking off the excess water. Note the time you save and the wattage shown on the dryer. Calculate your savings as follows:

$$\text{Percent energy savings} = \frac{\text{Time without shaking water} - \text{Time with shaking water}}{\text{Time without shaking water}}(100\%)$$

$$\text{Total energy saved (watt-sec)} = \frac{\text{Time saved by shaking water (sec)}}{} \times \frac{\text{Wattage shown on dryer}}{}$$

A watt-second is a joule, and 4.185 joules = 1 calorie.

EXPERIMENT 16 Transportation

Obtain a map of your local community. Mark the locations of your residence, your school, the grocery store, post office, bank, and the five other stores that you visit most frequently. Measure the distance from your home to each of these locations. What type of transportation do you usually use to travel from one to another? Discuss the efficiency in terms of fuel consumption, time, and cost of the transportation system you use. Would it be difficult to improve the fuel efficiency of this system?

EXPERIMENT 17 Solar Collectors

Remove the inner dividers from each of four identical ice cube trays and fill each tray with the *same* amount of water, until it is about two-thirds full. (Between 1½ and 2 cups will do for the average-size tray. Use a measuring cup.) Set all four trays in the freezer compartment of a refrigerator until the water is frozen. Cover two of the trays with clear plastic wrap, sealed around the edges with string or a large rubber band. Replace the trays in the freezer and let them stay overnight. Next day, about midmorning, prop the four trays on a small box on a waterproof surface outdoors so that two of them (a covered one and an open one) slope toward the north and the other two slope south (Fig. L–4). Let them sit in the sun and note the rate at which they melt. Which design was the most efficient collector of solar energy? Why? Plan

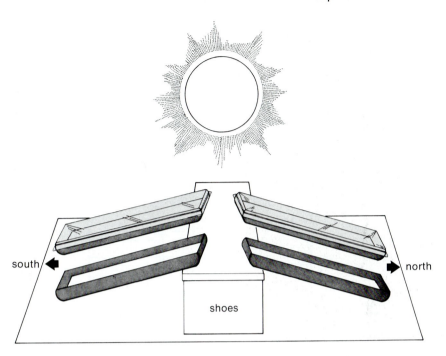

Figure L–4 Solar collector.

and carry out an experiment to test the effect of insulating the sides and bottom of your tray. (Use plastic foam or even dry crumpled newspaper as insulation.) Describe your results.

EXPERIMENT 18 Building a Solar Air Heater

Start with a cardboard box about 30 cm wide by 60 cm long by 10 cm deep. Remove the top flaps so that the box is open. Cut a hole about 3 cm in diameter at one end of the long side and another at the opposite end on the wide side, as shown in Figure L–5 *A*. Next, using additional cardboard and tape, construct a set of baffles and place the box upright, as shown in Figure L–5 *B*. Spray-paint the inside surfaces black and cover the open side with clear plastic. Set your solar heater in the sunlight and leave it for about half an hour. Measure the air temperature at the inlet port and at the outlet. How well does your heater work? Explain the purpose of the baffles, the black paint, the clear plastic.

EXPERIMENT 19 Radioactive Decay

You are not going to attempt any home-made bombs, but you can illustrate nuclear processes by other means. Shuffle a deck of cards. The exact number doesn't matter, but the more cards the better, so you may even combine two decks. However, count your cards before you start. Most card shuffling is far from thorough, so mix your cards at least a dozen

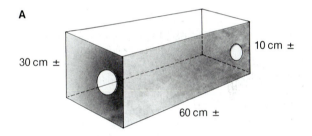

A

30 cm ±   10 cm ±

60 cm ±

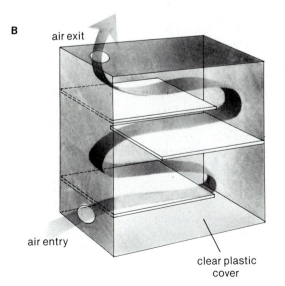

B

air exit

air entry

clear plastic cover

Figure L–5 Solar air heater.

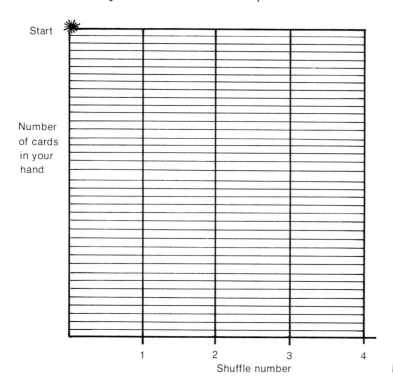

Start

Number
of cards
in your
hand

1   2   3   4

Shuffle number

Figure L–6 *Pattern of radioactive decay.*

times. Now turn the deck so that you see the faces of the cards. Fan them out, and discard any card *that follows a card of the same color*. If you do this correctly, you will be left with a set of cards of alternating colors—red, black, red, black, and so on. Count your remaining cards and record the number. Now shuffle your hand again (even *more* thoroughly than before) and repeat the process. Do this twice more, for a total of four shuffles. Now plot your results on a graph of the type shown in Figure L–6. Repeat the entire process as many times as you have patience for, so that there will be several points on the graph corresponding to each shuffle. Now draw a line that fits most of the points. This curve is the pattern of *first order decay*, showing a fall-off by "half-lives." Can you explain why the card procedure generated this type of curve?

Resources and Recycling

EXPERIMENT 20 Recycled Paper

In this experiment we will manufacture recycled paper from old newspaper. Cut a square of newspaper approximately 30 cm on a side and shred it into small pieces. Fill a large bowl one-quarter full of water, add the shredded paper, and let the mixture soak for an hour or two. Beat the mixture vigorously with an electric mixer or hand-operated egg beater until the paper breaks up into fibers and the mixture appears creamy and homogeneous. Next, dissolve two heaping tablespoons of starch or wallpaper paste in ½ L of warm water, add this solution to the creamy slurry, and stir. Take a piece of fine window screen and dip it into the solution. Lift the screen out horizontally, as shown in Figure L–7. If you have done this correctly, there will be a fine layer of paper fibers on the screen. Stir up the contents of the bowl and redip the screen carefully as many times as is needed to build up a layer about ¼ cm thick.

This layer must now be pressed and dried. To do this, place a cloth towel on a tabletop and lay the screen over the towel so that the fiber layer is facing upward. Cover the fibers with a piece of thin plastic

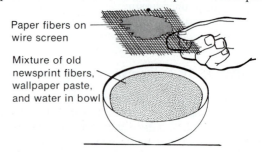

Paper fibers on wire screen

Mixture of old newsprint fibers, wallpaper paste, and water in bowl

Recycling: Making New Paper from Old

Figure L–7 *Recycling: making new paper from old.*

(a plastic bag is fine). Now squeeze down evenly on the plastic, using a block of wood for a press. Most of the water should be squeezed from between the fibers, through the screen, and on to the cloth towel. Set the screen out to dry for a day or two, peel off the newly fabricated recycled paper, and write a letter to a friend!

EXPERIMENT 21 Conserving Water

In this experiment you will measure how much water you can save in a shower. Using a stopwatch, measure the time required to take an ordinary shower. (Have someone outside the shower run the stopwatch so it doesn't get wet.) The next day, take a shower in the following manner: Turn the water on and wet yourself completely. Then turn the water *off* while you soap yourself. Turn the water back on to rinse. Measure the time that the water was running. What fraction (or percent) of the water used in an ordinary shower can be conserved? To determine the amount of water saved, measure the time needed for your shower to deliver 1 gal of water into a bucket. (First mark the 1-gal level in a bucket with a crayon.) Then,

$$\begin{array}{l} \text{Water saved} \\ \text{(gallons)} \end{array} = \dfrac{\begin{array}{c}\text{Time saved in the} \\ \text{conserving shower (sec)}\end{array}}{\begin{array}{c}\text{Time per gallon} \\ \text{in your shower (sec/gal)}\end{array}}$$

EXPERIMENT 22 Sand and Soil

For this experiment you will need a small shovelful of *dry* sand or sandy gravel. Mix one portion of the sand with an approximately equal weight of *dry* rotten leaves, coffee grounds, grass clippings, or other organic matter. Place about 100 g of the unmixed portion of sand in one small dish and 100 g of the sand-organic mixture in the other. Add about 20 mL of water to each, and then weigh both dishes accurately. Place them side by side on a table and weigh them every half-hour for several hours. Does the weight change? Why? Which sample loses water faster? Why? Discuss the implications of this experiment for agricultural practice.

EXPERIMENT 23 Soluble Minerals in Soil

Mix 25 to 30 g (about an ounce) of soil with enough rain water or distilled water so that you can pour the mixture into a filter. The filtered liquid should be clear. If it is not, simply pour it back through the same filter and let it run through again. Now evaporate all the water. You will notice a small residue in the bottom of your container. This residue is a mixture of various minerals that dissolve in water. For example, if the material is yellow-brown, iron may be present.

To make a further test, dissolve some washing soda (not bicarbonate of soda or baking soda; it must be washing soda, which is a form of sodium carbonate) in a little water. Dissolve your soil minerals in two or three drops of water and add a few drops of the carbonate solution. If salts of calcium, magnesium, or certain other metals are in the soil, the mixture will become cloudy.

EXPERIMENT 24 Dissolved Solids in Water

Use a clean jar to get a sample of water you wish to test. Filter the water if it contains undissolved solid matter. Then evaporate the filtered water to dryness. Test the following types of water samples: rain water or distilled water; tap water; water from a natural source, such as a river, stream, or pond; seawater, if available. Add to this list any other sample you consider to be of interest.

EXPERIMENT 25 Solid Waste Disposal

Sort through your waste and garbage containers at home. List the items that (1) need not have been purchased in the first place, (2) could have been reused, and (3) could be easily recycled. Weigh your garbage daily for a week. Then start a program to reduce the solid waste in your household and repeat the weighings. How much difference has your program made? How have your personal habits been changed? Locate the recycling centers in your area. Compile data on (1) the time required to prepare cans, bottles, and paper for recycling and to bring them to the center, and (2) the cost of transporting them.

EXPERIMENT 26 Iron and Its Oxides

For your experiments with iron, use unsoaped "steel" wool (really iron wool). If you cannot get any of the unsoaped variety in the grocery store, try the hardware store. You will also need a magnet and a tweezers or forceps.

Take a small wad (about the size of a grape) of iron wool in a forceps or tweezers and burn it in a gas flame on a stove or bunsen burner. Remove the burnt iron from the flame and examine it carefully. Note the small, rounded black globs. The rounding shows that the material melted in the flame and turned into droplets. The black product is the magnetic oxide of iron (Fe_3O_4), known as **magnetite.** The ancient name of this substance is **lodestone.** It was used as a compass magnet by sailors as early as the eleventh century. Use your magnet to demonstrate that these tiny black globs are magnetic. The melting point of both Fe and Fe_3O_4 is about 1500° C.

Now take another small portion of your iron wool and leave it in an open dish containing a little water until it is completely rusted. This will require some days or weeks. The product is the red oxide, Fe_2O_3. In mineral form it is known as **hematite.** Verify with your magnet that this oxide is not magnetic.

Both hematite and magnetite are important iron ores.

Environmental Disruption and Environmental Improvement

EXPERIMENT 27 Erosion

Find a place near your school where you can observe the effects of erosion. Describe what is happening, and, if possible, photograph the site. What agent (wind, ice, water, or other) is causing the breakdown and removal of material? Collect samples of the rock or soil as it appears before being disturbed, and then collect samples of the sediment that is being carried away. Compare the textures of the two materials. In your example is the erosion directly harmful to humans and their activities?

EXPERIMENT 28 Acids and Bases in Water

Acidic and basic impurities occur as pollutants in water. They can be detected conveniently by substances whose color depends on acidity or basicity; these substances are called **indicators.**

Litmus paper shows only whether a solution is acidic (the paper turns red) or basic (blue). Laboratory pH paper (see Appendix C for the pH scale) is more precise, ranging from deep blue (pH 10 or higher) through green (pH 8), all the way down to red (pH 2). Test the pH of tap water, rain water, club soda, lemon juice, vinegar, and solutions of mild soap, dishwasher soap, and bicarbonate of soda. Verify that bases and acids can neutralize each other.

You can also make your own indicator liquids from naturally colored foods or flowers. Purple grapes, red cabbage, or blueberries are among the best choices. Grind up one of these materials and extract the juice. This juice is your indicator. Place a few drops of your indicator in the acid solution and note the color. Repeat with the basic solution.

EXPERIMENT 29 Fermentation

Mix about 30 mL (about 1 fluid ounce) of molasses with about 250 mL (about a cup) of water and

about 1 g of dry yeast (about ⅛ of a packet) in a narrow-necked bottle. Close the bottle and shake up the mixture. Now remove the cap and cover the bottle with a piece of paper secured with a rubber band, and punch a few pinholes in the paper. DO NOT SEAL THE BOTTLE; PRESSURE WILL DEVELOP. Let the bottle stand in a moderately warm room for three days. Do you see any bubbles in the liquid? What is the gas? Remove the paper and pour some of the contents in a cup. Smell the liquid. What substance was produced?

EXPERIMENT 30 Automobile Exhaust

Go to your friendly neighborhood gasoline station and request the attendant's permission to let you ask a few questions. (Interview hint: Do not use a notebook, clipboard, or any other writing material. Know your questions by heart. Remember the answers—it takes a little practice but you can do it. When the interview is over, leave the gas station and only then should you fill out your questionnaire form. This informal method will be much more acceptable.) The questions are (1) Can you tell the difference between cars that take leaded gas and those that take non-leaded gas just by the smell of the exhaust? (If the answer is no, your interview is over. If the answer is yes, ask the second question.) (2) What do the two exhausts smell like? Remember the *descriptive* words that the attendant uses. Some of the probable chemical equivalents are

Like gas, Gasoline } Hydrocarbons

Pungent, Acidic, Tickling } Partially oxidized hydrocarbons

Cabbage, Skunk } Sulfur Compounds

What are your conclusions about the action of catalytic converters that are used on automobiles that require non-leaded gasoline?

EXPERIMENT 31 Observation of Smoke Shade

When fuel burns inefficiently, some of the unburned carbon particles are visible as a black or gray smoke. The density of such smoke coming from the stack of a factory or power plant can be estimated visually by comparing it with a series of printed grids. These grids, first suggested by Maximilian Ringelmann in 1898, are formed from squares of black lines on a white background, as follows:

RINGELMANN NO.	BLACK (%)	WHITE (%)
1	20	80
2	40	60
3	60	40
4	80	20

Select a suitable location for observing a smoke plume from a stack. Hold the chart in front of you and view the smoke while comparing it to the chart (Fig. L–8). The light shining on the chart should be the same as that shining on the smoke. For best results, the sun should be behind you.

Match the smoke with the corresponding Ringelmann smoke grid (1, 2, 3 or 4). Record your results and the time of your observation.

Calculate "observed smoke density" as follows:

Observed smoke density (percent) for a single observation

$$= \text{Ringelmann number} \times 20$$

Observed smoke density (average percent) for a number of observations

$$= \frac{\text{sum of all Ringelmann numbers} \times 20}{\text{number of observations}}$$

Check the air pollution regulations in your community and compare them with your findings.

EXPERIMENT 32 Dispersal of Gases in Air

Ask someone to pour a little fragrant liquid, such as cologne water or shaving lotion, into a saucer or other shallow open dish. The dish should be covered with plastic or aluminum foil and set in one corner of a quiet room. Now sit down in some other part of the room where you can keep busy with a quiet activity, such as reading a book. Ask your friend to remove the foil from the dish, while you note the time. Now read your book until you become aware of the smell, and note the time again. Repeat the experiment under various conditions, in different locations, perhaps with a fan blowing, and with different sources of odor. Note your findings with regard to the various factors that influence the dispersal of gases in air.

EXPERIMENT 33 Dustfall Measurement of Particulate Air Pollution

Get about three half-gallon wide-mouthed jars, such as restaurant-size mayonnaise jars, and wash them out. If you live in a hot, dry climate, fill the jars about one-fourth full of distilled water. In winter or rainy season, fill them only to a height of about ½ to 1 inch. If you expect freezing weather, use a 50-50 mixture of water and rubbing alcohol. (Distilled water may be available from your school, automobile service station, or drugstore. "Deionized" water will also do. You may use rainwater if you filter it first through a coffee filter or a paper towel.) Set the jars in *open* areas where you wish to measure dustfall, not under a tree or any part of a building. The jars should be elevated,

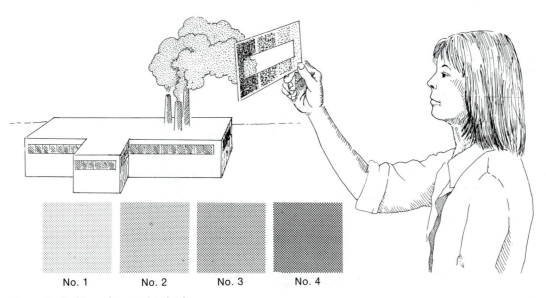

No. 1   No. 2   No. 3   No. 4

Figure L–8 *Observing smoke shade.*

preferably at least six feet, to avoid contamination from coarse windblown material, such as soil, that does not reflect general air pollution levels.

Leave the jars exposed for 30 days. Visit them at least once a week to replace any evaporated water by refilling the jars to the ¼-level mark. (If the water dries out completely, the test is invalid, because the wind may blow fine dusts out of the jar.) In rainy weather, check that the jar does not overflow. If the level approaches the top, stop the test.

Now evaporate all the liquid to dryness and collect and weigh the residue. (If your liquid contained alcohol, do not use a flame or electric heater because alcohol is flammable. Use a warm surface such as a radiator.) Calculate the dustfall pollution as follows:

1. Measure the diameter of each jar opening in centimeters.

2. Calculate the area of each opening:

$$\text{area} = \frac{\pi \times (\text{diameter})^2}{4}$$

3. Calculate the dustfall:

$$\text{Dustfall} = \frac{\text{wt of dust (mg)}}{\text{area (cm}^2) \times \text{time (days)}} \times 30 \frac{\text{days}}{\text{month}}$$

This result expresses dustfall in milligrams per square centimeter per month. To convert to tonnes of dust per square kilometer per month, multiply your answer by 10. To convert to tons of dust per square mile per month, multiply your answer by 28. Compare your results with reports from your local health or environmental agency.

EXPERIMENT 34 Water Purification with Activated Carbon

Food colors are usually available in small containers from which they can be dispensed in drops. Using such colors, prepare a set of lightly tinted solutions in ordinary drinking glasses, filled about three-quarters full. Now stir a little activated carbon powder, of the type used for aquariums, into each glass. The carbon may be purchased from a pet shop, drugstore, or hobby shop. Place a saucer over each glass and allow the carbon to settle overnight. Note the effectiveness with which the colored impurities are removed.

Can you design a series of experiments to determine how much carbon is needed to remove a given amount of dye? Or to determine which dyes are easier or harder to remove?

EXPERIMENT 35 Loudness

Carry this book around with you for a few days so you can refer to Table 10–4. Make a diary of various sounds you hear. Estimate the decibel levels as well as you can from the columns in the table that show the sound levels of various sources and their perceived loudness. Try to include the following sources: (1) a television, or radio at normal listening volume in your room or apartment; (2) the central study area in the school library; (3) your physical science classroom; (4) the street outside your classroom; (5) the background noise in your room at night; (6) the school cafeteria; (7) a local factory or construction job site; (8) your own activities. Which sounds are annoying? Offer suggestions for reducing the perceived loudness in each of the instances where the sound is annoying.

EXPERIMENT 36 Sound

Select a convenient constant source of sound, such as a ringing alarm clock, and try to reduce the loudness you hear from it by the following means: (1) stuff some cotton loosely into your ears; (2) hold your hands over your ears; (3) use both the cotton and your hands; (4) submerse your head in the bathtub (face up) until your ears are underwater; (5) if you can borrow a pair of earmuffs of the kind used in factories or at airports, try them. (*Safety Note:* Don't try to stuff any small hard objects in your ears; the results could be harmful.) Draw a straight line (of any length) in

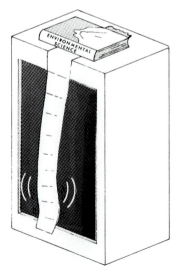

Figure L–9 Sound power.

your notebook, labeling one end "inaudible" and the other end "loudness to naked ear."

Inaudible
(zero loudness)

Loudness to
naked ear

Mark the positions of each of the sound-reducing methods on the line at a point that corresponds to the loudness you heard. For example, if you think that one of the methods reduced the loudness by half, mark its position halfway along the line. If you think

it reduced it by only 25 percent, mark it at one quarter of the length away from the "naked ear" end. Discuss the reasons for your findings.

EXPERIMENT 37 Sound Power

Hang a strip of tissue paper in front of the loudspeaker of a sound system (Fig. L–9). Shut off the sound, close the windows, and turn off any fan or air conditioner in the room. The paper should hang motionless. Now switch the sound on and turn up the volume. Describe the effect on the paper and explain your observations.

INDEX